Lecture Notes in Computer Science 12139

More information about this series at http://www.springer.com/series/7407

Valeria V. Krzhizhanovskaya ·
Gábor Závodszky · Michael H. Lees ·
Jack J. Dongarra · Peter M. A. Sloot ·
Sérgio Brissos · João Teixeira (Eds.)

Computational Science – ICCS 2020

20th International Conference
Amsterdam, The Netherlands, June 3–5, 2020
Proceedings, Part III

 Springer

Editors
Valeria V. Krzhizhanovskaya (iD)
University of Amsterdam
Amsterdam, The Netherlands

Michael H. Lees
University of Amsterdam
Amsterdam, The Netherlands

Peter M. A. Sloot (iD)
University of Amsterdam
Amsterdam, The Netherlands

ITMO University
Saint Petersburg, Russia

Nanyang Technological University
Singapore, Singapore

João Teixeira
Intellegibilis
Setúbal, Portugal

Gábor Závodszky (iD)
University of Amsterdam
Amsterdam, The Netherlands

Jack J. Dongarra (iD)
University of Tennessee
Knoxville, TN, USA

Sérgio Brissos
Intellegibilis
Setúbal, Portugal

ISSN 0302-9743   ISSN 1611-3349   (electronic)
Lecture Notes in Computer Science
ISBN 978-3-030-50419-9   ISBN 978-3-030-50420-5   (eBook)
https://doi.org/10.1007/978-3-030-50420-5

LNCS Sublibrary: SL1 – Theoretical Computer Science and General Issues

This Springer imprint is published by the registered company Springer Nature Switzerland AG
The registered company address is: Gewerbestrasse 11, 6330 Cham, Switzerland

Preface

Twenty Years of Computational Science

Welcome to the 20th Annual International Conference on Computational Science (ICCS – https://www.iccs-meeting.org/iccs2020/).

During the preparation for this 20th edition of ICCS we were considering all kinds of nice ways to celebrate two decennia of computational science. Afterall when we started this international conference series, we never expected it to be so successful and running for so long at so many different locations across the globe! So we worked on a mind-blowing line up of renowned keynotes, music by scientists, awards, a play written by and performed by computational scientists, press attendance, a lovely venue... you name it, we had it all in place. Then corona hit us.

After many long debates and considerations, we decided to cancel the physical event but still support our scientists and allow for publication of their accepted peer-reviewed work. We are proud to present the proceedings you are reading as a result of that.

ICCS 2020 is jointly organized by the University of Amsterdam, NTU Singapore, and the University of Tennessee.

The International Conference on Computational Science is an annual conference that brings together researchers and scientists from mathematics and computer science as basic computing disciplines, as well as researchers from various application areas who are pioneering computational methods in sciences such as physics, chemistry, life sciences, engineering, arts and humanitarian fields, to discuss problems and solutions in the area, to identify new issues, and to shape future directions for research.

Since its inception in 2001, ICCS has attracted increasingly higher quality and numbers of attendees and papers, and 2020 was no exception, with over 350 papers accepted for publication. The proceedings series have become a major intellectual resource for computational science researchers, defining and advancing the state of the art in this field.

The theme for ICCS 2020, "Twenty Years of Computational Science", highlights the role of Computational Science over the last 20 years, its numerous achievements, and its future challenges. This conference was a unique event focusing on recent developments in: scalable scientific algorithms, advanced software tools, computational grids, advanced numerical methods, and novel application areas. These innovative novel models, algorithms, and tools drive new science through efficient application in areas such as physical systems, computational and systems biology, environmental systems, finance, and others.

This year we had 719 submissions (230 submissions to the main track and 489 to the thematic tracks). In the main track, 101 full papers were accepted (44%). In the thematic tracks, 249 full papers were accepted (51%). A high acceptance rate in the thematic tracks is explained by the nature of these, where many experts in a particular field are personally invited by track organizers to participate in their sessions.

ICCS relies strongly on the vital contributions of our thematic track organizers to attract high-quality papers in many subject areas. We would like to thank all committee members from the main and thematic tracks for their contribution to ensure a high standard for the accepted papers. We would also like to thank Springer, Elsevier, the Informatics Institute of the University of Amsterdam, the Institute for Advanced Study of the University of Amsterdam, the SURFsara Supercomputing Centre, the Netherlands eScience Center, the VECMA Project, and Intellegibilis for their support. Finally, we very much appreciate all the Local Organizing Committee members for their hard work to prepare this conference.

We are proud to note that ICCS is an A-rank conference in the CORE classification.

We wish you good health in these troubled times and hope to see you next year for ICCS 2021.

June 2020

Valeria V. Krzhizhanovskaya
Gábor Závodszky
Michael Lees
Jack Dongarra
Peter M. A. Sloot
Sérgio Brissos
João Teixeira

Organization

Thematic Tracks and Organizers

Advances in High-Performance Computational Earth Sciences: Applications and Frameworks – IHPCES

Takashi Shimokawabe
Kohei Fujita
Dominik Bartuschat

Agent-Based Simulations, Adaptive Algorithms and Solvers – ABS-AAS

Maciej Paszynski
David Pardo
Victor Calo
Robert Schaefer
Quanling Deng

Applications of Computational Methods in Artificial Intelligence and Machine Learning – ACMAIML

Kourosh Modarresi
Raja Velu
Paul Hofmann

Biomedical and Bioinformatics Challenges for Computer Science – BBC

Mario Cannataro
Giuseppe Agapito
Mauro Castelli
Riccardo Dondi
Rodrigo Weber dos Santos
Italo Zoppis

Classifier Learning from Difficult Data – CLD^2

Michał Woźniak
Bartosz Krawczyk
Paweł Ksieniewicz

Complex Social Systems through the Lens of Computational Science – CSOC

Debraj Roy
Michael Lees
Tatiana Filatova

Computational Health – CompHealth

Sergey Kovalchuk
Stefan Thurner
Georgiy Bobashev

Computational Methods for Emerging Problems in (dis-)Information Analysis – DisA

Michal Choras
Konstantinos Demestichas

Computational Optimization, Modelling and Simulation – COMS

Xin-She Yang
Slawomir Koziel
Leifur Leifsson

Computational Science in IoT and Smart Systems – IoTSS

Vaidy Sunderam
Dariusz Mrozek

Computer Graphics, Image Processing and Artificial Intelligence – CGIPAI

Andres Iglesias
Lihua You
Alexander Malyshev
Hassan Ugail

Data-Driven Computational Sciences – DDCS

Craig C. Douglas
Ana Cortes
Hiroshi Fujiwara
Robert Lodder
Abani Patra
Han Yu

Machine Learning and Data Assimilation for Dynamical Systems – MLDADS

Rossella Arcucci
Yi-Ke Guo

Meshfree Methods in Computational Sciences – MESHFREE

Vaclav Skala
Samsul Ariffin Abdul Karim
Marco Evangelos Biancolini
Robert Schaback

Rongjiang Pan
Edward J. Kansa

Multiscale Modelling and Simulation – MMS

Derek Groen
Stefano Casarin
Alfons Hoekstra
Bartosz Bosak
Diana Suleimenova

Quantum Computing Workshop – QCW

Katarzyna Rycerz
Marian Bubak

Simulations of Flow and Transport: Modeling, Algorithms and Computation – SOFTMAC

Shuyu Sun
Jingfa Li
James Liu

Smart Systems: Bringing Together Computer Vision, Sensor Networks and Machine Learning – SmartSys

Pedro J. S. Cardoso
João M. F. Rodrigues
Roberto Lam
Janio Monteiro

Software Engineering for Computational Science – SE4Science

Jeffrey Carver
Neil Chue Hong
Carlos Martinez-Ortiz

Solving Problems with Uncertainties – SPU

Vassil Alexandrov
Aneta Karaivanova

Teaching Computational Science – WTCS

Angela Shiflet
Alfredo Tirado-Ramos
Evguenia Alexandrova

Uncertainty Quantification for Computational Models – UNEQUIvOCAL

Wouter Edeling
Anna Nikishova
Peter Coveney

Program Committee and Reviewers

Ahmad Abdelfattah
Samsul Ariffin
  Abdul Karim
Evgenia Adamopoulou
Jaime Afonso Martins
Giuseppe Agapito
Ram Akella
Elisabete Alberdi Celaya
Luis Alexandre
Vassil Alexandrov
Evguenia Alexandrova
Hesham H. Ali
Julen Alvarez-Aramberri
Domingos Alves
Julio Amador Diaz Lopez
Stanislaw
  Ambroszkiewicz
Tomasz Andrysiak
Michael Antolovich
Hartwig Anzt
Hideo Aochi
Hamid Arabnejad
Rossella Arcucci
Khurshid Asghar
Marina Balakhontceva
Bartosz Balis
Krzysztof Banas
João Barroso
Dominik Bartuschat
Nuno Basurto
Pouria Behnoudfar
Joern Behrens
Adrian Bekasiewicz
Gebrai Bekdas
Stefano Beretta
Benjamin Berkels
Martino Bernard

Daniel Berrar
Sanjukta Bhowmick
Marco Evangelos
  Biancolini
Georgiy Bobashev
Bartosz Bosak
Marian Bubak
Jérémy Buisson
Robert Burduk
Michael Burkhart
Allah Bux
Aleksander Byrski
Cristiano Cabrita
Xing Cai
Barbara Calabrese
Jose Camata
Mario Cannataro
Alberto Cano
Pedro Jorge Sequeira
  Cardoso
Jeffrey Carver
Stefano Casarin
Manuel Castañón-Puga
Mauro Castelli
Eduardo Cesar
Nicholas Chancellor
Patrikakis Charalampos
Ehtzaz Chaudhry
Chuanfa Chen
Siew Ann Cheong
Andrey Chernykh
Lock-Yue Chew
Su Fong Chien
Marta Chinnici
Sung-Bae Cho
Michal Choras
Loo Chu Kiong

Neil Chue Hong
Svetlana Chuprina
Paola Cinnella
Noélia Correia
Adriano Cortes
Ana Cortes
Enrique
  Costa-Montenegro
David Coster
Helene Coullon
Peter Coveney
Attila Csikasz-Nagy
Loïc Cudennec
Javier Cuenca
Yifeng Cui
António Cunha
Ben Czaja
Pawel Czarnul
Flávio Martins
Bhaskar Dasgupta
Konstantinos Demestichas
Quanling Deng
Nilanjan Dey
Khaldoon Dhou
Jamie Diner
Jacek Dlugopolski
Simona Domesová
Riccardo Dondi
Craig C. Douglas
Linda Douw
Rafal Drezewski
Hans du Buf
Vitor Duarte
Richard Dwight
Wouter Edeling
Waleed Ejaz
Dina El-Reedy

Amgad Elsayed
Nahid Emad
Chriatian Engelmann
Gökhan Ertaylan
Alex Fedoseyev
Luis Manuel Fernández
Antonino Fiannaca
Christos
    Filelis-Papadopoulos
Rupert Ford
Piotr Frackiewicz
Martin Frank
Ruy Freitas Reis
Karl Frinkle
Haibin Fu
Kohei Fujita
Hiroshi Fujiwara
Takeshi Fukaya
Wlodzimierz Funika
Takashi Furumura
Ernst Fusch
Mohamed Gaber
David Gal
Marco Gallieri
Teresa Galvao
Akemi Galvez
Salvador García
Bartlomiej Gardas
Delia Garijo
Frédéric Gava
Piotr Gawron
Bernhard Geiger
Alex Gerbessiotis
Ivo Goncalves
Antonio Gonzalez Pardo
Jorge
    González-Domínguez
Yuriy Gorbachev
Pawel Gorecki
Michael Gowanlock
Manuel Grana
George Gravvanis
Derek Groen
Lutz Gross
Sophia
    Grundner-Culemann

Pedro Guerreiro
Tobias Guggemos
Xiaohu Guo
Piotr Gurgul
Filip Guzy
Pietro Hiram Guzzi
Zulfiqar Habib
Panagiotis Hadjidoukas
Masatoshi Hanai
John Hanley
Erik Hanson
Habibollah Haron
Carina Haupt
Claire Heaney
Alexander Heinecke
Jurjen Rienk Helmus
Álvaro Herrero
Bogumila Hnatkowska
Maximilian Höb
Erlend Hodneland
Olivier Hoenen
Paul Hofmann
Che-Lun Hung
Andres Iglesias
Takeshi Iwashita
Alireza Jahani
Momin Jamil
Vytautas Jancauskas
João Janeiro
Peter Janku
Fredrik Jansson
Jirí Jaroš
Caroline Jay
Shalu Jhanwar
Zhigang Jia
Chao Jin
Zhong Jin
David Johnson
Guido Juckeland
Maria Juliano
Edward J. Kansa
Aneta Karaivanova
Takahiro Katagiri
Timo Kehrer
Wayne Kelly
Christoph Kessler

Jakub Klikowski
Harald Koestler
Ivana Kolingerova
Georgy Kopanitsa
Gregor Kosec
Sotiris Kotsiantis
Ilias Kotsireas
Sergey Kovalchuk
Michal Koziarski
Slawomir Koziel
Rafal Kozik
Bartosz Krawczyk
Elisabeth Krueger
Valeria Krzhizhanovskaya
Pawel Ksieniewicz
Marek Kubalcík
Sebastian Kuckuk
Eileen Kuehn
Michael Kuhn
Michal Kulczewski
Krzysztof Kurowski
Massimo La Rosa
Yu-Kun Lai
Jalal Lakhlili
Roberto Lam
Anna-Lena Lamprecht
Rubin Landau
Johannes Langguth
Elisabeth Larsson
Michael Lees
Leifur Leifsson
Kenneth Leiter
Roy Lettieri
Andrew Lewis
Jingfa Li
Khang-Jie Liew
Hong Liu
Hui Liu
Yen-Chen Liu
Zhao Liu
Pengcheng Liu
James Liu
Marcelo Lobosco
Robert Lodder
Marcin Los
Stephane Louise

Frederic Loulergue
Paul Lu
Stefan Luding
Onnie Luk
Scott MacLachlan
Luca Magri
Imran Mahmood
Zuzana Majdisova
Alexander Malyshev
Muazzam Maqsood
Livia Marcellino
Tomas Margalef
Tiziana Margaria
Svetozar Margenov
Urszula
    Markowska-Kaczmar
Osni Marques
Carmen Marquez
Carlos Martinez-Ortiz
Paula Martins
Flávio Martins
Luke Mason
Pawel Matuszyk
Valerie Maxville
Wagner Meira Jr.
Roderick Melnik
Valentin Melnikov
Ivan Merelli
Choras Michal
Leandro Minku
Jaroslaw Miszczak
Janio Monteiro
Kourosh Modarresi
Fernando Monteiro
James Montgomery
Andrew Moore
Dariusz Mrozek
Peter Mueller
Khan Muhammad
Judit Muñoz
Philip Nadler
Hiromichi Nagao
Jethro Nagawkar
Kengo Nakajima
Ionel Michael Navon
Philipp Neumann

Mai Nguyen
Hoang Nguyen
Nancy Nichols
Anna Nikishova
Hitoshi Nishizawa
Brayton Noll
Algirdas Noreika
Enrique Onieva
Kenji Ono
Eneko Osaba
Aziz Ouaarab
Serban Ovidiu
Raymond Padmos
Wojciech Palacz
Ivan Palomares
Rongjiang Pan
Joao Papa
Nikela Papadopoulou
Marcin Paprzycki
David Pardo
Anna Paszynska
Maciej Paszynski
Abani Patra
Dana Petcu
Serge Petiton
Bernhard Pfahringer
Frank Phillipson
Juan C. Pichel
Anna
    Pietrenko-Dabrowska
Laércio L. Pilla
Armando Pinho
Tomasz Piontek
Yuri Pirola
Igor Podolak
Cristina Portales
Simon Portegies Zwart
Roland Potthast
Ela Pustulka-Hunt
Vladimir Puzyrev
Alexander Pyayt
Rick Quax
Cesar Quilodran Casas
Barbara Quintela
Ajaykumar Rajasekharan
Celia Ramos

Lukasz Rauch
Vishal Raul
Robin Richardson
Heike Riel
Sophie Robert
Luis M. Rocha
Joao Rodrigues
Daniel Rodriguez
Albert Romkes
Debraj Roy
Katarzyna Rycerz
Alberto Sanchez
Gabriele Santin
Alex Savio
Robert Schaback
Robert Schaefer
Rafal Scherer
Ulf D. Schiller
Bertil Schmidt
Martin Schreiber
Alexander Schug
Gabriela Schütz
Marinella Sciortino
Diego Sevilla
Angela Shiflet
Takashi Shimokawabe
Marcin Sieniek
Nazareen Sikkandar
    Basha
Anna Sikora
Janaína De Andrade Silva
Diana Sima
Robert Sinkovits
Haozhen Situ
Leszek Siwik
Vaclav Skala
Peter Sloot
Renata Slota
Grazyna Slusarczyk
Sucha Smanchat
Marek Smieja
Maciej Smolka
Bartlomiej Sniezynski
Isabel Sofia Brito
Katarzyna Stapor
Bogdan Staszewski

Jerzy Stefanowski
Dennis Stevenson
Tomasz Stopa
Achim Streit
Barbara Strug
Pawel Strumillo
Dante Suarez
Vishwas H. V. Subba Rao
Bongwon Suh
Diana Suleimenova
Ray Sun
Shuyu Sun
Vaidy Sunderam
Martin Swain
Alessandro Taberna
Ryszard Tadeusiewicz
Daisuke Takahashi
Zaid Tashman
Osamu Tatebe
Carlos Tavares Calafate
Kasim Tersic
Yonatan Afework
    Tesfahunegn
Jannis Teunissen
Stefan Thurner

Nestor Tiglao
Alfredo Tirado-Ramos
Arkadiusz Tomczyk
Mariusz Topolski
Paolo Trunfio
Ka-Wai Tsang
Hassan Ugail
Eirik Valseth
Pavel Varacha
Pierangelo Veltri
Raja Velu
Colin Venters
Gytis Vilutis
Peng Wang
Jianwu Wang
Shuangbu Wang
Rodrigo Weber
    dos Santos
Katarzyna
    Wegrzyn-Wolska
Mei Wen
Lars Wienbrandt
Mark Wijzenbroek
Peter Woehrmann
Szymon Wojciechowski

Maciej Woloszyn
Michal Wozniak
Maciej Wozniak
Yu Xia
Dunhui Xiao
Huilin Xing
Miguel Xochicale
Feng Xu
Wei Xue
Yoshifumi Yamamoto
Dongjia Yan
Xin-She Yang
Dongwei Ye
Wee Ping Yeo
Lihua You
Han Yu
Gábor Závodszky
Yao Zhang
H. Zhang
Jinghui Zhong
Sotirios Ziavras
Italo Zoppis
Chiara Zucco
Pawel Zyblewski
Karol Zyczkowski

Contents – Part III

Biomedical and Bioinformatics Challenges for Computer Science

Advances in High-Performance Computational Earth Sciences: Applications and Frameworks

Data-Driven Approach to Inversion Analysis of Three-Dimensional Inner Soil Structure via Wave Propagation Analysis

Takuma Yamaguchi[1(✉)], Tsuyoshi Ichimura[1,3], Kohei Fujita[1], Muneo Hori[2], Lalith Wijerathne[1], and Naonori Ueda[3]

[1] Earthquake Research Institute and Department of Civil Engineering,
The University of Tokyo, Bunkyo, Tokyo, Japan
{yamaguchi,ichimura,fujita,lalith}@eri.u-tokyo.ac.jp
[2] Research Institute for Value-Added-Information Generation, Japan Agency
for Marine-Earth Science and Technology, Yokohama, Kanagawa, Japan
horimune@jamstec.go.jp
[3] Center for Advanced Intelligence Project, RIKEN, Tokyo, Japan
naonori.ueda@riken.jp

Abstract. Various approaches based on both computational science and data science/machine learning have been proposed with the development of observation systems and network technologies. Computation cost associated with computational science can be reduced by introducing the methods based on data science/machine learning. In the present paper, we focus on a method to estimate inner soil structure via wave propagation analysis. It is regarded as one of the parameter optimization approaches using observation data on the surface. This application is in great demand to ensure better reliability in numerical simulations. Typical optimization requires many forward analyses; thus, massive computation cost is required. We propose an approach to substitute evaluation using neural networks for most cases of forward analyses and to reduce the number of forward analyses. Forward analyses in the proposed method are used for producing the training data for a neural network; thereby they can be computed independently, and the actual elapsed time can be reduced by using a large-scale supercomputer. We demonstrated that the inner soil structure was estimated with the sufficient accuracy for practical damage evaluation. We also confirmed that the proposed method achieved estimating parameters within a shorter timeframe compared to a typical approach based on simulated annealing.

Keywords: Data-driven computing · Finite element analysis · Conjugate gradient method · GPU computing

© Springer Nature Switzerland AG 2020
V. V. Krzhizhanovskaya et al. (Eds.): ICCS 2020, LNCS 12139, pp. 3–17, 2020.
https://doi.org/10.1007/978-3-030-50420-5_1

1 Introduction

The demand for exploiting the power of Big Data has been increasing due to the rapid development of observation systems and network technologies (e.g., introduction of Internet of Things and 5th Generation networks). Thereafter, integration of computational science, data science, and machine learning has been proposed (Big Data & Extreme Computing (BDEC) [1]). The main purpose of this project is to provide the available data to computer systems and to produce new information with social value by data processing and computations. It is assumed that the data are supplied in real time; therefore, large amount of computer power will be required. Several large-scale computer systems are designed for data science and machine learning (e.g., AI Bridge Cloud Infrastructure (ABCI) [2]), and therefore, we can see that these approaches are being expanded. From the perspective of computational science, computation cost such as power consumption becomes a more important issue with the development of computation environments. Introduction of data science and machine learning can facilitate reducing the computation cost.

We focus on the estimation of the inner structure whose material properties vary by location. This problem has been discussed in various fields including biomedicine [4] and gas and oil exploration [12], and it is also important for damage evaluation in the case of earthquake disasters. Here, we estimate the inner soil structure (boundary surfaces of soil layers) required in ground shaking analysis: in this problem, material properties of each ground layer in the domain are relatively easy to estimate by geological survey. However, direct measurement of the position of boundaries between soil layers with different material properties is difficult. Reference [8] notes that the inner soil structure has significant effects on the distribution of displacement on the ground surface and strain in underground structures. The inner soil structure is not available with high resolution or appropriate accuracy, which deteriorates the accuracy of simulations even if numerical methods capable of modeling complex geometry is used.

One of the realistic ways to address this issue is to introduce an optimization method using the observation data on the ground surface in the case of a small earthquake. If we could generate many models and conduct wave propagation analysis for each of them, it would be possible to select a model capable of reproducing available observation data most accurately. The use of optimized models may help increase reliability of damage evaluation. This procedure requires a large number of forward analyses. Therefore, the challenge is an increase in the computation cost for many analyses with the large number of degrees of freedom.

In our previous study [14], we proposed a simulated annealing method with reduction in computation cost required in each forward analysis by overlapping model generation on CPUs and finite element solver computations on GPUs. We demonstrated that one boundary surface between two layers in the soil structure was estimated by 1,500 wave propagation analyses using finite element models with 3,000,000 degrees of freedom in 13 h. However, in this application, we have to estimate the three-dimensional inner soil structure using the data on two-dimensional ground surface. In cases when we estimate multiple boundary surfaces, the convergence of optimization methods can deteriorate, as the number of control parameters becomes much larger than the number of observation points. Taking this into account, we note that practical estimation cannot be materialized by just accelerating each forward analysis. Reduction in the number of forward analyses is another important issue to consider.

The approach to introduce data science or machine learning into computational science has been proposed aiming to reduce the number of forward analyses. For instance, [9] applied a machine learning-based methodology for microstructure optimization and materials design. Generally, evaluation using data science or machine learning is faster than forward analysis; therefore, more trials can be evaluated within the same timeframe.

We examine the applicability of the data-driven approach to estimation of the inner soil structure. To address this problem, we combine the wave propagation analysis and neural network methodology. We conduct many forward analyses and use these results as the training data for a neural network. By implementing the neural network that takes control parameters as input and outputs error levels of models, we can extract the parameters that are expected to reproduce observation data more accurately from many samples. Computation cost is reduced by replacing forward analysis with inference using neural networks. In addition, forward analyses can be computed independently; therefore, we can reduce the actual elapsed time by using large-scale supercomputers. In the application example, we demonstrate that the inner soil structure has a considerable effect on the strain distribution, and that the proposed method can estimate the soil structure with the sufficient accuracy for damage estimation within a shorter timeframe compared with a typical approach based on simulated annealing.

2 Methodology

In the present paper, we propose a method that estimates the inner soil structure of the target domain by conducting a large number of wave propagation analyses and choosing the inner structure with the maximum likelihood. Here, we estimate boundary surfaces of the domains with different material properties. For simplicity, we assume that the target domain has a stratified structure, and that the target parameters considered for optimization are elevations of the boundary surfaces on control points which are located at regular intervals in x and y directions. The set of control parameters is denoted by $\boldsymbol{\alpha}$. Boundary surfaces are generated in the target domain by interpolating elevations at the control points using bi-cubic functions. We employ the time history of waves on the ground surface $\mathbf{v}_{i_{\mathrm{obs}}}^{\mathrm{ref}}$ ($i_{\mathrm{obs}} = 1, ..., n_{\mathrm{obs}}$), where n_{obs} is the number of observation points. In addition, we assume that these waves do not contain noise. We define an error between a generated and a reference model as follows:

$$E = \sqrt{\frac{\sum_{i_{\mathrm{obs}}=1}^{n_{\mathrm{obs}}} \sum_{i_{\mathrm{t}}=1}^{n_{\mathrm{t}}} \left\| \mathbf{v}_{i_{\mathrm{obs}}, i_{\mathrm{t}}} - \mathbf{v}_{i_{\mathrm{obs}}, i_{\mathrm{t}}}^{\mathrm{ref}} \right\|^2}{\sum_{i_{\mathrm{obs}}=1}^{n_{\mathrm{obs}}} \sum_{i_{\mathrm{t}}=1}^{n_{\mathrm{t}}} \left\| \mathbf{v}_{i_{\mathrm{obs}}, i_{\mathrm{t}}}^{\mathrm{ref}} \right\|^2}}, \tag{1}$$

where i_t is each time step, and $\mathbf{v}_{i_{\mathrm{obs}},i_t}$ is the velocity on i_{obs}-th observation point at the i_t-th time step, which can be computed from the parameters $\boldsymbol{\alpha}$ and the known input waves obtained by pullback analysis. We assume that the models that reproduce the observation data closely have smaller error E. Therefore, we search $\boldsymbol{\alpha}$ that minimizes E. There are several gradient-based methods applied to optimization, for example, three-dimensional crustal structure optimization proposed in [11]. These methods have the advantage that the number of trials is small; however, it may be difficult to escape from a local solution if control parameters have a low sensitivity to the error function. Simulated annealing [7] is one of the most robust approaches; however, the convergence rate is not high for problems with many control parameters. If we obtain $\mathbf{v}_{i_{\mathrm{obs}},i_t}$ by forward analysis for each parameter, a large number of forward analyses are required. Reduction in the number of cases to compute and introduction of parallel computations of forward analyses are essential to conduct the target estimation within a realistic timeframe. Accordingly, we introduce an approach based on machine learning employing the results of forward analyses. We define the neural networks that can be used to estimate the error E based on the input parameters $\boldsymbol{\alpha}$. The details about the neural networks and forward analysis are described in the following subsections.

2.1   Introduction of Neural Networks

The proposed algorithm based on neural networks is described as Algorithm 1. Firstly, we conduct n_0 forward analyses to generate the adequate training data (Algorithm 1, lines 3–4). We have to use the parameter sets that are scattered in the control parameter space. In the present study, a random number retrieved from the normal distribution is added up to initial elevation of one control point. We fluctuate elevations under the constraint that the order of layers is consistent in all of the models. We obtain errors for all parameters by performing forward analyses with generated models. Each case can be computed independently; therefore, elapsed time is reduced when we use a large-scale computer system.

Next, we implement neural networks to estimate error E roughly (Algorithm 1, line 5). Here, input parameters of neural networks consist of the fluctuation amount of elevation at each control point. To improve the performance, we normalize these values, so that the value of -3σ is set equal to 0, and the value of 3σ is set equal to 1. Here, we generate the classifiers instead of regression models, as suggested by [6], due to the fact that the number of samples constrained by the massive computation cost is limited for modeling complex modes of the error function. We classify errors into 10 levels that are equally divided between the maximum and minimum errors in n_0 sets of parameters. We define one classifier for each observation point so that contribution of each point to the error becomes clearer. We define the point-wise error based on the original error in Eq. (1) as follows:

$$E_{i_{\mathrm{obs}}} = \sqrt{\frac{\sum_{i_t=1}^{n_t} \left\| \mathbf{v}_{i_{\mathrm{obs}},i_t} - \mathbf{v}_{i_{\mathrm{obs}},i_t}^{\mathrm{ref}} \right\|^2}{\sum_{i_t=1}^{n_t} \left\| \mathbf{v}_{i_{\mathrm{obs}},i_t}^{\mathrm{ref}} \right\|^2}}. \tag{2}$$

We divide the range of the possible values of $E_{i_{\text{obs}}}$ into ten levels (1–10) equally, and the implemented neural networks learn the classification of these levels. A final evaluation value for each parameter set is obtained by summation across levels in all neural networks. We assume that the parameters with smaller evaluation values provide smaller E. We have to define a neural network per observation point in this procedure; however, associated computation cost is insignificant, as each network can be trained independently, and the number of the training data elements is at most 10^3.

We perform inference using the generated neural networks. A dataset consisting of $n_1(\gg n_0)$ cases is inputted into the neural networks (Algorithm 1, lines 6–7). We extract n_0 cases that have the lowest estimated error levels and compute actual errors E by using forward analyses. Parameters that achieves the smallest E are chosen in Algorithm 1, lines 8–10. Further improvement can be achieved by iterating the same procedures using the estimated parameters as required (Algorithm 1, line 11).

Algorithm 1: Algorithm based on neural networks for estimation of the inner structure.

---

1 Data: initial parameters α_{base}, observation data $\mathbf{v}_{i_{\text{obs}}}^{\text{ref}}(i_{\text{obs}} = 1, ..., n_{\text{obs}})$, counter $i_{\text{trial}} = 1$

2 for $i_{\text{trial}} \leq MaxTrialNumber$ **do**

3   generate n_0 samples based on α_{base} to produce $TrainingDataset_{i_{\text{trial}}}$

4   **for** each α in $TrainingDataset_{i_{\text{trial}}}$ **do**
        $\lfloor$ compute actual error $E_{i_{\text{obs}}}$

5   **for** each observation point i_{obs} **do**
        $\lfloor$ produce classifier of $E_{i_{\text{obs}}}$ for given α

6   generate $n_1(\gg n_0)$ samples based on α_{base} to produce $InferenceDataset_{i_{\text{trial}}}$

7   **for** each α in $InferenceDataset_{i_{\text{trial}}}$ **do**
        $\lfloor$ estimate $E_{i_{\text{obs}}}$ using classifiers

8   choose n_0 samples from $InferenceDataset_{i_{\text{trial}}}$ which provides the lowest E to construct $SuperiorDataset_{i_{\text{trial}}}$

9   **for** each α in $SuperiorDataset_{i_{\text{trial}}}$ **do**
        $\lfloor$ compute actual error E

10   choose α_{best}, which provides the lowest E

11   update $\alpha_{\text{base}} \leftarrow \alpha_{\text{best}}, i_{\text{trial}} \leftarrow i_{\text{trial}} + 1$

2.2 Finite Element Analyses

In the proposed scheme, more than 10^3 finite element analyses are required; therefore, it is important to conduct them in a shorter time possible. We assume that we use the observation data in the case of an earthquake small enough to ignore nonlinearity for estimation of the inner structure. Therefore, we focus on linear wave propagation analysis. The target equation is defined as follows: $\left(\frac{4}{dt^2}\mathbf{M} + \frac{2}{dt}\mathbf{C} + \mathbf{K}\right)\mathbf{u}_{i_t} = \mathbf{f}_{i_t} + \mathbf{C}\mathbf{v}_{i_t-1} + \mathbf{M}\left(\mathbf{a}_{i_t-1} + \frac{4}{dt}\mathbf{v}_{i_t-1}\right)$, where $\mathbf{u}$, $\mathbf{v}$, $\mathbf{a}$, and $\mathbf{f}$ are displacement, velocity, acceleration, and force vector, respectively; and $\mathbf{M}$, $\mathbf{C}$, and $\mathbf{K}$ are mass, damping, and stiffness matrix, respectively. In addition, dt denotes the time increment, and i_t is the number of time steps. For the damping matrix $\mathbf{C}$, we apply Rayleigh damping and compute it by linear combination as follows: $\mathbf{C} = \alpha\mathbf{M} + \beta\mathbf{K}$. Coefficients α and β are set so that $\int_{f_{\min}}^{f_{\max}}(h - \frac{1}{2}(\frac{\alpha}{2\pi f} + 2\pi f\beta))^2 df$ is minimized, where $f_{\max}$, $f_{\min}$, and h are maximum/minimum targeting frequency and damping ratio. We apply the Newmark-β method with $\beta = 1/4$ and $\delta = 1/2$ for time integration. Vectors $\mathbf{v}_{i_t}$ and $\mathbf{a}_{i_t}$ can be described as follows: $\mathbf{v}_{i_t} = -\mathbf{v}_{i_t-1} + \frac{2}{dt}(\mathbf{u}_{i_t} - \mathbf{u}_{i_t-1})$ and $\mathbf{a}_{i_t} = -\mathbf{a}_{i_t-1} - \frac{4}{dt}\mathbf{v}_{i_t-1} + \frac{4}{dt^2}(\mathbf{u}_{i_t} - \mathbf{u}_{i_t-1})$. We obtain displacement vector $\mathbf{u}_{i_t}$ by solving the linear equation and updating vectors $\mathbf{v}_{i_t}$ and $\mathbf{a}_{i_t}$ accordingly. Generation of finite element models and computation of the finite element solver are the most computationally expensive parts. We automatically generate finite element models using the method proposed by [5]. This method applies CPU computations using OpenMP. In the solver part, we apply the OpenACC-accelerated solver based on the conjugate gradient method described in our previous study [13]. The solver combines the conjugate gradient method with adaptive preconditioning, geometric multigrid method, and mixed precision arithmetic to reduce the amount of arithmetic counts and the data transfer size. In the solver, sparse matrix vector multiplication is computed by using the Element-by-Element method. It is applied to compute the element matrix on-the-fly and allows reducing the memory access cost. Specifically, the multiplication of matrix $\mathbf{A}$ and vector $\mathbf{x}$ is defined as $\mathbf{y} = \sum_{i=1}^{n_{\text{elem}}}(\mathbf{Q}^{(i)T}(\mathbf{A}^{(i)}(\mathbf{Q}^{(i)}\mathbf{x})))$, where $\mathbf{y}$ is the resulting vector, and n_{elem} is the number of elements in the domain; $\mathbf{Q}^{(i)}$ is a mapping matrix to make a transition from the local node numbers in the i-th element to the global node numbers; and $\mathbf{A}^{(i)}$ is the i-th element matrix that satisfies the following: $\mathbf{A} = \sum_{i=1}^{n_{\text{elem}}}\mathbf{Q}^{(i)T}\mathbf{A}^{(i)}\mathbf{Q}^{(i)}$, where $\mathbf{A}^{(i)} = \frac{4}{dt^2}\mathbf{M}^{(i)} + \frac{2}{dt}\mathbf{C}^{(i)} + \mathbf{K}^{(i)}$. The solver proposed by [13] originally includes the procedure to extract parts with bad convergence to be extensively solved in the preconditioning part; however, we skip it for simplicity. In addition, only single and double precision numbers are used, and the custom data type FP21 is not used in the computations.

3   Application Example

3.1   Definition of the Considered Problem

We apply the developed method to estimate the soil structure, aiming to verify its efficiency. The supercomputer ABCI [2], operated by the National Institute of Advanced Industrial Science and Technology, is used for each forward analysis. Each compute node of ABCI has four NVIDIA Tesla V100 GPUs and two Intel Xeon Gold 6148 CPUs (20 cores). The GPUs in each compute node are connected via NVLink, with a bandwidth of 50 GB/s in each direction. We conduct each forward analysis using one compute node of ABCI. We assign one GPU for the finite element solver using MPI and use all CPU cores for generating models with OpenMP. Moreover, a GPU cluster composed of the IBM Power System AC922, which has four NVIDIA Tesla V100 GPUs and two IBM Power9 CPUs (16 cores) per compute node, is used for learning and inference of neural networks.

The target domain has four layers, and we define their boundary surfaces. In the case of this problem, material properties of the soil structure are deterministic. These properties are described in Table 1. The target domain is of the following size: $0\,\mathrm{m} \leq x \leq 500\,\mathrm{m}$, $0\,\mathrm{m} \leq y \leq 400\,\mathrm{m}$, and $-50\,\mathrm{m} \leq z \leq 1.62\,\mathrm{m}$. A finite element model with approximately 1,400,000 degrees of freedom is generated. Figure 1 represents one of the finite element models used in the analysis. The resolution was 5 m at maximum.

Table 1. Material properties of the target domain. V_p, V_s, and ρ are primary and secondary wave velocity, and density, respectively. h is the damping ratio.

	V_p(m/s)	V_s(m/s)	ρ (kg/m^3)	h
1st layer	1500	130	1700	0.02
2nd layer	1600	220	1800	0.02
3rd layer	1600	160	1700	0.02
4th layer	2000	400	2000	0.001

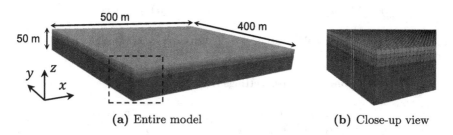

(a) Entire model   **(b)** Close-up view

Fig. 1. One of finite element models in the analysis.

Twelve observation points are located at points $(x, y) = (100i, 100j)$ ($i = 1 - 4$, $j = 1 - 3$) on the ground surface. We assume that the observation data at these points are available without noise. In fact, we target the low frequency band up to 2.5 Hz; therefore, the influence of noise is small. The control points are located at regular intervals in x and y directions. We denote the elevation of the k-th layer on points $(x, y) = (100i, 100j)$ ($i = 0 - 5$, $j = 0 - 4$) as α_{ij}^k(m). The points $x = 0$, $x = 500$, $y = 0$, and $y = 400$ are the edges of the domain, and we assume that α_{ij}^k are constant among these points for each layer. Boundary surfaces with reference parameters are represented in Fig. 2. We input the observed small earthquake wave for 80 s to the bottom surface of our target

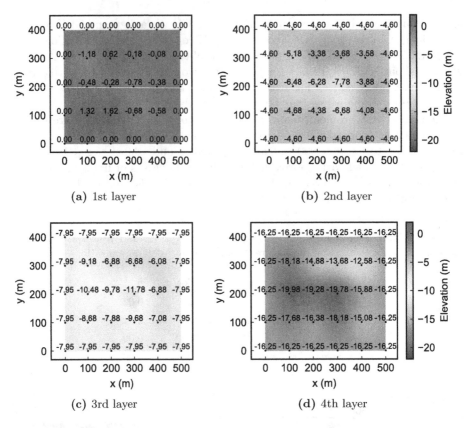

Fig. 2. Distribution of actual elevation (m) of each layer. The value of control parameters α_{ij}^k are also described.

Table 2. Problem settings for the control points.

	Control parameters	# of parameters
Case 1	$\alpha_{ij}^k(i = 1 - 4, j = 1 - 3, k = 4)$	12
Case 2	$\alpha_{ij}^k(i = 1 - 4, j = 1 - 3, k = 2 - 4)$	36

model. Within the current problem settings, the target frequency is as much as 2.5 Hz, so we apply a band-pass filter to remove unnecessary frequency bands. Time increment used in the analysis is 0.01 s; therefore, each wave propagation analysis comprises 8,000 times steps.

We perform testing on the two cases of the control parameters. The details are provided in Table 2. In case 1, we only estimate the elevation of the 4th layer. The elevations of other layers are known. In case 2, we estimate the elevations of the 2nd, 3rd, and 4th layers. The elevation of the 1st layer, which is the ground surface, is known. It should be noted that in case 2, there are more control parameters.

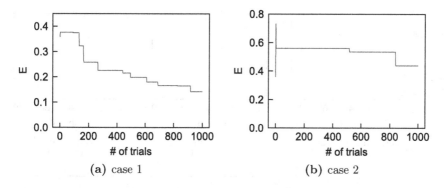

(a) case 1   (b) case 2

Fig. 3. History of the error E calculated by Eq. (1) in simulated annealing.

3.2   Estimation by a Typical Approach

First, we apply the approach proposed by [14] for parameter estimation, which is based on very fast simulated annealing [7]. Each control parameter is changed by multiplication of the difference between the lower and the upper limit values of the target parameter and the ratio r. r is computed as follows: $sgn(u - 0.5)T_i\left[(1 + 1/T_i)^{|2u-1|} - 1\right]$, where u is extracted from the uniform distribution between $[0.0, 1.0]$, and T_i is a temperature parameter in simulated annealing. The temperature at i-th trial is defined using initial temperature T_0 and the number of control points D as follows: $T_i = T_0\exp(-ci^{\frac{1}{D}})$, where parameter c is defined by T_0, D, lowest temperature T_f, and the number of trials i_f as $T_f = T_0\exp(-m)$, $i_f = \exp n$, and $c = m\exp(-\frac{n}{D})$. Here, $D = 12$ in case 1 and $D = 36$ in case 2. We set the number of trials $i_f = 1000$; and $c = 8.18$ in case 1 and $c = 12.00$ in case 2, respectively. These parameters satisfy the requirement that the parameters that increase the value of the error function by ΔE are adopted with the probability of 80% at the initial temperature, where ΔE is the value of the error function obtained in the initial model. In addition, the parameters which increase the value of the error function by $\Delta E \times 10^{-5}$ are adopted with the probability of 0.1% at the lowest temperature. These settings for c, T_0, and T_f are the same as those in [14].

The histories of the error value for both cases are presented in Fig. 3. In case 1, the parameters are updated at 100-trial intervals, on average. In contrast, the parameters are rarely updated in case 2. The final value of E is 39.7% of the initial parameters in case 1 and 122.2% in case 2. Figure 4 describes the time history of velocity on point $(x, y, z) = (400, 300, -0.08)$ when conducting wave propagation analysis with both estimated parameters. Analyzing these results, for case 1, we can see that the obtained wave has come close to that of the reference model to some extent. For case 2, the result obtained using the estimated model completely differs from that of the reference model. We assume that a much larger number of trials would be required for case 2.

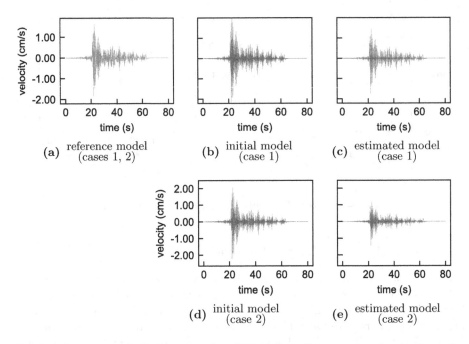

Fig. 4. x component of velocity at $(x, y) = (400, 300)$ on the ground surface in the analysis (blue lines). Differences with the results of the reference model are also represented in (b) to (e) as red lines. (Color figure online)

3.3   Estimation by the Proposed Method

Next, we apply the proposed method for case 2, which is presumed to be more difficult. We introduce neural networks for classification of error $E_{i_{\text{obs}}}$ in Eq. (2) according to the problem settings described above. We add random numbers following the normal distribution with $3\sigma = 5$ m to the control parameters α_{ij}^k. We generate $n_0 = 1000$ parameter sets and conduct forward analysis for each case. Here, we assign 800 cases for the training data and 200 cases for the test data. We produce twelve neural networks for each of the twelve observation points in the target domain. We employ Pytorch [10] to develop neural networks. Learning and inference are accelerated by GPUs. We consider fully connected layers for the neural networks. The number of units, the number of layers, and dropout rate, which are representative hyperparameters in neural networks, are optimized using optuna [3]. This framework searches hyperparameters that minimizes the summation of $\frac{|L^{\text{ref}} - L|}{L^{\text{ref}}}$ in the test data, where L is the level of error judged by a neural network ($1 \leq L \leq 10$), and L^{ref} is the actual level of error derived by forward analysis ($1 \leq L^{\text{ref}} \leq 10$). We use this index aiming to obtain more accurate classification for lower levels of error. The number of units, the number of layers, and dropout rate are fluctuated within the range of 6–72, 3–8, and 0.1–0.9, respectively. The number of trials is set to 50. We use a rectified linear unit for the activation function. In addition, we use Adam as the optimizer. Here, the batch size is set to 300, and the number of epochs is set to 1,000.

Optimized neural networks output correct levels with the probability of 71% and levels with an error of no more than plus/minus one level with the probability of 99% on average. We infer the error for $n_1 = 2,000,000$ parameter sets using the generated neural networks and extract $n_0 = 1,000$ cases that have lower estimated error levels. Next, we compute the actual error for the extracted parameter sets ($SuperiorDataset_1$ in Algorithm 1) by forward analyses. As a result, error E is generally reduced compared to the original dataset ($TrainingDataset_1$ in Algorithm 1), as represented in Fig. 5. This figure also includes the result obtained by repeating the same procedure once again ($SuperiorDataset_2$). The error distribution is shifted to the left side of the previous distribution. Analyzing this figure, we can confirm that the approach based on neural network is effective for extracting parameters that provide smaller errors. By using neural networks twice, the error E is reduced to 21% with respect to that of the initial parameters, as shown in Table 3.

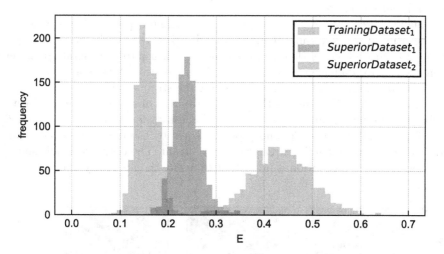

Fig. 5. Histogram of the error E for initial dataset ($TrainingDataset_1$ in Algorithm 1) and datasets extracted by using neural networks ($SuperiorDataset_1$, $SuperiorDataset_2$). Each dataset includes 1,000 sets of parameters.

Table 3. Error calculated by Eq. (1) in the proposed method. Results when generating AI once ($i_{\text{trial}} = 1$ in Algorithm 1) and twice ($i_{\text{trial}} = 2$) are listed.

	Initial parameters	Estimated parameters	
		$i_{\text{trial}} = 1$	$i_{\text{trial}} = 2$
E	0.358594	0.165238	0.077423

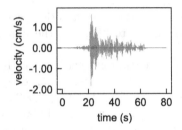

Fig. 6. x component of velocity at $(x, y) = (400, 300)$ on the ground surface using the model estimated by the proposed method (blue line). Differences with the results of the reference model are also represented by the red line. (Color figure online)

For confirmation of the accuracy of the estimated model, we conduct wave propagation analysis using the estimated parameters. Figure 6 describes the time history of velocity on point $(x, y, z) = (400, 300, -0.08)$, and Fig. 7 represents the distribution of displacement on the ground surface at time $t = 30.00$ s. We confirm that the results obtained by the estimated model and reference model are sufficiently consistent.

Finally, we computed the distribution of maximum principal strain, which can be utilized for screening of underground structures which might be damaged. Reference [8] notes that strain should be taken into consideration rather than velocity for damage estimation of buried pipelines, which is a typical type of underground structure. Figure 8 represents the strain distribution on the surface at $z = -1.2$ m. We confirm that the distribution of strain is estimated with sufficient accuracy by our proposed method. These strain distributions obtained by using the estimated model and those obtained by the initial model are considerably different, including areas with high strain. The obtained result shows that the proposed estimation is important to assure the reliability of evaluation.

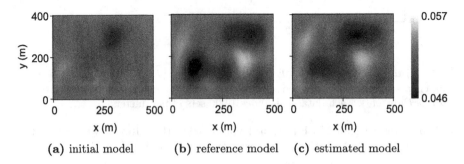

(a) initial model   (b) reference model   (c) estimated model

Fig. 7. Norm distribution of displacement (cm) on the ground surface at $t = 30.00$ s in the analysis.

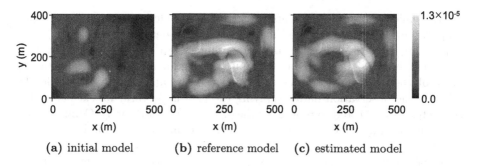

Fig. 8. Norm distribution of max principal strain on the plane $z = -1.2$ in the analysis.

3.4   Evaluation of Computation Cost

In this study, we evaluate the computation cost for case 2. In the approach based on simulated annealing, we computed 1,000 forward analyses sequentially. Each trial took twelve min to complete; and therefore, the total elapsed time was $12\,\text{min} \times 1,000\,\text{cases} = 12,000\,\text{min} \simeq 8$ days. In fact, the greater number of trials would be necessary, as the value of error remained high. In the proposed method, we set $MaxTrialNumber$ in Algorithm 1 to be 2 and iterated the generation of a neural network with 1,000 forward analyses and computation of top 1,000 cases extracted by neural networks twice. In total, we constructed neural networks twice and conducted 4,000 forward analyses. It required 30 min to define each neural network; therefore, one hour was required for training the neural networks in total. We used the supercomputer ABCI to perform wave propagation analysis. About 200 nodes were simultaneously available on average, although the number of available compute nodes depended on the utilization of the system. As we computed 200 cases in parallel, the elapsed time for forward analyses was $12\,\text{min} \times (4,000\,\text{cases}/200\,\text{cases}) = 240\,\text{min} = 4\,\text{h}$. Computation cost in other parts was negligible; therefore, the total elapsed time was approximately $1\,\text{h}+4\,\text{h} = 5\,\text{h}$. Although the number of the computed cases was larger than that of simulated annealing, the actual elapsed time was reduced owing to parallel computations. As a result of this comparison, we confirmed the effectiveness of the proposed method.

4   Conclusion

Introduction of data science and machine learning is one of the effective approaches to reduce the computation cost associated with computational science. In the present paper, we focus on estimation of the inner soil structure via wave propagation analysis. This estimation of the inner structure is important to improve reliability in numerical simulations. However, massive computation cost is required, as typical optimization requires many forward analyses.

We applied the data-driven approach to this problem aiming to reduce the computation cost. The proposed method combined neural networks and wave propagation analysis. We generated the training data by executing many forward analyses on a large-scale supercomputer. We implemented a neural network that took the parameters of inner structures as input and outputted error levels based on the observation data. Applying the neural network, we extracted the parameter sets expected to reproduce the observation data closely. Computation cost required in inference was negligible; therefore, many cases could be roughly evaluated in a shorter time.

In the application example, we estimated the soil structure using the observation data with a total of 4,000 wave propagation analyses. We confirmed that the estimated model had the sufficiently consistent results compared with the reference model via evaluation of a strain distribution. Each forward analysis required in the proposed method was computed in parallel; and thereby the actual elapsed time was reduced by using a large-scale supercomputer. We confirmed the effectiveness of the proposed method via performance comparison with the typical approach based on very fast simulated annealing. It is future task to examine validity of our proposed method for more complex models.

The demand for utilization of Big Data will continue to increase further, and the development of computation environments, observation systems, and network technologies will follow accordingly. In the present paper, we outlined the importance of developing an algorithm that enables capacity computing and maximizes utilization of computer resources for processing the large quantity of data and obtaining new information within a realistic timeframe.

Acknowledgments. Our results were obtained using computational resource of the AI Bridging Cloud Infrastructure (ABCI) at the National Institute of Advanced Industrial Science and Technology. We acknowledge support from Japan Society for the Promotion of Science (18H05239 and 18K18873).

References

1. Big data and extreme-scale computing. https://www.exascale.org/bdec/
2. Computing resources of AI bridging clound infrastructure. https://abci.ai/en/about_abci/computing_resource.html
3. Akiba, T., Sano, S., Yanase, T., Ohta, T., Koyama, M.: Optuna: a next-generation hyperparameter optimization framework. In: Proceedings of the 25th ACM SIGKDD International Conference on Knowledge Discovery & Data Mining, KDD 2019, pp. 2623–2631. Association for Computing Machinery, New York (2019). https://doi.org/10.1145/3292500.3330701
4. Clement, R., Schneider, J., Brambs, H.J., Wunderlich, A., Geiger, M., Sander, F.G.: Quasi-automatic 3D finite element model generation for individual single-rooted teeth and periodontal ligament. Comput. Methods Programs Biomed. **73**(2), 135–144 (2004)
5. Ichimura, T., et al.: An elastic/viscoelastic finite element analysis method for crustal deformation using a 3-D island-scale high-fidelity model. Geophys. J. Int. **206**(1), 114–129 (2016)

6. Ichimura, T., Fujita, K., Yamaguchi, T., Hori, M., Lalith, M., Ueda, N.: Fast multi-step optimization with deep learning for data-centric supercomputing. In: 4th International Conference on High Performance Compilation, Computing and Communications (2020, accepted)

7. Ingber, L.: Very fast simulated re-annealing. Math. Comput. Model. **12**(8), 967–973 (1989)

8. Liang, J., Sun, S.: Site effects on seismic behavior of pipelines: a review. J. Press. Vessel Technol. **122**(4), 469–475 (2000)

9. Liu, R., Kumar, A., Chen, Z., Agrawal, A., Sundararaghavan, V., Choudhary, A.: A predictive machine learning approach for microstructure optimization and materials design. Sci. Rep. **5**(1), 1–12 (2015)

10. Paszke, A., et al.: PyTorch: an imperative style, high-performance deep learning library. In: Wallach, H., Larochelle, H., Beygelzimer, A., d'Alché Buc, F., Fox, E., Garnett, R. (eds.) Advances in Neural Information Processing Systems 32, pp. 8024–8035. Curran Associates, Inc. (2019). http://papers.nips.cc/paper/9015-pytorch-an-imperative-style-high-performance-deep-learning-library.pdf

11. Quinay, P.E.B., Ichimura, T., Hori, M.: Waveform inversion for modeling three-dimensional crust structure with topographic effects. Bull. Seismol. Soc. Am. **102**(3), 1018–1029 (2012)

12. Warner, M., et al.: Anisotropic 3D full-waveform inversion. Geophysics **78**(2), R59–R80 (2013)

13. Yamaguchi, T., Fujita, K., Ichimura, T., Naruse, A., Lalith, M., Hori, M.: GPU implementation of a sophisticated implicit low-order finite element solver with FP21-32-64 computation using OpenACC. In: Sixth Workshop on Accelerator Programming Using Directives (2019). https://waccpd.org/program/

14. Yamaguchi, T., Ichimura, T., Fujita, K., Hori, M., Wijerathne, L.: Heuristic optimization with CPU-GPU heterogeneous wave computing for estimating three-dimensional inner structure. In: Rodrigues, J., et al. (eds.) ICCS 2019. LNCS, vol. 11537, pp. 389–401. Springer, Cham (2019). https://doi.org/10.1007/978-3-030-22741-8_28

Data Assimilation in Volcano Deformation Using Fast Finite Element Analysis with High Fidelity Model

Sota Murakami[1]([⊠]), Takuma Yamaguchi[1], Kohei Fujita[1], Tsuyoshi Ichimura[1], Maddagedara Lalith[1], and Muneo Hori[2]

[1] Earthquake Research Institute and Department of Civil Engineering, The University of Tokyo, Bunkyo, Tokyo, Japan
{souta,yamaguchi,fujita,ichimura,lalith}@eri.u-tokyo.ac.jp
[2] Research Institute for Value-Added-Information Generation, Japan Agency for Marine-Earth Science and Technology, Yokohama, Japan
horimune@jamstec.go.jp

Abstract. Estimation of the inner state of volcanoes are important for understanding the mechanism of eruption and reduction of disaster risk. With the improvement in observation networks, data assimilation of internal magma state using time-history crustal deformation data observed at the surface is expected to be suitable for solving such problems. Using finite-element methods capable of modeling complex geometry is desirable for modeling the three-dimensional heterogeneous crust structure, and nonlinear time-history analysis is required for considering the change in material properties due to the movement of magma. Thus, many cases of large-scale finite-element analysis is required, and the computational cost incurred is expected to become a bottleneck. As a basic study towards data assimilation of internal magma state considering change in material properties of the crust, we demonstrated that many case analyses of volcano deformation problems can be conducted in a reasonable time frame by development of a crustal deformation analysis method accelerated by GPUs. For verification of the data assimilation method, we estimated the magma trend in an actual three-dimensional heterogeneous crust structure without temporal change in material properties. We confirmed that the magma movement trend can be reproduced using the model considering crust heterogeneity, while models disregarding three-dimensional crust structure resulted in wrong estimations. Thus, we can see that using finite-element methods capable of modeling three-dimensional heterogeneity for crustal deformation analysis is important for accurate magma state estimation.

Keywords: Data assimilation · Kalman filter · Finite element analysis · GPU computation

V. V. Krzhizhanovskaya et al. (Eds.): ICCS 2020, LNCS 12139, pp. 18–31, 2020.
https://doi.org/10.1007/978-3-030-50420-5_2

1   Introduction

Estimating the state of magma inside volcanoes are important for understanding the physical mechanism of volcanoes and predicting eruptions for disaster mitigation purposes. Available observation data is increasing with the improvement of observation networks by advance in sensor technology. For example, ground surface deformation data is available by Global Navigation Satellite Systems (GNSS) and Interferometric Synthetic Aperture Radar (InSAR) observations conducted by a variety of satellites. GNSS data have high temporal resolution, while InSAR data have high spatial resolution; thus, we can expect obtaining high resolution crustal deformation information by using both of these observation data. On the other hand, there is room for improvement in the analysis methods used for inner state estimation using these observation data. Using time-history observation data of crustal deformation, we target to estimate three-dimensional movement of magma in the crust. Here, the crust has three-dimensional heterogeneity with material property changing with the movement of magma. For such problems, data assimilation of magma state based on crustal deformation analysis using finite-element analysis which is capable of considering three-dimensional heterogeneity is expected to be effective. However, the finite-element analysis cost becomes huge and thus realization of such analysis is considered challenging. In such problems, cost for generating finite-element models with 10^{6-8} degrees of freedom and analyzing its linear response for 10^{3-5} cases is required. Thus, most methods approximate the heterogeneous three-dimensional crust with a homogeneous half space in crustal deformation analysis of volcanoes [10]. Crustal deformation analysis based on the finite-element method is started to be used; however, the scale of the analysis remains up to 2×10^5 degrees-of-freedom for 20 cases [13]. We can see that many-case finite-element analyses required for reflecting the change in material properties is challenging.

Fast finite-element solvers capable of solving island-scale crustal deformation analysis with fault slip at inter-plate boundaries have been developed [3,14]. We may be able to reduce the computational cost for volcanic crustal deformation by extending this method. Thus, we improve the large-scale seismic finite-element solver for application to volcano problems, and verify its applicability in this study. Together, we measure the computational cost required in many-case analyses. Furthermore, using the developed method, we conduct data assimilation for a hypothetical problem with a three-dimensional heterogeneous crust structure of an actual site to verify the estimation accuracy of time-history underground magma distribution estimation. This paper targets basic study on development of a method to estimate the movement of magma in the crust by using time-history crustal deformation observation at the surface, with appropriate modeling of the time-history change of the three-dimensional heterogeneous crust structure. The main target is development of a crustal deformation analysis method for volcanoes, and confirmation of its validity for many case analyses and applicability to data-assimilation. The change in material property of crust is slow in the target problem; thus, the characteristics of the problem is known to be constant in terms of the data assimilation. Considering the above, we neglect temporal change in crust material properties in the hypothetical problem in this study.

The rest of this paper is organized as follows. In Sect. 2, we explain the data assimilation method used in this study, and then explain the fast finite-element method developed for volcano problems. In Sect. 3, we verify the accuracy of the solver and measure the computational cost. In Sect. 4, we validate the method using a hypothetical problem with three-dimensional heterogeneous crust structure of an actual volcano Shinmoedake using GNSS and InSAR data. In Sect. 5, we summarize the paper with discussions on future prospects.

2   Methodology

2.1   Data Assimilation

We model the change in excessive pressure and the resulting expansion and contraction of the magma chamber using a spherical pressure source. In previous studies [12], it has been shown that the surface response for a spherical source can be approximated using a point source if the sphere is located in a depth more than three times its radius. In a point source, magma can be expressed as body forces. Thus, if the magma source is small compared to its located depth, we can express the distribution of excessive pressure using a body force density distribution without considering the material properties of the magma. Here, the excessive pressure of the magma source can be expressed as a linear combination of basis spatial distribution functions $B_k(x)$ as

$$f(x,t) = \sum_{k=1}^{M} B_k(x)c_k(t). \tag{1}$$

Here, x, t denote spatial coordinate and time, M is the number of basis functions, and $c_k(t)$ is the coefficient for each basis function. Using Eq. (1), we model the surface displacement $u_r(x,t)$ as

$$u_r(x,t) = G_r(f(x,t)) + e, \tag{2}$$

where G_r is the Green's function of surface displacement for excessive pressure at magma source, and e is the observation error. Here, r denotes the coordinate axis, where $r = 1, 2, 3$ corresponds to the North, East, and vertical directions. By neglecting abrupt volcanic activities such as eruptions or change in material properties due to movement of magma, we only need to consider long term quasi-static deformation for analyzing crustal deformation due to underground magma activity. Following this assumption, we approximate the crust as a linear elastic body. Due to the linear properties of Green's functions, applying Eq. (1) to Eq. (2) leads to

$$u_r(x,t) = \sum_{k=1}^{M} H_{rk}(x)c_k(t) + e, \tag{3}$$

where

$$H_{rk}(x) = G_r(B_k(x)). \tag{4}$$

In this study, we assimilate data observed by GNSS and InSAR. Three component displacement is obtained at each GNSS observation point; thus, the observation vector y_1 obtained by GNSS is expressed as

$$y_{1i}(t_j) = u_r(x_{n_1}, t_j), \quad i = 3(n_1 - 1) + r, \tag{5}$$

where t_j is the time at j-th time step, and x_{n_1} ($n_1 = 1, ..., N_1$) are the coordinates of the GNSS observation points. This observation vector consists of $3 \times N_1$ displacement components. We can rewrite Eq. (4) by substituting $x = x_{n_1}$ as

$$H_{ik}^1 = G_r(B_k(x_{n_1})). \tag{6}$$

On the other hand, the InSAR data only consists of the difference between two observations, with only a single displacement component parallel to the view direction. Here, we assume that we obtain one displacement component at each observation point via preprocessing. Thus, the observation vector y_2 obtained by InSAR is expressed as

$$y_{2i}(t_j) = n_r^T u_r(x_{n_2}, t_j), \quad i = n_2, \tag{7}$$

where x_{n_2} ($n_2 = 1, ..., N_2$) are the coordinates of the InSAR observation points. This observation vector consists of N_2 displacement components. Here, $n_r^T = (\cos\theta \sin\phi, \sin\theta \sin\phi, \cos\phi)$ is the normal vector parallel to the observation direction, where θ is the angle from the vertical direction to the satellite observation direction, and ϕ is the counter clockwise angle from the moving direction of the satellite measured from the North direction. In the case of InSAR, Eq. (6) is expressed as

$$H_{ik}^2 = n_r^T G_r(B_k(x_{n_2})). \tag{8}$$

By defining a combined observation vector $y = \{y_1, y_2\}$,

$$y_i(t_j) = \sum_{k=1}^{M} H_{ik} c_k(t) + e. \tag{9}$$

where we define a combined observation matrix $H = \{H^1, H^2\}$. Although GNSS and InSAR observation data includes observation errors (e.g., local GNSS benchmark motion, GNSS reference frame errors, InSAR planar correction) that must be considered during data assimilation, we assume these are removed by preprocessing.

A simple and high accuracy model for time-history evolution of excessive pressure distribution of magma is not yet known. Thus, assuming that occurrence of abrupt change is rare, we use the trend model

$$\frac{dX}{dt} = X + v_t, \tag{10}$$

in this paper. Here, $\boldsymbol{X}(t) = (c_1(t), \cdots, c_M(t))$ and $\boldsymbol{v}_t$ is system error following the Gauss distribution. Temporal discretization of Eq. (10) with the finite-difference method leads to

$$\boldsymbol{X}_{it} = \boldsymbol{X}_{it-1} + \boldsymbol{v}_{it}, \quad \text{where} \quad \boldsymbol{v}_{it} \sim N(\boldsymbol{0}, \boldsymbol{Q}_{it}). \tag{11}$$

Here, it indicates the time step number. The observation equation becomes

$$\boldsymbol{y}_{it} = \boldsymbol{H}\boldsymbol{X}_{it} + \boldsymbol{w}_{it}, \quad \text{where} \quad \boldsymbol{w}_{it} \sim N(\boldsymbol{0}, \boldsymbol{R}_{it}). \tag{12}$$

Using the scaling parameter α^2, the covariance of the evolution equation can be expressed as

$$\boldsymbol{Q}_{it} = \alpha^2 \Delta t \boldsymbol{I}, \tag{13}$$

where $\boldsymbol{I}$ is the identity matrix and $\Delta t = t_{it} - t_{it-1}$. The covariance of observation noise becomes

$$\boldsymbol{R}_{it} = \begin{pmatrix} \sigma_1^2 \boldsymbol{\Sigma}_1 & \boldsymbol{0} \\ \boldsymbol{0} & \sigma_2^2 \boldsymbol{\Sigma}_2 \end{pmatrix}, \tag{14}$$

where $\sigma_1^2, \boldsymbol{\Sigma}_1$ and $\sigma_2^2, \boldsymbol{\Sigma}_2$ are the covariance and correlation matrices of GNSS and InSAR observation errors, respectively. As the state-space Eqs. (10) and (11) are expressed as linear models following the Gauss distribution, we can use the Kalman filter algorithm for data assimilation. We use a Kalman smoother with length 200.

In order to conduct data assimilation of the above, computations of Green's functions for surface ground displacement response for each basis function is required. We use the finite-element method for this computation in this study.

2.2   Computation of Green's Functions Using the Finite-Element Method

We target the linear elastic crustal deformation under volcano expansion. The governing equation is

$$\sigma_{ij,j} + f_i = 0, \tag{15}$$

where

$$\sigma_{ij} = \lambda \epsilon_{kk} \delta_{ij} + 2\mu \epsilon_{ij}. \tag{16}$$

Here, σ, f are stress and external force, $\lambda, \epsilon, \delta, \mu$ are the first Lame coefficient, strain, the Kronecker delta and the shear modulus, respectively. By spatial discretization with the finite-element method, the governing equation becomes a linear system of equations

$$\boldsymbol{K}\boldsymbol{u} = \boldsymbol{f}, \tag{17}$$

where $\boldsymbol{K}, \boldsymbol{u}, \boldsymbol{f}$ are the global stiffness matrix, nodal displacement, and external force vectors, respectively.

We use the method in [2] for expressing external force, where an equivalent displacement field for excessive pressure in a void in the magma chamber is expressed using infinitesimal spheroidal pressure sources. This method expresses body force as a sum of three perpendicular dipole moments. For an expansion source with dipole moment of f_0 at coordinate $\boldsymbol{x}'$, the body force $\boldsymbol{f}_c$ can be expressed as

$$\boldsymbol{f}_c = f_0 \nabla_x \delta(\boldsymbol{x} - \boldsymbol{x}'), \tag{18}$$

where $\delta(\boldsymbol{x})$ is the Dirac's delta function. The sizes of the dipole moments are given by Bonafede and Ferrari [1] as

$$f_0 = a^3 \Delta P \frac{\lambda(\boldsymbol{x}') + \mu(\boldsymbol{x}')}{\mu(\boldsymbol{x}')} \pi, \tag{19}$$

where a and ΔP are the radius and the pressure increment of the spherical pressure source. The external force vector in the finite-element analysis becomes

$$\boldsymbol{f} = \sum_e \int_{\Omega_e} [\boldsymbol{N}]_e^T \boldsymbol{f}_c d\Omega, \tag{20}$$

where $\boldsymbol{N}$ and Ω are the shape functions and element domains. By applying Eq. (18) to Eq. (20), the external force vector becomes

$$\boldsymbol{f} = \sum_e f_0 \nabla [\boldsymbol{N}(\boldsymbol{x}')]_e. \tag{21}$$

The outer force for a basis function for spatial distribution $\boldsymbol{B}_k(\boldsymbol{x})$ becomes

$$\boldsymbol{f}_k = \int \sum_e \boldsymbol{B}_k(\boldsymbol{x}) \nabla [\boldsymbol{N}_k(\boldsymbol{x})]_e dV. \tag{22}$$

By using this as the right-hand side vector of Eq. (17) and solving the linear system of equations, we can obtain **u** which corresponds to the Green's functions.

Most of the computation cost involved in finite-element analysis is generation of the finite-element model and the solver cost. We use the method of [8] to robustly generate finite-element mesh from digital elevation maps of crust structures. In our method, the spherical pressure sources are expressed without explicit modeling of the void regions in magma chambers. Thus, we do not require mesh that reflects the magma chamber geometry; we can use a single mesh reflecting the crustal structure for analyzing any pressure source. On the other hand, we need to solve the linear equation for each basis function. Thus, we can expect that the number of solver computations becomes more than the number of models to be generated. Thus, reduction of the time required for solving each of the linear set of equations becomes important for reduction of the cost of the whole procedure.

We use the method developed by Yamaguchi et al. [14] for the finite-element solver. This solver was originally developed for island-scale crustal deformation analysis for a given inter-plate fault-slip on CPU based large-scale computing environments, and was ported to GPU environments using OpenACC. The solver is based on the Adaptive Conjugate Gradient method, where the preconditioning equation is solved roughly using another Conjugate Gradient solver [6]. Here we call the iterations of the original solver as the outer loop, and the iterations of the solver used in solving the preconditioning equation as the inner loop. As the inner loop is only used as a preconditioner, we can solve it roughly. Thus, we can implement methods that can reduce computational cost in the inner loop. In this solver, we used mixed precision arithmetic and the multi-grid method. While the outer loop is computed in double precision, the inner loop is computed using single precision. By assigning suitable thresholds for the inner loop solvers, we can move most of the computation to the low-precision arithmetic parts which required less computational cost. The geometric multi-grid method and the algebraic multi-grid method are used for the multi-grid preconditioning. These methods lead to reduction in computation and data transfer; thus, it is effective for reducing computational cost on both CPUs and GPUs.

The sparse matrix-vector product Ku is the most time-consuming kernel in the preconditioned conjugate gradient method. In the solver, we use the Element-by-Element (EBE) method for the sparse matrix vector products except algebraic multi-grid. The multiplication of matrix K and vector u is obtained by adding up element-wise matrix-vector products as

$$f \leftarrow \sum_e Q_e K_e Q_e^T u, \tag{23}$$

where f is the resulting vector, Q_e is the mapping matrix from the local node number to the global node number, and K_e is the e-th element stiffness matrix. By computing the element stiffness on the fly instead of storing it in memory, we can reduce the amount of memory access, and thus improve computational performance. As the matrix for algebraic multi-grid is generated algebraically, the EBE method cannot be applied. Thus, we use 3×3 block compressed row storage for this part. Although Q_e is shown in matrix form in Eq. (23), it corresponds to random access in the actual code. Random read access to memory is involved for the right-hand side vector u part, and random write access to memory is involved for the left-hand side vector f part. Thus, the EBE method consists of element multiplications and random memory access. Generally, the performance on GPUs deteriorate with random memory access. Thus, in order to reduce the randomness of data access, we solve several equations at the same time. If the matrix is common, the EBE computations for several vectors can be expressed as

$$[f^1, ..., f^n] \leftarrow \sum_e Q_e K_e Q_e^T [u^1, ..., u^n]. \tag{24}$$

Here, n vectors are solved at the same time. Randomness of memory access is reduced as the read/write are conducted in consecutive address with the length of the number of vectors. This is expected to remove the performance bottleneck of the EBE kernel.

The approach of solving several vectors at the same time was originally developed for solving the crustal deformation to inter-plate fault-slip in a short time on GPUs. Although the target problem of this paper is different, it is the same from the viewpoint that displacement response to several sources are solved. Both of the discretized equations have the same stiffness matrix as the sources are only reflected to the external force vectors. Thus, we can expect applying the multiple vector method to our problem. Thus, we extended the multiple vector method such that point sources can be computed for the volcano expansion problem. By computing multiple Green's functions for multiple point sources, we reduce the time required for computation of each Green's function.

3   Verification

Using the developed method, we check the convergence of numerical solution and performance of the solver. Here we use an IBM Power System AC922 with 2 16-core IBM POWER9 2.60 GHz CPUs and 4 NVIDIA Tesla V100 GPUs. We use MPI to use all GPUs during computation, and use MPI/OpenMP to use all cores in the case of CPU computation.

We compared the numerical solution with Mogi's analytical solution [10] corresponding to response of elastic half space for a spherical pressure source. The compute domain is $-50\,\mathrm{km} \leq x \leq 50\,\mathrm{km}$, $-50\,\mathrm{km} \leq y \leq 50\,\mathrm{km}$, $-100\,\mathrm{km} \leq z \leq 0\,\mathrm{km}$, with the input source at $(0\,\mathrm{km}, 0\,\mathrm{km}, -4\,\mathrm{km})$. The dipole moment size is $1.0 \times 10^{16}\,\mathrm{Nm}$, with crust material properties of $V_p = 5.0\,\mathrm{km/s}$, $V_s = 2.9\,\mathrm{km/s}$, and density of $2.60\,\mathrm{g/cm^3}$. The finite-element model is generated with resolution of 500 m. From Fig. 1, we can see that the obtained solution follows the analytical solution, with relative error less than 0.02 at coordinate $(0, 0, 0)$ km. We can see that the analysis method is converged to the correct solution.

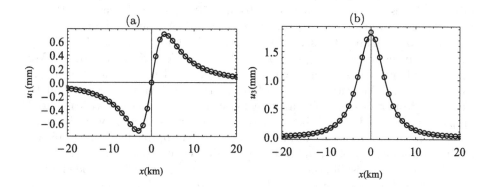

Fig. 1. Comparison of analytical solution (solid line) and finite-element results (circles) at horizontal and vertical directions at coordinate $(x, 0, 0)$.

Next, we measure performance for solving 12 point sources using GPUs. For comparison, we measured elapsed time for solving 1 vector at a time on GPUs, and 1 vector at a time on CPUs. Fig. 2 shows the elapsed time per vector. We can see that with the use of GPUs and the method for computing multiple vectors at the same time, we can accelerate computation by 35.5-fold from the CPU version. The elapsed time for the developed method was 23 s per vector, which corresponds to $23\,\text{s} \times 10^4 = 2.66$ days for computing 10^4 equations. We can see that many case analyses can be conducted in practical time by use of the developed method.

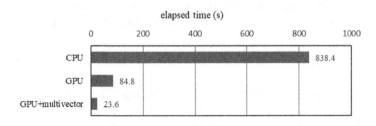

Fig. 2. Performance comparison of the elapsed time for solving 1 vector

4   Data Assimilation of Magma Source of Shinmoedake

We conducted data assimilation of Shinmoedake mountain using the developed method. Shinmoedake is an active volcano in the Kirishima-mountain range in Kyushu region of Japan. Recently, supply of magma and contraction due to eruption is repeated, and thus the state of magma chamber is changing. Thus, the estimation of the state of magma source is of high demand. In this section, we generate an artificial observation data of a magma eruption process, and use it to confirm that data assimilation of magma source of Shinmoedake is possible. The target region is of size $-50\,\text{km} \leq x \leq 50\,\text{km}$, $-50\,\text{km} \leq y \leq 50\,\text{km}$, $-100\,\text{km} \leq z \leq 1.4\,\text{km}$. Coordinate $(x, y, z) = (0, 0, 0)$ is set to the position of Shinmoedake according to the map of Geospatial Information Authority of Japan [5] with elevation set to the reference ellipsoid of the World Geodetic System (GRS80). We use the underground structure geometry and material properties given in the Japan Integrated Velocity Structure Model by the Headquarters for Earthquake Research Promotion [7]. The target region consists of 14 layers with material properties in Fig. 3. Figure 4 shows the generated finite-element model with minimum element size of $ds = 500\,\text{m}$. The total degrees-of-freedom was 4,516,032 and the total number of second-order tetrahedral elements was 2,831,842.

We express the distribution of magma source $B_k(x)$ using a 12 noded interpolation function shown in Fig. 5. The coordinates of nodes are given in Fig. 6. Nodes 5, 6, 7, and 8 models the volcanic vent, and the magma distribution in these sections are interpolated with one-dimensional linear functions. The other nodes at plane $z = -6\,\text{km}$ model the magma chamber; these sections are interpolated with two-dimensional linear functions. Each basis function is normalized such that the integration $\int B_k(x)dx$ over the domain becomes 1 N m. Horizontal resolution is set based on resolution of GNSS station. We assume a hypothetical trend in the magma source, and use this as the target for estimation. The target trend is shown in Fig. 8, where only nodes 2, 5, 6, 7, and 8 have non-zero values. This trend is generated such that the magma pressure at the bottom of the vent is increasing, and the magma is ascending through the vent. Considering that displacement of 4 cm is observed in EBINO-MAKIZONO GNSS baseline length in one year in 2011 [9], we set the trend such that the maximum displacement becomes the order of 10 cm. We use time step increment of $\Delta t = 0.25$ day, and total number of time steps as $T = 1000$. The scaling parameter for the system noise in Eq. (13) is set to $\alpha^2 = 4 \times 10^{32}\,\mathrm{N}^2\mathrm{m}^2/\mathrm{day}$. This value is set such that it becomes similar to the value of covariance at the abrupt change in the target trend.

We obtain the observation data by applying the Green's function to the magma source. We assume that GNSS data is available every 6 h at 8 observation points located near Shinmoedake (GEONET GPS-based Control Stations of Geospatial Information Authority of Japan [4]), and assume that InSAR data is available every 14 days at 60×60 km region with 500 m mesh resolution (total of 2001 observation channels). We assume that GNSS data is available every 6 h at 8 observation points located near Shinmoedake (GEONET GPS-based Control Stations of Geospatial Information Authority of Japan [4]). The coordinates of GNSS reference stations are given in Fig. 7. We also assume that InSAR data is available every 14 days at 60×60 km region with 500 m mesh resolution (total of 2001 observation channels). The observation error is assumed to be $\sigma_1 = 1\,\mathrm{mm}^2$, $\sigma_2 = 5\,\mathrm{mm}^2$. The correlation matrix $\boldsymbol{\Sigma}_1$, $\boldsymbol{\Sigma}_2$ are assumed to be identity matrices with noise applied to the artificial observation data. We use the same observation data and correlation matrix in the data assimilation process. We use $\theta = 30°$ and $\phi = 10°$ assuming a typical polar orbit satellite for the InSAR observation directions. Although InSAR data is known to involve correlation between data, we neglect this correlation for simplicity. In this section, we conduct estimation for three cases with different observation data: case (a) using GNSS, case (b) using InSAR, and case (c) using both GNSS and InSAR.

Figure 9 shows the estimated results for each case. While node 2 representing the magma chamber is estimated for all cases, estimation at nodes 5 and 6 corresponding to the vent differs among the cases. We can see that the movement of magma in the vertical direction is less sensitive compared to its horizontal movement, and thus estimation of vertical distribution is more challenging. In order to compare the results quantitatively, we use the root mean square error RMSE $= \sqrt{\frac{1}{TM}\sum_{t=1}^{T}\|\mathbf{X}_t - \hat{\mathbf{X}}_t\|^2}$, where $\hat{\mathbf{X}}_{1:T}$ is the estimation obtained by data assimilation, and M is the number of basis functions ($M = 12$). From values in Fig. 9, we can see that RMSE is the smallest for case (c). From this, we can see that the accuracy of inner state estimation is improved with combination of available data when compared to using a single observation source.

In order to discuss the effect of the crust structure, we finally conducted data assimilation for the case disregarding the three-dimensional crust structure. A finite-element model with flat surface and homogeneous material properties is generated, and the same data assimilation procedure was conducted. Here, material number 14 in Fig. 3 was used, and the surface elevation was set to the

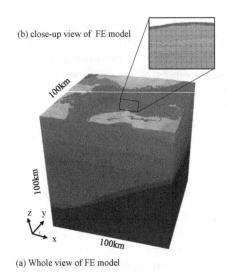

(b) close-up view of FE model

(a) Whole view of FE model

Layer	V_p (km/s)	V_s (km/s)	ρ (g/cm^3)
1	1.7	0.35	1.80
2	1.8	0.5	1.95
3	2.0	0.6	2.00
4	2.1	0.7	2.05
5	2.2	0.8	2.07
6	2.3	0.9	2.10
7	2.4	1.0	2.15
8	2.7	1.3	2.20
9	3.0	1.5	2.25
10	3.2	1.7	2.30
11	3.5	2.0	2.35
12	4.2	2.4	2.45
13	5.0	2.9	2.60
14	5.5	3.2	2.65

Fig. 3. Material properties of layers in the target domain. V_p is P wave velocity, V_s is S wave velocity, and ρ is density.

Fig. 4. Overview of the finite element model.

mean elevation of the domain. The estimation results using both GNSS and InSAR is shown in Fig. 9 (d). Even if both of the observation data is used, the estimated trend was completely different from the input trend when disregarding the crust structure. In addition, the estimation results vary with the observation data; thus, periodic peaks appear when InSAR data is included. Note that the homogeneous Green's function works for qualitative estimation of single source, as described in [11]. However, the error in Fig. 9 (d) was significantly larger than the case considering the crustal structure; we can see that consideration of the crust structure is important for high resolution estimation of the target problem. From here, we can see that the use of Green's functions computed by finite-element modeling proposed in this paper is important.

5   Closing Remarks

In this paper, we conducted basic study towards magma state estimation using time-history crust deformation data observed at the surface. As the crust involves three-dimensional heterogeneity and that the material properties of the crust changes with the state of magma, data assimilation using many-case three-dimensional finite-element analysis is expected to be suitable. In this study, we developed a fast GPU-accelerated solver for computing crustal deformation for excessive pressure sources of the magma. The elapsed time for solving a single Green's function was 23 s; corresponding to solving 10^4 equations in 2.6 days. We can see that many case finite-element analysis has become possible within a reasonable time frame. In the application example, we showed that the

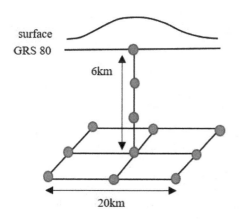

node	coordinate
1	$(-10, -10, -6)$
2	$(-10, 0, -6)$
3	$(-10, 10, -6)$
4	$(0, -10, -6)$
5	$(0, 0, -6)$
6	$(0, 0, -4)$
7	$(0, 0, -2)$
8	$(0, 0, 0)$
9	$(0, 10, -6)$
10	$(10, -10, -6)$
11	$(10, 0, -6)$
12	$(10, 10, -6)$

Fig. 5. Location of the input points. The magma distribution is interpolated with one-dimensional or two-dimensional linear functions, and input points represent control point of the magma distribution.

Fig. 6. Location of nodes in km shown in Fig. 5

station name	coordinate
KAGOSHIMAOOKUCHI	$(-27.2, 16.4)$
EBINO	$(-2.0, 15.3)$
NOJIRI	$(18.2, 6.2)$
AIRA	$(-27.1, -9.4)$
MAKIZONO	$(-12.0, -6.0)$
MIYAKONOJOU2	$(8.6, -3.5)$
HAYATO	$(-14.3, -18.4)$
MIYAKONOJOU	$(12.8, -18.8)$

Fig. 7. Coordinates of GNSS reference stations in km. z is set to the surface level.

magma source trend in a three-dimensional heterogeneous crust structure can be estimated using an artificial observation data set. Data assimilation based on a homogeneous half space model resulted in wrong results; thus, we can see the importance of considering the heterogeneous crust structure and the use of finite-element method capable of modeling three-dimensional heterogeneity. As the data assimilation results were shown to differ with the crust structure, we plan to incorporate subsurface structure and topography with much more heterogeneity than the stratified structure used in Sect. 4, and plan to estimate the effect of uncertainty in the underground crust structure in our future work. Furthermore, we plan to extend the method for application to problems with material nonlinearity of the crust.

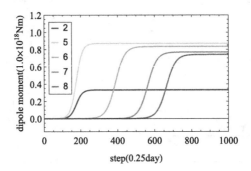

Fig. 8. Target input trend in strength of each dipole. Nodes that are not indicated in the figure are set to have zero dipoles.

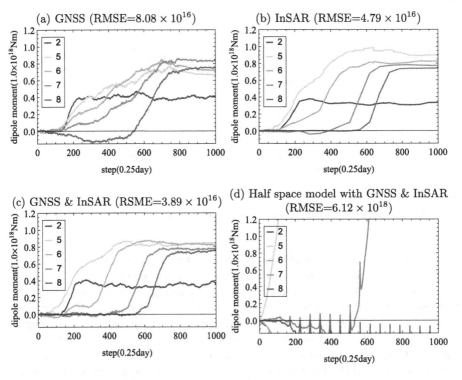

Fig. 9. Estimated strength of each dipole (N m) with the observation data (a) GNSS alone (b) InSAR alone (c) GNSS & InSAR on the three-dimensional heterogeneous models. (d) is the result using Half space model with GNSS & InSAR. The numbers in the legend correspond to nodes shown in the Fig. 6. A result closer to the input trend in Fig. 8; a lower RMSE is preferable.

Acknowledgements. We thank Dr. Takane Hori and Dr. Ryoichiro Agata of JAM-STEC for preparation of the DEM (digital elevation model) data for constructing the FE model. We acknowledge support from Japan Society for the Promotion of Science (18H05239 and 18K18873).

References

1. Bonafede, M., Ferrari, C.: Analytical models of deformation and residual gravity changes due to a Mogi source in a viscoelastic medium. Tectonophysics **471**(1–2), 4–13 (2009)
2. Charco, M., Galán del Sastre, P.: Efficient inversion of three-dimensional finite element models of volcano deformation. Geophys. J. Int. **196**(3), 1441–1454 (2014)
3. Fujita, K., Ichimura, T., Koyama, K., Inoue, H., Hori, M., Maddegedara, L.: Fast and scalable low-order implicit unstructured finite-element solver for earth's crust deformation problem. In: Proceedings of the Platform for Advanced Scientific Computing Conference, pp. 1–10 (2017)
4. Geospatial Information Authority of Japan: GNSS earth observation network system. http://terras.gsi.go.jp/geo_info/geonet_top.html
5. Geospatial Information Authority of Japan: GSI Maps. http://www.gsi.go.jp
6. Golub, G.H., Ye, Q.: Inexact preconditioned conjugate gradient method with inner-outer iteration. Soc. Ind. Appl. Math. J.: Sci. Comput. **21**(4), 1305–1320 (1999)
7. Headquarters for Earthquake Research Promotion: Underground model. https://www.jishin.go.jp/evaluation/seismic_hazard_map/underground_model/
8. Ichimura, T., et al.: An elastic/viscoelastic finite element analysis method for crustal deformation using a 3-D island-scale high-fidelity model. Geophys. J. Int. **206**(1), 114–129 (2016)
9. Kato, K., Yamasato, H.: The 2011 eruptive activity of Shinmoedake volcano, Kirishimayama, Kyushu, Japan—overview of activity and volcanic alert level of the Japan meteorological agency—. Earth Planets Space **65**(6), 2 (2013). https://doi.org/10.5047/eps.2013.05.009
10. Kiyoo, M.: Relations between the eruptions of various volcanoes and the deformations of the ground surfaces around them. Bull. Earthq. Res. Inst. **36**, 99–134 (1958)
11. Masterlark, T.: Magma intrusion and deformation predictions: sensitivities to the Mogi assumptions. J. Geophys. Res.: Solid Earth **112**(B6) (2006))
12. McTigue, D.: Elastic stress and deformation near a finite spherical magma body: resolution of the point source paradox. J. Geophys. Res.: Solid Earth **92**(B12), 12931–12940 (1987)
13. Ronchin, E., Masterlark, T., Dawson, J., Saunders, S., Martì Molist, J.: Imaging the complex geometry of a magma reservoir using FEM-based linear inverse modeling of InSAR data: application to Rabaul Caldera, Papua New Guinea. Geophys. J. Int. **209**(3), 1746–1760 (2017)
14. Yamaguchi, T., Fujita, K., Ichimura, T., Hori, M., Lalith, M., Nakajima, K.: Implicit low-order unstructured finite-element multiple simulation enhanced by dense computation using OpenACC. In: Chandrasekaran, S., Juckeland, G. (eds.) WACCPD 2017. LNCS, vol. 10732, pp. 42–59. Springer, Cham (2018). https://doi.org/10.1007/978-3-319-74896-2_3

Optimization and Local Time Stepping of an ADER-DG Scheme for Fully Anisotropic Wave Propagation in Complex Geometries

Sebastian Wolf[1][(✉)], Alice-Agnes Gabriel[2], and Michael Bader[1]

[1] Technical University of Munich, Munich, Germany
wolf.sebastian@in.tum.de
[2] Ludwig-Maximilians-Universität München, Munich, Germany

Abstract. We present an extension of the earthquake simulation software SeisSol to support seismic wave propagation in fully triclinic anisotropic materials. To our best knowledge, SeisSol is one of the few open-source codes that offer this feature for simulations at petascale performance and beyond. We employ a Discontinuous Galerkin (DG) method with arbitrary high-order derivative (ADER) time stepping. Here, we present a novel implementation of fully physical anisotropy with a two-sided Godunov flux and local time stepping. We validate our implementation on various benchmarks and present convergence analysis with respect to analytic solutions. An application example of seismic waves scattering around the Zugspitze in the Bavarian Alps demonstrates the capabilities of our implementation to solve geophysics problems fast.

1 Introduction

To successfully model earthquakes and perform seismic simulations, accurate models for the source dynamics and the propagation of seismic waves are needed. For seismic wave propagation, acoustic, isotropic and anisotropic elastic, attenuating and poroelastic materials are the most important rheologies [21]. Seismic anisotropy affects speed and scattering of seismic waves depending on the direction of propagation and can be found on all scales in the solid Earth. Anisotropy can stem from finely layered or cracked materials, the internal crystal structure of minerals or the alignment of ice crystals in glaciers. Anisotropic material behavior is observed in fault zones [16,17] and accounted for on global scale in refinements of the Preliminary Reference Earth Model [8]. Effective anisotropy on the scales of seismic wavelengths can be modeled by assuming homogeneous materials with directional dependent properties.

Anisotropy is one of the key seismic properties next to velocity and intrinsic attenuation. Locally, at the scale of earthquake fault zones, large variations in anisotropy reflect the strong material contrasts, extreme strains, and high dissipation of damaged rock. At the global scale, variations in anisotropy provide snapshots of our planet's interior that inform our understanding of plate

© Springer Nature Switzerland AG 2020
V. V. Krzhizhanovskaya et al. (Eds.): ICCS 2020, LNCS 12139, pp. 32–45, 2020.
https://doi.org/10.1007/978-3-030-50420-5_3

tectonics. Imaging of anisotropy is also crucial in industry contexts such as exploration or geothermal reservoir development and maintenance. All these applications require efficient forward solvers, ideally accounting for physical anisotropy together with the geometrical complexity of the geological subsurface. High-order accuracy is crucial to resolve small variations of anisotropy, which are often within only a few percent variation of isotropic material, depending on tectonic context.

Anisotropic material behavior has been successfully included in Finite Difference schemes [11,22], pseudo-spectral methods [5], Spectral Element codes [13] and Discontinuous Galerkin (DG) schemes for the velocity–stress formulation [20] and for the velocity–strain formulation [26]. Only few open-source codes exist which are able to simulate seismic wave propagation in anisotropic materials and which are also tailored to run efficiently on supercomputers. The DG ansatz on unstructured grids allows us to include full physical anisotropy as we do not encounter geometrical restrictions. The DG software SeisSol has undergone end-to-end performance optimization over the last years [3,10,25]. However, anisotropic effects have been neglected thus far.

In this paper, we present a novel implementation of fully anisotropic wave propagation that exploits SeisSol's high-performance implementation of element-local matrix operations and supports local time stepping. We first lay out the physical model and state the governing equations (Sect. 2). In Sect. 3 these equations are discretized using the DG method combined with *arbitrary high-order derivative* time stepping (ADER-DG). Our main numerics contribution is to introduce a two-sided numerical flux of the Godunov type in conjunction with a free-surface boundary condition based on solving an inverse Riemann problem. We highlight details of how we implemented theses features into the existing code base, and extended it to make use of local time stepping. Here, the key novelty is a general approach to integrate a numerical eigenvalue solver in SeisSol that replaces analytically derived formulas in the respective precomputation steps. In Sect. 5 we verify our implementation against various analytical solutions and community benchmark problems. We also present an updated reference solution for the AHSP1 benchmark [18], since our implementation revealed physical inconsistencies in the previous community reference solution. To demonstrate the capability of our code to solve real-world geophysical problems we model anisotropically scattering seismic waves radiating from a point source under the strong topography contrasts of Mount Zugspitze in the Bavarian Alps.

2   Physical Model

Linear elastic materials are characterized by a stress–strain relation in the form

$$\sigma_{ij} = c_{ijkl}\epsilon_{kl} \qquad \text{for } i,j \in \{1,2,3\}, \tag{1}$$

with stress and strain tensors denoted by $\sigma, \epsilon \in \mathbb{R}^{3 \times 3}$. Symmetry considerations reduce the 81 parameters c_{ijkl} to only 21 independent values [2]. Employing Voigt notation we can write the relation in a matrix–vector manner:

$$
\begin{pmatrix} \sigma_{11} \\ \sigma_{22} \\ \sigma_{33} \\ \sigma_{23} \\ \sigma_{13} \\ \sigma_{12} \end{pmatrix} = \underbrace{\begin{pmatrix} c_{11} & c_{12} & c_{13} & c_{14} & c_{15} & c_{16} \\ & c_{22} & c_{23} & c_{24} & c_{25} & c_{26} \\ & & c_{33} & c_{34} & c_{35} & c_{36} \\ & & & c_{44} & c_{45} & c_{46} \\ & sym & & & c_{55} & c_{56} \\ & & & & & c_{66} \end{pmatrix}}_{=: \mathcal{H}} \begin{pmatrix} \epsilon_{11} \\ \epsilon_{22} \\ \epsilon_{33} \\ 2\epsilon_{23} \\ 2\epsilon_{13} \\ 2\epsilon_{12} \end{pmatrix} . \tag{2}
$$

This constitutive relation can be combined with the equations of motion of continuum mechanics which we write in the velocity–strain formulation, where the vector $Q = \left(\sigma_{11}, \sigma_{22}, \sigma_{33}, \sigma_{13}, \sigma_{23}, \sigma_{13}, u_1, u_2, u_3\right)^T$ defines the quantities of interest. The combined equation reads:

$$
\frac{\partial Q_p}{\partial t} + A_{pq}^1 \frac{\partial Q_q}{\partial x_1} + A_{pq}^2 \frac{\partial Q_q}{\partial x_2} + A_{pq}^3 \frac{\partial Q_q}{\partial x_3} = 0. \tag{3}
$$

The Jacobian matrices $A^d, d = 1, 2, 3$, can be deduced from the stress–strain relation and have the form

$$
A^d = \begin{pmatrix} 0 & \mathcal{C}^d \\ \mathcal{R}^d & 0 \end{pmatrix} \text{ with, e.g., } \mathcal{C}^1 = \begin{pmatrix} -c_{11} & -c_{16} & -c_{15} \\ -c_{21} & -c_{26} & -c_{25} \\ -c_{31} & -c_{36} & -c_{35} \\ -c_{61} & -c_{66} & -c_{65} \\ -c_{41} & -c_{46} & -c_{45} \\ -c_{51} & -c_{56} & -c_{55} \end{pmatrix} . \tag{4}
$$

We observe that the second index is constant for each column. To construct the matrices $\mathcal{C}^2$ and $\mathcal{C}^3$ we replace these by 6, 2, 4 and 5, 4, 3 respectively. The blocks $\mathcal{R}^d$ are the same as for the isotropic case and are detailed in [20]. The material parameters can vary in space. For better readability the space dependence has been dropped in Eq. (3). Isotropic material behavior can be seen as a specialization of anisotropy, where $c_{11} = c_{22} = c_{33} = \lambda + 2\mu$, $c_{12} = c_{13} = c_{23} = \lambda$, $c_{44} = c_{55} = c_{66} = \mu$ and all other parameters are zero.

3   Numerical Approximation

De la Puente et al. [20] presented the numerics of including anisotropic material effects into ADER-DG seismic wave propagation simulations. We here improve the numerical scheme by a two-sided Godunov flux and a free-surface boundary condition, as well as adaptions necessary for local time stepping with anisotropy. A two-sided flux is physically more accurate and allows for coupling between different rheologies. Local time stepping improves performance drastically.

3.1 Spatial Discretization

To solve Eq. (3), we follow a DG ansatz [7]. The underlying geometry is approximated by a mesh of tetrahedral elements τ^m. For the discretization polynomial ansatz functions Φ_l are defined on a reference element τ_{ref}. On each element the numerical solution Q_p^m is expanded in terms of the basis functions:

$$Q_p^m(x, t) = \hat{Q}_{pl}^m(t)\Psi_l^m(x) = \hat{Q}_{pl}^m(t)\Phi_l(\Xi^m(x)). \tag{5}$$

Here the function $\Xi^m : \tau^m \to \tau_{\text{ref}}$ is an affine linear coordinate transformation. By Ψ_l^m we denote the lth basis function transformed to the mth element.

On each element Eq. (3) is multiplied by a test function and integration by parts is applied leading to a semi-discrete formulation:

$$\int_{\tau^m} \Psi_k^m \frac{\partial \hat{Q}_{pl}^m}{\partial t} \Psi_l^m dV + \int_{\partial \tau^m} \Psi_k^m (n_d A_{pq}^d Q_q)^* dS - \int_{\tau^m} \frac{\partial \Psi_k^m}{\partial x_d} A_{pq}^d \hat{Q}_{ql}^m \Psi_l^m dV = 0. \tag{6}$$

The Jacobians A_{pq}^d are element-wise constant. Also $\hat{Q}$ and its time derivative are constant on each cell. This allows us to pull these quantities out of the integrals. Applying a change of variables to the reference element the integrals can be precomputed. Together with an appropriate flux formulation this leads to a quadrature-free numerical scheme.

3.2 Flux and Boundary Conditions

In DG schemes continuity across element boundaries is only enforced in a weak sense, via the flux term $(n_d A_{pq}^d Q_q)^*$ in Eq. (6). Hence, a proper numerical flux, which also takes the underlying physics into account, is essential. De la Puente et al. [20] demonstrated anisotropy with one-sided Rusanov flux and discussed an extension to Godunov fluxes for ADER-DG. Two-sided fluxes capture the correct jump conditions of the Rankine-Hugoniot condition on both sides of the inter-element boundaries. They have been introduced to SeisSol in [23] for acoustic and (visco)elastic materials. In the following we give an overview over the most important aspects of using two-sided flux formulations and on generalizing the isotropic flux to a two-sided formulation for the anisotropic case.

In Eq. (6), the surface integral over $\partial \tau^m$ can be dispatched into four integrals over the four triangular faces of each element. We evaluate the flux for each face individually, so we need to transform the quantities Q as well as the stress–strain relation into face-aligned coordinate systems. For anisotropic materials the stress–strain relation is represented by the matrix $\mathcal{H}$, see Eq. (2). We can express the constitutive behavior in the face-aligned coordinate system via the matrix $\tilde{\mathcal{H}} = N \cdot \mathcal{H} \cdot N^T$ (cf. [4]), where N is the so-called Bond matrix (cf. [20]). We define the matrix $\tilde{A}$ to have the same structure as the matrix A^1 but with entries $\tilde{c}_{ij}$ from the matrix $\tilde{\mathcal{H}}$. At each face we have the Jacobians and approximations on the inside $\tilde{A}^-, Q^-$ and on the outside $\tilde{A}^+, Q^+$.

The Godunov flux approximates the solution at the element boundary by solving a Riemann problem across the element interfaces. First the equations are transformed to a face-aligned coordinate system. The Rankine-Hugoniot condition states that discontinuities travel with wave speeds given via the eigenvalues of the Jacobian. The differences between the quantities are the corresponding eigenvectors. A detailed derivation can be found in [23, 26].

We observe that the eigenvectors of the Jacobian $\tilde{A}$ have the form

$$
R = \begin{pmatrix}
r_1^1 & r_2^1 & r_3^1 & 0 & 0 & 0 & r_3^1 & r_2^1 & r_1^1 \\
r_1^2 & r_2^2 & r_3^2 & 1 & 0 & 0 & r_3^2 & r_2^2 & r_1^2 \\
r_1^3 & r_2^3 & r_3^3 & 0 & 1 & 0 & r_3^3 & r_2^3 & r_1^3 \\
r_1^4 & r_2^4 & r_3^4 & 0 & 0 & 0 & r_3^4 & r_2^4 & r_1^4 \\
r_1^5 & r_2^5 & r_3^5 & 0 & 0 & 1 & r_3^5 & r_2^5 & r_1^5 \\
r_1^6 & r_2^6 & r_3^6 & 0 & 0 & 0 & r_3^6 & r_2^6 & r_1^6 \\
r_1^7 & r_2^7 & r_3^7 & 0 & 0 & 0 & -r_3^7 & -r_2^7 & -r_1^7 \\
r_1^8 & r_2^8 & r_3^8 & 0 & 0 & 0 & -r_3^8 & -r_2^8 & -r_1^8 \\
r_1^9 & r_2^9 & r_3^9 & 0 & 0 & 0 & -r_3^9 & -r_2^9 & -r_1^9
\end{pmatrix}. \tag{7}
$$

The eigenvectors and corresponding eigenvalues resemble three incoming and outgoing waves. To take different material values on the inside and outside into account an eigendecomposition of both Jacobians $\tilde{A}^-$ and $\tilde{A}^+$ is performed and the matrix R is constructed taking the first three columns from R^- and the last three columns from R^+. With indicator matrices $I^- = \mathrm{diag}(1,1,1,0,0,0,0,0,0)$ and $I^+ = \mathrm{diag}(0,0,0,0,0,0,1,1,1)$, we can then compute the flux as:

$$
F = \frac{1}{2} \underbrace{T\tilde{A}^-(RI^+R^{-1})T^{-1}Q^-}_{G^+} + \frac{1}{2} \underbrace{T\tilde{A}^-(RI^-R^{-1})T^{-1}Q^+}_{G^-}. \tag{8}
$$

Here T is a matrix that rotates Q from the global coordinate system to the face-aligned coordinate system. We take into account that the first six components of $Q \in \mathbb{R}^9$ represent a symmetric tensor and the last three components represent a vector. Both parts can be rotated independently, so T combines the rotation of tensorial and vectorial quantities.

Analogous to inter-element boundaries, we also impose boundary conditions via a specialized flux. For a free surface boundary we want to impose $s = \sigma n = 0$. In Eq. (8) the term $RI^+R^{-1}T^{-1}Q^-$ is identified with the state at the inside of the inter-element boundary. We can use this fact to construct a flux which will yield the free surface boundary. To do so we set the traction s to zero and compute the velocity u consistently:

$$
\begin{pmatrix} s^b \\ u^b \end{pmatrix} = \begin{pmatrix} 0 & 0 \\ -R_{21}R_{11}^{-1} & I \end{pmatrix} \begin{pmatrix} s^- \\ u^- \end{pmatrix}. \tag{9}
$$

Superscripts b denote values at the boundary. The values s^- and u^- are the traction and velocity in the face-aligned coordinate system. The matrices R_{11} and R_{21} slice out the first three columns of R and the rows corresponding to the traction respectively the velocity components. The flux is obtained as $F = T\tilde{A}^-Q^b$. Note that we did not specify the non-traction components of σ. As these lie in the null space of the Jacobian $\tilde{A}$ the flux is not altered by their value.

3.3   Time Discretization

To integrate Eq. (6) in time SeisSol employs the ADER method [6]. ADER time stepping expands each element solution locally in time as a Taylor series up to a certain order. The time derivatives are replaced by the discretized spatial derivatives following the Cauchy–Kowalewski theorem. To update one element we therefore only need the values of the element itself and its four neighbors, which fosters efficient parallelization.

ADER time stepping inherently supports local time stepping [7]: Smaller elements or elements with high wave speeds will be updated more often than large elements or elements with low wave speeds. Local time stepping is thus crucial for performant applications that use strong adaptive mesh refinement or where meshes suffer from badly shaped elements. Setups with a heterogeneous material can also benefit substantially from local time stepping.

Each element has to satisfy the stability criterion $\Delta t^m < \frac{1}{2N+1} \frac{l^m}{v^m}$ for the time step size Δt^m, where l^m and v^m denote the in-sphere diameter and maximum wave speed of element τ^m, N is the order of the method.

In anisotropic materials the wave speeds depend on the direction of propagation. This has not been considered in previous work (e.g. [7]). For a fixed direction d we define the matrix $M(d)_{ij} = d_k c_{iklj} d_l$. We calculate the wave speeds in direction d from the eigenvalues λ_i of the matrix $M(d)$ via $v_i = \sqrt{\lambda_i / \rho}$ resulting in a primary and two secondary waves (cf. [4]). The element-wise maximum wave speed is the maximum of these speeds over all directions d.

4   Implementation

SeisSol's ADER-DG discretization is implemented via element-local matrix chain multiplications, which allows for high performance on modern CPUs [10,25]. All required matrices are precomputed in the initialization phase and optimized kernels are generated for the matrix chain operations [24]. In the following we present the most important choices we made for our implementation of anisotropy. Concerning the matrices, the compute kernels of the isotropic case can be reused, just the assembly of the Jacobians and flux matrices differs.

4.1   Storage Requirements

For each element we store the material values. In comparison to isotropic materials 22 instead of 3 values have to be stored. This overhead is negligible compared to the storage required for the degrees of freedoms. For example, a discretization of order 6 requires 504 degrees of freedom per element.

Concerning the storage of the precomputed matrices, only the Jacobians A and the flux matrices G^+ and G^- change between the isotropic and the anisotropic case. Based on the sparsity pattern of a matrix, the biggest rectangular non-zero block of the matrix is stored. We store the matrices A, G^+ and G^- as full 9×9 matrices for the isotropic as well as for the anisotropic case, thus no overhead is produced. The underlying data structures are not changed.

4.2   Calculation of the Flux Term

Unlike as with isotropic materials the eigenstructure of the Jacobians given in Eq. (7) is hard to express analytically. In the applied scheme the eigendecomposition has to be calculated once for every element. We use the open source software package Eigen3 [9] to obtain the eigenvectors numerically. We chose this approach for three reasons:

(i)   Even if an analytic expression for the eigenvalues is available it is lengthy and hence the implementation is error-prone.

(ii) From a software engineering point of view the use of the numerical eigenvalue solver replaces a lot of code that was previously needed for each material model individually. We unified the formulation of the Riemann solver for all material models (isotropic, anisotropic and viscoelastic). We use templating to distinguish the assembly of the Jacobians for each model. From then on we can use the same code to precalculate G^+ and G^-. We expect that these software engineering choices make it easy to include additional material models into SeisSol in the future. Also coupling between different material models within the same simulation can be obtained with little overhead.

(iii) The question of accuracy and stability of the numerical solver may arise. But stability of an analytically derived formula is also not guaranteed. Round-off errors and cancellation could drastically influence the accuracy of the derived eigenvectors. With our choice for using a stable numerical solver instead, we circumvent this problem.

4.3   Maximal Wave Speeds for Local Time Stepping

Local time stepping is implemented with a clustered scheme to meet the requirements of modern supercomputers [3]. To cluster the elements the required time step for each element has to be known in advance. To obtain the maximum wave speed for one element, we would have to find the maximum wave speed over all directions. This boils down to solving an optimization problem which involves the calculation of eigenvalues of arbitrary matrices. Solving this optimization problem analytically results in lengthy calculations. In practice, the time step is relaxed by a security factor to meet the CFL condition, so the maximum wave speed does not have to be computed exactly. We sample the wave speeds for several directions d and take their maximum as the maximum wave speed v^m.

5   Validation and Performance

5.1   Convergence Analysis

Planar wave analytic descriptions are widely used in wave propagation problems. Here we present a numerical convergence study to analyze the correct implementation to confirm its expected convergence properties. To this end, we verify our implementation solving the 3-D, anisotropic, seismic wave equations in the form of periodic, sinusoidal waves in a unit-cube as explained in [19]. The computational domain is the unit cube $[-1, 1]^3$ with periodic boundary conditions. The ansatz for our plane-wave solution is

$$Q_p(x, t) = \text{Re}\left(Q_p^0 e^{i(\omega t - k \cdot x)}\right) \tag{10}$$

where ω denotes the frequency and k is the vector of wave numbers. When we combine this with Eq. (3) we see that the initial condition Q^0 has to be a solution of the eigenvalue problem

$$A_{pq}^d k_d Q_q^0 = \omega Q_p^0. \tag{11}$$

In the case of linear elasticity there is a zero eigenvalue with multiplicity 3. The other eigenvalues appear pairwise with different signs and correspond to the P wave and two S waves. For isotropic materials the two S waves coincide, whereas for anisotropic media a slow and a fast S wave can be distinguished.

For linear PDEs a linear combination of several solutions is a solution again. To take the directional dependence of anisotropic materials into account we superimpose three planar waves with wave number vectors $k^1 = (\pi, 0, 0)$, $k^2 = (0, \pi, 0)$ and $k^3 = (0, 0, \pi)$. For each direction a P wave traveling in the direction of k^l and an S wave traveling in the opposite direction has been chosen. When we denote the eigenvectors of the matrix $A^d k_d^l$ with R^l and the corresponding eigenvalues with ω^l, the analytic solution can be written as

$$Q_p(x, t) = \sum_{l=1}^{3} \text{Re}\left(R_{p2}^l e^{i(\omega_2^l t - k^l \cdot x)} + R_{p9}^l e^{i(\omega_9^l t - k^l \cdot x)}\right). \tag{12}$$

The computational domain is discretized into cubes of edge length $h = \frac{1}{2}, \frac{1}{4}, \frac{1}{8}, \frac{1}{16}$. Each cube is split up into five tetrahedrons. The material is given by density $\rho = 1.00\,\frac{kg}{m^3}$ and the elastic tensor $c_{ij} = 0$ except for

$$\begin{aligned} c_{11} &= 192\,\text{Pa} & c_{12} &= 66.0\,\text{Pa} & c_{13} &= 60.0\,\text{Pa} \\ c_{22} &= 160\,\text{Pa} & c_{23} &= 56.0\,\text{Pa} & c_{33} &= 272\,\text{Pa} \\ c_{44} &= 60.0\text{Pa} & c_{55} &= 62.0\,\text{Pa} & c_{66} &= 49.6\,\text{Pa}. \end{aligned} \tag{13}$$

We compare the numerical solution to the analytic solution at time $t = 0.02$. Figure 1 shows the convergence behavior for the stress component σ_{11} in the L^2-norm. We clearly observe the expected convergence orders. The plots for the L^1- and L^∞-norm are comparable. All other quantities also show the expected convergence rates.

5.2 Isotropy via Anisotropy: Layer over Halfspace (LOH1)

The community benchmark LOH1 [18] is designed for isotropic elastic materials. We here use it to validate backwards compatibility, as isotropic elasticity is a special case of anisotropic elasticity. The setup consists of a layered half space

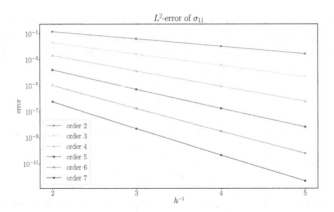

Fig. 1. 3D unit cube plane wave test case. Visualization of convergence behavior of the stress component σ_{11}. The L^2 error is shown versus the mesh spacing h.

with a free surface boundary on top. The top layer is 1 km thick with material parameters $\rho = 2600\,\frac{\text{kg}}{\text{m}^3}$, $\mu = 10.4\,\text{GPa}$ and $\lambda = 20.8\,\text{GPa}$. The half space below has material parameters $\rho = 2700\,\frac{\text{kg}}{\text{m}^3}$, $\mu = \lambda = 32.4\,\text{GPa}$. The source is a double couple point source with the only non-zero moment M_{xy} located in 2 km depth. The moment rate time history is given by the function $M_{xy}(t) = M_0 \frac{t}{T^2} \exp\left(-\frac{t}{T^2}\right)$ with maximal moment $M_0 = 10^{18}\,\text{Nm}$ and onset time $t = 0.1\,\text{s}$.

We compare the velocities at the free surface to the given reference solutions. There are nine receivers located along three different lines through the origin. The domain is the cuboid $[-40000, 40000] \times [-40000, 40000] \times [0, 32500]\,\text{m}^3$. On all other surfaces than the free surface ($x_3 = 0$) we impose absorbing boundary conditions. The mesh is refined around the source and coarsened away from it. The characteristic length is approximately 300 m in the vicinity of the source and grows up to 2000 m towards the boundary. In total the mesh consists of 2.57 million cells. The mesh is large enough that waves do not leave the domain during the computational time and we do not observe problems with artificial reflections. At the same time we keep the computational effort reasonable due to the coarsening towards the absorbing boundaries.

We ran the simulation with convergence order 6 up to a final time of 5 s and compared our solutions to the reference solution using envelope and pulse misfit [14]. We recorded the velocities at all nine receivers. In Table 1 we present the results for the third, sixth and ninth receiver, which are farthest away from the source (the longer the wave is propagated, the less accurately it is typically resolved due to numerical dissipation and dispersion). With a maximal envelope misfit of 1.32% and a maximal phase misfit of 0.20% we are well within the highest level of accuracy defined in the LOH1 benchmark description. To investigate the computational overhead of incorporating anisotropic effects, we compare the execution times for the setup executed with the isotropic and the anisotropic implementation. We ran each simulation 5 times on 100 nodes of SuperMUC-NG (Intel Skylake Xeon Platinum 8174, 48 cores per node) and averaged the

Table 1. Envelope and pulse misfits for the different velocity components

Receiver	EM			PM		
	x	y	z	x	y	z
3	1.32%	0.06%	0.06%	0.15%	0.00%	0.00%
6	0.79%	0.79%	1.04%	0.12%	0.10%	0.12%
9	0.99%	1.04%	0.83%	0.14%	0.10%	0.10%

total wall time. The average runtime of the anisotropic implementation was 185.0 s, which is only a little slower than the isotropic implementation, which averaged at 184.6 s. The difference between the runtimes may be explained by the more complicated initialization phase for the anisotropic case. We also point out that the deviation in run times in between different runs was larger than the difference between the averages. The computations achieved an average of 990.3 GFlop/s in the isotropic case and 994.7 GFlop/s in the anisotropic case.

5.3   Anisotropic Homogeneous Space (AHSP, SISMOWINE)

The SISMOWINE test suite [18] proposes the following test case for seismic wave propagation in anisotropic materials: The geometry is a homogeneous full space. The homogeneous material has a density of $\rho = 2700 \frac{\text{kg}}{\text{m}^3}$. The elastic response is characterized by the elastic tensor $c_{ij} = 0$ expect for

$$c_{11} = 97.2\,\text{GPa} \quad c_{12} = 10.0\,\text{GPa} \quad c_{13} = 30.0\,\text{GPa}$$
$$c_{22} = 97.2\,\text{GPa} \quad c_{23} = 30.0\,\text{GPa} \quad c_{33} = 70.0\,\text{GPa} \tag{14}$$
$$c_{44} = 32.4\,\text{GPa} \quad c_{55} = 32.4\,\text{GPa} \quad c_{66} = 43.6\,\text{GPa}.$$

The source is identical to the LOH1 benchmark. We again refine the mesh around the source with a characteristic edge length of about 300 m and coarsen away from the source. This results in a total of 3.98 million cells. To simulate a full space, absorbing boundary conditions are imposed on all six surfaces. Just as for the LOH1 test case we do not encounter artificial reflections from the boundaries.

Figure 2 shows an exemplary comparison between our new implementation and the reference solution of SISMOWINE. One can clearly see that the reference solution does not feature a second shear wave whereas our implementation does. Since shear wave splitting is a well-known physical feature of wave propagation in anisotropic media [2] we assume an error in the proposed reference solution. The correctness of our calculation is confirmed in comparison to an analytical reference: We used the open source tool christoffel [12] to compute the wave speeds for the given material depending on the direction of wave propagation. The shown receiver 6 is located at $(7348, 7348, 0)$, 10392 m away from the source. The calculated arrival time for the P wave is 1.73 s, the slow and the fast S wave arrive after 2.59 s and 3.00 s respectively. We observe that the arrival times align very well with the results calculated by the new SeisSol implementation.

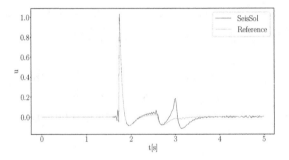

Fig. 2. Comparison of velocity component in x direction for receiver 6

In agreement with the original authors and the maintainers of the SISMOWINE project, our here presented solution has been accepted as the new reference solution due to its physical plausibility.

5.4 Tilted Transversally Isotropic Medium

For transversally isotropic materials analytical solutions can be found along the axis of symmetry [5]. As mentioned earlier the representation of the elastic tensor depends on the chosen coordinate system. By tilting the symmetry axis of the transversally isotropic material we can generate an almost densely filled elastic tensor. We take the material characterized by the tensor $c_{ij} = 0$ except for

$$
\begin{array}{lll}
c_{11} = 66.6\,\text{GPa} & c_{12} = 19.7\,\text{GPa} & c_{13} = 39.4\,\text{GPa} \\
c_{22} = 66.6\,\text{GPa} & c_{23} = 39.4\,\text{GPa} & c_{33} = 39.9\,\text{GPa} \\
c_{44} = 10.9\,\text{GPa} & c_{55} = 10.9\,\text{GPa} & c_{66} = 23.6\,\text{GPa}.
\end{array} \tag{15}
$$

and density $\rho = 2590\,\frac{\text{kg}}{\text{m}^3}$ and tilt it around the x axis about 30°. We consider the computational domain $\Omega = [0, 2500]^3$. The source is placed at $(1250, 1562.5, 937.5)$ and a receiver is placed at $(1250, 1198.05, 1568.75)$, which is along the symmetry axis of the tilted material 728.9 m away from the source. The source either acts along the axis of symmetry or orthogonal to the axis. The time history is a Ricker wavelet with dominant frequency $f_0 = 16.0\,\text{Hz}$ and onset time $t_0 = 0.07\,\text{s}$. The whole simulation was run for 0.6 s on a mesh which is refined in a sphere around the source and in a cylinder along the axis of symmetry. In the most refined region the characteristic length is 5 m and grows towards the boundary. In total the mesh consisted of 6.10 million elements.

We compare our solution obtained at the receiver with the analytic solution using envelope and pulse misfit. For the horizontal source we obtained a maximal envelope misfit of 2.09% and a pulse misfit of 0.49%. For the vertical source the misfits were 2.16% and 0.28% respectively. For both source types the numerical solution fits the analytic solution very well.

Fig. 3. Isotropic (left) vs. anisotropic (right) seismic wave field scattered at the strong topography free surface of Mount Zugspitze at $t = 3\,$s simulation time.

5.5   Application Example: Mount Zugspitze

Accurate numerical simulation of scattering seismic waves by complex geometries are critical for assessing and quantifying seismic hazards. In the context of regional scale seismology wave propagation simulations, many numerical methods are challenged by geometric restrictions or low-order accuracy. We here spatially discretize the surface topography around Mount Zugspitze [1] in an unstructured tetrahedral computational mesh of size 90 km × 90 km up to 70 km depth with a resolution of 600 m at the surface. The mesh contains 1.47 million cells. We chose a discretization of order 6 which results in 740 million degrees of freedom. This means we can resolve frequencies up to 4.2 Hz with an estimated envelope misfit smaller than 1% [15]. A kinematic point source with the same parameters as for the LOH1 test case is placed in the center of the domain at 10 km depth. We visually compare the wave field scattered by topography in an isotropic material with parameters $\rho = 2670\,\frac{kg}{m^3}, \lambda = 36.4\,GPa, \mu = 29.8\,$GPa with an anisotropic material. The elastic tensor is chosen such that in EW-direction the P wave speed is $6000\,\frac{m}{s}$ for both materials. To illustrate the effects of anisotropy the P wave speed in NS-direction of the anisotropic material is 5% lower.

In Fig. 3 snapshots of the vertical velocity field on the free surface are plotted. A circular shape for the isotropic example and an elliptic shape for the anisotropic part illustrate the effects of anisotropic materials on wave propagation under strong velocity contrast. We compare this simulation with and without local time stepping: moving from global to local time stepping drastically reduced the execution time from 5990 s to 210 s. The simulation was run on 50 nodes of SuperMUC-NG. The computational more intense version with global time stepping achieved 1.414 TFlop/s, the version with local time stepping achieved 1.015 TFlop/s. This shows that local time stepping is crucial to obtain fast simulations when the element sizes vary a lot, such as in the case of surface topography.

6 Conclusion

The earthquake simulation code SeisSol has been successfully extended to take general anisotropic materials into account. A two-sided Godunov flux for anisotropic media has been derived and implemented. Together with the formulation of the free-surface boundary condition as the solution of an inverse Riemann problem it fits well with the other rheological models. Necessary changes to include local time stepping have been described and implemented.

The scheme has been validated against various benchmarks. The expected convergence rates are demonstrated in comparison to analytic solutions. The mismatch between our results on a community benchmark have been discussed with the maintainers and led to an update of the reference solution.

As anisotropy is non-neglectable to describe the Earth's subsurface structure we expect a wide range of applications. Besides the importance of seismic anisotropy for exploration purposes, earthquake fault zones may be characterised by pervasive anisotropy. Earthquake monitoring and forecasting can be built upon this observation.

Acknowledgments. The research leading to these results has received funding from European Union Horizon 2020 research and innovation program (ENERXICO, grant agreement No. 828947), KAUST-CRG (FRAGEN, grant no. ORS-2017-CRG6 3389.02) and the European Research Council (TEAR, ERC Starting grant no. 852992). Computing resources were provided by the Leibniz Supercomputing Centre (project no. pr45fi on SuperMUC-NG).

References

1. Copernicus EU-DEM. https://land.copernicus.eu/pan-european/satellite-derived-products/eu-dem/eu-dem-v1.1. Accessed 14 Mar 2018
2. Aki, K., Richards, P.: Quantitative Seismology, 2nd edn. University Science Books, Sausalito (2002)
3. Breuer, A., Heinecke, A., Bader, M.: Petascale local time stepping for the ADER-DG finite element method. In: IPDPS, pp. 854–863 (2016)
4. Carcione, J.M.: Wave Fields in Real Media, 3rd edn. Elsevier, Oxford (2015)
5. Carcione, J.M., Kosloff, D., Behle, A., Seriani, G.: A spectral scheme for wave propagation simulation in 3-D elastic-anisotropic media. Geophysics **57**(12), 1593–1607 (1992)
6. Dumbser, M., Käser, M.: An arbitrary high-order discontinuous Galerkin method for elastic waves on unstructured meshes - II. The three-dimensional isotropic case. Geophys. J. Int. **167**(1), 319–336 (2006)
7. Dumbser, M., Käser, M., Toro, E.F.: An arbitrary high-order discontinuous Galerkin method for elastic waves on unstructured meshes - V. Local time stepping and p-adaptivity. Geophys. J. Int. **171**(2), 695–717 (2007)
8. Dziewonski, A.M., Anderson, D.L.: Preliminary reference earth model. Phys. Earth Planet. Inter. **25**, 297–356 (1981)
9. Guennebaud, G., Jacob, B., et al.: Eigen v3 (2010)
10. Heinecke, A., et al.: Petascale high order dynamic rupture earthquake simulations on heterogeneous supercomputers. In: Proceedings of SC 2014, pp. 3–14 (2014)

11. Igel, H., Mora, P., Riollet, B.: Anisotropic wave propagation through finite-difference grids. Geophysics **60**(4), 1203–1216 (1995)
12. Jaeken, J.W., Cottenier, S.: Solving the christoffel equation: phase and group velocities. Comput. Phys. Commun. **207**, 445–451 (2016)
13. Komatitsch, D., Barnes, C., Tromp, J.: Simulation of anisotropic wave propagation based upon a spectral element method. Geophysics **65**(4), 1251–1260 (2000)
14. Kristeková, M., Kristek, J., Moczo, P.: Time-frequency misfit and goodness-of-fit criteria for quantitative comparison of time signals. Geophys. J. Int. **178**(2), 813–825 (2009)
15. Käser, M., Hermann, V., de la Puente, J.: Quantitative accuracy analysis of the discontinuous Galerkin method for seismic wave propagation. Geophys. J. Int. **173**(3), 990–999 (2008)
16. Leary, P.C., Li, Y.G., Aki, K.: Observation and modelling of fault-zone fracture seismic anisotropy - I. P, SV and SH travel times. Geophys. J. Int. **91**(2), 461–484 (1987)
17. Licciardi, A., Eken, T., Taymaz, T., Piana Agostinetti, N., Yolsal-Çevikbilen, S.: Seismic anisotropy in central north Anatolian fault zone and its implications on crustal deformation. Phys. Earth Planet. Inter. **277**, 99–112 (2018)
18. Moczo, P., et al.: Comparison of numerical methods for seismic wave propagation and source dynamics - the SPICE code validation. In: 3rd International Symposium on the Effects of Surface Geology on Seismic Motion, pp. 1–10 (2006)
19. de la Puente, J., Dumbser, M., Käser, M., Igel, H.: Discontinuous Galerkin methods for wave propagation in poroelastic media. Geophysics **73**(5), T77–T97 (2008)
20. de la Puente, J., Käser, M., Dumbser, M., Igel, H.: An arbitrary high-order discontinuous Galerkin method for elastic waves on unstructured meshes - IV. Anisotropy. Geophys. J. Int. **169**(3), 1210–1228 (2007)
21. Stein, S., Wysession, M.: An Introduction to Seismology, Earthquakes, and Earth Structure, 1st edn. Wiley-Blackwell, Malden (2002)
22. Sun, Y.C., Zhang, W., Chen, X.: 3D seismic wavefield modeling in generally anisotropic media with a topographic free surface by the curvilinear grid finite-difference method 3D seismic wavefield modeling in generally anisotropic media. Bull. Seismol. Soc. Am. **108**(3A), 1287–1301 (2018)
23. Uphoff, C.: Flexible model extension and optimisation for earthquake simulations at extreme scale. PhD thesis, Technical University of Munich (2020)
24. Uphoff, C., Bader, M.: Yet another tensor toolbox for discontinuous Galerkin methods and other applications. Submitted to ACM TOMS (2019)
25. Uphoff, C., et al.: Extreme scale multi-physics simulations of the tsunamigenic 2004 Sumatra megathrust earthquake. In: Proceedings of SC 2017, pp. 21:1–21:16 (2017)
26. Zhan, Q., et al.: Full-anisotropic poroelastic wave modeling: a discontinuous Galerkin algorithm with a generalized wave impedance. Comput. Methods Appl. Mech. Eng. **346**, 288–311 (2019)

The Challenge of Onboard SAR Processing: A GPU Opportunity

Diego Romano[1]([✉])(iD), Valeria Mele[2](iD), and Marco Lapegna[2](iD)

[1] Institute for High Performance Computing and Networking (ICAR), CNR, Naples, Italy
diego.romano@cnr.it
[2] University of Naples Federico II, Naples, Italy
{valeria.mele,marco.lapegna}@unina.it

Abstract. Data acquired by a Synthetic Aperture Radar (SAR), onboard a satellite or an airborne platform, must be processed to produce a visible image. For this reason, data must be transferred to the ground station and processed through a time/computing-consuming focusing algorithm. Thanks to the advances in avionic technology, now GPUs are available for onboard processing, and an opportunity for SAR focusing opened. Due to the unavailability of avionic platforms for this research, we developed a GPU-parallel algorithm on commercial off-the-shelf graphics cards, and with the help of a proper scaling factor, we projected execution times for the case of an avionic GPU. We evaluated performance using ENVISAT (Environmental Satellite) ASAR Image Mode level 0 on both NVIDIA Kepler and Turing architectures.

Keywords: Onboard SAR focusing · GPU-parallel · Range-Doppler algorithm

1 Introduction

In the domain of environmental monitoring, Synthetic Aperture Radar (SAR) plays an important role. It is an active microwave imaging technology for remote sensing, which can be employed for observations in all-day and all-weather contexts. Satellites and aircraft have limited space for a radar antenna, therefore a SAR sensor creates a synthetic aperture by exploiting their motion. As a platform moves along a direction (called *azimuth* direction), the sensor transmits pulses at right angles (along *range* direction) and then records their echo from the ground (see Fig. 1).

Thanks to its synthetic aperture, SAR systems can acquire very long land swaths organized in proper data structures. However, to form a comprehensible final image, a processing procedure (*focusing*) is needed.

The focusing of a SAR image can be seen as an inherently space-variant two-dimensional correlation of the received echo data with the impulse response of the system. Radar echo data and the resulting Single-Look Complex (SLC)

V. V. Krzhizhanovskaya et al. (Eds.): ICCS 2020, LNCS 12139, pp. 46–59, 2020.
https://doi.org/10.1007/978-3-030-50420-5_4

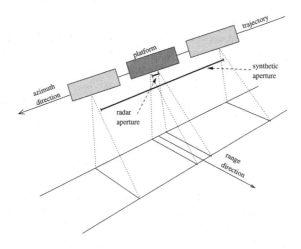

Fig. 1. Representation of the synthetic aperture created by a moving platform provided with a sensor.

image are stored in matrices of complex numbers representing the in-phase and quadrature (i/q) components of the SAR signal. Several processors are available, based on three main algorithms: Range-Doppler, ωk, and Chirp Scaling [7].

Usually, this processing takes time and needs HPC algorithms in order to process data quickly. Heretofore, considering the limited computing hardware onboard, data had been transmitted to ground stations for further processing. Nevertheless, the vast amount of acquired data and the severely limited downlink transfer bandwidth imply that any SAR system also needs an efficient raw data compression tool. Because of structures with apparent higher entropy, a quasi-independence of in-phase and quadrature components showing histograms with nearly Gaussian shape and identical variance, conventional image compression techniques are ill-suited, and resulting compression rates are low.

Thanks to advances in the development of avionic specialized computing accelerators (GPUs) [1,12], now the onboard SAR processing with real-time GPU-parallel focusing algorithms is possible. These could improve sensor data usability on both strategic and tactical points of view. For example, we can think of an onboard computer provided with a GPU directly connected to both a ground transmitter and a SAR sensor through GPUDirect [13] RDMA [5] technology.

Several efforts have been made to implement GPU SAR processors for different raw SAR data using CUDA Toolkit. In [4], the focusing of an ERS2 image with $26,880 \times 4,912$ samples on an NVIDIA Tesla C1060 was obtained in 4.4 s using a Range-Doppler algorithm. A similar result is presented in [14], where a COSMO-SkyMed image of $16,384 \times 8,192$ samples has been processed employing both Range-Doppler and ωk algorithms in 6.7 s. Another implementation of the ωk algorithm, described in [20], focused a Sentinel-1 image with $22,018 \times 18,903$ in 10.87 s on a single Tesla K40, and 6.48 s in a two GPUs configuration. In [15], a ωk-based SAR processor implemented in OpenCL and run on four Tesla K20 has been used to focus an ENVISAT ASAR IM image of $30,000 \times 6,000$ samples in 8.5 s and a Sentinel-1 IW image of $52,500 \times 20,000$ samples in 65 s. All these results have accurately analyzed the ground station case, where one or more Tesla GPU products have been used.

Our idea is to exploit the onboard avionic GPU computing resources, which are usually more limited than the Tesla series. For example, on the one hand, the avionic EXK107 GPU of the Kepler generation is provided with 2 Streaming Multiprocessors (SMs), each with 192 CUDA core. On the other hand, the Tesla K20c, of the same architecture generation, has 13 SMs, also with 192 CUDA core each.

Historically, the development of SAR processors has been characteristic of the industrial sector, and therefore there is little availability of open-source processors. This work is based on the *esarp* processor within the GMTSAR processing system [17], a focuser written in C and implementing a Range-Doppler algorithm (Fig. 2) for ERS-1/2, ENVISAT, ALOS-1, TerraSAR-X, COSMOS-SkyMed, Radarsat-2, Sentinel-1A/B, and ALOS-2 data. For testing convenience, the GPU-parallel processor herein presented is limited to ENVISAT ASAR Image Mode level 0 data [18], but with a reasonably little effort, it can be adapted to other sensors raw data.

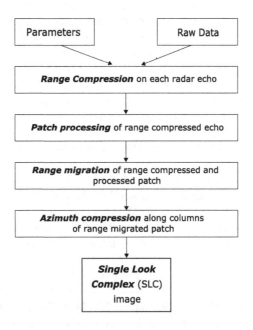

Fig. 2. Range-Doppler Algorithm flow in *esarp* processor

This paper shares the experiences gathered during the testing of a prototype HPC platform, whose details are subject to a non-disclosure agreement and therefore excluded from this presentation. However, several insights can be useful to discuss new approaches in the design of SAR processing procedures and strategies. Indeed, from previous experiences in GPU computing, which also included special devices ([8–11,16]), we can make some assumptions. Furthermore, the reasoning made when dealing with an off-the-shelf hardware solution can be in some way translated to an avionic product, accepting that the algorithmic logic does not change. In order to develop and test our algorithm, with the intent to exploit the massive parallelism of GPUs, we applied the approach proposed in [2].

In the next section, we provide a schematic description of the Range-Doppler algorithm, and we focus on data-parallel kernels that can be efficiently implemented on a GPU. Section 3 presents the actual kernels implemented and their relative footprint in the perspective of avionic hardware. Testing is presented in Sect. 4, with an estimation of the execution time on an avionic GPU. Finally, we discuss results and conclude in Sect. 5.

2 Range-Doppler Algorithm and Identification of Data-Parallel Kernels

The GMTSAR processing system relies on precise orbits (sub-meter accuracy) to simplify the processing algorithms, and techniques such as *clutterlock* and *autofocus* are not necessary to derive the orbital parameters from the data.

In the *esarp* focusing component, data are processed by patches in order not to overload the computing platform. Each patch contains all the samples along the range direction and a partial record along the azimuth direction. Several patches are concatenated to obtain the image for the complete strip.

1. **Range Compression** – In the ENVISAT signal, there are 5681 points along the range direction that must be recovered in a sharp radar pulse by deconvolution with the chirp used during signal transmission. The operation is done in the frequency domain: firstly, the chirp is transformed, then the complex product of each row with the conjugate of the chirp is computed. A Fast Fourier Transform (FFT) is therefore needed before and after the product. In order to take advantage of the speed of radix-2 FFT, data are zero-padded to the length of 8192. This procedure allows obtaining phase information for a longer strip, which will be later reduced to 6144 points for further processing.

2. **Patch Processing** – In order to focus the image in the azimuth direction, data must be transformed in the range-doppler domain, which means in the frequency domain for the azimuth direction, by applying an FFT on the transposed matrix representing the range compressed image. For the ENVISAT radar, the synthetic aperture is 2800 points long. Again, to exploit the speed of radix-2 FFT, 4096 rows are loaded and processed, consisting of a patch. The last 1296 rows are overlapped with the following patch.

3. **Range Migration** – As the platform moves along the flight path, the distance between the antenna and a point target changes, and that point appears as a hyperbolic-shaped reflection. To compensate for this effect, we should implement a remapping of samples in the range-doppler domain through a sort of interpolator. Such a migration path can be computed from the orbital information required by the GMTSAR implementation and must be applied to all the samples in the range direction.

4. **Azimuth Compression** – To complete the focusing in the azimuth direction, a procedure similar to the Range Compression is implemented. In the range-doppler domain, a frequency-modulated chirp is created to filter the phase shift of the target. This chirp depends on: the pulse repetition frequency, the range, and the velocity along the azimuth direction. As before, after the complex product, the result is inversely Fourier transformed back to the spatial domain to provide the focused image.

In the four steps described above, many operations can be organized appropriately, respecting their mutual independence [3]. As shown in Fig. 3, each subalgorithm corresponds to a GPU kernel exploiting possible data parallelism. The several planned FFTs can be efficiently implemented through cuFFT batching. If the raw data matrix is memorized in a 1-dimensional array with row-major order, all the FFTs in range direction can be executed in efficient batches [19]. When the FFTs runs in the azimuth direction, a pre- and post-processing matrix transpose becomes necessary.

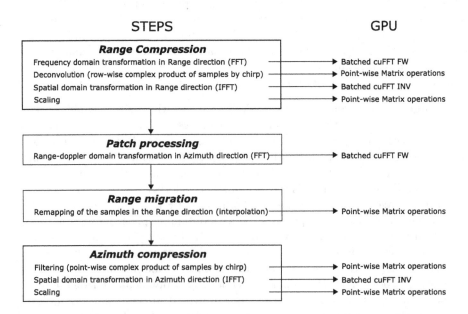

Fig. 3. Steps of the Range-Doppler Algorithm and correspondence with possible data-parallel GPU operations

The filtering sub-algorithms can be easily organized as point-wise matrix operations, assuming that the chirps are available in the device memory for reading. This step is efficiently achievable by building the range chirp directly on the GPU, as it consists of a mono-dimensional array with spatial properties, and by subsequently transforming it in the frequency domain through a proper FFT. Similarly, the azimuth chirp can be built and transformed directly on the GPU, but this time it is a 2-D array.

About the mapping of the samples in the Range direction, assuming enough memory is available for storing the migrated samples, it can be seen as point-wise matrix operation, as each sample corresponds to a previously patch processed data subject to operations involving orbital information.

3 GPU Kernels and Memory Footprint

In order to evaluate the feasibility of onboard processing, we present an analysis of the resources needed.

Firstly, let us observe that cuFFT proposes a convenient function to get an accurate estimate of the additional work area size needed to run a batched plan. Since the dimensions used in the Range-Doppler algorithm for ENVISAT data are a power of 2, that is 8192 complex numbers of 8 bytes each in the range direction for 4096 rows, the additional work area consists of 256 MBytes for the batches in the range direction. Similarly, in the azimuth direction, the batches are organized in 6144 columns of 4096 points, and the additional work area required is about 192 MBytes.

In Algorithm 1, a GPU-parallel pseudo-code presents the kernels and the cuFFT runs of the GPU-parallel version of *esarp*. In the following, we analyze the kernels with their possible sources of Algorithmic Overhead [3] and their memory footprint.

- **d_orbit_coef**: in order to remap the range samples and to compensate platform movement within the range migration step, for each sample in the range, there are 8 parameters describing the orbit characteristics and their influence on the migration. These parameters are the same for each row of the patch, and they are scaled considering the position in the synthetic aperture, that is the position in the azimuth direction. They are also useful to put up the chirp in the azimuth direction. To save useless recomputing, this kernel precomputes 8 arrays of 6144 elements with a corresponding memory footprint of 384 KBytes. Their values can be computed independently by 6144 threads in an appropriate thread-block configuration that takes into account the number of SMs in the GPU.
- **d_ref_rng**: this kernel populates an array with the chirp in range direction based on the pulse emitted by the sensor. The array is also zero-padded to the length of the nearest power of 2 to exploit subsequent radix-2 FFT efficiency. For the ENVISAT data, the array consists of 8192 complex numbers of 8 bytes each, i.e., 64 Kbytes. The workload of this kernel is proportional to the number of elements in the array. Moreover, each element can be processed

---

Algorithm 1: esarp on GPU

Result: SAR focused image

initialization;
d_orbit_coef(coef) ;   // kernel to create arrays with orbital info
d_ref_rng(r_ref) ;   // kernel to set up range chirp
cuFFT(r_ref,FW) ;   // transform range chirp in frequency domain
d_ref_az(a_ref) ;   // kernel to set up azimuth chirp
cuFFT(a_ref,FW) ;   // transform azimuth chirp in frequency domain

while *patches to be focused* **do**
>   receive patch;
>
>   // Range compression
>   cuFFT(patch,FW) ;   // transf. freq. in range direction
>   d_mul_r(patch,r_ref) ;   // kernel for deconvolution in range dir.
>   cuFFT(patch,INV) ;   // transform back in spatial domain
>   d_scale(patch) ;   // kernel for scaling partial results
>
>   // Patch processing
>   d_trans_mat(patch) ;   // kernel to transpose patch
>   cuFFT(patch,FW) ;   // transf. freq. in azimuth direction
>
>   // Range migration
>   d_intp_tot(patch,coef) ;   // kernel to remap samples in range dir.
>
>   // Azimuth Compression
>   d_mul_a(patch,a_ref) ;   // kernel for filtering in azimuth dir.
>   cuFFT(patch,INV) ;   // transform back in spatial domain
>   d_scale(patch) ;   // kernel for scaling results
>   d_trans_mat(patch) ;   // kernel to transpose patch

end

---

independently of the others, meaning that the workload can be split among threads. If those are organized in a number of blocks, which is multiple of the number of SMs present in the GPU, we can have a good occupancy of the devices. Also, the divergence induced by the zero-padding can be minimized during thread-block configuration.

– **d_ref_az**: by using previously calculated orbital parameters, a 2-D array of the same size of the patch is populated with the chirp in the azimuth direction, which is different for each column. Hence, the memory footprint is $6144 \cdot 4096 \cdot 8 = 192$ MBytes. Beforehand, the array is reset to zero values since not all the samples are involved in the filtering. To limit divergence, each element in the array can be assigned to a thread that populates the array if necessary, or it waits for completion. Since the same stored orbital parameters are used for each row, the threads can be arranged in blocks with column-wise memory access in mind in order to limit collisions among different SMs. Hence, the execution configuration can be organized in a 2-D memory grid with blocks of threads on the same column.

- **d_mul_r**: implements a point-wise multiplication of each row of the patch by the conjugate of the chirp in the frequency domain. The workload can be assigned to independent threads with coalescent memory accesses. Following reasoning similar to d_ref_az, with the idea of limiting memory collisions, each thread in a block can compute one column of the patch in a *for* cycle, realizing a coalesced write of the results with the other threads in the same warp. This kernel does not require additional memory occupation.
- **d_scale**: after the inverse FFT needed to transform the patch back to the spatial domain, a point-wise scaling is needed. As before, independent threads can work with coalescent memory accesses, and efficient workload assignments can be configured.
- **d_trans_mat**: this kernel follows the highly efficient sample proposed in [6]. In this case, the memory footprint corresponds to a new array with the same dimension of the patch, i.e., 192 MBytes.
- **d_intp_tot**: the remapping of the samples is carried on in a point-wise procedure. The output patch must be in a different memory location, and therefore the memory footprint consists again of an additional 192 MBytes. Making similar reasoning on the memory accesses as we did for the d_ref_az kernel, we can configure the execution to minimize global memory collisions, optimizing block dimensions for occupancy.
- **d_mul_a**: this kernel filters the patch to focus the final image in the frequency domain. The operations consist of element-wise matrix products and do not need additional work area in memory. An efficient thread-block configuration can follow the reasoning made for the previous kernel.

To summarize the analysis of the memory footprint for the whole procedure to focus a patch: 192×2 MBytes are necessary to swap the patch for transposing and remapping data in several kernels, 256 MBytes are necessary for the most demanding FFT, and the preliminary computing of chirps and orbit data require $\approx$192.5 MBytes. The total is less than 1 GByte of memory, which is a fair amount available on every GPU.

4 Testing on Workstation and Reasoning on Avionic Platform

As mentioned in the introduction, we had access to a prototype avionic platform for testing purposes, and we had the opportunity to run our algorithm repeatedly. Even if we cannot disclose details about platform architecture and testing outcomes due to an NDA, we can refer to the GPU installed, which is an Nvidia EXK107 with Kepler architecture.

In this section, we will present the results collected on a workstation with a Kepler architecture GPU (see Table 1), to propose some reasoning on the avionic platform with the help of a scale factor, and on another workstation with a Turing architecture GPU (see Table 2) to evaluate the running time on a more recent device.

Table 1. Workstation used for testing on Kepler architecture

	Workstation Kepler
OS	Ubuntu 18.04
CPU	Intel Core i5 650 @3.20 GHz
RAM	6 GB DDR3 1333 MT/s
GPU	GeForce GTX 780 (12 SMs with 192 cores each)

Let us consider the execution time of our GPU version of the esarp processor, excluding any memory transfer between host and device, i.e., considering data already on the GPU memory. Such is a fair assumption since all the focusing steps are executed locally without memory transfers between host and device. In an avionic setting, only two RDMA transfers happen: the input of a raw patch from the sensor, the output of a focused patch to the transmitter (Fig. 4).

If we call t_{wk} the execution time for focusing a patch on the *Workstation Kepler*, and t_a the execution time to focus a patch on an avionic platform provided with an EXK107 GPU, from our testing we noticed a constant scale factor:

$$s_f = \frac{t_{wk}}{t_a} = 0.23$$

It should not be considered a universal scale factor for whatever kernel run on both devices. However, it is a constant behavior on the total execution time to focus whatever patch from ENVISAT ASAR IM data using our GPU-parallel version of the esarp processor. Therefore s_f is useful to estimate the time needed to focus a swath on an avionic platform using such application.

To verify the functionalities of the focusing algorithm, we used data freely available from http://eo-virtual-archive4.esa.int. Measures presented in this section are relative to the processing of the image in Fig. 5 subdivided in 9 patches.

Table 2. Workstation used for testing on Turing architecture

	Workstation Turing
OS	CentOS 7.6
CPU	Gold Intel Xeon 5215
RAM	94 GB
GPU	Quadro RTX 6000 (72 SMs with 64 cores each)

In Table 3 we present the execution times of the GPU-esarp software, relatively to the steps of the Range-Doppler algorithm, on Workstation Kepler. The preliminary processing step, which includes the creation of arrays containing orbital information and chirps in both range and azimuth direction, is executed just for the first patch, as the precomputed data do not change for other patches within the same swath. The total execution time needed to focus the whole image is $t_{wk} = 1.12$ s, excluding input-output overhead and relative memory transfers between host and device.

We can, therefore, expect that the execution time needed on the avionic platform is:

$$t_a = \frac{t_{wk}}{s_f} = 4.87 \text{ s}$$

which is less than the ENVISAT stripmap acquisition time $t_{in} \approx 16$ s for the relative dataset. Moreover, if each sample of the resulting image consists of a complex number of 16 bits, the total size of the output is ≈ 295 MBytes. In a pipelined representation of a hypothetical avionic system, as pictured in Fig. 4, all data transfers are subject to their respective connection bandwidth. Considering that the payload communication subsystem of the ENVISAT mission had a dedicated bandwidth for SAR equipment of 100 Mbit/s, the time necessary to transmit the result to the ground would be $t_{out} \approx 24$ s. That is, we can suppose that:

$$t_a < t_{in} < t_{out}$$

hence, we have an expected GPU-parallel focusing algorithm able to satisfy real-time requirements on an EXK107 device.

If we consider the execution times on Workstation Turing (Table 4), we see that the total time needed to focus the whole image is $t_{wt} = 0.208$ s, excluding input-output transfers, which is very promising for the next generation of avionic GPUs. Moreover, considering the spare time available for further processing during down-link transmission, we can think about computing Azimuth FM rate and Doppler Centroid estimators. Those algorithms are useful to provide parameters for Range Migration, and Azimuth Compression steps in case of non-uniform movements of the platform, as it happens on airborne SAR.

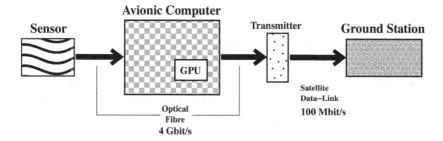

Fig. 4. Transfer data rates in an avionic system: sensors are usually connected to the computer unit through Optical Fibre, which allow rates of the Gbit/s magnitude or more; within the Avionic Computer, GPUs allows transfers at rates with a magnitude of Gbit/s; at the end of this pipeline, a data-link connection to the ground station can transfer with a maximum rate of 100 Mbit/s with current technology.

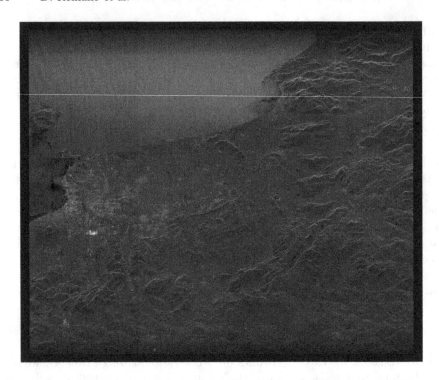

Fig. 5. Focused SAR image of Napoli area, consisting of 6144 samples in the range direction and 25200 samples in the azimuth direction. The sampled area is 106 × 129 Km², with an Azimuth resolution of 5 m. For rendering purposes, here the image is proposed with vertical range direction and with the azimuth direction squeezed to map on square pixels.

Table 3. Execution times in milliseconds for each step of the GPU-esarp software on the Workstation Kepler

	Execution time in milliseconds								
Preliminary processing	21.7								
Range compression	46.9	46.2	46.2	46.1	46.3	46	45.8	45.9	46
Patch processing	4.8	4.8	4.8	4.8	4.8	4.8	4.8	4.7	4.7
Range migration	47.8	47.1	46.9	46.9	47.2	46.9	47.1	47.2	47.3
Azimuth compression	24.4	24.1	24.1	24.2	24.3	24.2	24.8	24.1	24.1
Total (excl. I/O)	145.4	122.2	122	122	122.6	121.9	122.5	121.9	122.1
Patch	1	2	3	4	5	6	7	8	9

Table 4. Execution times in milliseconds of the GPU-esarp software on the Workstation Turing

	Execution time in milliseconds								
Total (excl. I/O)	28.5	24.3	22.9	22.3	22.3	22.2	22	22	22
Patch	1	2	3	4	5	6	7	8	9

5 Conclusions

When thinking about SAR sensing, a common approach is to consider it as an instrument for delayed operational support. Usually, SAR raw data are compressed, down-linked, and processed in the ground stations to support several earth sciences research activities, as well as disaster relief and military operations. In some cases, timely information could be advisable, and onboard processing is becoming an approach feasible thanks to advances in GPU-technology with reduced power consumption.

In this work, we developed a GPU-parallel algorithm based on the Range-Doppler algorithm as implemented in the open-source GMTSAR processing system. The results, in terms of execution time on off-the-shelf graphics cards, are encouraging if scaled to proper avionic products. Even if we did not present actual results on an avionic GPU, thanks to some insights acquired during testing of a prototype avionic computing platform and a constant scale factor, we showed that onboard processing is possible when an efficient GPU-parallel algorithm is employed.

Since this result is based on the algorithmic assumption that orbital information is available, some processing techniques such as *clutterlock* and *autofocus* have been avoided. That is the case for many satellite SAR sensors, but further experiments must be carried on to verify the feasibility of onboard processing on airborne platforms, where parameters like altitude and velocity may slightly change during data acquisition. In this sense, as future work, we plan to implement a GPU-parallel algorithm for parameters estimation.

References

1. GRA112 graphics board, July 2018. https://www.abaco.com/products/gra112-graphics-board
2. D'Amore, L., Laccetti, G., Romano, D., Scotti, G., Murli, A.: Towards a parallel component in a GPU–CUDA environment: a case study with the L-BFGS harwell routine. Int. J. Comput. Math. **92**(1), 59–76 (2015). https://doi.org/10.1080/00207160.2014.899589
3. D'Amore, L., Mele, V., Romano, D., Laccetti, G.: Multilevel algebraic approach for performance analysis of parallel algorithms. Comput. Inform. **38**(4), 817–850 (2019). https://doi.org/10.31577/cai_2019_4_817
4. di Bisceglie, M., Di Santo, M., Galdi, C., Lanari, R., Ranaldo, N.: Synthetic aperture radar processing with GPGPU. IEEE Signal Process. Mag. **27**(2), 69–78 (2010). https://doi.org/10.1109/MSP.2009.935383

5. Franklin, D.: Exploiting GPGPU RDMA capabilities overcomes performance limits. COTS J. **15**(4), 16–20 (2013)
6. Harris, M.: An efficient matrix transpose in CUDA C/C++, February 2013. https://devblogs.nvidia.com/efficient-matrix-transpose-cuda-cc/
7. Hein, A.: Processing of SAR Data Fundamentals, Signal Processing, Interferometry, 1st edn. Springer, Heidelberg (2010)
8. Laccetti, G., Lapegna, M., Mele, V., Montella, R.: An adaptive algorithm for high-dimensional integrals on heterogeneous CPU-GPU systems. Concurr. Comput.: Pract. Exper. **31**(19), e4945 (2019). https://doi.org/10.1002/cpe.4945. https://onlinelibrary.wiley.com/doi/abs/10.1002/cpe.4945, e4945 cpe.4945
9. Laccetti, G., Lapegna, M., Mele, V., Romano, D.: A study on adaptive algorithms for numerical quadrature on heterogeneous GPU and multicore based systems. In: Wyrzykowski, R., Dongarra, J., Karczewski, K., Waśniewski, J. (eds.) PPAM 2013. LNCS, vol. 8384, pp. 704–713. Springer, Heidelberg (2014). https://doi.org/10.1007/978-3-642-55224-3_66
10. Marcellino, L., et al.: Using GPGPU accelerated interpolation algorithms for marine bathymetry processing with on-premises and cloud based computational resources. In: Wyrzykowski, R., Dongarra, J., Deelman, E., Karczewski, K. (eds.) PPAM 2017. LNCS, vol. 10778, pp. 14–24. Springer, Cham (2018). https://doi.org/10.1007/978-3-319-78054-2_2
11. Montella, R., Giunta, G., Laccetti, G.: Virtualizing high-end GPGPUs on ARM clusters for the next generation of high performance cloud computing. Cluster Comput. **17**(1), 139–152 (2014). https://doi.org/10.1007/s10586-013-0341-0
12. Munir, A., Ranka, S., Gordon-Ross, A.: High-performance energy-efficient multicore embedded computing. IEEE Trans. Parallel Distrib. Syst. **23**(4), 684–700 (2012). https://doi.org/10.1109/TPDS.2011.214
13. NVIDIA Corporation: Developing a Linux Kernel Module Using RDMA for GPUDirect (2019). http://docs.nvidia.com/cuda/gpudirect-rdma/index.html, version 10.1
14. Passerone, C., Sansoè, C., Maggiora, R.: High performance SAR focusing algorithm and implementation. In: 2014 IEEE Aerospace Conference, pp. 1–10, March 2014. https://doi.org/10.1109/AERO.2014.6836383
15. Peternier, A., Boncori, J.P.M., Pasquali, P.: Near-real-time focusing of ENVISAT ASAR Stripmap and Sentinel-1 TOPS imagery exploiting OpenCL GPGPU technology. Remote Sens. Environ. **202**, 45–53 (2017). https://doi.org/10.1016/j.rse.2017.04.006. Big Remotely Sensed Data: Tools, Applications and Experiences
16. Rea, D., Perrino, G., di Bernardo, D., Marcellino, L., Romano, D.: A GPU algorithm for tracking yeast cells in phase-contrast microscopy images. Int. J. High Perform. Comput. Appl. **33**(4), 651–659 (2019). https://doi.org/10.1177/1094342018801482
17. Sandwell, D., Mellors, R., Tong, X., Wei, M., Wessel, P.: GMTSAR: an InSAR processing system based on generic mapping tools (2011)
18. Schättler, B.: ASAR level 0 product analysis for image, wide-swath and wave mode. In: Proceedings of the ENVISAT Calibration Review. Citeseer (2002)

19. Střelák, D., Filipovič, J.: Performance analysis and autotuning setup of the cuFFT library. In: Proceedings of the 2nd Workshop on AutotuniNg and ADaptivity AppRoaches for Energy Efficient HPC Systems, ANDARE 2018. Association for Computing Machinery, New York (2018). https://doi.org/10.1145/3295816.3295817

20. Tiriticco, D., Fratarcangeli, M., Ferrara, R., Marra, S.: Near real-time multi-GPU ωk algorithm for SAR processing. In: Agency-Esrin, E.S. (ed.) Big Data from Space (BiDS), pp. 277–280, October 2014. https://doi.org/10.2788/1823

High-Resolution Source Estimation of Volcanic Sulfur Dioxide Emissions Using Large-Scale Transport Simulations

Mingzhao Liu[1], Yaopeng Huang[1], Lars Hoffmann[5], Chunyan Huang[6], Pin Chen[2,3], and Yi Heng[2,3,4(✉)]

[1] School of Chemical Engineering and Technology, Sun Yat-sen University, Guangzhou, China
[2] School of Data and Computer Science, Sun Yat-sen University, Guangzhou, China
hengyi@mail.sysu.edu.cn
[3] National Supercomputing Center in Guangzhou (NSCC-GZ), Guangzhou, China
[4] Guangdong Province Key Laboratory of Computational Science, Guangzhou, China
[5] Jülich Supercomputing Centre, Forschungszentrum Jülich, Jülich, Germany
[6] School of Statistics and Mathematics, Central University of Finance and Economics, Beijing, China

Abstract. High-resolution reconstruction of emission rates from different sources is essential to achieve accurate simulations of atmospheric transport processes. How to achieve real-time forecasts of atmospheric transport is still a great challenge, in particular due to the large computational demands of this problem. Considering a case study of volcanic sulfur dioxide emissions, the codes of the Lagrangian particle dispersion model MPTRAC and an inversion algorithm for emission rate estimation based on sequential importance resampling are deployed on the Tianhe-2 supercomputer. The high-throughput based parallel computing strategy shows excellent scalability and computational efficiency. Therefore, the spatial-temporal resolution of the emission reconstruction can be improved by increasing the parallel scale. In our study, the largest parallel scale is up to 1.446 million compute processes, which allows us to obtain emission rates with a resolution of 30 min in time and 100 m in altitude. By applying massive-parallel computing systems such as Tianhe-2, real-time source estimation and forecasts of atmospheric transport are becoming feasible.

Keywords: Source estimation · High-throughput computing · Transport simulations · Volcanic emissions

1 Introduction

Model simulations and forecasts of volcanic aerosol transport are of great importance in many fields, e.g., aviation safety [1], studies of global climate change [2, 3] and atmospheric dynamics [4]. However, existing observation techniques, e.g., satellite measurements, cannot provide detailed and complete spatial-temporal information due to their own limitations. With appropriate initial conditions, numerical simulations can provide

© Springer Nature Switzerland AG 2020
V. V. Krzhizhanovskaya et al. (Eds.): ICCS 2020, LNCS 12139, pp. 60–73, 2020.
https://doi.org/10.1007/978-3-030-50420-5_5

relatively complete and high-resolution information in time and space. Model predictions can help to provide early warning information for air traffic control or input to studies of complex global or regional atmospheric transport processes.

In order to achieve accurate atmospheric transport simulations, it is necessary to first combine a series of numerical techniques with limited observational data to achieve high-resolution estimates of the emission sources. These techniques include backward-trajectory methods [5], empirical estimates [6] and inverse approaches. Among them, the inverse approaches are universal and systematic in the identification of atmospheric emission sources due to their mathematical rigor.

For instance, Stohl et al. [7] used an inversion scheme to estimate the volcanic ash emissions related to the volcanic eruptions of Eyjafjallajökull in 2010 and Kelut in 2014. They utilized Tikhonov regularization to deal with the ill-posedness of the inverse problem. Flemming and Inness [8] applied the Monitoring Atmospheric Composition and Climate (MACC) system to estimate sulfur dioxide (SO_2) emissions by Eyjafjallajökull in 2010 and Grimsvötn in 2011, in which the resolution of the emission rates is about 2–3 km in altitude and more than 6 h in time. Due to limitations in computational power and algorithms, the spatial-temporal resolution of the reconstructed source obtained in previous studies is relatively low.

The main limitations of real-time atmospheric transport forecasts are the great computational effort and data I/O issues. Some researchers tried to employ graphics processing units to reduce the computational time and got impressive results [9–11]. Lagrangian particle dispersion models are particularly well suited to distributed-memory parallelization, as each trajectory is calculated independently of each other. To reduce the computational cost, Larson et al. [12] applied a shared- and distributed-memory parallelization to a Lagrangian particle dispersion model and achieved nearly linear scaling in execution time with the distributed-memory version and a speed-up factor of about 1.4 with the shared-memory version. In the study of Müller et al. [11], the parallelization of the Lagrangian particle model was implemented in the OpenMP shared memory framework and good strong scalability up to 12 cores was achieved.

In this work, we implement the Lagrangian particle dispersion model Massive-Parallel Trajectory Calculations (MPTRAC) [5] on the Tianhe-2 supercomputer, along with an inverse modeling algorithm based on the concept of sequential importance resampling [13] to estimate time- and altitude-dependent volcanic emission rates. In order to realize large-scale SO_2 transport simulations on a global scale, high-resolution emission reconstructions and real-time forecasts, the implementation is based on state-of-the-art techniques of supercomputing and big-data processing. The computing performance is assessed in the form of strong and weak scalability tests. Good scalability and computational efficiency of our codes make it possible to reconstruct emission rates with unprecedented resolution both in time and altitude and enable real-time forecasts.

The remainder of this manuscript is organized as follows: Sect. 2 introduces the forward model, the inverse modeling algorithm and the parallelization strategies. Section 3 presents the parallel performance of the forward and inverse code on the Tianhe-2 supercomputer. In Sect. 4, the results of the emission reconstruction and forward simulation are presented for a case study. Discussions and conclusions are provided in Sect. 5.

2 Data and Methods

2.1 Lagrangian Particle Dispersion Model

In this work, the forward simulations are conducted with the Lagrangian particle dispersion model MPTRAC, which has been successfully applied for volcanic eruption cases of Grímsvötn, Puyehue-Cordón Caulle and Nabro [5]. Meteorological fields of the ERA-Interim reanalysis [14] provided by the European Centre for Medium-Range Weather Forecasts (ECMWF) are used as input data for the transport simulations. The trajectory of an individual air parcel is calculated by

$$\frac{d\mathbf{x}(t)}{dt} = \mathbf{v}(\mathbf{x}(t), t), \tag{1}$$

where $\mathbf{x} = (x, y, z)$ denotes the spatial position and $\mathbf{v} = (u, v, w)$ denotes the velocity of the air parcel at time t. Here, x and y coordinates refer to latitude and longitude whereas the z coordinate refers to pressure. The horizontal wind components u and v and the vertical velocity $w = dp/dt$ are obtained by 4-D linear interpolation from the meteorology data, which is common in Lagrangian particle dispersion models [15]. Small-scale diffusion and subgrid-scale wind fluctuations are simulated based on a Markov model following Stohl et al. [16].

In our previous work [17], truncation errors of different numerical integration schemes of MPTRAC have been analyzed in order to obtain an optimal numerical solution strategy with accurate results and minimum computational cost. The accuracy of the MPTRAC trajectory calculations has been analyzed in different studies, including [18], which compared trajectory calculations to superpressure balloon tracks.

2.2 Evaluation of Goodness-of-Fit of Forward Simulation Results

Atmospheric InfraRed Sounder (AIRS) satellite observations are used to detect volcanic SO_2 based on a brightness temperature differences (BTD) algorithm [19]. To evaluate the goodness-of-fit of the forward simulation results obtained by MPTRAC, the critical success index (CSI) [20] is calculated by

$$\text{CSI} = C_x / (C_x + C_y + C_z). \tag{2}$$

Here, the number of positive forecasts with positive observations is C_x, the number of negative forecasts with positive observations is C_y, and the number of positive forecasts with negative observations is C_z. The CSI, representing the ratio of successful predicts to the total number of predicts that were either made ($C_x + C_z$) or needed (C_y), is commonly used for the assessment of the simulation results of volcanic eruptions and other large-scale SO_2 transport problems. Basically, it provides a measure of the overlap of a simulated volcanic SO_2 plume from the model with the real plume as found in the satellite observations. CSI time series are calculated using the AIRS satellite observations and MPTRAC simulation results mapped on a discrete grid, which are essential to the inverse modeling algorithm presented in the next section.

2.3 Inverse Source Estimation Algorithm

The strategy for the inverse estimation of time- and altitude-dependent emission rates is shown in Fig. 1 and Algorithm 1. The time- and altitude-dependent emissions are considered for the domain $E := [t_0, t_f] \times \Omega$, which is discretized with n_t and n_h uniform intervals into $N = n_t \cdot n_h$ subdomains. For each subdomain, a forward calculation of a set of air parcel trajectories is conducted with MPTRAC, which is referred to here as 'unit simulation' for a given time and altitude. Each unit simulation is assigned a certain amount of SO_2, where we assume that the total SO_2 mass over all unit simulations is known a-priori. During the inversion, a set of importance weights $w_i (i = 1, \cdots, N)$, which satisfy $\sum_{i=1}^{N} w_i = 1$, are estimated to represent the relative posterior probabilities of the occurrence of SO_2 emission mass.

At first, the subdomains are populated with SO_2 emissions (air parcels) according to an equal-probability strategy. N parallel unit simulations with a certain amount of air parcels are performed in an iterative process and the corresponding time series (CSI_k^i) with $k = 1, \cdots, n_k$ at different times t_k and $i = 1, \cdots, N$ are calculated to evaluate the agreement of the simulations with the satellite observations. Then, the importance weights are updated according to the following formulas:

$$w_i = m_i \Big/ \sum_{a=1}^{N} m_a, \tag{3}$$

$$m_i = \left(\sum_{1}^{n_k} CSI_k^i \right) \Big/ n_k. \tag{4}$$

During the iteration, the m_i represents the probability of emitted source air parcels that fall in the ith temporal and spatial subdomain. Finally, after the termination criterion is satisfied, the emission source is obtained based on the final importance weight distribution. To define the stopping criterion, we calculate the relative difference d by

$$d\left(\mathbf{W}^{r+1}, \mathbf{W}^r \right) = \frac{\left\| \mathbf{W}^{r+1} - \mathbf{W}^r \right\|}{\max \left(\left\| \mathbf{W}^{r+1} \right\|, \left\| \mathbf{W}^r \right\| \right)}, r \geq 1, \tag{5}$$

where r denotes the iterative step and the norm is defined by

$$\left\| \mathbf{W}^r \right\| = \sqrt{\sum_{i=1}^{N} \left| w_i^r \right|^2}, \mathbf{W}^r = \left(w_i^r \right)_{i=1\cdots N}. \tag{6}$$

As a stopping criterion, threshold of the relative difference d is chosen to be 1%.

In practice, in order to deal with the complexity of the SO_2 air parcel transport, a so-called "product rule" is utilized in the resampling process, in which the average CSI time series is replaced by the product of two average CSI time series in subsequent and separate time periods:

$$m_i = \left(\sum_{k=1}^{n_k'} CSI_k^i \Big/ n_k' \right) \cdot \left(\sum_{k=n_k'+1}^{n_k} CSI_k^i \Big/ (n_k - n_k') \right), 1 \leq n_k' < n_k, \tag{7}$$

where n'_k is a "split point" of the time series. This strategy can better eliminate some low-probability local emissions when reconstructing source terms, thus leading to accurate final forward simulation results locally and globally. A detailed description of the inverse algorithm and the improvements due to applying the product rule can be found in [21].

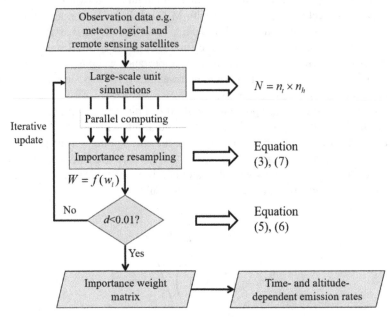

Fig. 1. Flow chart of the inverse modeling strategy

Algorithm 1 Importance sampling algorithm in the inverse modeling

Initialize: Set r = 1, and initially set all importance weights $w_i^0 = 1/N$;

repeat

    for $i = 1 : N$ **do** in parallel

        Perform forward simulation with w_i^{r-1};

        Calculate $CSI_k^i = C_x/(C_x + C_y + C_z)$;

        Update m_i^r, w_i^r with Equation (3), (7);

    end

    Calculate $d(\mathbf{W}^r, \mathbf{W}^{r-1})$ with Equation (5), (6);

    Set r=r+1;

until $d \leq 0.01$

Output: $\mathbf{W}^r$.

2.4 Parallel Implementation

The Tianhe-2 supercomputer at the National Supercomputing Center of Guangzhou (NSCC-GZ) consists of 16000 compute nodes, with each node containing two 12-core Intel Xeon E5-2692 CPUs with 64 GB memory [22]. The advanced computing performance and massive computing resources of Tianhe-2 provide the possibility to conduct more complex mathematical research and simulations on much larger scales than before. Based on off-line simulations in previous work, we expect that Tianhe-2 will facilitate applications of real-time forecasts for larger-scale problems. The computational efficiency of the high-precision inverse reconstruction of emission source will directly determine whether the atmospheric SO_2 transport process can be predicted in real time.

To our best knowledge, few studies focus on both, direct inverse source estimation and forecasts, at near-real-time. Fu et al. [23] conducted a near-real-time prediction study on volcanic eruptions based on the LOTOS-EUROS model and an ensemble Kalman filter. Santos et al. [10] developed a GPU-based code to process the calculations in near real time. In this work, we attempt to further develop the parallel inverse algorithm for reconstruction of volcanic SO_2 emission rates based on sequential importance resampling methods, utilizing the computational power of the Tianhe-2 supercomputer to achieve large-scale SO_2 transport simulations and real-time or near-real-time predictions.

The parallelization of MPTRAC and the inverse algorithm is realized by means of a hybrid scheme based on the Message Passing Interface (MPI) and Open Multi-Processing (OpenMP). Since each trajectory can be computed independently, the ensembles of unit simulations are distributed to different compute nodes using the MPI distributed memory parallelization. On a particular compute node, the trajectory calculations of the individual unit simulations are distributed using the OpenMP shared memory parallelization. Theoretically, the calculation time will decrease near linearly with an increasing number of compute processes. Therefore, sufficient computational performance can greatly reduce the computational costs and enable simulations of hundreds of millions of air parcels on the supercomputer system.

The implementation of the inverse algorithm is designed based on a high-throughput computing strategy. At each iterative step, the time- and altitude-dependent domain $E := [t_0, t_f] \times \Omega$ is discretized with n_t and n_h uniform intervals $N = n_t \cdot n_h$, which leads to $N = n_t \cdot n_h$ unit simulations that are calculated in parallel as shown in Fig. 1. Theoretically, the high-throughput parallel computing strategy can greatly improve the resolution of the inversion in time and altitude through increasing the value of N. Only little communication overhead is needed to distribute the tasks and gather the results. With more computing resources, it is possible to operate on more detailed spatial-temporal grids and to obtain more accurate results. In this work, we have achieved a resolution of 30 min in time and 100 m in altitude for the first time with our modeling system.

In summary, the goal of this work is to develop an inverse modeling system using parallel computing on a scale of millions of cores, including high-throughput submission, monitoring, error tolerance management and analysis of results. A multi-level task scheduling strategy has been employed, i.e., the computational performance of each sub-task was analyzed to maximize load balancing. During the calculation, every task is monitored by a daemon with an error tolerance mechanism being established to avoid accidental interruptions and invalid calculations.

3 Parallel Performance Analysis

In this section, we evaluate the model parallel performance on the Tianhe-2 supercomputer based on the single-node performance of MPTRAC for the unit simulations and the multi-node performance of the sequential importance resampling algorithm.

Since our parallel strategy is based on high-throughput computing to avoid communication across the compute nodes, the single-node computing performance is essential in determining the global computing efficiency. To test the single-node computing performance, we employ the Paratune Application Runtime Characterization Analyzer to measure the floating-point speed. An ensemble of 100 million air parcels was simulated on a single node and the gigaflops per second (Gflops) turned out to be 13.16. The strong scalability test on a single node is conducted by simulating an ensemble of 1 million air parcels. The results on strong scaling are listed in Table 1 and the results on weak scaling are listed in Table 2. Referred to a single process calculation, the strong and weak scaling efficiency using 16 computing processes reach 84.25% and 85.63%, respectively.

Table 1. Strong scaling of a single-node MPTRAC simulation

Problem size	Number of processes $N_{process}$	Clock time/s	Strong speed-up ratio R_s	Efficiency ($R_s/N_{process}$)
1 million parcels	1	1496	1x	100%
	4	399	3.75x	93.75%
	16	111	13.48x	84.25%

Table 2. Weak scaling of a single-node MPTRAC simulation

Problem size	Number of processes $N_{process}$	Clock time/s	Weak speed-up ratio R_w	Efficiency ($R_w/N_{process}$)
1 million parcels	1	1496	1x	100%
4 million parcels	4	1597	3.75x	93.75%
16 million parcels	16	1753	13.7x	85.63%

Since the ensemble simulations with MPTRAC covering multiple unit simulations are conducted independently on each node, the scaling efficiency of the MPI parallelization is mostly limited by I/O issues rather than communication or computation. Nevertheless, we tested the strong and weak scalability of the high-throughput based inverse calculation process, with a maximum computing scale of up to 38400 computing processes. The results are shown in Tables 3 and 4. The scaling is nearly linear with respect to the number of compute nodes. Especially for weak scaling, the efficiency is close to ideal conditions, except for little extra costs related to the calculation of the CSI and I/O issues.

Table 3. Strong scaling of the multi-node inverse algorithm (Each unit simulation cover 1 million air parcels. The same goes for Table 4)

Problem size	Number of processes $N_{process}$	Clock time/s	Strong speed-up ratio R_s	Efficiency $(R_s/N_{process})$
1600 unit simulations	2400	16775	1x	100%
	9600	4452	3.77x	94.25%
	38400	1138	14.74x	92.13%

Table 4. Weak scaling of multi-node inverse algorithm

Problem size	Number of processes $N_{process}$	Clock time/s	Weak speed-up ratio R_w	Efficiency $(R_w/N_{process})$
100 unit simulations	2400	1123	1x	100%
400 unit simulations	9600	1132	3.97x	99.25%
1600 unit simulations	38400	1138	15.79x	98.69%

In summary, the high-throughput based hybrid MPI/OpenMP parallel strategy of MPTRAC and the inverse algorithm show good strong and excellent weak scalability on the Tianhe-2 supercomputer. That means the inverse modeling system has high potential in massive parallel applications, meeting the requirements of real-time forecasts. However, the forward calculation still has some potential for optimization. In future work, we will investigate the possibility of cross-node computing with MPTRAC and try to further improve the single-node computing performance by using hyper threading. Besides, some further improvements may also be possible for the multi-node parallelization, in particular for the I/O issues and the efficiency of temporary file storage.

4 Case Study of the Nabro Volcanic Eruption

Following Heng et al. [21], we choose an eruption of the Nabro volcano, Eritrea, as a case study to test the inverse modeling system on the Tianhe-2 supercomputer. The Nabro volcano erupted at about 20:30 UTC on 12 June 2011, causing a release of about 1.5×10^9 kg of volcanic SO_2 into the troposphere and lower stratosphere. The volcanic activity lasted over 5 days with varying plume altitudes. The simulation results obtained for the Nabro volcanic eruption are of particular interest for studies of the Asian monsoon circulation [4, 24].

4.1 Reconstructed Emission Results with Different Resolutions

In general, with increasing resolution of the initial emissions, the forward simulation results are expected to become more accurate, but the calculation cost will also be much larger. In this work, the resolution of the volcanic SO_2 emission rates has been raised to 30 min of time and 100 m of altitude for the first time. The largest computing scale employs 60250 compute nodes on Tianhe-2 simultaneously. Each node calculated the kinematic trajectories of 1 million air parcels using a total of 24 cores. On such a computing scale, the inverse reconstruction and final forward simulation take about 22 min and require about 530,000 core hours in total.

Based on the inverse algorithm and parallel strategy described in Sects. 2 and 3, the SO_2 emission rates are reconstructed at different temporal and spatial resolutions, as shown in Fig. 2. The resolutions are (a) 6 h of time, 2.5 km of altitude, (b) 3 h of time, 1 km of altitude, (c) 1 h of time, 250 m of altitude, and (d) 30 min of time, 100 m of altitude. More fine structures in the emission rates become visible at higher resolution in Figs. 2a to 2c. However, the overall result in Fig. 2d at the highest resolution appears to be unstable with oscillations occurring between 12 to 16 km of altitude. The reason of this is not clear and will require further study, e.g., in terms of regularization of the inverse problem. For the time being, we employ the results in the Fig. 2c for the final forward simulation.

Compared with our previous work performed on the JuRoPA supercomputer at the Jülich Supercomputing Centre [21], the simulation results of this work performed on Tianhe-2 are rather similar. The reconstructed emissions show that the Nabro volcano had three strong eruptions on June 13, 14 and 16. For validation, Table 5 compares altitude and time of the major eruptions obtained with observations from different satellite sensors, which shows that the emission data constructed by the inverse modeling approach qualitatively agree with the measurements. Here, we also refer to the time series of the 2011 Nabro eruption based on Meteosat Visible and InfraRed Imager (MVIRI) infrared imagery (IR) and water-vapor (WV) measurements, which were used as validation data sets in [21] as shown in Fig. 3.

Table 5. Major eruption altitudes of the Nabro volcano on different days

	June 13	June 14	June 16
CALIOP and MIPAS data	19 km	9–13 km	–
Fromm et al. (2013) [25]	15–19 km	–	–
Fromm et al. (2014) [26]	–	–	17.4 km
Result from this work	15–17 km	9–13 km	17 km

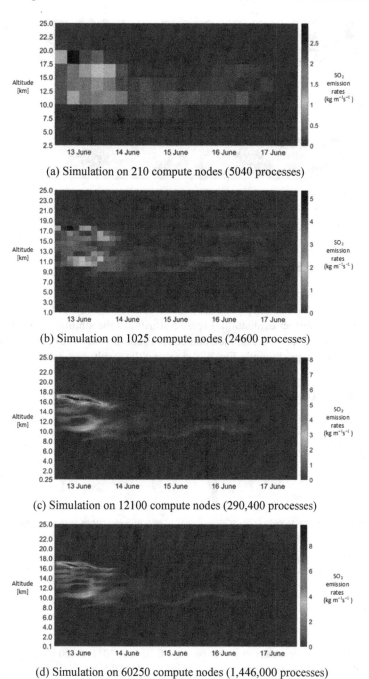

(a) Simulation on 210 compute nodes (5040 processes)

(b) Simulation on 1025 compute nodes (24600 processes)

(c) Simulation on 12100 compute nodes (290,400 processes)

(d) Simulation on 60250 compute nodes (1,446,000 processes)

Fig. 2. Reconstructed volcanic SO_2 emission rates of the Nabro eruption in June 2011. The x-axis refers to time, the y-axis refers to altitude (km), and the color bar refers to the emission rate $(kg\ m^{-1}\ s^{-1})$. (Color figure online)

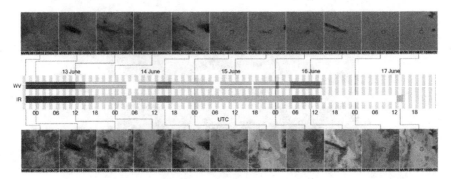

Fig. 3. Time line of the 2011 Nabro eruption based on MVIRI IR and WV measurements. Here white is none, light blue is low level, blue is medium level, dark blue is high level [21] (Color figure online)

4.2 Final Forward Simulation Results

Based on the reconstructed emission data with 1 h in time and 250 m in altitude resolution, the final simulation applying the product rule was conducted on the Tianhe-2 supercomputer for further evaluation. Figure 5 illustrates the simulated SO_2 transport, providing information on both, altitude and concentration, which are comparable to the AIRS observation maps shown in Fig. 4, suggesting the results are stable and accurate.

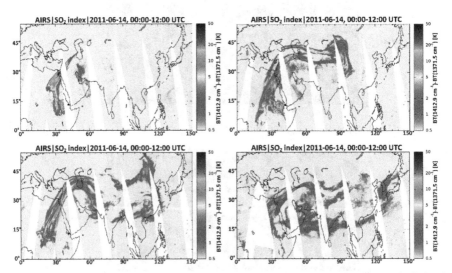

Fig. 4. The AIRS satellite observations on 14, 16, 18, 20 June 2011, 06:00 UTC (SO_2 index is a function of column density obtained from radiative transfer calculations. Here we refer to [19] for more detailed description of detection of volcanic emissions based on brightness temperature differences (BTDs) technique.)

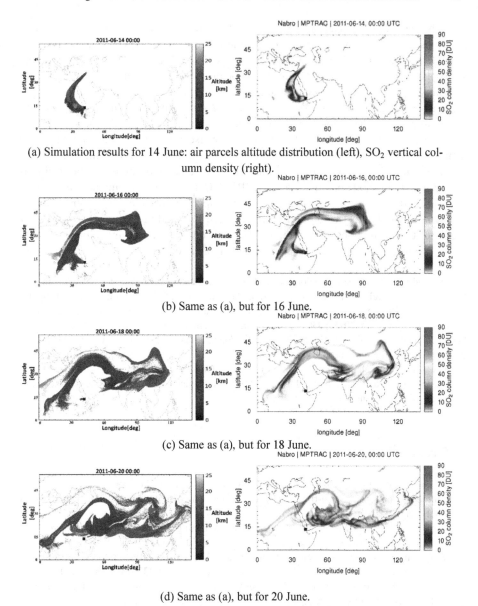

(a) Simulation results for 14 June: air parcels altitude distribution (left), SO_2 vertical column density (right).

(b) Same as (a), but for 16 June.

(c) Same as (a), but for 18 June.

(d) Same as (a), but for 20 June.

Fig. 5. Final forward simulation results of volcanic SO_2 released by the Nabro eruption. The black square indicates the location of the Nabro volcano.

5   Conclusions and Outlook

The high-resolution reconstruction of source information is critical to obtain precise atmospheric aerosol and trace gas transport simulations. The work we present in this paper has potential applications for studying the effects of large-scale industrial emissions, nuclear leaks and other pollutions of the atmosphere and environment. The computational costs and efficiency of the inverse model will directly determine whether the atmospheric pollutant transport process can be predicted in real time or near real time. For this purpose, we implemented and assessed a high-throughput based inverse algorithm using the MPTRAC model on the Tianhe-2 supercomputer. The good scalability demonstrates that the algorithm is well suited for large-scale parallel computing. In our case study, the computational costs for the inverse reconstruction and final forward simulation at unprecedented resolution satisfy the requirements of real-time forecasts.

In the future work, we will study further improvements of the computational efficiency, e.g., multi-node parallel usage of MPTRAC, mitigation of remaining I/O issues, post-processing overhead, efficient storage of temporary files, etc. Also, some stability problems at the highest resolution problems need to be addressed, e.g., by means of regularization techniques. Nevertheless, we think that the inverse modeling system in its present form is ready to be tested in further applications.

Acknowledgements. The corresponding author Yi Heng acknowledges support provided by the "Young overseas high-level talents introduction plan" funding of China, Zhujiang Talent Program of Guangdong Province (No.2017GC010576) and the Natural Science Foundation of Guangdong (China) under grant no. 2018A030313288. Chunyan Huang is supported by the National Natural Science Foundation of China (No. 11971503), the Young Talents Program (No. QYP1809) and the disciplinary funding of Central University of Finance and Economics. Pin Chen is supported by the Key-Area Research and Development Program of Guangdong Province (2019B010940001). Part of this work was funded by the Deutsche Forschungsgemeinschaft (DFG, German Research Foundation) – Projektnummer 410579391.

References

1. Brenot, H., Theys, N., Clarisse, L., et al.: Support to Aviation Control Service (SACS): an online service for near-real-time satellite monitoring of volcanic plumes. Nat. Hazards Earth Syst. Sci. **14**(5), 1099–1123 (2014)
2. Sigl, M., Winstrup, M., McConnell, J.R., et al.: Timing and climate forcing of volcanic eruptions for the past 2,500 years. Nature **523**(7562), 543–549 (2015)
3. Solomon, S., Daniel, J.S., Neely, R.R., et al.: The persistently variable "background" stratospheric aerosol layer and global climate change. Science **333**(6044), 866–870 (2011)
4. Bourassa, A.E., Robock, A., Randel, W.J., et al.: Large volcanic aerosol load in the stratosphere linked to Asian monsoon transport. Science **337**(6090), 78–81 (2012)
5. Hoffmann, L., Rößler, T., Griessbach, S., et al.: Lagrangian transport simulations of volcanic sulfur dioxide emissions: impact of meteorological data products. J. Geophys. Res.-Atmos. **121**(9), 4651–4673 (2016)
6. Lacasse, C., Karlsdottir, S., Larsen, G., et al.: Weather radar observations of the Hekla 2000 eruption cloud, Iceland. Bull. Volcanol. **66**(5), 457–473 (2004). https://doi.org/10.1007/s00445-003-0329-3
7. Stohl, A., Prata, A.J., Eckhardt, S., et al.: Determination of time- and height-resolved volcanic ash emissions and their use for quantitative ash dispersion modeling: the 2010 Eyjafjallajokull eruption. Atmos. Chem. Phys. **11**(9), 4333–4351 (2011)

8. Flemming, J., Inness, A.: Volcanic sulfur dioxide plume forecasts based on UV satellite retrievals for the 2011 Grimsvotn and the 2010 Eyjafjallajokull eruption. J. Geophys. Res.-Atmos. **118**(17), 10172–10189 (2013)
9. Molnar, F., Szakaly, T., Meszaros, R., et al.: Air pollution modelling using a Graphics Processing Unit with CUDA. Comput. Phys. Commun. **181**(1), 105–112 (2010)
10. Santos, M.C., Pinheiro, A., Schirru, R., et al.: GPU-based implementation of a real-time model for atmospheric dispersion of radionuclides. Prog. Nucl. Energy **110**, 245–259 (2019)
11. Müller, E.H., Ford, R., Hort, M.C., et al.: Parallelisation of the Lagrangian atmospheric dispersion model NAME. Comput. Phys. Commun. **184**(12), 2734–2745 (2013)
12. Larson, D.J., Nasstrom, J.S.: Shared- and distributed-memory parallelization of a Lagrangian atmospheric dispersion model. Atmos. Environ. **36**(9), 1559–1564 (2002)
13. Gordon, N.J., Salmond, D.J., Smith, A.F.M.: Novel approach to nonlinear/non-Gaussian Bayesian state estimation. IEE Proc. F (Radar Signal Process.) **140**(2), 107–113 (1993)
14. Dee, D.P., Uppala, S.M., Simmons, A.J., et al.: The ERA-Interim reanalysis: configuration and performance of the data assimilation system. Q. J. Roy. Meteorol. Soc. **137**(656), 553–597 (2011)
15. Bowman, K.P., Lin, J.C., Stohl, A., et al.: Input data requirements for Lagrangian trajectory models. Bull. Am. Meteorol. Soc. **94**(7), 1051–1058 (2013)
16. Stohl, A., Forster, C., Frank, A., et al.: Technical note: the Lagrangian particle dispersion model FLEXPART version 6.2. Atmos. Chem. Phys. **5**, 2461–2474 (2005)
17. Rößler, T., Stein, O., Heng, Y., et al.: Trajectory errors of different numerical integration schemes diagnosed with the MPTRAC advection module driven by ECMWF operational analyses. Geosci. Model Dev. **11**(2), 575–592 (2018)
18. Hoffmann, L., Hertzog, A., Rößler, T., et al.: Intercomparison of meteorological analyses and trajectories in the Antarctic lower stratosphere with Concordiasi superpressure balloon observations. Atmos. Chem. Phys. **17**, 8045–8061 (2017)
19. Hoffmann, L., Griessbach, S., Meyer, C.I.: Volcanic emissions from AIRS observations: detection methods, case study, and statistical analysis. In: Proceedings of SPIE, vol. 9242 (2014)
20. Schaefer, J.: The critical success index as an indicator of warning skill. Weather Forecast. **5**, 570–575 (1990)
21. Heng, Y., Hoffmann, L., Griessbach, S., et al.: Inverse transport modeling of volcanic sulfur dioxide emissions using large-scale simulations. Geosci. Model Dev. **9**(4), 1627–1645 (2016)
22. Che, Y., Yang, M., Xu, C., et al.: Petascale scramjet combustion simulation on the Tianhe-2 heterogeneous supercomputer. Parallel Comput. **77**, 101–117 (2018)
23. Fu, G.L., Lin, H.X., Heemink, A., et al.: Accelerating volcanic ash data assimilation using a mask-state algorithm based on an ensemble Kalman filter: a case study with the LOTOS-EUROS model (version 1.10). Geosci. Model Dev. **10**(4), 1751–1766 (2017)
24. Wu, X., Griessbach, S., Hoffmann, L.: Equatorward dispersion of a high-latitude volcanic lume and its relation to the Asian summer monsoon: a case study of the Sarychev eruption in 2009. Atmos. Chem. Phys. **17**(21), 13439–13455 (2017)
25. Fromm, M., Nedoluha, G., Charvat, Z.: Comment on "large volcanic aerosol load in the stratosphere linked to Asian monsoon transport". Science **339**(6120), 647 (2013)
26. Fromm, M., Kablick, G., Nedoluha, G., et al.: Correcting the record of volcanic stratospheric aerosol impact: Nabro and Sarychev Peak. J. Geophys. Res.-Atmos. **119**(17), 10343–10364 (2014)

Granulation-Based Reverse Image Retrieval for Microscopic Rock Images

Magdalena Habrat[ID] and Mariusz Młynarczuk[(✉)][ID]

AGH University of Science and Technology, Mickiewicza 30, 30-059 Kraków, Poland
mlynar@agh.edu.pl

Abstract. The paper presents a method of object detection on microscopic images of rocks, which makes it possible to identify images with similar structural features of the rock. These features are understood as the sizes and shapes of its components and the mutual relationships between them. The proposed detection methodology is an adaptive and unsupervised method that analyzes characteristic color clusters in the image. It achieves good detection results for rocks with clear clusters of colored objects. For the analyzed data set, the method finds in the rock image sets with high visual similarity, which translates into the geological classification of rocks at a level of above 78%. Considering the fact that the proposed method is based on segmentation that does not require any input parameters, this result should be considered satisfactory. In the authors' opinion, this method can be used in issues of rock image search, sorting, or e.g. automatic selection of optimal segmentation techniques.

Keywords: Objects retrieval · Classification of rocks · Microscopic analysis of rock · CBIR · CBVIR

1 Introduction

Recent years we have seen a dynamic technological progress in the field of computer processing and recognition of images. The developed methods find further applications, which include earth sciences. The active development of IT methods, initiated in the nineties of the last century, resulted in a situation where the basis of many measurements in geology are images and digital image sequences obtained from e.g. optical, electron, confocal microscopes, etc. [1–5]. Carrying out automatic measurements on images means that the researchers have access to constantly increasing image databases. It necessitates their automatic interpretation and that, in turn, requires their automatic indexation. Currently in geology, such data sets are managed manually using specialized knowledge. However, in the case of very large collections that often contain hundreds of thousands of images, this approach is difficult to implement due to the huge amount of time that it requires. This situation necessitates the use of methods that would allow for automatic management of image data sets [6, 7]. These include, among others, image search techniques, which have been intensively developed in recent years [8–12]. These methods are also the subject of the research described in this work. The paper presents

© Springer Nature Switzerland AG 2020
V. V. Krzhizhanovskaya et al. (Eds.): ICCS 2020, LNCS 12139, pp. 74–86, 2020.
https://doi.org/10.1007/978-3-030-50420-5_6

research for an unsupervised retrieval system, where the query is in the form of an image (or its fragment) and no additional conditions are explicitly defined. This approach is defined as image search, and also referred to as reverse image search [13].

1.1   The Idea of Image Retrieval Systems

Image retrieval systems belong to the group of techniques that support the automatic management of image data sets [6–8, 10, 11, 14, 15]. Two main approaches to image search have become particularly popular in the literature.

The first (which is not the subject of this work) involves finding images based on the so-called metadata, which can be both technical parameters of the image as well as verbal descriptions, e.g. regarding the content of the image. This type of image search systems is often referred to as TBIR (Text Based Image Retrieval).

The other mainstream image search is based on image content analysis and is often referred to as CBIR, CBVIR (Content Based (Visual) Image Retrieval) [16, 17]. This approach largely reflects the way in which images are compared by the human mind, referring to the content of images that differ in colors, texture and content. Such an image can be characterized by certain characteristics. In order to detect these features, it is necessary to determine optimal and possible universal methods for obtaining descriptors (understood as a numerical representation of image features). In the search issue, both the search query and the set of searched images are processed to extract the feature vector. Then, these features are compared and adjusted according to established criteria. In a nutshell, one can distinguish two main trends in the construction of image search systems. These trends differ in the way they determine the similarity between the data. The first is the so-called unsupervised search, otherwise referred to as "without interaction", based on unsupervised machine learning methods [18–20]. The other trend is the so-called supervised search, otherwise defined - "with interaction" or feedback, based on supervised machine learning methods (relevance feedback) [21–23]. Regardless of the selection of the similarity determination methodology, in the reverse image search problem it is important that the feature detection process be as effective as possible.

1.2   Initial Description of the Presented Method

Due to the cognitive nature of geological research and the constantly growing resources of digital archives of rock images, the application of methods supporting the work of experts seems to be fully justified. Prerequisites for the system that may be created based on the methodology proposed in the work should be introduced:

- the user has no knowledge of the searched image archive;
- the data in the archive are not described, they only have their IDs given during the algorithm operation;
- the user can indicate the search key in the form of an image (or fragment), and the search is based only on the analysis of the features of the selected query and the available archive.

The adoption of such assumptions means that the reverse search method based on the analysis of the shapes of objects recorded in the pictures should be based on universal steps that would result in acceptable effectiveness for images of various rocks. On this basis, the main stages of the method can be distinguished:

1. feature detection stage, i.e. automatic processing of the query key and images from the database to extract the characteristic features, including:

   a. extraction of values describing selected features both for the query image and for the entire available image database - this stage is described in detail in this work;

   b. creating a common feature vector that includes query feature values and an image database. As part of the work, its normalization is also carried out;

2. the stage of determining similarity between the values of the described features in order to determine the search results.

In this paper, the authors present a stepwise description of the unsupervised reverse search methodology, placing particular emphasis on the description and results of the object detection method that suggest similar features of rock granulation.

2   Method of Conglomerates Detection

In order to quantify the groups of visually similar grains (called here conglomerates), it is necessary to segment the rock image. The research proposed a detection method consisting of four main steps, i.e.

1. selection of data for analysis;
2. image generalization - that is, determining the number of dominant colors, and then dividing the image into areas coherent in terms of color;
3. estimation of the number of color clusters of detected regions (in step No. 2);
4. segmentation of the input image (from step 1) by color clustering, according to the cluster number values determined in step 3; final image preparation.

2.1   Selection of Data for Analysis

The research used images of 8 different groups of rocks. Rocks were selected in such a way that they differ in the grain size (i.e. granulation). The study analyzed three groups of grain sizes: fine-grained: dolomite and quartzite, medium-grained: crystalline slate, metamorphic shale, limestone and coarse-grained: anhydrite, granite and granodiorite (see Fig. 1).

The images were recorded using an optical microscope with polarized light, with optimal lighting and 100x magnification, which did not change during the recording of all photos. The image database on which the analyzes were carried out comprised 800 digital images, i.e. 100 images for each rock.

anhydrite   dolomite   granite

metamorphic shale   crystalline slate   limestone

Fig. 1. Sample images for rocks analyzed in searching for structurally similar granulation.

The method described in the work was developed and tested in the RGB color space, where each layer is a gray image with values of gray levels and in the range from 0 to 255. In the first step, it is recommended to set a temporary parameter for the maximum number of colors taken into account during further analysis. It can be a division of the entire color space into 256 clusters, referring to the number of gray levels of the image. Choosing larger values allows for more detailed results but it takes more time to run the algorithm.

2.2 Determining the Areas of the Image Dominating in Terms of Color

The next stage of the method is the initial generalization of the image by dividing it into fragments with possibly consistent color representation. It can be done through:

- manual selection - a very large fixed number of areas can be assumed that can be generated by the growth segmentation method or, for example, the method of super pixel detection [24];
- adaptive selection - by calculating the number of real dominant colors in the image. For this purpose, the number of unique colors is selected that most often occur in the image. Additionally, the minimum number of occurrences of the detected color is checked. The more dominant colors there are, the smaller their area of occurrence. For images with different resolutions and sizes, normalization should be carried out.

Knowing the initial number of dominant colors, one can go to the generalization stage. The research used the SLIC image super pixel detection method [24]. Such generalization results in the creation of the so-called image mosaics, as shown in Fig. 2. It is worth noting that the human eye distinguishes only a few colors on the obtained mosaics, while the histograms indicate the existence of many close but separated clusters of colors with similar levels of gray. The use of adaptive binarization results in the extraction of many invalid conglomerates. This can be seen on the example of granite in Figs. 3 and 4.

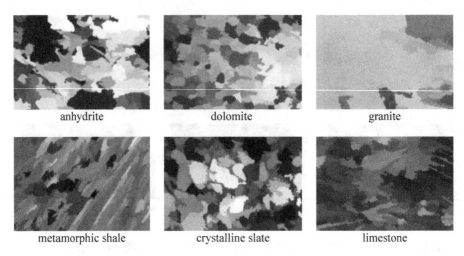

Fig. 2. Sample image mosaics for selected rock groups obtained by SLIC super pixel detection.

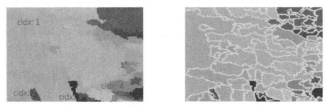

Fig. 3. Comparison of mosaic appearance (4 main colors-*cidx*) (left) and imposed limits after binarization (right).

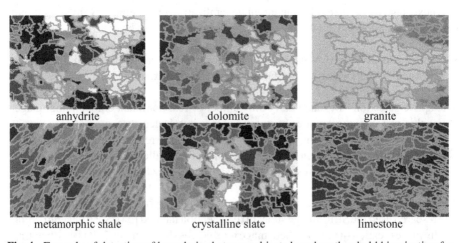

Fig. 4. Example of detection of boundaries between objects based on threshold binarization for mosaic colors.

The issue of excess objects caused by the closeness of color is the subject of the next step of the proposed method (step 3). The goal of this step is to answer the question of how to binarize obtained mosaics properly so that the fragments of images with visually similar and numerically different colors become one final cluster (conglomerate).

2.3 Estimation of the Number of Color Clusters of Detected Image Regions

It is suggested that in order to binarize mosaics, they should be clustered with respect to the number of major dominant colors. To determine this number automatically, it is proposed to use the method of estimating the number of clusters. The feature vector for this grouping were the colors of the image after generalization. The Elbow method was used, which analyzes the percentage of stability (variance) within a given cluster. It requires the input of the condition of stabilization of variance. It says at what degree of variance of parameter values, the resulting number of clusters is considered optimal. By submitting the image color vector after generalization as a method parameter, one can observe a graph of variance stabilization (y axis) and the number of clusters (x axis) indicating the number of dominant colors in the mosaic (Fig. 5). In all cases, stabilization within clusters is achieved for values higher than 0,9, but it is not constant for all rocks.

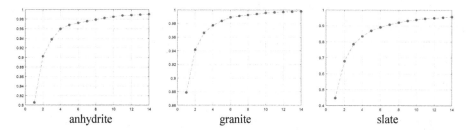

anhydrite   granite   slate

Fig. 5. Graphs of variance stabilization for selected images of different rocks (x axis – number of clusters, y axis - variance stabilization).

In order to determine the value of the optimal number of clusters, one can proceed in various ways, e.g.:

- the interesting stabilization value can be assumed a priori to be at a constant level for all rocks, e.g. 0,95;
- the last significant change in differences between increasing the value of cluster stabilization can be detected. It comes down to the detection of the so-called elbow (factor) on the Elbow method chart. Analyzing the values of subsequent differences, it can be seen that if for each iteration a factor indicates large changes, then in the analyzed cases it obtained values lower than 0,99. Therefore, it can be assumed that the values of this coefficient lower than 0,99 indicate disproportionate stabilization of variance and the need to increase the number of iterations. Values equal to 0,99 or higher indicate stabilization and the stop condition. Thus, the maximum value and for the stabilization condition may indicate a given number of clusters.

2.4   Input Image Segmentation

Having obtained the number of main dominant colors in previous step, it is proposed
to use it as the number of clusters for the image segmentation method. One can then
specify the number of result groups that the image should be divided into so that each of
the pixels belonging to the image is combined into homogeneous clusters of gray levels
(ultimately colors) - see Fig. 6 and 7. Both original images and mosaics can be clustered.

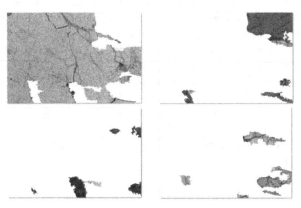

Fig. 6. Groups of the most characteristic rock fragments with a coherent grain color (granite from
the Giant Mountains) - estimation consistent with visual assessment.

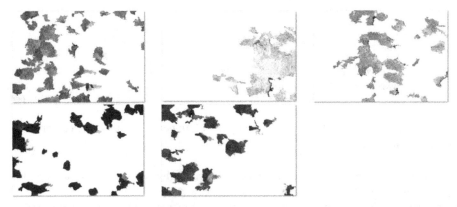

Fig. 7. Groups of the most characteristic rock fragments with a coherent grain color (dolomite
from the Beskid Mountains) - estimation consistent with visual assessment.

Figure 8 presents the detected objects (grain conglomerates) for sample images for
each of the 8 rocks studied. Input images were clustered (so as not to lose information
omitted by the generalization method). A morphological gradient by erosion was used
to detect the conglomerate boundary.

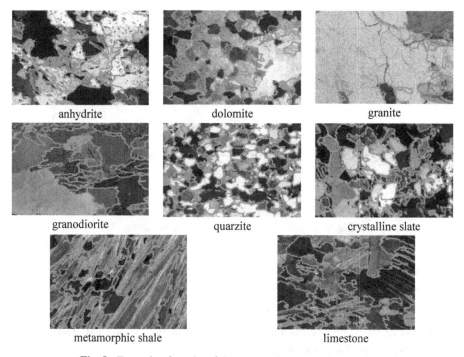

anhydrite

dolomite

granite

granodiorite

quarzite

crystalline slate

metamorphic shale

limestone

Fig. 8. Example of results of the proposed segmentation method.

2.5 Final Algorithm of Conglomerates Detection

The methods described in this chapter lead to segmentation of groups of visually similar grains - called conglomerates. Figure 9 is a graphical presentation of the proposed algorithm of conglomerate segmentation. It should be emphasized that the literature contains descriptions of algorithms for the segmentation of only certain types of rocks. However, these methods work very well for those rocks for which they were developed, and for others they usually fail. The method we proposed was developed to properly segment grains or grain groups for most grained rocks. This method does not always lead to very correct segmentation of grains. However, in our opinion, it is sufficient for the correct classification of rocks based on similar structural features.

3   Results for Similarity Determination Stage

Binary images with detected conglomerates can become the basis for the creation of feature vectors. Each object detected in the image can be described by a set of parameters. In this research the objects parameters were used: surface area, circular coefficient, longest/shortest diameter, equivalent diameter, Feret diameters. As a result, the feature vector of each object in the defined feature space is obtained. One can stop at this approach when the system aims to find all similar objects in the database.

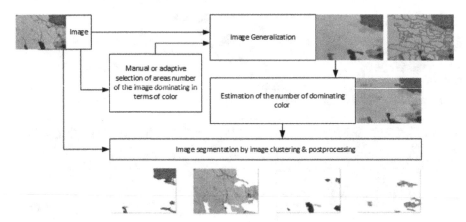

Fig. 9. Algorithm of conglomerate segmentation.

However, if it is necessary to find images of rocks similar in shape and mutual quantitative relationships, then statistical measures describing the features of the objects can be used. In this work, the following were analyzed:

- skewness of the feature value for all objects within the image; it determines the asymmetry of the distribution of the analyzed variable;
- coefficient of variation of the feature value for all objects within the image; it defines the degree of the variation in the distribution of features;
- spread of the feature value for all objects within the image; determines the quantitative diversity between objects in the image;
- arithmetic mean of the feature values for all objects within the image; determines the average quantitative measure between objects in the image;

In addition, the parameters of mean, coefficient of variation, dispersion and skewness of the orientation of objects in the image were introduced. The number of objects and the ratio of conglomerate surface area coverage to their number was also taken into account. In this way, a vector consisting of 42 features (X42) was defined. Additionally, the usefulness of different variants of the 42 dimensional subspace was assessed:

- X2 - descriptor described by two features (number of objects in the image - the ratio of conglomerate surface area coverage to their number),
- X10m - a descriptor based only on average values,
- X10cv - descriptor based only on bottom coefficients of variation,
- X10r - a descriptor based only on dispersion of values,
- X10s - a descriptor based on the skewness of parameters values.

The average results of the geological correctness of detection of kNN for various variants are presented in Table 1. It can be seen that the results deteriorate with the increase in the number of analyzed neighbors k. The best results, on average for all rock groups, were obtained for the X10m space. The lowest results were obtained for searches based on the analysis of only the number of conglomerates and the ratio of the surface area of conglomerates and their number.

Table 1. Comparison of the average (for all groups of images) values of the geological fit for different numbers of the most similar images (kNN method) for different feature spaces [%].

k	10	20	30	40	50	60	70	80	90	100
X42	67,98	60,96	56,67	53,23	50,58	48,31	46,28	44,45	42,69	41,10
X2	39,06	31,68	29,21	27,47	26,66	26,00	25,48	25,13	24,93	24,73
X10cv	45,98	39,88	36,96	34,78	33,32	32,13	31,08	30,25	29,42	28,64
X10r	38,24	32,79	29,85	28,17	26,96	25,94	25,02	24,31	23,68	**23,16**
X10s	48,34	42,01	38,99	36,61	34,98	33,35	31,9	30,64	29,52	28,50
X10m	**78,31**	73,04	68,98	66,18	63,79	61,75	59,87	58,18	56,34	54,45

Table 2 presents a summary of the average value of correct geological fit for individual groups of images. It can be noted that, as in the case of Table 1, the highest results are obtained for descriptors of average values of shape parameters and based on the entire unreduced X42 space.

Table 2. Comparison of the average (for all groups of images) values of the geological fit for ten the most similar images (kNN method) for different feature spaces [%].

	X42	X2	X10cv	X10r	X10s	X10m
Anhydrite	54,00	48,40	31,50	41,60	32,80	61,50
Dolomite	68,60	34,40	52,80	41,90	52,60	76,10
Granite	62,80	61,90	39,00	42,20	46,00	72,80
Granodiorite	59,10	45,50	30,30	32,70	37,70	73,30
Quarzite	68,90	35,00	58,30	35,80	46,40	75,20
Crys. slate	75,00	30,00	48,70	39,40	45,30	88,70
Met. shale	87,10	28,00	62,90	43,80	82,50	**96,20**
Limestone	68,30	29,30	44,30	**28,50**	43,40	82,70

Examples of matching results are shown in Figs. 10 and 11. Searching with a key in the form of an image of a rock with a coarse or fine-grained structure results in obtaining the most similar images with such a structure. Thus, it seems possible that image search would make it possible to determine the structural similarity of images of different rocks.

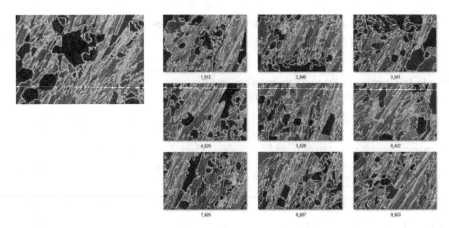

Fig. 10. Geologically and visually correct search results (key – left, results – right).

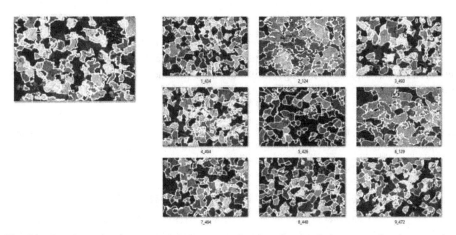

Fig. 11. Search results that are correct in terms of grains characteristics aspect, but incorrect in terms of geology (key – left, results – right).

4   Recapitulation

The paper presents a method of detecting conglomerates on microscopic images of rocks that allows searching for images of rocks with similar structural features. These features are understood as the size and shape of its components and the relationship between them. The proposed detection methodology is an unsupervised and adaptive method. It analyzes the number of characteristic color clusters on the examined images. The method returns good results for rocks with clear clusters of colorful rock-forming objects. It is not sensitive to individual small color charges (e.g. minerals stuck in the binder, e.g. clay) and treats them as noise, which it skips at the stage of estimating the number of the most characteristic clusters. It seems that the method cannot be a segmentation method used to accurately describe the rock, because it does not produce results accurate enough to become the basis for quantitative analysis of the rock. However, this method is a good starting point for automatic analysis and interpretation of the content of petrographic images. In the authors' opinion, this method can be successfully used for rock image retrieval. It can also be used in image sorting for given geological features, when the input data is unknown, not described, and thus training the system and the use of supervised classification methods (e.g. for segmenting rock grains) is impossible.

Acknowledgements. This work was financed by the AGH-University of Science and Technology, Faculty of Geology, Geophysics and Environmental Protection as a part of statutory project.

References

1. Zhang, Y., Wang, G., Li, M., Han, S.: Automated classification analysis of geological structures based on images data and deep learning model. Appl. Sci. **8**(12), 2493 (2018)
2. Aligholi, S., Khajavi, R., Razmara, M.: Automated mineral identification algorithm using optical properties of crystals. Comput. Geosci. **85**, 175–183 (2015)
3. Shu, L., McIsaac, K., Osinski, G.R., Francis, R.: Unsupervised feature learning for autonomous rock image classification. Comput. Geosci. **106**, 10–17 (2017)
4. Izadi, H., Sadri, J., Bayati, M.: An intelligent system for mineral identification in thin sections based on a cascade approach. Comput. Geosci. **99**, 37–49 (2017)
5. Młynarczuk, M., Habrat, M., Skoczylas, N.: The application of the automatic search for visually similar geological layers in a borehole in introscopic camera recordings. Measurement **85**, 142–151 (2016)
6. Espinoza-Molina, D., Datcu, M.: Earth-observation image retrieval based on content, semantics, and metadata. IEEE Trans. Geosci. Remote Sens. **51**(11), 5145–5159 (2013)
7. Castelli, V., Bergman, L.D.: Image Databases: Search and Retrieval of Digital Imagery. Wiley, New York (2004)
8. Fergus, R., Fei-Fei, L., Perona, P., Zisserman, A.: Learning object categories from Google's image search. In: Tenth IEEE International Conference on Computer Vision, ICCV 2005, vol. 2, pp. 1816–1823 (2005)
9. Ładniak, M., Młynarczuk, M.: Search of visually similar microscopic rock images. Comput. Geosci. **19**(1), 127–136 (2014). https://doi.org/10.1007/s10596-014-9459-2
10. Liu, Y., Zhang, D., Lu, G., Ma, W.Y.: A survey of content-based image retrieval with high-level semantics. Pattern Recogn. **40**(1), 262–282 (2007)
11. Najgebauer, P., et al.: Fast dictionary matching for content-based image retrieval. In: Rutkowski, L., Korytkowski, M., Scherer, R., Tadeusiewicz, R., Zadeh, L., Zurada, J. (eds.) ICAISC 2015. LNCS (LNAI), vol. 9119, pp. 747–756. Springer, Cham (2015). https://doi.org/10.1007/978-3-319-19324-3_67
12. Habrat, M., Młynarczuk, M.: Evaluation of local matching methods in image analysis for mineral grain tracking in microscope images of rock sections. Minerals **8**, 182 (2018)
13. Gaillard, M., Egyed-Zsigmond, E.: Large scale reverse image search. In: XXXVème Congrès INFORSID, p. 127 (2017)
14. Habrat, M., Młynarczuk, M.: Object retrieval in microscopic images of rocks using the query by sketch method. Appl. Sci. **10**, 278 (2020)
15. Wang, X.J., Zhang, L., Li, X., Ma, W.Y.: Annotating images by mining image search results. IEEE Trans. Pattern Anal. Mach. Intell. **30**(11), 1919–1932 (2008)
16. Sivic, J., Zisserman, A.: Video Google: a text retrieval approach to object matching in videos. In: Proceedings of the Ninth IEEE International Conference on Computer Vision (ICCV 2003), vol. 2, pp. 1470–1477 (2003)

17. Aigrain, P., Zhang, H., Petkovic, D.: Content-based representation and retrieval of visual media: a state-of-the-art review. In: Zhang, H., Aigrain, P., Petkovic, D. (eds.) Representation and Retrieval of Visual Media in Multimedia Systems, pp. 3–26. Springer, Boston (1996). https://doi.org/10.1007/978-0-585-34549-9_2

18. Chen, Y., Wang, J.Z., Krovetz, R.: An unsupervised learning approach to content-based image retrieval. In: Seventh International Symposium on Signal Processing and Its Applications, vol. 1, pp. 197–200. IEEE (2003)

19. Dy, J.G., Brodley, C.E.: Feature selection for unsupervised learning. J. Mach. Learn. Res. **5**, 845–889 (2004)

20. Zakariya, S., Ali, R., Ahmad, N.: Combining visual features of an image at different precision value of unsupervised content based image retrieval. In: IEEE International Conference on Computational Intelligence and Computing Research, ICCIC 2010, pp. 1–4. IEEE (2010)

21. Minka, T.P., Picard, R.W.: Interactive learning with a "society of models". Pattern Recogn. **30**, 565–581 (1997)

22. Rui, Y., Huang, T.S., Ortega, M., Mehrotra, S.: Relevance feedback: a power tool for interactive contentbased image retrieval. IEEE Trans. Circuits Syst. Video Technol. **8**(5), 644–655 (1998)

23. Zhou, X.S., Huang, T.S.: Relevance feedback in image retrieval: a comprehensive review. Multimedia Syst. **8**(6), 536–544 (2003)

24. Achanta, R., Shaji, A., Smith, K., Lucchi, A., Fua, P., Susstrunk, S.: SLIC superpixels compared to state-of-the-art superpixel methods. IEEE Trans. Pattern Anal. Mach. Intell. **34**(11), 2274–2282 (2012)

Hybrid SWAN for Fast and Efficient Practical Wave Modelling - Part 2

Menno Genseberger[1]([⊠]) and John Donners[2]

[1] Deltares, P.O. Box 177, 2600 MH Delft, The Netherlands
Menno.Genseberger@deltares.nl
[2] Atos, Munich, Germany

Abstract. In the Netherlands, for coastal and inland water applications, wave modelling with SWAN on structured computational grids has become a main ingredient. However, computational times are relatively high. Benchmarks showed that the MPI version of SWAN is not that efficient as the OpenMP version within a single node.

Therefore, in a previous paper [5] a hybrid version of SWAN was proposed for computations on structured computational grids. It combines the efficiency of the OpenMP version on shared memory with the capability of the MPI version to distribute memory over nodes. In the current paper we extend this approach by an improved implementation, verification of the model performance with a testbed, and extensive benchmarks of its parallel performance. With these benchmarks for important real life applications we show the significance of this hybrid version. We optimize the approach and illustrate the behavior for larger number of nodes. Parallel I/O will be subject of future research.

Keywords: Wave modelling · Hybrid method · Distributed and shared memory

1 Introduction

In the Netherlands, for assessments of the primary water defences (for instance [8]), operational forecasting of flooding [7,9], and water quality studies in coastal areas and shallow lakes (for instance [4]) waves are modelled with the third generation wave simulation software SWAN [1]. These are applications with SWAN on structured computational grids. However, computational times of SWAN are relatively high. Operational forecasting of flooding and water quality studies require a faster SWAN, at the moment this is a major bottleneck. Assessments of the primary water defences require both a fast and efficient SWAN.

We acknowledge the Dutch Ministry of Infrastructure and the Environment, PRACE, Fortissimo, and NWO for supporting a part of this research. We acknowledge PRACE for awarding us access to resource Cartesius based in The Netherlands at SURFsara.

© Springer Nature Switzerland AG 2020
V. V. Krzhizhanovskaya et al. (Eds.): ICCS 2020, LNCS 12139, pp. 87–100, 2020.
https://doi.org/10.1007/978-3-030-50420-5_7

Therefore, in a previous paper [5] we studied the current MPI version [13] and OpenMP version [2] of SWAN. That study was the basis of a hybrid version of SWAN.

In the present paper ("part 2"), with results of extensive benchmarks for important real life applications, we show the significance of the hybrid version. For a proper understanding of the underlying principles we recapitulate the main ingredients from [5] of SWAN and its parallel implementations in Sect. 2 and Sect. 3, respectively. The setup of the benchmarks is outlined in Sect. 4. In Sect. 5 we indicate how we verified the model performance of SWAN for the hybrid version. Then we optimize the approach in Sect. 6. Section 7 ends with two illustrations of the behavior of the hybrid version for larger number of nodes. Parallel I/O will be subject of future research.

2   SWAN

The simulation software package SWAN (Simulating WAves Near-shore) developed at Delft University of Technology [1], computes random, short-crested wind-generated waves in coastal areas and inland water systems. It solves a spectral action balance equation that incorporates spatial propagation, refraction, shoaling, generation, dissipation, and nonlinear wave-wave interactions. The coupling of wave energy via the spectral action balance equation is global over the entire geographical domain of interest. Compared to spectral methods for oceanic scales that can use explicit schemes, SWAN has to rely on implicit upwind schemes to simulate wave propagation for shallow areas in a robust and economic way. This is because typical scales (both spatial, temporal, and spectral) may have large variations when, for instance, waves propagate from deep water towards the surf zone in coastal areas.

For spectral and temporal discretization fully implicit techniques are applied. As a consequence the solution procedure of SWAN is computationally intensive. For typical applications these computations dominate other processes like memory access and file I/O.

In the present paper we consider SWAN for structured computational grids (both rectangular and curvilinear) that cover the geographical domain. The spectral space is decomposed into four quadrants. In geographical space a Gauss-Seidel iteration, or sweep technique is applied for each quadrant. This serial numerical algorithm is based on the Strongly Implicit Procedure (SIP) by Stone [12]. Figure 1 illustrates the sweep technique.

3   Parallel Implementation

Given a serial numerical algorithm, in general two parallelization strategies can be followed [3]: type (1): change the algorithm for a high degree of parallelism or type (2): do not change the algorithm but try to implement it in parallel as much as possible. For the serial numerical algorithm of SWAN based on implicit schemes with sweep technique, a strategy of type (2) has an upperbound of

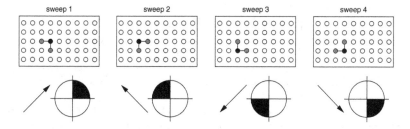

Fig. 1. Illustration of the sweep technique for the four quadrants. For every quadrant the arrow indicates the sweep direction and the black bullet represents a computational grid point that is being processed for which information comes from the two grey bullets via the upwind coupling stencil.

computational grid	natural ordering	hyperplane ordering
	37 38 39 40 41 42 43 44 45	5 6 7 8 9 10 11 12 13
	28 29 30 31 32 33 34 35 36	4 5 6 7 8 9 10 11 12
	19 20 21 22 23 24 25 26 27	3 4 5 6 7 8 9 10 11
	10 11 12 13 14 15 16 17 18	2 3 4 5 6 7 8 9 10
	1 2 3 4 5 6 7 8 9	1 2 3 4 5 6 7 8 9

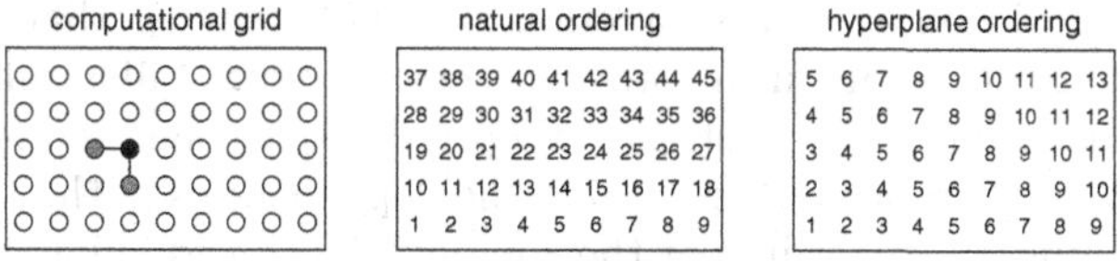

Fig. 2. Natural ordering and corresponding hyperplane ordering for sweep 1 of the sweep technique. The black bullet represents a point in the computational grid that is being processed for which information comes from the two grey bullets via the upwind coupling stencil.

maximal parallelism for the computations. For each of the four sweeps of the sweep technique this upperbound is related to a hyperplane ordering [3, § 4.1]. This depends on the stencil that couples the points in the computational grid: a new value at a point in the computational grid cannot be computed before values are known at neighbouring points that are coupled via this stencil. If computations proceed via some ordering (for instance the natural ordering for sweep 1 as shown in Fig. 2) then the corresponding hyperplane ordering shows those points in the computational grid for which new values can be computed simultaneously (i.e. concurrent computations, in parallel, with opportunity for fine-grained synchronization). These points have the same number in the ordering (a hyperplane), points on which they depend via the coupling stencil have a lower number (data dependency).

3.1 Distributed Memory

To reduce computational times of SWAN, Zijlema [13] considered parallelization approaches for distributed memory architectures. The current MPI version of SWAN is based on this work. The approach followed is of type (2): a block wavefront approach for which the author of [13] was inspired by a parallelization of an incomplete LU factorization.

In fact, it is based in a more coarse-grained way on the hyperplane ordering for the sweeps of the sweep technique from Sect. 3. For this purpose, the

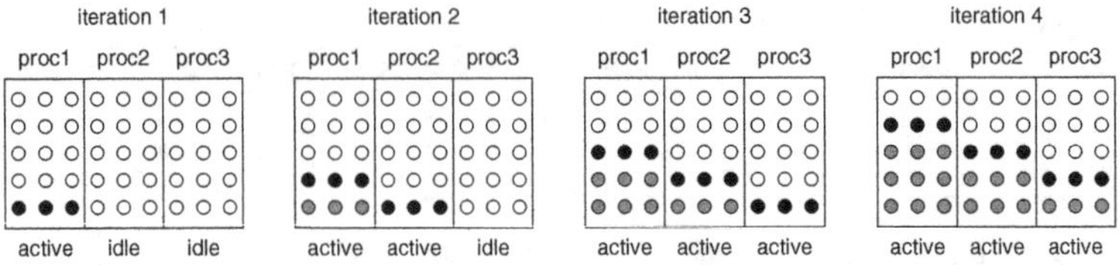

Fig. 3. Illustration of the block wavefront approach for sweep 1 of the sweep technique. Shown are succeeding iterations in case of three parallel processing units (proc1, proc2, and proc3). The black bullets represent computational grid points that are being updated in the current iteration. The grey bullets were updated in a previous iteration.

computational grid is decomposed into strips in one direction. The number of computational grid points in this direction is equal or higher than the number of computational grid points in the other direction. Figure 3 illustrates the block wavefont approach for sweep 1 of the sweep technique (the idea for the other sweeps is similar). In iteration 1, following the dependencies of the upwind stencil, processor 1 updates the values at the computational grid points in the lowest row of strip 1. All other processors are idle in iteration 1. When sweep 1 arrives at the right-most point in the lowest row of strip 1, after the update the corresponding value is communicated to strip 2. Then processor 2 is activated. In iteration 2, processor 1 performs sweep 1 on the next row of strip 1, processor 2 performs sweep 1 on the lowest row of strip 2. Etcetera. Note that not all processors are fully active during start and end phase of this approach. However, for a larger number of computational grid points (compared to the number of processors) this becomes less important. The block wavefront approach is implemented in the current editions of SWAN with MPI. Data is distributed via the decomposition in strips. For each sweep, at the end of every iteration communication between adjacent strips is needed to pass updated values. This global dependency of data may hamper good parallel performance on distributed memory architectures. Note that the MPI version can run on shared memory multi-core architectures too. Furthermore, this approach can be seen as a block (or strip) version of the approach that will be discussed next.

3.2 Shared Memory

In [2], Campbell, Cazes, and Rogers considered a parallelization strategy of type (2) for SWAN. The approach is based in a fine-grained way on the hyperplane ordering for the given sweep from Sect. 3. This ordering determines the data dependency and enables concurrent computations with maximal parallelism for type (2). For the implementation with fine-grained synchronization, [2] uses pipelined parallel steps in one direction of the computational grid. Lines with computational grid points in the other direction are assigned to the available processors in a round-robin way. Figure 4 illustrates the pipelined parallel approach based on the hyperplane ordering for sweep 1 of the sweep technique (the idea for the other sweeps is similar). In the current editions of SWAN, this approach

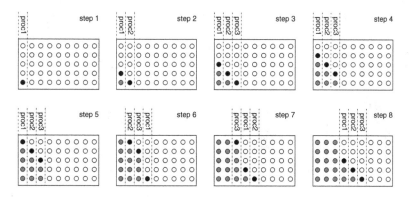

Fig. 4. Illustration of the pipelined parallel approach based on the hyperplane ordering for sweep 1 of the sweep technique. Shown are succeeding steps in case of three parallel processing units (proc1, proc2, and proc3). The black bullets represent computational grid points that are being updated in the current step. The grey bullets were updated in a previous step.

is implemented on shared memory multi-core architectures (or SM-MIMD, [11, § 2.4, 2012] with OpenMP.

3.3 Hybrid Version

Further inspection of the approaches used by the MPI and OpenMP versions of SWAN learned us that, conceptually, a combination should be quite straightforward. We illustrated the conceptual approaches for both versions in Fig. 3 and Fig. 4, respectively. Both illustrations were for the same computational grid. Let us reconsider the situation for the approach of the OpenMP version in Sect. 3.2. As the approach is based on the hyperplane ordering for the given sweep, the approach also holds for the transpose of the situation shown in Fig. 4. (In fact [2] uses this transposed situation as illustration.) In this transposed situation lines with computational grid points perpendicular to the other direction are assigned to the available processors in a round-robin way. Now, the point is that this transposed situation for the pipelined parallel approach fits nicely in one strip of the block wavefront approach of Sect. 3.1. The block wavefront approach distributes the strips and for each strip the grid lines are processed efficiently by the pipelined parallel approach within shared memory. In this way the part of the sweeps inside the strips are built up by the pipelined parallel approach and the block wavefront approach couples the sweeps over the strips. Again, the parallelization strategy is of type (2): all computations can be performed without changing the original serial numerical algorithm. Therefore, except for rounding errors, the hybrid version gives identical results to the original serial numerical algorithm.

Note that, for one strip the hybrid approach reduces to the pipelined parallel approach, whereas for one processor per strip it reduces to the block wavefront approach.

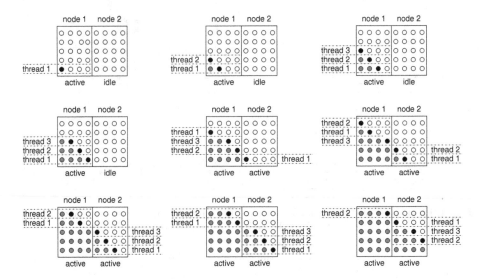

Fig. 5. Illustration of the hybrid approach based on a combination of the block wavefront approach and the pipelined parallel approach for sweep 1 of the sweep technique. Shown are succeeding steps in case of three OpenMP threads (thread 1, thread 2, and thread 3) within two MPI processes (node 1 and node 2). The black bullets represent computational grid points that are being updated in the current step. The grey bullets were updated in a previous step.

In Fig. 5 we illustrate this hybrid version for sweep 1 of the sweep technique. To make the link with the actual implementation we give a short description in terms of OpenMP threads and MPI processes. Shown are succeeding steps in case of three OpenMP threads (thread 1, thread 2, and thread 3) within two MPI processes. The black bullets represent computational grid points that are being updated in the current step. The grey bullets were updated in a previous step. Note that, the OpenMP threads on node 1 and node 2 are different threads. With MPI two strips are created: strip 1 is located on node 1, strip 2 on node 2. On node 1, OpenMP starts with the pipelined parallel approach for strip 1. Lines (horizontal for this example) with computational grid points are assigned to the three OpenMP threads in a round-robin way. Node 2 stays idle until sweep 1 arrives at the right-most point in the lowest row of strip 1, after the update the corresponding value is communicated to node 2. Then on node 2 OpenMP starts with the pipelined parallel approach on strip 2. Etcetera.

The hybrid version required some subtle modifications in the source code of SWAN for structured computational grids. They are essential to accommodate the combination of OpenMP and MPI. To make the hybrid version available for general use, we handed these modifications to the maintainers of the official SWAN version at Delft University of Technology [1].

4   Setup of Benchmarks

As a central case for the benchmarks a SWAN model is used that has been developed for the assessment of the primary water defences in the northern part of the Netherlands [8]. The model covers the Dutch part of the Wadden Sea, a complex area of tidal channels and flats sheltered by barrier islands from the North Sea. See Fig. 6 in [5] for the bathymetry. The model is relatively large compared to other SWAN models, with a 2280 × 979 curvilinear computational grid for the geographical domain, resulting in more than 2 million active computational grid points and a required working memory of about 6 GB.

In addition, benchmarks are performed for SWAN models of Lake IJssel (454 × 626 rectangular computational grid with full simulation period and I/O) and Lake Marken (195 × 204, 586 × 614, and 975 × 1024 curvilinear computational grids with shortened simulation period and no I/O). See Fig. 6 in [5] for the locations. These models are incorporated too as they differ in size and concern other important application areas. The first model has been developed for operational forecasting of flooding near the Dutch major lakes [7]. The second model has been developed for water quality studies in Lake Marken and is used in combination with a shallow water and advection diffusion solver for modelling resuspention and sedimentation, light penetration, and related ecological effects [4,6].

Here we present results of benchmarks that were performed on "2690 v3" nodes of the Cartesius supercomputer (Mellanox ConnectX-3 InfiniBand adapter providing 4 × FDR resulting in 56 Gbit/s inter-node bandwidth, Intel MPI, Bull B720 bullx system, SURFsara, the Netherlands). Each node contains 2 Intel twelve-core Xeon E5-2690 v3 processors resulting in 24 cores per node with 2.60 GHz per core. There is 30 MB cache per processor, no hyperthreading is used.

Benchmarks have been performed for MPI, OpenMP, and hybrid implementations of Deltares[1] SWAN versions 40.72ABCDE (Wadden Sea and Lake IJssel cases) and 40.91AB.8 (Lake Marken case) for Linux 64 bits platforms. Note that for one computational process with one thread, the OpenMP, MPI, and hybrid version are functionally identical to the serial version of SWAN. Standard compiler settings are used as supplied with the Fortran source code at the SWAN website [1] resulting in level 2 optimization for the Intel Fortran 14 compiler as used on the Intel processors.

Timings of the wall-clock time have been performed three times. Results presented here are averages of these timings. To have an indication of the variance (i.e. measurement error), also the average minus the standard deviation and the average plus the standard deviation are included. Shown are double logarithmic plots for wall-clock time as a function of the number of computational cores. In case of linear parallel scaling, lines will have a downward slope of 45°.

---

[1] This Deltares version is in use for the applications mentioned in Sect. 1. It has some small but subtle additional functionalities compared to the official version at Delft University of Technology (see website [1]) to enable interaction with a shallow water solver and wave growth in depth-limited situations like Lake IJssel and Lake Marken.

5   Verification of Model Performance

As mentioned before, the MPI, OpenMP, and hybrid versions do not change the original serial numerical algorithm for parallelization. Therefore, except for rounding errors, these versions give identical results to the original serial numerical algorithm, i.e. the model performance of SWAN stays the same.

During the benchmarks we verified this aspect by checking for all benchmark cases that the different combinations (MPI version, OpenMP version, hybrid version, hardware, number of processes/threads) show the same convergence behavior of the numerical algorithm of SWAN. Furthermore, for the hybrid implementation of Deltares SWAN version 40.91AB.8 the Deltares SWAN testbed was run to verify this aspect too for all testcases in the testbed. The Deltares SWAN testbed originates from the ONR testbed for SWAN [10]. It runs analytical, laboratory, and field testcases for SWAN for typical functionality and compares results with previous tested versions on different platforms and measured wave characteristics. Based on statistical postprocessing results of the testbed runs can be accumulated in numbers that indicate the model performance on which it can be decided to accept a new SWAN version. For the hybrid implementation of Deltares SWAN version 40.91AB.8 no significant differences were observed.

6   Further Optimization and Behavior Inside a Node

Current hardware trends show an increase in the number of computational cores per processor whereas multiple processors share memory inside a node. Therefore, here we first try to further optimize the MPI, OpenMP, and hybrid version before we scale up to larger number of nodes. Note that, by doing so, no new discoveries are expected, the only aim is to have a good basis on the hardware before scaling up. For this purpose, with some numerical experiments we investigate the effect of the position of the computational processes on specific locations (cores, processors) inside a node. Considered are the iterations of the Wadden Sea case (i.e. no I/O).

Figure 6 shows the parallel performance of the serial, MPI, and OpenMP versions on one Cartesius 2690 v3 node as a function of the number of cores used. For the MPI and OpenMP version two ways of pinning the processes/threads to the cores are shown:

> **compact**: computational processes of neighbouring strips (MPI) or lines (OpenMP) are placed as close as possible to each other and
> **spread**: computational processes of neighbouring strips (MPI) or lines (OpenMP) are placed in corresponding order but spread over the free cores as much as possible.

For example: if only 6 cores are used then for compact the computational processes 1, 2, 3, 4, 5, and 6 are placed on physical cores 0, 1, 2, 3, 4, and 5, respectively. For spread they are placed on physical cores 0, 4, 8, 12, 16, and 20, respectively. From the figure we may conclude that, as might be expected,

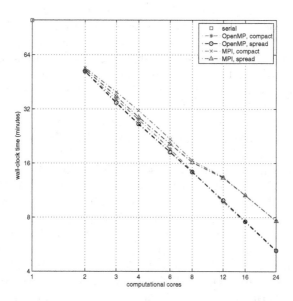

Fig. 6. Parallel performance of the serial, MPI, and OpenMP versions of SWAN for the Wadden Sea case on one Cartesius 2690 v3 node as a function of the number of cores used. For the MPI and OpenMP version two ways of pinning the processes/threads to the cores are shown: compact and spread. See Sect. 6 for further explanation. Shown is the wall-clock time in minutes for the iterations.

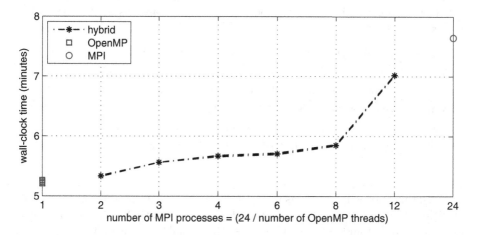

Fig. 7. Parallel performance of the hybrid version of SWAN for the Wadden Sea case on one Cartesius 2690 v3 node using all 24 cores as a function of the number of MPI processes. Each MPI process uses (24/number of MPI processes) OpenMP threads. See Sect. 6 for further explanation. Shown is the wall-clock time in minutes for the iterations.

when using not all cores inside a node, it is beneficial for the wall-clock time to "spread" the computational processes over the cores. Note that differences between "compact" and "spread" are not that large, this may be explained from the intensive computations of SWAN that dominate in the wall-clock time (see Sect. 2). Furthermore, in Fig. 6 it can be seen that this effect disappears when using more than 8 cores and that then also the difference between MPI and OpenMP becomes more prominent.

A similar observation can be made from Fig. 7. This figure shows the parallel performance of the hybrid version of SWAN on one Cartesius 2690 v3 node using all 24 cores as a function of the number of MPI processes. Each MPI process uses (24/number of MPI processes) OpenMP threads. Note that for one MPI process the hybrid version reduces to the original OpenMP version with 24 OpenMP threads (most left), whereas for 24 MPI processes it reduces to the original MPI version (most right). For each MPI process of the hybrid version the OpenMP threads are pinned compact to the cores, for the next MPI process in the ordering of the algorithm the OpenMP threads are pinned compact to the next cores in the ordering of the node. For the MPI and OpenMP versions the MPI processes respectively OpenMP threads are pinned compact to the cores of the node. Shown is the wall-clock time in minutes for the iterations.

Based on the previous numerical experiments we conclude that, when using all cores of a Cartesius 2960 v3 node, optimal settings for the MPI, OpenMP, and hybrid versions is pinning the MPI processes and/or OpenMP threads compact to the cores. We will use these settings for the remainder of this paper.

Table 1. Wall-clock time in minutes of the MPI, OpenMP, and hybrid versions on 1 node (top), 2 nodes (middle), and 4 nodes (bottom) for Cartesius 2690 v3 nodes for the Wadden Sea case.

1 node, 24 processes/threads	MPI version (t1)	OpenMP version (t2)	t1/t2
Wall-clock time iterations (m)	7.645 ± 0.013	5.236 ± 0.026	1.4601
2 nodes, 48 processes/threads	MPI version (t1)	Hybrid version (t2)	t1/t2
Wall-clock time iterations (m)	4.530 ± 0.002	2.994 ± 0.002	1.5130
4 nodes, 96 processes/threads	MPI version (t1)	Hybrid version (t2)	t1/t2
Wall-clock time iterations (m)	2.540 ± 0.012	1.668 ± 0.005	1.5228

With this knowledge/settings we extend the numerical experiments in § 4.2 of [5] with results on 1, 2, and 4 Cartesius nodes for the Wadden Sea case (iterations, no I/O) in Table 1 and the Lake IJssel case (full simulation with I/O) in Table 2. These results confirm the trends observed in the previous paper (the same cases are used in the benchmarks there) that the hybrid version improves the parallel performance of the current MPI version for larger number of cores per node and/or more nodes.

Table 2. Wall-clock time in minutes of the MPI, OpenMP, and hybrid versions on 1 node (top), 2 nodes (middle), and 4 nodes (bottom) for Cartesius 2690 v3 nodes for the Lake IJssel case.

1 node, 24 processes/threads	MPI version (t1)	OpenMP version (t2)	t1/t2
Wall-clock time full simulation (m)	117.229 ± 0.153	60.600 ± 0.350	1.9345
Wall-clock time iterations (m)	115.493 ± 0.165	57.736 ± 0.410	2.0004
Wall-clock time I/O at end (m)	1.736 ± 0.012	2.863 ± 0.060	0.6064
2 nodes, 48 processes/threads	MPI version (t1)	Hybrid version (t2)	t1/t2
Wall-clock time full simulation (m)	65.509 ± 0.051	45.906 ± 0.119	1.4270
Wall-clock time iterations (m)	64.373 ± 0.055	43.197 ± 0.151	1.4902
Wall-clock time I/O at end (m)	1.136 ± 0.004	2.709 ± 0.032	0.4193
4 nodes, 96 processes/threads	MPI version (t1)	Hybrid version (t2)	t1/t2
Wall-clock time full simulation (m)	38.282 ± 0.025	27.480 ± 0.052	1.3931
Wall-clock time iterations (m)	37.351 ± 0.052	25.101 ± 1.071	1.4880
Wall-clock time I/O at end (m)	0.931 ± 0.027	2.379 ± 1.019	0.3913

7   Behavior for Large Number of Nodes

We end with two numerical experiments in which we increase the number of nodes.

First, to compare MPI and hybrid implementations, we consider the Wadden Sea case (iterations, no I/O) on Cartesius 2690 v3 nodes. So in fact we extend Table 1 to larger number of nodes. Per node all 24 cores are used and MPI processes and/or OpenMP threads are pinned compact as described in Sect. 6. Figure 8 shows the resulting wall-clock times in minutes for the OpenMP, MPI, and hybrid versions. It can be seen that the gap between the MPI and hybrid version stays constant up to 16 nodes. From this point on the wall-clock time for the MPI version increases as the strips are becoming very thin. The MPI version divides the computational work in only one direction. For this it chooses the direction with most computational grid points, for the Wadden Sea case with 2280×979 computational grid this is the first grid direction. For 32 nodes the computational work is divided in $32 \times 24 = 768$ strips whereas there are only 2280 grid points in this direction, resulting in only 2 to 3 grid points per strip. For 128 nodes there are not enough grid points anymore to have at least one grid point per strip. In that case, the current software implementation of the MPI version (as provided by the maintainers of the official SWAN version at Delft University of Technology [1]) crashes. For the hybrid version, however, the number of MPI processes is a factor 24 lower and the OpenMP threads work in the other grid direction. In Fig. 8 the wall-clock time for the hybrid version still decreases after 16 nodes. The lowest value occurs between 64 and 128 nodes. After 128 nodes the wall-clock time increases again. In case of 95 nodes the hybrid version has strips of width $2280/95 = 24$. Then, with 24 OpenMP threads per node, precisely all points in the computational grid for which new values can be computed simultaneously are processed at the same time. So for this situation we obtain the maximal parallelism that we can obtain for the given algorithm (see also § 3 from [5]). This corresponds with the observation that the lowest value of the wall-clock time occurs between 64 and 128 nodes.

Second, to study the effect of increasing the grid size on the parallel performance of the hybrid version of SWAN, we consider the Lake Marken case (iterations, no I/O) on Cartesius 2690 v3 nodes. For this purpose the original 195×204 curvilinear computational grid of [4,6] is uniformly refined with a factor of 3, respectively 5, in both horizontal grid directions. This resulted in a 586×614 and 975×1024 curvilinear computational grid. The corresponding number of computational grids points is 39 780 (original grid), 359 804 (3 × 3 refined grid), and 998 400 (5 × 5 refined grid). (For the Wadden Sea case these numbers are $2280 \times 979 = 2\,232\,120$.) The study of the effect of the refinement on computational times of SWAN is important as the original grid is quite coarse for local ecological impact assessments like in [6]. We did not take into account the coupling with the shallow water solver nor the advection diffusion solver as we only want to know a lower bound of the contribution of SWAN to the computational times. Furthermore we restricted the simulation period to 1 day (instead of a typical full simulation period of 373 days). Figure 9 shows the resulting wall-clock times in minutes for the different grid sizes. As can be seen the behavior is similar as for the Wadden Sea case in Fig. 8. For larger grids, more nodes can be used to lower computational times. Again, lowest values of the wall-clock time occur for the maximal parallelism that we can obtain for the given algorithm: for $204/24 \approx 9$ nodes, $614/24 \approx 26$ nodes, and $1024/24 \approx 43$ nodes (for 195×204, 586×614, and 975×1024 grid, respectively).

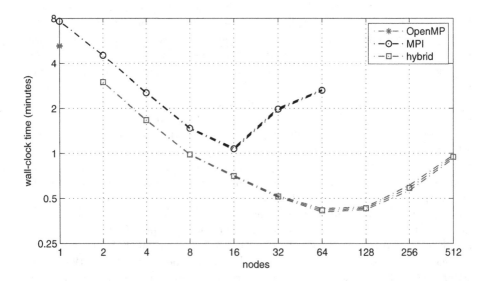

Fig. 8. Parallel performance of the OpenMP, MPI, and hybrid versions of SWAN for the Wadden Sea case on Cartesius 2690 v3 nodes for large numbers of nodes. Shown is the wall-clock time in minutes for the iterations.

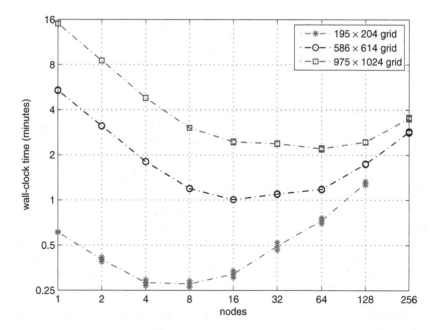

Fig. 9. Parallel performance of the hybrid version of SWAN for the Lake Marken case on Cartesius 2690 v3 nodes for large numbers of nodes and different grid sizes. Shown is the wall-clock time in minutes for the iterations.

8   Conclusions and Outlook

Because of the importance for real life applications in the Netherlands, we investigated the parallel efficiency of the current MPI and OpenMP versions of SWAN for computations on structured computational grids. In a previous paper [5] we proposed a hybrid version of SWAN that naturally evolves from these versions. It combines the efficiency of the OpenMP version with the capability of the MPI version to use more nodes.

In the current paper we extended this approach. With extensive benchmarks for important real life applications we showed the significance of this hybrid version. We optimized the approach. Numerical experiments showed that the hybrid version improves the parallel performance of the current MPI version even more for larger number of cores per node and/or larger number of nodes. Given the current trends in hardware this is of great importance. Parallel I/O will be subject of future research.

References

1. Booij, N., Ris, R.C., Holthuijsen, L.H.: A third-generation wave model for coastal regions, part i, model description and validation. J. Geophys. Res. **104**(C4), 7649–7666 (1999). (Software (GNU GPL) can be downloaded from http://swanmodel. sourceforge.net)

2. Campbell, T., Cazes, V., Rogers, E.: Implementation of an important wave model on parallel architectures. In: Oceans 2002 MTS/IEEE Conference, pp. 1509–1514. IEEE (2002). http://www7320.nrlssc.navy.mil/pubs/2002/Campbell.etal.pdf

3. Chan, T.C., van der Vorst, H.A.: Approximate and incomplete factorizations. In: Parallel Numerical Algorithms, ICASE/LaRC Interdisciplinary Series in Science and Engineering, vol. 4, pp. 167–202. Kluwer Academic (1997)

4. Donners, J., et al.: Using high performance computing to enable interactive design of measures to improve water quality and ecological state of lake Marken. In: Proceedings 15th World Lake Conference (2014). http://www.unescowaterchair. org/activities/publications

5. Genseberger, M., Donners, J.: A hybrid SWAN version for fast and efficient practical wave modelling. Procedia Comput. Sci. **51**, 1524–1533 (2015)

6. Genseberger, M., Noordhuis, R., Thiange, C.X.O., Boderie, P.M.A.: Practical measures for improving the ecological state of lake Marken using in-depth system knowledge. Lakes Reserv. Res. Manag. **21**(1), 56–64 (2016)

7. Genseberger, M., Smale, A.J., Hartholt, H.: Real-time forecasting of flood levels, wind driven waves, wave runup, and overtopping at dikes around Dutch lakes. In: 2nd European Conference on FLOODrisk Management, pp. 1519–1525. Taylor & Francis Group (2013)

8. Groeneweg, J., Beckers, J., Gautier, C.: A probabilistic model for the determination of hydraulic boundary conditions in a dynamic coastal system. In: International Conference on Coastal Engineering (ICCE 2010) (2010)

9. Kleermaeker, S.H.D., Verlaan, M., Kroos, J., Zijl, F.: A new coastal flood forecasting system for the Netherlands. In: Hydro 2012 Conference. Hydrographic Society Benelux (2012). http://proceedings.utwente.nl/246

10. Ris, R.C., Holthuijsen, L.H., Smith, J.M., Booij, N., van Dongeren, A.R.: The ONR test bed for coastal and oceanic wave models. In: International Conference on Coastal Engineering (ICCE 2002) (2002)

11. van der Steen, A.J.: Overview of recent supercomputers. Technical report, NWO-NCF (2008, 2010, 2011, 2012). http://www.euroben.nl/reports.php. 2010 version appeared in J. J. Dongarra and A. J. van der Steen. High-performance computing systems. Acta Numerica, 21:379–474, 2012

12. Stone, H.L.: Iterative solution of implicit approximations of multidimensional partial differential equations. SIAM J. Numer. Anal. **5**, 530–558 (1968)

13. Zijlema, M.: Parallelization of a nearshore wind wave model for distributed memory architectures. In: Parallel Computational Fluid Dynamics - Multidisciplinary Applications, pp. 207–214. Elsevier Science (2005)

Agent-Based Simulations, Adaptive Algorithms and Solvers

An Agent-Based Simulation of the Spread of Dengue Fever

Imran Mahmood[1,2(✉)], Mishal Jahan[2], Derek Groen[1], Aneela Javed[3],
and Faisal Shafait[2,4]

[1] Department of Computer Science, College of Engineering Design and Physical
Sciences, Brunel University, Kingston Lane, Uxbridge, London UB8 3PH, UK
imran.mahmoodqureshihashmi@brunel.ac.uk
[2] School of Electrical Engineering and Computer Science (SEECS),
National University of Sciences and Technology (NUST), H-12,
Islamabad 44000, Pakistan
[3] Atta-ur-Rahman School of Applied Biosciences (ASAB), National University
of Sciences and Technology (NUST), H-12, Islamabad 44000, Pakistan
[4] Deep Learning Laboratory, National Center of Artificial Intelligence (NCAI),
H-12, Islamabad 44000, Pakistan

Abstract. Vector-borne diseases (VBDs) account for more than 17% of
all infectious diseases, causing more than 700,000 annual deaths. Lack
of a robust infrastructure for timely collection, reporting, and analysis
of epidemic data undermines necessary preparedness and thus posing
serious health challenges to the general public. By developing a simulation framework that models population dynamics and the interactions
of both humans and mosquitoes, we may enable epidemiologists to analyze and forecast the transmission and spread of an infectious disease
in specific areas. We extend the traditional SEIR (Susceptible, Exposed,
Infectious, Recovered) mathematical model and propose an Agent-based
model to analyze the interactions between the host and the vector using:
(i) our proposed algorithm to compute vector density, based on the reproductive behavior of the vector; and (ii) agent interactions to simulate
transmission of virus in a spatio-temporal environment, and forecast the
spread of the disease in a given area over a period of time. Our simulation results identify several expected dengue cases and their direction of
spread, which can help in detecting epidemic outbreaks. Our proposed
framework provides visualization and forecasting capabilities to study
the epidemiology of a certain region and aid public health departments
in emergency preparedness.

Keywords: Dengue epidemiology · Agent-based modeling ·
Validation · Anylogic · Host-vector interaction

1 Introduction

Each year, millions of people are exposed to serious health risks due to emerging
infectious and communicable diseases. This poses a severe threat to the public

The original version of this chapter was revised: the missing funding information has been added. The correction to this chapter is available at
https://doi.org/10.1007/978-3-030-50420-5_49

health security at local, regional or national level; especially in underdeveloped countries. This is primarily due to the lack of infrastructure for timely collection, reporting, and analysis of epidemic data; non-existent early warning and forecasting systems; inadequate preparedness and emergency response management [22].

3.9 billion people in 128 countries are at risk of contracting dengue, with estimated 96 million cases annually. The worldwide incidence of dengue has risen 30-fold in the past 30 years. Pakistan is among the 110 countries in the world which are badly affected by the mosquito-borne dengue virus. The first outbreak of dengue fever (DF) in Pakistan was confirmed in 1994. The country is currently experiencing among worst-ever dengue outbreaks, recording about 45,000 confirmed cases [27]. Even today, an effective dengue vaccine offering balanced protection is still elusive. Unfortunately, existing dengue vaccines are known to have limited efficacy and cure [8]. This underscores the critical need of preventing dengue transmission and eventual outbreak by: (i) investigating favorable conditions for the dengue epidemic to occur [21]; (ii) plummeting the vector population [23]; and averting the vector-human contact [7], all of which are perceived as daunting challenges of Epidemiology. Due to the lack of ICT enabled governance, the existing infectious disease surveillance systems in Pakistan are unable to perform epidemiological spread analyses and effective emergency response planning [11,18]. Researchers have started exploring latest techniques like Artificial Intelligence for diagnostic screen of dengue suspects [12], but these methods are still in their infancy.

To mitigate these challenges, we need to develop a reliable *health surveillance* and rapid *emergency response* infrastructure, that monitors and responds to known endemic diseases in the country, and evolves to cater potential new outbreaks. This infrastructure should have the capability to collect spatio-temporal epidemiological data, analyze it using computational methods, forecast possible outbreaks and generate early warnings for rapid emergency response management. Due to the sparsity of data, the dynamics of the vector population and its interaction with the human population is quite difficult to capture. We therefore need to build models that incorporate the vector population, disease transmission and the spread direction of the infectious disease, and that support the application of preemptive strategies and countermeasures.

In this study we present a simulation approach to analyze and predict the mosquito population density and its consequence on dengue spread. We also simulate the pathogen transmission, and observe the dynamic interaction of human and mosquito population, both of which we use to forecast outbreaks in a spatial environment. Our proposed agent-based simulation framework allows modeling of both human and vector population dynamics, using separate layers. Human agents evolve between different states, from Susceptible to Exposed, Infected and eventually Recovered states while Mosquito agents evolve from Egg to Larva, Pupa and Adult state. Both population layers are spatially distributed and we model the interaction between both layers using our proposed algorithm. In addition, we expose both layers to exogenous variables such a temperature, humidity, rainfall and the permeable water surfaces in the region. From our simulation results we can reproduce predicted dengue cases across age groups and highlight them using spatio-temporal visualizations. We validate our results using data of an existing study of the local region [4]. Our proposed framework can be re-used for simulation, visualization and forecasting of any region and aid Public health departments in emergency preparedness.

The rest of the paper is organized as follows: Sect. 2 outlines the background concepts used in this paper and the literature review. Section 3 discusses our proposed framework. Section 4 provides simulation results and model validation and Sect. 5 provides conclusions and future work.

2 Background and Literature Review

2.1 Dengue Epidemiology

This section provides key concepts used in our modeling approach. Epidemiology is the study of models, causes, effects, risk factors, transmission, spread and outbreaks of infectious diseases in a particular population [25]. DENV is transmitted to humans through the bite of infected female Aedes mosquitoes primarily Aedes aegypti and Aedes albopictus. A susceptible mosquito can acquire infection from an infectious person and transmit it further, or human gets infection from the bite of already Infectious Aedes aegypti. It spends its lives around or in side houses, becomes adult and typically flies up to 400 m [26]. Adees aegypti life cycle is composed of four stages as shown in Fig. 2. Aedes Aygepti proliferates around 30–32 °C. The number of eggs laid per batch depends on the weight of mosquito and other factors. Details of the dengue epidemiology can be viewed at [9].

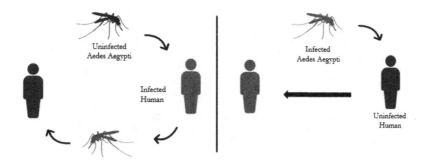

Fig. 1. Mosquito to human dengue transmission

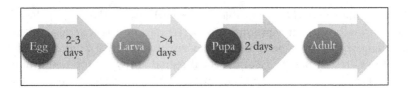

Fig. 2. Life-cycle of Aedes aegypti

2.2   Environmental Factors

Various meteorological factors also influence the growth of the vector population and the spread of dengue fever. Temperature and humidity influence the incidence of dengue fever by modifying adult feeding behavior, larval development and mosquito survival. The average life span of Aedes mosquito is 25 days, with a range from one day to 76 days. Population density of Aedes Aegypti rapidly increases in summers as rising temperatures shortens the incubation period of mosquitoes and they take less time to emerge from eggs to adults. This increase the overall risk of dengue transmission [14]. The mortality rate of mosquitoes increases at high temperatures. For temperatures 15 °C, the feeding frequency of mosquitoes increases resulting in greater risk of viral transmission. Mosquito life cycle consists of two stages one of which is aquatic i.e they require stagnant water in order to develop and reproduce. Rainfall plays a vital role in the development of mosquitoes and Dengue transmission. There is an increase in dengue cases during and after rainy seasons. Pakistan is unfortunately experiencing severe forms of above stated changes. Relative humidity, temperature and rain remained noteworthy prognosticators of dengue occurrence in Pakistan. Surge of cases occurred from September to October [24], as shown in Fig. 3.

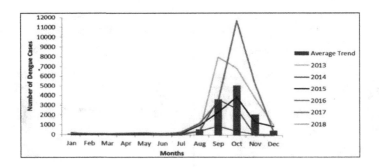

Fig. 3. Dengue cases in Pakistan (2013–2018)

2.3   Mathematical Modeling in Epidemiology

Traditional non-spatial epidemiology models have been used use to represent epidemics of communicable diseases [3]. A *Susceptible, Exposed, Infectious, Recovered (SEIR)* is an epidemiological model that describes the transmission process of an infectious disease [19]. The individuals enter or leave the compartment based on flow rates defined. There is a decrease in Susceptible individuals when individuals get infection, or they die. Number of individuals in the Exposed compartment increases when they get infection and decreases as soon as they show the symptoms of infection and decreases with the death rate. The infection compartment increases by the infection events of Exposed and decreases by the death and recovery rate. There is an increase in Recovered compartment with the recovery rate and decreases with the natural death. Another extension to the SIS model was proposed in [13]. It contains three population components: Humans (H), Vectors (M) and Eggs (E). 'L' represents the number of Latent mosquitoes that are in incubation period, as shown in Fig. 4. At first a Susceptible human gets infection from a vector moves to the Infectious compartment. While at infectious state, a human may get recovered and move to the Recovered state or may die. An Infectious human may spread infection to a Susceptible vector which would move to the Latent compartment, where it stays for the time between getting Infection and becoming infectious, called latency period; the vector then moves to the Infectious compartment. A susceptible mosquito if lays eggs increases the population of susceptible vectors, whereas an Infectious mosquito if lays eggs would increase the population of Infectious vectors. It is also possible that an Infectious mosquito lays normal eggs. The main goal of modeling the life cycle of mosquitoes is to estimate the growth of the vector density at a place and time; and the interaction with the hosts to predict the rate of spread of disease. Our proposed approach is based on the foundation of this principle.

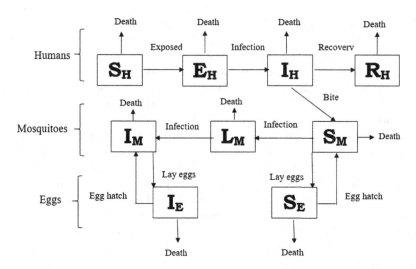

Fig. 4. Humans, mosquitoes and eggs. S = Susceptible, E = Exposed, I = Infected, R = Recovered, L = Latent, H = Human, M = Mosquitoes, E = Egg

2.4   Agent-Based Modeling and Simulation

Most existing mathematical models focus on non-spatial macro-level aspects of the system. Because we are strongly interested in the micro-level aspects of our agents (a level where we can more easily extract rules and behaviors), we therefore use agent-based modelling. An agent-based model consists of autonomous decision-making entities: *Agents*, each able to perceptively assess situations and make decision according to a set of predefined rules. ABM is decentralized and produces collective behavior by agents communicating and interacting with each other in their environment [16]. The interest in using ABM has been recently renewed due to their ability to model complex geospatial structures and interacting networks [5].

2.5   Literature Review

Many researchers have proposed different dengue epidemic simulation models. However, selection of the methodology and platform greatly depends upon the purpose of the study. Jacintho, et al. [16] propose an agent-based model of the dengue spread, using the Swarm platform that models the micro-level behavior of agents in the spread and transmission of dengue fever using a rule-based approach, however it lacks GIS based spatial representation. Almeida, et al. [1] propose an individual-based approach to model Aedes aegypti population considering vectors, humans and objects as agents, using repast framework. Kang and Aldstadt [17] proposed an approach to validate spatially explicit ABM for several specifications of vector-borne disease transmission models, using multiple scale spatio-temporal patterns. Hunter et al. [15] developed a data driven agent-based SEIR model that resulted in the emergence of patterns and behaviors that are not directly programmed into the model. Lima, et al. [20] developed DengueME, a collaborative open source platform to simulate dengue disease and its vector's dynamics. It supports compartmental and individual-based models, implemented over a GIS database, that represent Aedes aegypti population dynamics, human demography, human mobility, urban landscape and dengue transmission mediated by human and mosquito encounters. Yein Ling Hii [14] proposed a machine learning approach to study dengue fever considering climatic factors: temperature and rainfall using Poisson multivariate regression and validated through multiple statistical models. Guo, et al. [10] presented a comparison of various machine learning algorithms for the accurate prediction of dengue in China, and shows that support vector regression achieved a superior performance in comparison with other forecasting techniques assessed in this study.

Our proposed framework uses the mathematical model proposed by [13], as underlying foundation and extend it using an ABM approach, which consists of two population layers: (i) human and (ii) vectors (mosquitoes and eggs). Both layers express the population dynamics, mobility and microscopic behavior of human and vector agents separately, yet provide a common spatial environment for their interactions, to study the spread of disease within the desired spatiotemporal resolution.

3   Proposed Framework

This section discusses the details of our proposed framework which is composed
of three layers: (i) Host Layer (Human population); (ii) Vector Layer (Mosquito
population); and (iii) Pathogen Layer (Dengue parameters), as shown in Fig. 5.
The framework further integrates: (i) GIS based spatial environment for agent
distribution and mobility; (ii) time step using a user defined temporal resolution (from Min to Year); (iii) simulation engine to handle change of states and
dynamic event processing; (iv) database for importing input data and exporting
simulation results; (v) exogenous variables for weather and climate data; and
(vi) validation data for model validation. The framework also provide a visualization dashboard for viewing the change of states of the distributed population
of agents in real-time (or virtual time a.k.a faster than real-time) and for the
graph visualization of the simulation results.

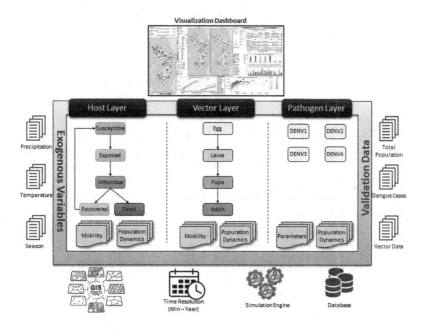

Fig. 5. Agent-based simulation framework

3.1   Host Layer

This layer concerns with the structure, behavior and the interactions of host
agent and allows the modelers to initialize a host population (e.g., human or
animals) using variable configurations. The behavior of the host agent is represented by a state chart, as shown in Fig. 6 which consists of six states: Susceptible, Exposed, Infected (or Immune), Recovered (or Dead). In this paper

we assume the entire host population is initialized in Susceptible state. In the future, we intend to support a distribution of initial states so that we can use more realistic model configurations of a study area. When at *'Susceptible'* state, a host agent receives an incoming message from an infectious mosquito, representing the 'infectious bite', it transits to an *'Exposed'* state. After a certain period, defined by the incubation period parameter (3–8 days [28]) either the host will transit to *'Infected'* state or go to immune state (if resolved immune by the immunity probability). When at infected state, an incoming 'bite' message causes a self-loop transition, which is used to transmit the virus to an uninfected mosquito, with a given probability. After a certain duration, defined by the illness duration parameter (20 days) the host will transit to 'Recovered' state, if resolved true by the survival probability (0.9) or 'Dead' otherwise. We implement the interactions between Host and Vector agents using message passing and a distance based network type [2], which implies that a vector is connected with multiple hosts that are situated within a given range (36 m), and can communicate i.e. bite with infection or acquire an infection through a bite, as shown in Fig. 8. The detailed bite algorithm is presented later in this section.

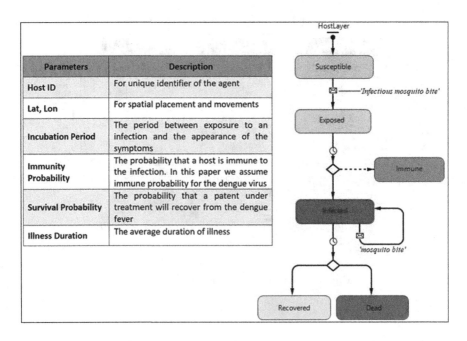

Fig. 6. Host agent (human)

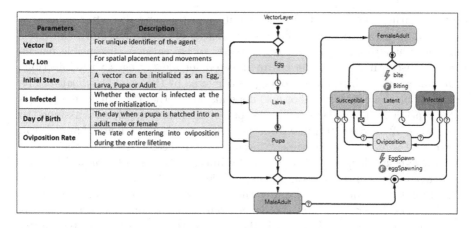

Parameters	Description
Vector ID	For unique identifier of the agent
Lat, Lon	For spatial placement and movements
Initial State	A vector can be initialized as an Egg, Larva, Pupa or Adult
Is Infected	Whether the vector is infected at the time of initialization.
Day of Birth	The day when a pupa is hatched into an adult male or female
Oviposition Rate	The rate of entering into oviposition during the entire lifetime

Fig. 7. Vector agent (mosquito)

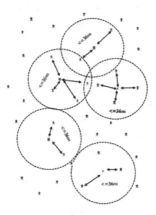

Fig. 8. Bite mechanism

3.2   Vector Layer

This layer concerns with the structure, behavior and the interactions of vector agent. The behavior of the vector agent is represented by a state chart, as shown in Fig. 7. It consists of states: *Egg, Larva, Pupa* and *Adult* (male or female). A vector agent can be initialized in any of these states. In this paper we assume initialize the vector population with the probability of 0.2 eggs, 0.2 larvae, 0.2 pupas and 0.4 adults (0.5 male and 0.5 female). The vector agents are uniformly distributed over the space, however this distribution can be fine tuned if the data of the concentrations of the mosquitoes nests is available. From egg, the agent moves on to the Larva state between 2–3 days. From larva it takes 4 days to transit to Pupa state, where it resides for 2 days and moves to the Adult stage, with a 0.5 probability to be a male or a female [6]. The male Aedes aegypti neither makes a bite nor transmits infection and dies after a lifespan of 14 days and then go to the final state '*dead*'. A female mosquito can enter into the '*Oviposition*' state when it is ready to lay eggs. A susceptible female will lay susceptible eggs whereas an infected female lay eggs with infectious probability = 0.3. The spawning of eggs is triggered by the event '*EggSpawn*'. The initial state of an egg (infected or not) depends on whether the female mosquito is in Susceptible or infected state. The lifespan of the female mosquito is assumed to be between 42–56 days. Inspired from the Helmersson's mathematical model [13], shown in Fig. 4, the 'susceptible', 'latent' and 'Infected' states of the female mosquito are implemented. If not initially infected (i.e., *IsInfected = false*), an adult female mosquito enters into 'susceptible' state. It goes to 'latent' state if bites an infected human and acquires the virus (with a transmission probability of 0.18). The latent period is the delay from transmission to infection and is assumed to be 10 days, after which the agent enters into the 'Infected' state. While at 'susceptible' or 'Infected' states, a female mosquito bites human and have her fill of blood after each meal, at the rate of a bite after 2 days. Aedes aegypti is an intermittent biter and prefers to bite more than one person during the feeding period, therefore we assume it bites all the connected hosts. This bite is triggered by the event 'bite' at random intervals until the mosquito is dead When a 'bite' occurs the interaction between Host and Vector is implemented as shown in Algorithm 1. It takes VectorID and a list of ConnectedHosts as input. A vector is connected to all the hosts that lie within the range of 36 m. This algorithm determines infectivity of the vector after the bite using a boolean variable 'Infected'. Line1 assigns the existing state (infected or not) of the vector. Line 2–14 iterates a list of all the connected hosts and evaluate two scenarios as shown in Fig. 1. If the vector is infected it will transmit an '*infectious bite*' through message passing, and cause the host to transit from '**susceptible**' to '**exposed**' state. Else, if the host is infected the vector will acquire infection (with a 0.018 probability), and transit from '**susceptible**' to '**Latent**' state. Otherwise it will just '*bite*' the host without the transmission of any infection. The number of times mosquito will bite human is defined by bite rate [13], which is temperature dependent and is calculated as:

$$BiteRate = \frac{(0.03 \times T + 0.66)}{7}$$

Algorithm 1 Bite algorithm

    Input: VectorID:*int*, ConnectedHosts[]:*list*
    Output: Infected:*bool*

```
1   Infected ←VectorAgent.IsInfected (ID = VectorID)
2   for i ← 1 to ConnectedHosts n do
3   │   HostAgent ← ConnectedHost[i]
4   │   if Infected = true then
5   │   │   SendMessage(HostAgent, "InfectiousBite")
6   │   else
7   │   │   if HostAgent.Infected = true then
8   │   │   │   SendMessage(HostAgent, "Bite")
9   │   │   │   if RandomTrue(0.3) then
10  │   │   │   │   Infected ←true
11  │   │   else
12  │   │   │   SendMessage(HostAgent, "Bite")
13  │   i ← i + 1
14  │   return Infected
```

3.3 Pathogen Layer

This layer deals with the modalities of the pathogen under study i.e., 'Dengue' in our case. In this paper the structure and the behaviour of the pathogen is limited only to initialize the serotype (i.e., DENV1, DENV2, DENV3 or DENV4) and key parameters such as infectivity, transmisiability, survivability and incubation period. In future, we aim to extend this layer for dealing with the complex logic of the cross-immunity with different serotypes.

4   Simulation and Results

Our simulation is implemented using AnyLogic University Edition, and performed for a population of 50,000 persons and 1,000 mosquito agents, for a period of 90 days, on an Intel Core i7-8700 CPU@3.20 GHz, 16.0 GB RAM, and a 64-bit Windows Operating System. The human and mosquito populations are randomly distributed within a selected region in the city of Islamabad. A simulation run is shown using spatial and temporal visualization in Fig. 9 and 10. Initially the person agent population is susceptible, as soon as it is bitten by infectious mosquito it becomes Exposed (orange), after completing its incubation period of 4–7 days in exposed state it either becomes Infectious (red) or Immune (gray). From Infectious state after completing 4–12 days it either becomes Recovered (green) or Dead (black).

The simulated results obtained are compared with the actual results of confirmed cases obtained from the dengue outbreak occurred in the local region

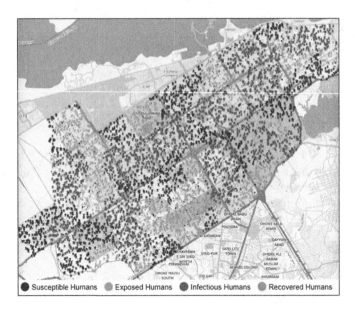

Fig. 9. Simulation results - spatial visualization (Color figure online)

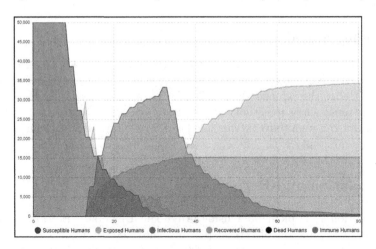

Fig. 10. Simulation results temporal visualization (Color figure online)

in 2013. The actual cases were 9,036 out of the total population of 1,257,602 while the simulated cases were 400 out of the total agents' population of 50,000. In order to compare the data results of actual and simulated dengue cases, we calculated the prevalence of both data sets:

$$\text{Prevalence} = \frac{\text{No. of Dengue Cases}}{\text{Total Population}} \times 100 \tag{1}$$

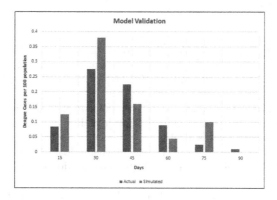

Fig. 11. Comparison of dengue cases - Source of actual data [4]

For comparison we present a graph containing both cases in Fig. 11. We observe a similar trend in the prevalence of dengue cases between the actual and simulated results. The Root Mean Square Error (RMSE) of both actual and simulated prevalence for the 90 days is *0.064*. The results show that the dengue transmission is temperature dependent. It is to be noted that with the increase in temperature, the biting rate of Aedes aegypti increases that give rise to the dengue cases.

5   Conclusion and Future Work

In this paper we presented a framework for modeling, simulation, visualization and forecasting of dengue spread, using an agent-based modeling approach. Our framework incorporates the structure, behaviour and interactions of key entities of disease epidemiology: (i) Host (Humans); (ii) Vector (mosquitoes); and (iii) Pathogen (dengue virus), using a separation of concern through independent layers. Our open-ended framework offers greater flexibility to the modelers for modification and extensibility of the disease under study, and can be used as a research tool by stakeholders (e.g., clinicians, microbiologists and public health professionals) to study the epidemiology of a region. The unique feature of our framework is its ability to model the life-cycle and population dynamics of both host and vector population, while incorporating extended mathematical models for the study of the epidemic spread. In the absence of vector population data, this tool provides means to synthesize vector population, and help improve the understanding of the spread of diseases. In the future, we plan to improve our proposed framework further by adding the mobility layer to incorporate movements, both in the Host and Vector layers. We also aim to extend our framework to support populations of a million agents or more.

Funding Information. This work was supported by the HiDALGO project, which has received funding from the European Union Horizon 2020 research and innovation programme under grant agreement No 824115.

References

1. de Almeida, S.J., Ferreira, R.P.M., Eiras, Á.E., Obermayr, R.P., Geier, M.: Multi-agent modeling and simulation of an Aedes aegypti mosquito population. Environ. Model. Softw. **25**(12), 1490–1507 (2010). https://doi.org/10.1016/j.envsoft.2010.04.021
2. Anylogic: Standard agent contacts network (2019). https://help.anylogic.com/index.jsp?topic=/com.anylogic.help/html/agentbased/Connections+and+Networks.html
3. Brauer, F., Castillo-Chavez, C., Feng, Z.: Mathematical Models in Epidemiology. TAM, vol. 69. Springer, New York (2019). https://doi.org/10.1007/978-1-4939-9828-9
4. Chaudhry, M., Ahmad, S., Rashid, H.B., Ud Din, I.: Dengue epidemic in postconflict Swat District. Am. J. Trop. Med. Hyg. **96**(4), 899–902 (2017). https://doi.org/10.4269/ajtmh.16-0608
5. Crooks, A., Malleson, N., Manley, E., Heppenstall, A.: Agent-Based Modelling and Geographical Information Systems: A Practical Primer. SAGE Publications Limited, Thousand Oaks (2018)
6. DengueVirusNet: Life cycle of aedes aegypti (2019). http://www.denguevirusnet.com/life-cycle-of-aedes-aegypti.html
7. Eisen, L., Barry, B.J., Morrison, A.C., Scott, T.W.: Proactive vector control strategies and improved monitoring and evaluation practices for dengue prevention. J. Med. Entomol. **46**(6), 1245–1255 (2009). https://doi.org/10.1603/033.046.0601
8. Gubler, D.J., Halstead, S.B.: Is Dengvaxia a useful vaccine for dengue endemicareas? BMJ **367** (2019). https://doi.org/10.1136/bmj.l5710
9. Gubler, D.J., Ooi, E.E., Vasudevan, S., Farrar, J.: Dengue and Dengue Hemorrhagic Fever. CABI, UK (2014)
10. Guo, P., et al.: Developing a dengue forecast model using machine learning: a case study in China. PLoS Negl. Trop. Dis. **11**(10), e0005973 (2017)
11. Hafeez, F., Akram, W., Suhail, A., Arshad, M.: Knowledge and attitude of the public towards dengue control in urban and rural areas of Punjab. Pak. J. Zool. **44**(1), 15–21 (2012)
12. Haneef, M., Mahmood, I., Younis, S., Shafait, F.: Artificial intelligence based diagnostic screening for dengue fever. In: International Conference on Dengue Prevention and Control, Lahore, Pakistan (2020)
13. Helmersson, J.: Mathematical modeling of dengue - temperature effect on vectorial capacity. Master's thesis (2012)
14. Hii, Y.L.: Climate and dengue fever: early warning based on temperature and rainfall. Ph.D. thesis, UmeÅ University, Epidemiology and Global Health (2013)
15. Hunter, E., Namee, B.M., Kelleher, J.: An open-data-driven agent-based model to simulate infectious disease outbreaks. PLoS One **13**(12), 0208775 (2018). https://doi.org/10.1371/journal.pone.0208775
16. Jacintho, L.F.O., Batista, A.F.M., Ruas, T.L., Marietto, M.G.B., Silva, F.A.: An agent-based model for the spread of the dengue fever: a swarm platform simulation approach. In: Proceedings of the 2010 Spring Simulation Multiconference. Society for Computer Simulation International, San Diego (2010). https://doi.org/10.1145/1878537.1878540
17. Kang, J.Y., Aldstadt, J.: Using multiple scale spatio-temporal patterns for validating spatially explicit agent-based models. Int. J. Geogr. Inf. Sci. **33**(1), 193–213 (2018). https://doi.org/10.1080/13658816.2018.1535121

18. Khan, J., Khan, I., Ghaffar, A., Khalid, B.: Epidemiological trends and risk factors associated with dengue disease in Pakistan (1980–2014): a systematic literature search and analysis. BMC Public Health **18**(1), 745 (2018). https://doi.org/10.1186/s12889-018-5676-2

19. Kretzschmar, M., Wallinga, J.: Mathematical models in infectious disease epidemiology. Modern Infectious Disease Epidemiology. SBH, pp. 209–221. Springer, New York (2009). https://doi.org/10.1007/978-0-387-93835-6_12

20. Lima, T.F.M.D., et al.: Dengueme: a tool for the modeling and simulation of dengue spatiotemporal dynamics. Int. J. Environ. Res. Public Health **13**(9), 920 (2016)

21. Masui, H., Kakitani, I., Ujiyama, S., Hashidate, K., Shiono, M., Kudo, K.: Assessing potential countermeasures against the dengue epidemic in non-tropical urban cities. Theor. Biol. Med. Model. **13**, 12 (2016). https://doi.org/10.1186/s12976-016-0039-0

22. Racloz, V., Ramsey, R., Tong, S., Hu, W.: Surveillance of dengue fever virus: a review of epidemiological models and early warning systems. PLoS Negl. Trop. Dis. **6**(5), e1648 (2012)

23. Regis, L.N., et al.: Sustained reduction of the dengue vector population resulting from an integrated control strategy applied in two Brazilian cities. PLoS ONE **8**(7), 920 (2013). https://doi.org/10.1371/journal.pone.0067682

24. Shaikh, S., Kazmi, S.J.H., Qureshi, S.: Monitoring the diversity of malaria and dengue vector in Karachi: studying variation of genera and subgenera of mosquitoes under different ecological conditions. Ecol. Process. **3**, 12 (2014). https://doi.org/10.1186/s13717-014-0012-y

25. Szklo, M., Nieto, F.J.: Epidemiology: Beyond the Basics. Jones & Bartlett Publishers, Burlington (2014)

26. WHO: Dengue control (2016). https://www.who.int/denguecontrol/disease/en/

27. WHO: Who scales up response to worldwide surge in dengue (2019). https://www.who.int/news-room/feature-stories/detail/who-scales-up-response-to-worldwide-surge-in-dengue

28. World Health Organization: Dengue and severe dengue- fact sheet (2019). http://www.who.int/mediacentre/factsheets/fs117/en/

Hypergraph Grammar-Based Model of Adaptive Bitmap Compression

Grzegorz Soliński[1], Maciej Woźniak[1]([⊠]) [iD], Jakub Ryzner[1], Albert Mosiałek[1], and Anna Paszyńska[2] [iD]

[1] Department of Computer Science, AGH University of Science and Technology, Kraków, Poland
macwozni@agh.edu.pl
[2] Faculty of Physics, Astronomy and Applied Computer Science, Jagiellonian University, Kraków, Poland
anna.paszynska@uj.edu.pl

Abstract. JPEG algorithm defines a sequence of steps (essential and optional) executed in order to compress an image. The first step is an optional conversion of the image color space from RBG (red-blue-green) to YCbCr (luminance and two chroma components). This step allows to discard part of chrominance information, a useful gain due to the fact, that the chrominance resolution of the human eye is much lower than the luminance resolution. In the next step, the image is divided into 8×8 blocks, called MCUs (Minimum Coded Units). In this paper we present a new adaptive bitmap compression algorithm, and we compare it to the state-of-the-art of JPEG algorithms. Our algorithm utilizes hypergraph grammar model, partitioning the bitmap into a set of adaptively selected rectangles. Each rectangle approximates a bitmap using MCUs with the size selected according to the entire rectangular element. The hypergraph grammar model allows to describe the whole compression algorithm by a set of five productions. They are executed during the compression stage, and they partition the actual rectangles into smaller ones, until the required compression rate is obtained. We show that our method allows to compress bitmaps with large uniform areas in a better way than traditional JPEG algorithms do.

Keywords: Hypergraph grammar · Bitmap compression · Adaptive projection-based interpolation

1 Introduction

Although baseline JPEG is still the most commonly used compression algorithm, a number of algorithms have emerged as an evolution to JPEG. The most prevalent one is JPEG2000 standard (ITU-T T.800 — ISO/IEC 15444-1), which introduces usage of the Discreet Wavelet Transform in place of the Discreet Cosine Transform used in traditional JPEG, as well as usage of a more sophisticated entropy encoding scheme [1]. The standard also introduces an interesting feature,

ⓒ Springer Nature Switzerland AG 2020
V. V. Krzhizhanovskaya et al. (Eds.): ICCS 2020, LNCS 12139, pp. 118–131, 2020.
https://doi.org/10.1007/978-3-030-50420-5_9

which gives a potential boost in compression ratio or quality on some images - Region of Interest (ROI) coding [2]. Region of Interest is a predetermined portion of the image of an arbitrary shape, that is coded before the rest of the image. ROI part of the image and the background can also be compressed with different qualities, which makes JPEG2000 possess some traits of adaptivity. The gain in compression performance of JPEG2000 in comparison to JPEG is mainly denoted by the lack of DCT-specific block artifacts of the compressed images, as well as generally better-quality images for compression ratios exceeding 20:1 [3].

Another standard meant as an evolution to baseline JPEG, originally developed by Microsoft as HD Photo, eventually got published as (ITU-T T.832 — ISO/IEC 29199-2), commonly referred to as JPEG XR. The new standard supports higher compression ratios than baseline JPEG for encoding an image with equivalent quality by introducing lossless color space transformation and integer transform employing a lifting scheme in place of JPEG's slightly lossy RGB to YCbCr linear transformation and the DCT, respectively. It also introduces a different organization of the blocks and another level of frequency transformation [4]. As the experiments carried out by the JPEG committee have shown, the compression performance of JPEG XR was typically very close to the performance of JPEG 2000, with the latter slightly outperforming JPEG XR. However, the difference was generally marginal in the scope of bit rates meaningful for digital photography [5].

Recent work performed by the JPEG committee and its subsidiaries, as well as other contributors, resulted in a series of JPEG related standards/extensions. One of them is JPEG XT (ISO/IEC 18477), a standard that specifies a number of backwards compatible extensions to the legacy JPEG standard. It addresses several plain points, that have stuck to JPEG over the years, such as support for compression of images with higher bit depths (9 to 16 bits-per-pixel), high dynamic range imaging and floating point coding, lossless compression and alpha channels coding [6]. Another recently presented standard is JPEG XS (ISO/IEC 21122). Unlike former standards, the focus JPEG XS was to provide a robust, cost-efficient codec for video-over-IP solutions, suitable for parallel architectures and ensuring a minimal end-to-end compression latency as well as minimal incremental quality degradation on re-compression [7]. As objectives of developing JPEG XT and JPEG XS standards were somewhat different than their e.g. that of JPEG2000, they rather introduce sets of features useful for some particular applications - such as HDR photography or video streaming, than contributing any particular gain in compression performance or visual fidelity.

In this paper we employ the adaptive algorithm for image compression. We use the hypergraph grammar-based approach to model the adaptive algorithm. The hypergraph grammar was originally created by A. Habel and H. Kreowski [8,9]. They were used to model adaptive finite element method computations [10,11], multi-frontal solver algorithm [12–14]. They define the hypergraph data structure and the abstraction of graph grammar productions. The process of adaptive compression of a bitmap can be expressed by a sequence of graph grammar productions. In our model we define five graph grammar productions to describe the adaptive algorithm. We compare our approach with the JPEG standards described above. Alternative graph grammars include the CP-graph grammar, also used for modeling adaptive finite element method [15–19], but for the CP-graph grammar generates hierarchical structures with the history of refinements, while for the bitmap compression application the flat graph seems to be more applicable.

2   Hypergraph and Hypergraph Grammar

In this section the main definitions connected with hypergraphs and hypergraph grammar are presented.

A hypergraph is a special kind of graph consisting of nodes and so called hyperedges joining the nodes. One hyperedge can join two or more nodes. The nodes as well as hyperedges can be labeled with the use of a fixed alphabet. Additionally, the sets of attributes can be assigned to nodes and hyperedges. Figure 1 presents an exemplary hypergraph consisting of five nodes labeled by v, two hyperedges labeled by I and one hyperedge labeled by $F1$. The hyperedges labeled by I have attribute *break*. One hyperedge has attribute *break* equal to 1 and the second one has the attribute *break* equal to 0.

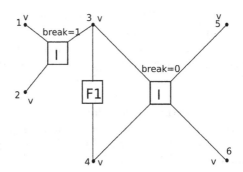

Fig. 1. An exemplary hypergraph with six nodes denoted by v, two hyperedges denoted by I and one hyperedge denoted by $F1$.

Definition 1. An undirected attributed labelled hypergraph over label alphabet C and attribute set A is defined as a system $G = (V, E, t, l, at)$, where:

- V is a finite set of nodes,
- E is a finite set of hyperedges,
- $t : E \rightarrow V^*$ is a mapping assigning sequences of target nodes to hyperedges,
- $l : V \cup E \rightarrow C$ is a node and hyperedge labelling function,
- $at : V \cup E \rightarrow 2^A$ is a node and hyperedge attributing function.

Hypergraph G is a subhypergraph of hypergraph g if its sets of nodes and edges are subsets of sets of nodes and edges of the hypergraph g, respectively and the corresponding nodes and edges of both graphs have the same labels and attributes.

Definition 2. Let $G = (V_G, E_G, t_G, l_G, at_G)$ and $g = (V_g, E_g, t_g, l_g, at_g)$ be two hypergraphs. G is a subhypergraph of g if:

1. $V_G \subset V_g, E_G \subset E_g$,
2. $\forall e \in E_G \; t_G(e) = t_g(e)$,
3. $\forall e \in E_G \; l_G(e) = l_g(e), \forall v \in V_G \; l_G(v) = l_g(v)$,
4. $\forall e \in E_G \; at_G(e) = at_g(e), \forall v \in V_G \; at_G(v) = at_g(v)$.

In order to derive complex hypergraphs from the simpler ones the so-called graph grammar productions can be used. This approach allows to replace subhypergraph of the hypergraph by new hypergraph. The operation of replacing the subhypergraph of the hypergraph by another hypergraph is allowed only under the assumption, that for each new hypergraph so called sequence of its external nodes is specified and that both hypergraphs have the same number of external nodes (the same so-called type of a graph). These nodes correspond to target nodes of a replaced hypergraph.

Definition 3. A hypergraph of type k is a system $H = (G, ext)$, where:

- $G = (V, E, t, l, at)$ is a hypergraph over C and A,
- ext is a sequence of nodes from the node set V, called external nodes, with $|ext| = k$.

Figure 1 presents an exemplary hypergraph of type 6 with denoted external nodes.

Definition 4. A hypergraph production is a pair $p = (L, R)$, where both L and R are hypergraphs of the same type. Applying a production $p = (L, R)$ to a hypergraph H means to replace a subhypergraph h of H isomorphic with L by a hypergraph R and replacing external nodes from graph h with the corresponding external nodes of R.

Figure 6 presents an exemplary hypergraph production. Figure 2 presents an exemplary starting graph and Fig. 4 presents this graph after application of the production from Fig. 3.

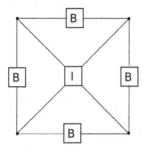

Fig. 2. An exemplary starting hypergraph.

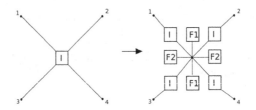

Fig. 3. An exemplary hypergraph production.

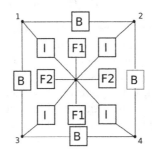

Fig. 4. The graph from Fig. 2 after application of the production from Fig. 3

The application of a production $p = (L, R)$ to a hypergraph H consists of replacing a subhypergraph of H isomorphic with L by a hypergraph isomorphic with R and replacing nodes of the removed subhypergraph isomorphic with external nodes of L by the corresponding external nodes of R.

Definition 5. Let P be a fixed set of hypergraph productions. Let G and G' be two hypergraphs.

We say that hypergraph H is directly derived from a hypergraph G ($G \Rightarrow H$) if there exists $p = (L, R) \in P$ such that:

- There exists h - a subhypergraph of G isomorphic with L,
- Let ext_h be a sequence of nodes of h consisting of nodes isomorphic with nodes of the sequence of external nodes of L (ext_L). The replacement of $h = (V_h, E_h, t_h, l_h, at_h)$ in $G = (V_G, E_G, t_G, l_G, at_G)$ by $R = (V_R, E_R, t_R, l_R, at_R)$ generates the the hypergraph $G' = (V'_G, E'_G, t'_G, l'_G, at'_G)$, where:

  - $V'_G = V_H - V_h \cup V_R$,
  - $E'_G = E_H - E_h \cup E_R$,
  - $\forall e \in E_R \; t'_G(e) = t_R(e)$,
  - $\forall e \in E_G - E_h$ with $t_G(e) = t_1, ..., t_n, t'_G(e) = t'_1, .., t'_n$, where each $t'_i = t_i$ if t_i does not belong to the sequence ext_h or $t'_i = v_j$ (v_j is an j-th element of the sequence ext_R) if t_i is an j-th element of the sequence ext_h,
  - $\forall e \in E_G - E_h \; l'_G(e) = l_G(e)$, $at'_G(e) = at_G(e)$, $\forall e \in E_R \; l'_G(e) = l_R(e)$, $at'_G(e) = at_R(e)$,

- $\forall v \in V_G - V_h \; l'_G(v) = l_G(v), \; at'_G(v) = at_G(v), \; \forall v \in V_R \; l'_G(v) = l_R(v),$
  $at'_G(v) = at_R(v).$

– H is isomorphic with the hypergraph G'.

Definition 6. A hypergraph grammar is a quadruple $G = (V, E, P, X)$, where:

– V is a finite set of labelled nodes,
– E is a finite set of labelled hyperedges,
– P is a finite set of hypergraph productions of the form $p = (L, R)$, where L and R are hypergraphs of the same type composed of nodes of V and hyperedges of E,
– X is an initial hypergraph called axiom of G.

3 Hypergraph Grammar for Bitmap Compression

In the section the hypergraph grammar productions modelling the bitmap compression based on projection based interpolation algorithm is presented. The finite element mesh is represented by hypergraph. The changes in the structure of the mesh are modeled by corresponding graph grammar productions.

A hypergraph modeling a mesh with rectangular elements is defined with the following sets of graph node and edge labels C:

$$C = \{C_1 \cup C_2 \cup C_3\} \tag{1}$$

where

– $C_1 = \{v\}$ is a singleton containing the node label which denotes a node of finite element,
– $C_2 = \{F1, F2, B\}$ is a set of edge labels which denote edges of finite elements,
– $C_3 = \{I\}$ is a singleton containing the edge label which denotes interior of finite element.

The following attributing functions are defined to attribute nodes and edges of the hypergraphs modeling the mesh.

– $geom : V \times C_1 \rightarrow R \times R$ is a function attributing nodes, which assigns the coordinates x, y to each node of the element,
– $break : E \times C_3 \rightarrow \{0, 1\}$ is a function attributing interiors, which assigns the value denoting if the adaptation should be performed, where 1 means that element should be broken, 0 means that element should not be broken.

The graph grammar for bitmap compression **(G1)** is defined as a set of hypergraph grammar productions: production **(P1)** for generation of the first element of the mesh, productions **(P2)**, **(P2')**, **(P2")**, **(P2"')** breaking the interior of the element, production **(P3)** for breaking of the boundary edge, production **(P4)** for breaking shared edge, productions **(P5)**, **(P5')**, **(P5")**, **(P5"')** for making the decision about the adaptation, and the starting graph consisting of one node labeled by S. Figure 5 presents production **(P1)** for generation of the first element. Productions **(P2)**, **(P2')**, **(P2")** and **(P2"')** from Fig. 6 allow for breaking the interior of the element for the case of all adjacent edges having the same "level of adaptation" as the interior (production **(P2)**), three adjacent edges having the same "level of adaptation" as the interior ((**(P2')**)), two adjacent edges having the same "level of adaptation" as the interior (production **(P2")**) and finally only one adjacent edge having the same "level of adaptation" as the interior (production **(P2"')**). Our model distinguishes the edges of element being in the boundary of the mesh (hyperdees with label B), and shared edges of elements (labeled by $F1$ and $F2$). Thus we have two separate productions for breaking edges: production **(P3)** for breaking boundary edges and production **(P4)** for breaking shared edges (see Fig. 7, 8). The similar productions to productions from Fig. 7, 8 are defined for hyperedges labeled by $F2$ instead of $F1$ and $F1$ instead of $F2$. The last set of productions, presented in Fig. 9, allows for making decision about the adaptation. Figure 10 presents an exemplary derivation in our graph grammar.

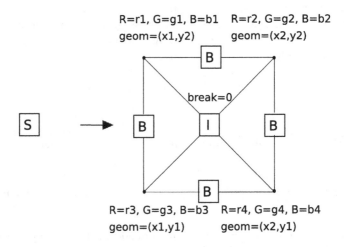

Fig. 5. Production P1 for generation of the first element

4   Results

In order to examine the validity of the idea of the presented hypergraph grammar-based Adaptive JPEG algorithm, we compare the compression ratio with the standard JPEG algorithm on three exemplary bitmaps, presented in Fig. 11. Compressed bitmaps with only luminance component's MCUs outlined in the images are presented in Fig. 12.

The Table 2 presents the compression ratio for different luminance and chrominance parameter values.

The main observation from looking at the Table 2 is that the algorithm (without taking into account visual quality) indeed gives higher compression ratios than baseline JPEG, with higher compression obtained from running the algorithm with larger quality parameter values. This is hardly a surprise, as theoretically the algorithm could give the same compression ratio as baseline

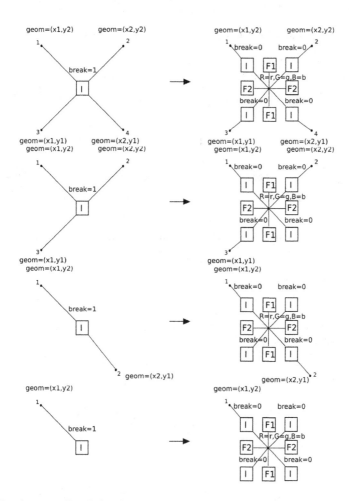

Fig. 6. Productions P2, P2', P2" and P2"' for breaking the interior of the element.

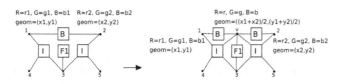

Fig. 7. Production P3 for breaking the boundary edge

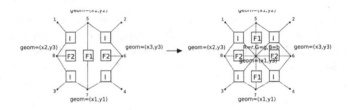

Fig. 8. Production P4 for breaking common edge

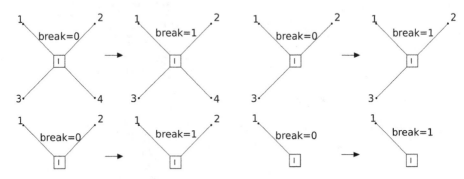

Fig. 9. Productions P5, P5', P5" and P5"' for making virtual adaptation

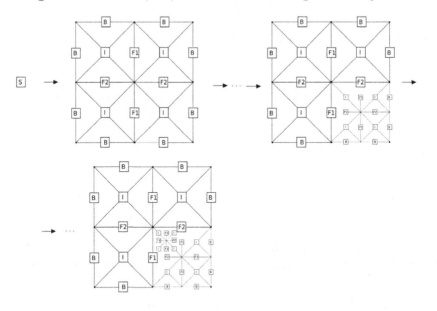

Fig. 10. Exemplary derivation

JPEG in a worst-case scenario, but usually the non-zero quality parameter values leave some room for error and allow some larger areas to stay undivided, resulting in greater compression.

Another observation from examining the graphs more carefully, is that the compression ratios for different test images are quite sparse: depending on the image, for a satisfactory quality, the compression ratios came down to as little as 0.69% (field image for parameter values 15×30) or to only as much as 8.5% (moon image for parameter values 6×60). This shows, that the algorithm clearly works better for certain kind of images than for others. This kind of images are very large images with large relatively uniform areas.

In order to examine the validity of the idea of the presented hypergraph grammar-based Adaptive JPEG algorithm, an effort to compare the performance with the standard JPEG and newer, JPEG successor standards was made: JPEG 2000 and JPEG XR. However, such comparison is not trivial, as the algorithms use different types of parameters to specify the level of compression, directly reflected in the resulting image quality. The parameter types used in JPEG2000 and JPEG XR do not necessarily coincide with the quality parameter type used in JPEG, that is 1–100 scaled quality parameter used to modify the quantization tables. Therefore, some assumptions were made when choosing the parameter values for JPEG2000 and JPEG XR. The JPEG2000 images were generated using the OpenJPEG library, which accepts a quality parameter denoting a target PSNR value in dB. As it was not possible to use the same type of parameter as JPEG accepts, the JPEG2000 images were compressed using PSNR value parameter equal to the PSNR of the corresponding baseline JPEG image. The JPEG XR images were generated using the jxrlib library, which, as a quality parameter, accepts a value ranging between 0.0 and 1.0. In this case, the images were compressed using the parameter value 0.8, to be as consistent with the JPEG compressed test images as possible. The actual impact of the respective parameter values however might differ considerably from JPEG in terms of resulting image quality to file size ratio. The presented Adaptive JPEG seems to work particularly well with large, relatively uniform images - as evident in the case of the field image (Table 1).

Table 1. Comparison of compression ratios for different luminance and chrominance parameters for JPEG and Adaptive JPEG.

Bitmap	JPEG	2×4	5×5	5×10	10×20	15×15	15×30	20×40	25×50	30×30	60×60
Field	3.19%	2.58%	2.26%	2.16 %	1.17%	0.72%	0.69%	0.46%	0.34%	0.28%	0.13%
Lake	5.68%	5.06%	4.53%	3.92%	3.35%	2.70%	2.62%	1.93%	1.55%	1.29%	0.54%
Moon	11.83%	11.24%	11.24%	11.24%	11.24 %	11.24%	11.24%	11.24%	11.24%	11.16%	8.50%

Fig. 11. Bitmap compressions. JPEG compresson on the left, adaptive JPEG on the right with error tolerance parameter values - 15/30.

Fig. 12. Compressed bitmaps with only luminance component's MCUs outlined.

Table 2. Comparison of compression ratios for different image standards.

Bitmap	ADAPTIVE JPEG	JPEG-XR	JPEG2000	JPEG
Field	0.72%	3.33%	2.60%	2.29%
Lake	2.70%	6.37%	1.47%	5.68%
Moon	11.24%	16.23%	5.94%	11.83%

5   Conclusions

There seems to be a class of images for which the algorithm presented in this paper achieves higher compression ratios than baseline JPEG, while retaining subjective image quality. Even though this class of images is not clearly defined, some common characteristics can be extracted: large size of the image - images with higher resolution seem to gain more from the adaptive extension introduced by the algorithm, and areas of low frequency - images containing areas that are relatively uniform seem to tolerate larger MCU sizes better, resulting in higher compression ratios. The maximum values of luminance and chrominance quality parameters, consumed by the algorithm, for which the images seemed to retain subjective image quality, were found to be around 15/15 (luminance/chrominance). In some cases, the image quality was still retained for the parameter values of 15/30, proving the less impactful nature of the chrominance component. Future work may include implementing the adaptive algorithm based on an existing improvement over JPEG, e.g. JPEG2000. Both JPEG2000 and JPEG XR, although very different, operate much the same in regard to MCU size, which is globally fixed. If not an implementation, a thorough performance comparison between the presented Adaptive JPEG algorithm and other lossy compression algorithms ought to be conducted, as only JPEG was exhaustively addressed in this work. Finally, we only address the problem of compression, completely leaving out the issue of decompression. Decompression of an image compressed in such non-deterministic way is a very complex problem, deserving a separate research.

Acknowledgement. The research presented in this paper was partially supported by the funds assigned to AGH University of Science and Technology by the Polish Ministry of Science and Higher Education.

References

1. International Organization for Standardization. ISO/IEC 15444–1:2016 - Information technology - JPEG 2000 image coding system: Core coding system (2016)
2. Askelöf, J., Carlander, M.L., Christopoulos, C.: Region of interest coding in jpeg 2000. Signal Process. Image Commun. **17**(1), 105–111 (2002)
3. Ebrahimi, F., Chamik, M., Winkler, S.: Jpeg vs. jpeg2000: an objective comparison of image encoding quality, pp. 55–58 (2004)

4. International Telecommunication Union. ITU-T T.832 - Information technology - JPEG XR image coding system - Image coding specification (2016)
5. Dufaux, F., Sullivan, G.J., Ebrahimi, T.: The jpeg xr image coding standard [standards in a nutshell]. IEEE Signal Process. Mag. **26**(6), 195–204 (2009)
6. Artusi, A., et al.: Jpeg xt: a compression standard for hdr and wcg images [standards in a nutshell]. IEEE Signal Process. Mag. **33**(2), 118–124 (2016)
7. Richter, T., Keinert, J., Descampe, A., Rouvroy, G.: Entropy coding and entropy coding improvements of jpeg xs. In: 2018 Data Compression Conference, pp. 87–96 (2018)
8. Habel, A., Kreowski, H.-J.: May we introduce to you: hyperedge replacement. In: Ehrig, H., Nagl, M., Rozenberg, G., Rosenfeld, A. (eds.) Graph Grammars 1986. LNCS, vol. 291, pp. 15–26. Springer, Heidelberg (1987). https://doi.org/10.1007/3-540-18771-5_41
9. Habel, A., Kreowski, H.-J.: Some structural aspects of hypergraph languages generated by hyperedge replacement. In: Brandenburg, F.J., Vidal-Naquet, G., Wirsing, M. (eds.) STACS 1987. LNCS, vol. 247, pp. 207–219. Springer, Heidelberg (1987). https://doi.org/10.1007/BFb0039608
10. Slusarczyk, G., Paszyńska, A.: Hypergraph grammars in hp-adaptive finite element method. Procedia Comput. Sci. **18**, 1545–1554 (2013)
11. Gurgul, P., Jopek, K., Pingali, K., Paszyńska, A.: Applications of a hyper-graph grammar system in adaptive finite-element computations. Int. J. Appl. Math. Comput. Sci. **28**(3), 569–582 (2018)
12. Gurgul, P., Paszyński, M., Paszyńska, A.: Hypergrammar based parallel multifrontal solver for grids with point singularities. Comput. Sci. **16**(1), 75–102 (2015)
13. Paszyński, M., Jopek, K., Paszyńska, A., Hassan, M.A., Pingali, K.: Hypergraph grammar based multi-thread multi-frontal direct solver with Galois scheduler. Comput. Sci. **18**(2), 27–55 (2019)
14. Paszyńska, A., et al.: Quasi-optimal elimination trees for 2d grids with singularities. Sci. Program. (2015)
15. Paszyńska, A., Paszyński, M., Grabska, E.: Graph Transformations for modeling hp-adaptive finite element method with mixed triangular and rectangular elements. In: Allen, G., Nabrzyski, J., Seidel, E., van Albada, G.D., Dongarra, J., Sloot, P.M.A. (eds.) ICCS 2009. LNCS, vol. 5545, pp. 875–884. Springer, Heidelberg (2009). https://doi.org/10.1007/978-3-642-01973-9_97
16. Goik, D., Jopek, K., Paszyński, M., Lenharth, A., Nguyen, D., Pingali, K.: Graph grammar based multi-thread multi-frontal direct solver with Galois scheduler. Procedia Comput. Sci. **29**, 960–969 (2014)
17. Paszyński, M., Paszyńska, A.: Graph transformations for modeling parallel hp-adaptive finite element method. In: Wyrzykowski, R., Dongarra, J., Karczewski, K., Wasniewski, J. (eds.) PPAM 2007. LNCS, vol. 4967, pp. 1313–1322. Springer, Heidelberg (2008). https://doi.org/10.1007/978-3-540-68111-3_139
18. Paszyński, M.: On the parallelization of self-adaptive hp-finite element methods part I. composite programmable graph grammarmodel. Fundam. Inf. **93**(4), 411–434 (2009)
19. Paszyński, M.: On the parallelization of self-adaptive hp-finite element methods part II partitioning communication agglomeration mapping (PCAM) analysis. Fundam. Inf. **93**(4), 435–457 (2009)

Simulation of Neurotransmitter Flow in Three Dimensional Model of Presynaptic Bouton

Andrzej Bielecki$^{(\boxtimes)}$ and Maciej Gierdziewicz

Chair of Applied Computer Science, Faculty of Automation, Electrical Engineering,
Computer Science and Biomedical Engineering, AGH University of Science
and Technology, Al. Mickiewicza 30, 30-059 Cracow, Poland
bielecki@agh.edu.pl

Abstract. In this paper a geometrical model for simulation of the nerve impulses inside the presynaptic bouton is designed. The neurotransmitter flow is described by using partial differential equation with nonlinear term. The bouton is modeled as a distorted geosphere and the mitochondrion inside it as a highly modified cuboid. The quality of the mesh elements is examined. The changes of the amount of neurotransmitter during exocytosis are simulated.

Keywords: Presynaptic bouton · PDE model · Finite elements method · Neurotransmitter flow

1 Introduction

The activity of neurons [16], including those taking place in the presynaptic bouton of the neuron, may be described theoretically with differential equations [1,2,4] or, more specifically, with partial differential equations (PDE) [3,11,13]. Numerical methods have been commonly used to solve PDE [9,10,12,13], which may be a proof that scientists are constantly interested in this phenomenon and, on the other hand, that results are variable and depend strongly on the biological assumptions and on the way of modeling. To obtain the solution, the appropriate design of the mesh of the studied object (in this paper: of a presynaptic bouton of the neuron) is necessary, which may be a demanding task [19,20].

Some experiments in the cited papers have already been performed in order to answer the question: how the neurotransmitter (NT) mediates the process of conducting nerve impulses, which is connected with the distribution of NT in the synapse and with its exocytosis. This paper is intended to expand that knowledge by using more complicated mathematical model in three dimensions together with a geometric model which is more complex than it was in some previous

The calculations have been supported by using computing grant *neuron2019* in PL-Grid infrastructure (the supercomputer *Prometheus*).

© Springer Nature Switzerland AG 2020
V. V. Krzhizhanovskaya et al. (Eds.): ICCS 2020, LNCS 12139, pp. 132–143, 2020.
https://doi.org/10.1007/978-3-030-50420-5_10

works [8, 13]. There are also examples of the description of a realistic geometric model [7], but without performing simulations. However, the NT flow in the realistically shaped model of the presynaptic bouton affected by the presence of a mitochondrion inside it has not been studied yet. Therefore, the objective of this paper was to examine the process of NT flow in a realistic model of the presynaptic bouton with a mitochondrion partly occupying its volume.

2   Theoretical Foundations

In this section the model based on partial differential equations is presented briefly. Such approach allows us to study the dynamics of the transport both in time and in spatial aspect. The model is nonlinear, diffusive-like - see the paper [3] for details. There are a few assumptions about the model. The bouton location is a bounded domain $\Omega \subset \mathbf{R}^3$. The boundary (the membrane) of the bouton is denoted as $\partial \Omega$. The total amount of neurotransmitter in the bouton is increasing when new vesicles are supplied inside the bouton and decreasing when their contents are released through the bouton membrane. The proper subset of the bouton, Ω_1, is the domain of vesicles supply. The release site, Γ_d, is a proper subset of the membrane. The function $f : \Omega \to \mathbf{R}$ models the synthesis of the vesicles that contain neurotransmitter. The value of this function is $f(x) = \beta > 0$ on Ω_1 and $f(x) = 0$ on $\Omega \setminus \Omega_1$. The neurotransmitter flow in the presynaptic bouton was modeled with the following partial differential equation:

$$\varrho_t(x,t) = a\Delta\varrho(x,t) + f(x)(\bar{\varrho} - \varrho(x,t))^+ \quad \text{in} \quad \Omega \times (0,T), \tag{1}$$

where $\varrho(x,t)$ is the density of neurotransmitter at the point x and time t and a is the diffusion coefficient. The last term contains the function $f(x)$, presented before, and the threshold NT density, above which synthesis dose not take place, denoted by $\bar{\varrho}$. For any $x \in \mathbf{R}$ the "x^+" symbol means $\max(0, x)$. The boundary conditions are

$$-\frac{\partial \varrho(x,t)}{\partial \nu} = 0 \quad \text{on} \quad (\partial\Omega \setminus \Gamma_d) \times (0,T),$$

$$-\frac{\partial \varrho(x,t)}{\partial \nu} = \eta(t)a\varrho(x,t) \quad \text{on} \quad \Gamma_d \times (0,T), \tag{2}$$

where $\partial/\partial\nu$ is a directional derivative in the outer normal direction ν. The function $\eta(t)$ depends on the time t and takes the value of 1 for $t \in [t_n, t_n + \tau]$ (with the action potential arriving at t_n and with the release time τ), and $\eta(t) = 0$ otherwise [5, 8].

Multiplying (2) by a smooth function $v : \Omega \to \mathbf{R}$, and integrating by parts gives (see [3] for details):

$$\int_\Omega \varrho_t(x,t)v(x)\,dx + a\int_\Omega \nabla\varrho(x,t) \cdot \nabla v(x)\,dx$$
$$+ a\eta(t)\int_{\Gamma_d} \varrho(x,t)v(x)\,dS = \beta\int_{\Omega_1} (\bar{\varrho} - \varrho(x,t))^+ v(x)\,dx, \tag{3}$$

provided that ϱ is sufficiently smooth.

Piecewise linear C^0 tetrahedral finite elements are used in the numerical scheme. The unknown function ϱ is approximated by ϱ_h, computed with the formula

$$\varrho_h(x,t) = \sum_{i=1}^{K} \varrho_i(t)v_i(x), \tag{4}$$

in which the basis functions of the finite element space are denoted by v_i. Using the substitution $v = v_j$ in (2) yields

$$\sum_{i=1}^{K} \varrho_i'(t) \int_\Omega v_i(x)v_j(x)\,dx + a\sum_{i=1}^{K} \varrho_i(t) \int_\Omega \nabla v_i(x) \cdot \nabla v_j(x)\,dx$$

$$+ \alpha\eta(t)\sum_{i=1}^{K} \varrho_i(t) \int_{\Gamma_d} v_i(x)v_j(x)\,dS = \beta \int_{\Omega_1} \left(\bar{\varrho} - \sum_{i=1}^{K} \varrho_i(t)v_i(x) \right)^+ v_j(x)\,dx. \tag{5}$$

When we denote $\varrho_h(t) = [\varrho_1(t),\ldots,\varrho_K(t)]$, then the mass matrix $\mathcal{G} = \{g_{ij}\}_{i,j=1}^{K}$, the stiffness matrix $\mathcal{A}_1 = \{a_{ij}^1\}_{i,j=1}^{K}$, and the release matrix $\mathcal{A}_2 = \{a_{ij}^2\}_{i,j=1}^{K}$ are given by

$$g_{ij} = \int_\Omega v_i(x)v_j(x)\,dx, \tag{6}$$

$$a_{ij}^1 = \int_\Omega \nabla v_i(x) \cdot \nabla v_j(x)\,dx,$$

$$a_{ij}^2 = \int_{\Gamma_d} v_i(x)v_j(x)\,dS.$$

Let $\mathcal{A}(t) = \mathcal{A}_1 + \eta(t)\mathcal{A}_2$. Let us also introduce the nonlinear operator $\mathcal{B} : \mathbf{R}^K \to \mathbf{R}^K$ in the following way

$$\mathcal{B}(\varrho_1,\ldots,\varrho_K)_j = \beta \int_{\Omega_1} \left(\bar{\varrho} - \sum_{i=1}^{K} \varrho_i v_i(x) \right)^+ v_j(x)\,dx. \tag{7}$$

The Eq. (3) may be rewritten as

$$\mathcal{G}\varrho_h'(t) + \mathcal{A}(t)\varrho_h(t) = \mathcal{B}(\varrho_h(t)).$$

The time step is defined as $t_k = k\Delta t$ where Δt is at least several times smaller than τ. If we assume that ϱ_h^k is approximated by $\varrho_h(t_k)$ we can use the Crank–Nicolson approximative scheme

$$\mathcal{G}\frac{\varrho_h^k - \varrho_h^{k-1}}{\Delta t} + \frac{1}{2}\mathcal{A}(t_k)\varrho_h^k + \frac{1}{2}\mathcal{A}(t_{k-1})\varrho_h^{k-1} = \frac{1}{2}\mathcal{B}(\varrho_h^k) + \frac{1}{2}\mathcal{B}(\varrho_h^{k-1}), \tag{8}$$

valid for $k = 1, 2, \ldots$. In the finite element basis $\{v_k\}_{k=1}^K$, ϱ_0 is approximated by ϱ_h^0. The scheme must be solved for ϱ_h^k. The problem of the nonlinearity in $\mathcal{B}$ is dealt with by the iterative scheme:

$$\varrho_h^{k(0)} = \varrho_h^{k-1}, \tag{9}$$

$$\mathcal{G}\frac{\varrho_h^{k(m+1)} - \varrho_h^{k-1}}{\Delta t} + \frac{1}{2}\mathcal{A}(t_k)\varrho_h^{k(m+1)} + \frac{1}{2}\mathcal{A}(t_{k-1})\varrho_h^{k-1} = \frac{1}{2}\mathcal{B}(\varrho_h^{k(m)}) + \frac{1}{2}\mathcal{B}(\varrho_h^{k-1}),$$

In each step of the iteration the linear system (9) is solved. The iterative procedure is stopped for the value m for which the correction of the solution due to one step of the scheme is sufficiently small.

Note that the non-linear term is calculated at the previous iteration (m) and the linear terms at iterations (m + 1). Such scheme reduces to solving the linear system in each step of the iteration. It is an alternative to the Newton method, and it has the advantage of not needing to cope with the non-differentiable non-linearity introduced with the last term (taking the positive part) of the equation.

From physiological point of view, the time step for the Eq. (9) should be less than the time scale of the modeled phenomenon. We do not have the proof of the scheme stability but the numerical experiments show that the iterations of the fixed point scheme with respect to index m converge, and the limit is the solution of the Crank–Nicolson scheme. This scheme is known to be unconditionally stable. Still, to avoid numerical artifacts, the time-step needs to be sufficiently small. Our numerical experiments show that no numerical artifacts are seen in the obtained solution for $\Delta t = 0.0001\,\mu s$ and for $\Delta t = 0.00005\,\mu s$.

For the numerical simulations, the bouton is modeled as a bounded polyhedral domain $\Omega \subset \mathbf{R}^3$. The boundary of the bouton, $\partial\Omega$, is represented by a set of flat polygons. The proper polyhedral subset of the bouton, Ω_1, is the domain of vesicles supply. The release site Γ_d is modeled as a finite sum of flat polygons.

Various geometrical models of the presynaptic bouton have been used by the authors to simulate the process of conducting nerve impulses so far. The natural reference point for assessing their quality is to compare them to the images of real boutons which have been thoroughly examined, for example, several years ago [17,18]. One of the models was based of the skeleton made up of two concentric spheres, the outer one having the diameter of 2–3 μm and the inner one free from synaptic vesicles, and also with numerous release sites [13]. The other structure consisting of two concentric globes and a single release site was filled with a tetrahedral mesh and it has been used to examine the distribution of neurotransmitter inside the bouton [6]. A very similar structure has been utilized to find the connection between the number and location of synthesis domains and the amount of neurotransmitter in different locations in the bouton [8]. Another, more complicated, surface model intended for the discrete simulation of vesicle movement consisted of a realistically shaped bouton and a mitochondrion partly occupying its space [7].

The amplitudes of evoked excitatory junctional potentials and the estimated number of the synaptic vesicles released during exocytosis were examined in [13] with the use of the model with empty central region. The study discussed, among others, the way the synaptic depression was reflected in the results of the simulations.

The model made up from two concentric spheres filled with a tetrahedral mesh was used to demonstrate synaptic depression [6]. The volume of the model was about $17{,}16\,\mu m^3$ and the surface about $32{,}17\,\mu m^2$. The diffusion coefficient was $3\,\mu m^2/s$, synthesis rate and the exocytosis rate around $21\,\mu m^3/s$. The results confirmed that during 0.5 s of 40 Hz stimulation, with the assumed values of parameters, the amount of neurotransmitter in the presynaptic bouton decreased, though only by the fraction of about 2%.

The next studies with the globe model [8] were performed to examine the influence of the location and number of synthesis zones on the amount of neurotransmitter in particular locations of the bouton. The simulation parameters were similar as in [6] but the number of synthesis zones was 1 or 2 (however, with the same total volume) and the location was also variable. The chief conclusion was that the closer the synthesis (or supply) domain to the release site, the faster the bouton becomes depleted of neurotransmitter.

3   The Realistic Model of the Bouton with Mitochondrion

The geometric aspects of the models of the bouton used so far, mentioned in the previous section, were extremely simplified. In this paper, the simulations have been conducted on the basis of the realistic model of the bouton in which the mitochondrion is modeled as well.

3.1   The Input Surface Mesh

The structure used to model the presynaptic bouton in this paper was based on the real shape of such a bouton and it was composed of geometric shapes modified so that it resembled the original. Therefore, the outer of the bouton was modeled as a moderately modified sphere, whereas a part of the mitochondrion inside it was a highly distorted cuboid, as it has been presented in [7]. The parameters of the surface mesh are described therein in detail. It should be stressed that the mesh described in this paper was designed to shorten the computing time of simulations of neurotransmitter flow, and therefore contains, among others, relatively large elements. The number of tetrahedra in the mesh was less then 10^5 and the number of the surface triangles (faces) was less than 5×10^4.

3.2   Three-Dimensional Model

The parameters of the input surface mesh were as follows. The total surface of the bouton was $S \approx 6.7811\,\mu m^2$, and the area of the active zone (of the release site) was $S_{AZ} \approx 0.2402\,\mu m^2$. The surface mesh is presented in Fig. 1.

The tetrahedral mesh was generated with the TetGen program [14,15]. The result is depicted in Fig. 2.

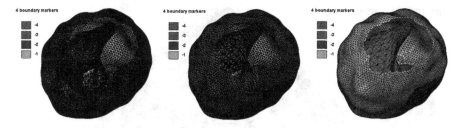

Fig. 1. The surface mesh of the presynaptic bouton - the view with the top cut off. The bouton would fit in a $1.33 \times 1.16 \times 1.23\,\mu$m cuboid. Skeleton - left, partly colored - middle, colored surface - right. The meaning of the colors is: red(-4) - mitochondrion, green(-3) - artificial cut, blue(-2) - release site, yellow(-1) - the rest. (Color figure online)

The input parameters for `TetGen` were chosen in order to minimize the number of mesh elements. As a result of this, the bouton three-dimensional mesh contained 77418 tetrahedra with the total volume $V \approx 0.9029\,\mu\mathrm{m}^3$. The volume of the part of the bouton located in the direct neighborhood of the mitochondrion (where the endings of the microtubules are sometimes found), assumed as the theoretical NT supply zone ("synthesis domain") was $V_s \approx 0.0198\,\mu\mathrm{m}^3$. The quality of the mesh was assessed by computing several measures for each tetrahedra and by taking, for each measure, its maximal value i.e. the value for the worst mesh element. The values are collected in Table 1. The relatively high values of the mesh quality parameters may suggest a cautious approach to the analysis of the results. However, further stages of the experiment revealed that the mesh proved sufficient for performing the planned simulation.

Table 1. The parameters of the quality of three dimensional mesh of the presynaptic bouton with a mitochondrion.

Parameter	Formula	Ideal value	Maximal value
SV	$\sqrt{S}/\sqrt[3]{V}$	$\sqrt{3}\sqrt[6]{72} \approx 2.6843$	7.146
ER	E_{max}/R	$2\sqrt{6} \approx 4.8989$	212.5
EH	E_{max}/H_{min}	$\sqrt{3}/\sqrt{2} \approx 1.2247$	104.42
MX	S_{max}/S	$1/4 = 0.2500$	0.491
MN	S_{min}/S	$1/4 = 0.2500$	0.0058

The initial density of the neurotransmitter was calculated by using the formula

$$\varrho(t = 0) = \varrho_0 = Ae^{-br^2}\,[\text{vesicles}/\mu\mathrm{m}^3] \tag{10}$$

where $A = 300[\text{vesicles}/\mu\mathrm{m}^3]$ is the theoretical maximal value of the function, $b = 0.28[1/\mu\mathrm{m}^2]$, and $r[\mu\mathrm{m}]$ is the distance from the "center" of the bouton i.e. from the center of the minimal cuboid containing the model.

4 boundary markers

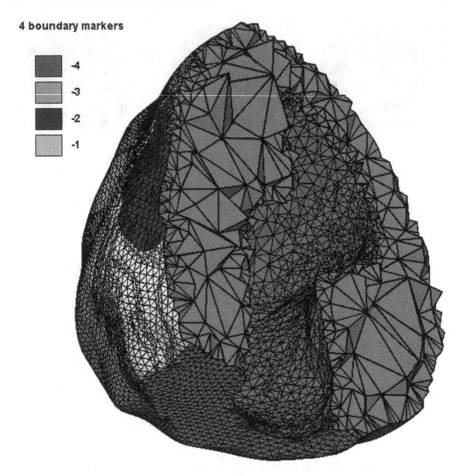

Fig. 2. The cross section of the tetrahedral mesh of the bouton used in calculations. The bouton would fit in a 1.33 × 1.16 × 1.23 μm cuboid. The meaning of the colors: red(-4) - inner surface of the mitochondrion, green(-3) - artificial cut, blue(-2) - release site, yellow(-1) - the rest of the bouton surface, purple - the elements of the tetrahedral mesh. (Color figure online)

4   Simulation Results

The simulation time was 0.1 s and during that time there were 4 impulses. The results are depicted in the following figures: Fig. 3 and Fig. 4, and in Table 2.

The program was designed primarily for validating the accuracy of the calculations; therefore the algorithm was implemented as a single threaded code in Python, with numeric and graphic modules included, and the calculations were relatively slow. The approximate speed was 30–40 iterations per hour. Therefore one simulation run lasted about 2–3 days.

Table 2. The amount of neurotransmitter in vesicles during 4 consecutive nerve impulses. The columns contain: Time - simulation time, Total - the total amount of NT in the bouton, Rel - the amount of NT released in a single time step. The NT amount unit is the amount of NT in a single synaptic vesicle.

Impulse 1			Impulse 2			Impulse 3			Impulse 4		
Time [s]	Total	Rel	Time [s]	Total	Rel	Time [s]	Total	Rel	Time [s]	Total	Rel
0.0123	250.67	0.12	0.0371	250.24	0.11	0.0621	249.63	0.11	0.0871	249.03	0.11
0.0124	250.52	0.20	0.0372	250.05	0.19	0.0622	249.44	0.18	0.0872	248.85	0.18
0.0125	250.40	0.15	0.0373	249.90	0.14	0.0623	249.31	0.14	0.0873	248.71	0.14
0.0126	250.35	0.12	0.0374	249.79	0.12	0.0624	249.19	0.11	0.0874	248.61	0.11
0.0127	250.35	0.06	0.0375	249.73	0.05	0.0625	249.14	0.05	0.0875	248.56	0.05

The total amount of neurotransmitter in the presynaptic bouton calculated during simulation was almost constant. The relative decrease of its value did not exceed 1%. Those values refer to the situation when the neuron is not very intensively stimulated, and synaptic depression is not likely to occur. From the analysis of Fig. 3 it may be concluded that the activity of the release site is very moderate. However, closer examination of Fig. 4 reveals that the spatial distribution of neurotransmitter does change during stimulation. In the region directly adjacent to the release site the amount of neurotransmitter drops a little, and also in its vicinity a slight decrease may be noticed, which is visible in the last picture in Fig. 4. The amount of neurotransmitter increased between stimuli, though at a very low speed, which may be attributed to the low value of exocytosis rate. The changes in these parameters can result in neurotransmitter amount sinking rapidly during exocytosis, thus leading to synaptic depression.

To verify the reliability of the results, two control simulations were run. The first one, with two times larger mesh, confirmed that the relative differences in the results did not exceed 0.14%. For the second one, with two times larger mesh (as before) and with halved time step, the results were almost the same; the relative difference between two control runs did not exceed 10^{-10}.

The process described throughout this paper is similar to which has been found before [1] where, in Fig. 4, one can notice the almost constant number of released vesicles in every time step at the end of the simulation. Taking into account the fact that the authors of the cited paper assumed that the total number of synaptic vesicles in the bouton was more than 10^4, and that the number of synaptic vesicles released in each time step was about 40, we may conclude that if with $M \approx 250$ vesicles in the bouton (the value assumed in this paper) the amount of neurotransmitter released during one time step is approximately 0.5, our results are similar to those found in literature; the proportions of the released amount to the total vesicle pool are of the same order of magnitude. In another study [2] the equilibrium state has been achieved, though at a lower level, indicating synaptic depression.

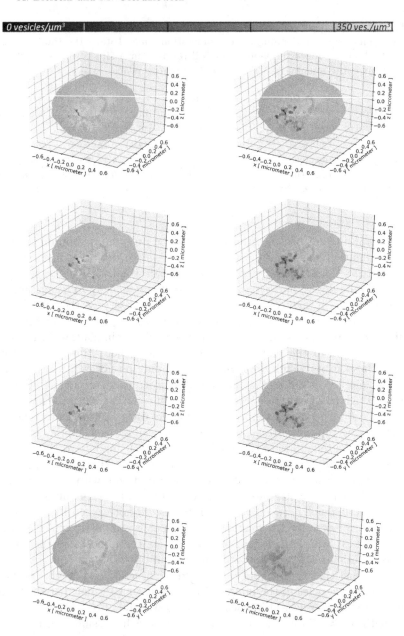

Fig. 3. The spatial distribution of neurotransmitter density $ND(x, y, z)$ in the presynaptic bouton after the arrival of 3 consecutive action potentials. The total time was 0.1 s and the impulses, presented in the rows, arrived at time intervals 0.0123 s–0.0127 s, 0.0373 s–0.0377 s, and 0.0623 s–0.0627 s. In the last row the first (0.0 s) and the last (0.1 s) moment of simulation is depicted. The tick marks in all axes are placed from $-0.6\,\mu m$ to $+0.6\,\mu m$, at $0.2\,\mu m$ intervals. The meaning of colors, corresponding to the range of ND ($0\div350$ vesicles/μm^3) is explained by the color bar at the top. (Color figure online)

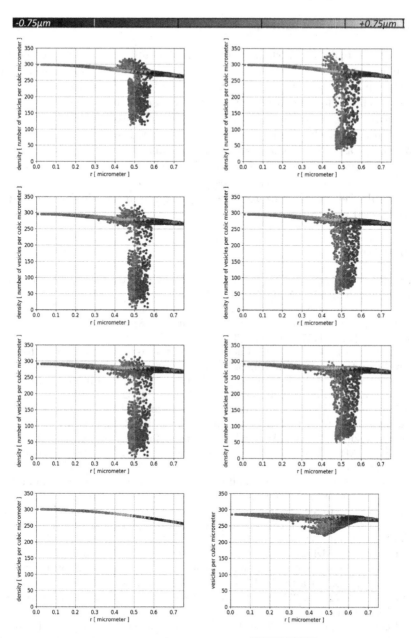

Fig. 4. The radial distribution ($\varrho(r)$, where $r = \sqrt{x^2 + y^2 + z^2}$) of NT density in the presynaptic bouton after 3 consecutive action potentials. Three impulses are in the first 3 rows, and in the last row the plots for the beginning (left) and for the end (right) of the simulation are drawn. The total time was 0.1 s and the impulses time intervals 0.0123 s–0.0127 s, 0.0373 s–0.0377 s, and 0.0623 s–0.0627 s. The tick marks in the r axis are in the range $0 \div 0.75\,\mu$m, placed each $0.1\,\mu$m, and in the ϱ axis they are placed from 0 to 350 vesicles/μm^3 at the intervals of 100 vesicles/μm^3. The meaning of colors, reflecting the z coordinate ($-0.75 \div +0.75\,\mu$m) is explained by the color bar at the top. (Color figure online)

5   Concluding Remarks

The model of realistically shaped presynaptic bouton with a mitochondrion partly blocking its volume, presented in this paper, proved its ability to be used in simulation of synaptic depression. It should be stressed that the values of the parameters chosen for the initial tests of the proposed structure refer to the situation of a very regular activity, not threatened by depression. However, *in silico* tests may reveal the changes in distribution of neurotransmitter in a presynaptic bouton during a very frequent stimulation, thus allowing us to study depression in detail, the more so because the results of the experiments found in literature, whether or not a strong synaptic depression was detected, confirm that our simulation results reflect the real processes taking place in the presynaptic bouton of a stimulated neuron.

References

1. Aristizabal, F., Glavinovic, M.I.: Simulation and parameter estimation of dynamics of synaptic depression. Biol. Cybern. **90**, 3–18 (2004)
2. Bui, L., Glavinović, M.I.: Temperature dependence of vesicular dynamics at excitatory synapses of rat hippocampus. Cogn. Neurodyn. **8**(4), 277–286 (2014). https://doi.org/10.1007/s11571-014-9283-3
3. Bielecki, A., Kalita, P.: Model of neurotransmitter fast transport in axon terminal of presynaptic neuron. J. Math. Biol. **56**, 559–576 (2008)
4. Bielecki, A., Kalita, P.: Dynamical properties of the reaction-diffusion type model of fast synaptic transport. J. Math. Anal. Appl. **393**, 329–340 (2012)
5. Bielecki, A., Gierdziewicz, M., Kalita, P.: Three-dimensional model of signal processing in the Presynaptic Bouton of the neuron. In: Rutkowski, L., Scherer, R., Korytkowski, M., Pedrycz, W., Tadeusiewicz, R., Zurada, J.M. (eds.) ICAISC 2018. LNCS (LNAI), vol. 10841, pp. 3–14. Springer, Cham (2018). https://doi.org/10.1007/978-3-319-91253-0_1
6. Bielecki, A., Gierdziewicz, M., Kalita, P.: Three-dimensional simulation of synaptic depression in axon terminal of stimulated neuron. In: KUKDM 2016, Ninth ACC Cyfronet AGH HPC Users' Conference, pp. 73–74. ACC Cyfronet AGH, Kraków (2016). http://www.cyfronet.krakow.pl/kdm16/prezentacje/s7_04_Gierdziewicz.pdf
7. Bielecki, A., Gierdziewicz, M., Kalita, P., Szostek, K.: Construction of a 3D geometric model of a Presynaptic Bouton for use in modeling of neurotransmitter flow. In: Chmielewski, L.J., Datta, A., Kozera, R., Wojciechowski, K. (eds.) ICCVG 2016. LNCS, vol. 9972, pp. 377–386. Springer, Cham (2016). https://doi.org/10.1007/978-3-319-46418-3_33
8. Bielecki, A., Gierdziewicz, M., Kalita, P.: A study on efficiency of 3D partial differential diffusive model of presynaptic processes. Biocybernetics Biomed. Eng. **40**(1), 100–110 (2020). https://doi.org/10.1016/j.bbe.2019.02.001
9. Bielecki, A., Kalita, P., Lewandowski, M., Siwek, B.: Numerical simulation for a neurotransmitter transport model in the axon terminal of a presynaptic neuron. Biol. Cybern. **102**, 489–502 (2010)
10. Bielecki, A., Kalita, P., Lewandowski, M., Skomorowski, M.: Compartment model of neuropeptide synaptic transport with impulse control. Biol. Cybern. **99**, 443–458 (2008)

11. Friedman, A., Craciun, G.: A model of intracellular transport of particles in an axon. J. Math. Biol. **51**, 217–246 (2005)
12. Lobos C., González E.: Mixed-element octree: a meshing technique toward fast and real-time simulations in biomedical applications. Int. J. Numer. Methods Biomed. Eng. **31**(12), article no. e02725, 24 p. (2015)
13. Knödel, M.M., et al.: Synaptic bouton properties are tuned to best fit the prevailing firing pattern. Front. Comput. Neurosci. **8**, 101 (2014)
14. Si, H.: TetGen: a quality tetrahedral mesh generator and 3D Delaunay triangulator, version 1.4 user manual. WIAS - Weierstrass Institute for Applied Analysis and Stochastics (WIAS) (2006)
15. Si, H.: TetGen, a Delaunay-based quality tetrahedral mesh generator. ACM Trans. Math. Softw. **41**(2), 1–36 (2015)
16. Tadeusiewicz, R.: New trends in neurocybernetics. Comput. Methods Mater. Sci. **10**, 1–7 (2010)
17. Wilhelm, B.: Stoichiometric biology of the synapse. Dissertation, Georg August University Göttingen (2013)
18. Wilhelm, B.G., et al.: Composition of isolated synaptic boutons reveals the amounts of vesicle trafficking proteins. Science **344**(6187), 1023–1028 (2014)
19. Yu, Z., Holst, M.J., McCammon, J.A.: High-fidelity geometric modeling for biomedical applications. Finite Elem. Anal. Des. **44**, 715–723 (2008)
20. Yu, Z., Wang, J., Gao, Z., Xu, M., Hoshijima, M.: New software developments for quality mesh generation and optimization from biomedical imaging data. Comput. Methods Programs Biomed. **113**, 226–240 (2014)

Scalable Signal-Based Simulation of Autonomous Beings in Complex Environments

Mateusz Paciorek, Agata Bogacz, and Wojciech Turek[✉]

AGH University of Science and Technology, Kraków, Poland
`wojciech.turek@agh.edu.pl`

Abstract. Simulation of groups of autonomous beings poses a great computational challenge in terms of required time and resources. The need to simulate large environments, numerous populations of beings, and to increase the detail of models causes the need for parallelization of computations. The signal-based simulation algorithm, presented in our previous research, prove the possibility of linear scalability of such computations up to thousands of computing cores. In this paper further extensions of the signal-based models are investigated and new method for defining complex environments is presented. It allows efficient and scalable simulation of structures which cannot be defined using two dimensions, like multi-story buildings, anthills or bee hives. The solution is applied for defining a building evacuation model, which is validated using empirical data from a real-life evacuation drill.

Keywords: Agent-based simulation · Autonomous beings simulation · Scalability · High performance computing

1 Introduction

The research on modeling and simulation of groups of autonomous beings, like crowds of pedestrians, cars in traffic or swarms of bees, has a great need for efficient simulation methods. The need for simulating numerous groups of beings in large environment, the growing complexity of models and the desire to collect the results fast increase the complexity of the computational task. Therefore, new, scalable simulation methods are constantly pursued by the researchers.

In our previous research [2], a novel method for parallelization of autonomous beings simulation has been presented. It is based on the concept of information propagation, which replaces the need for searching the required data by individual beings. The signal is modeled in a 2D grid of cells, it is directional and autonomously spreads to adjacent cells in the grid. This simple concept removes the need for analyzing remote locations during model update, and therefore, allows splitting the simulated environment between many computing processes, which communicate with fixed number of neighbors once in a single time step, which is the crucial requirement of super-scalable algorithm defined in [5]. The

© Springer Nature Switzerland AG 2020
V. V. Krzhizhanovskaya et al. (Eds.): ICCS 2020, LNCS 12139, pp. 144–157, 2020.
https://doi.org/10.1007/978-3-030-50420-5_11

basic version of the method allows representing two-dimensional environments as grids of cells, which are split into equal fragments and updated in parallel. After each time step, the processes updating neighbor fragments exchange information about signal and beings located in common borders. The method requires defining dedicated simulation models, however, in return it can provide linear scalability of simulation. Linear scalability has been achieved in several tests performed on HPC hardware, which involved up to 3456 computing cores.

Linearly scalable method for autonomous beings simulation encouraged further research on signal-based models and their validation. In [2] three simple models have been presented in order to demonstrate the capabilities of the method. In [9] a model of pedestrian behaviour based on proxemics rules have been developed and tested. Further work on signal-based models for pedestrians led to the problem considered in this paper: the need for modeling complex, multi-level buildings, which cannot be represented in two dimensions. Similar problem can be encountered in modeling of other species habitations, like anthills, bee hives or mole tunnel systems. This class of modeling problems can be addressed by using three-dimensional models, however, efficiency of such approach would be doubtful. Only a small part of the 3D space is accessible and significant so the 3D model would introduce a huge waste in memory and computing resources.

In this paper we propose the extension of the signal propagation method addressing the requirements of environments that cannot be represented by 2D grid. The extension introduces an abstraction over the relation of adjacency, which enables flexibility in defining the shape of the environment while preserving values of distance and direction. This concept is further explored by implementing a evacuation scenario of a multi-story building as an example of such an environment. Finally, we compare the metrics collected during the simulation with the available real-life data to validate the resulting model.

2   Scalability of Autonomous Beings Simulation

The problem of autonomous beings simulation and computer-aided analysis of phenomena taking place in large groups of such beings has been studied for many decades. For example, first significant result of traffic simulation using computers can be found in the fifties of XX century [6]. Since then countless reports on methods for modeling and simulation of different types of beings and different phenomena have been published. Specific problems, like urban traffic or pedestrian dynamics, attracted so much attention, that several different classifications of models can be found in the literature [10,12,15]. The taxonomy is based on the considered level of detail [12] can be found in several different problems, where macro-, mezo- and mico-scale models are being distinguished. In recent years the vast majority of research focus on micro-scale models, which distinguish individual entities and allow differences in their characteristics and behavior.

One of the basic decisions which has to be made while defining the model is the method of representing the workspace of the beings. The environment model can be discrete or continuous, it can represent 2 or 3 dimensions. The decision to discretize the workspace significantly simplifies the algorithms for model execution, which allows simulating larger groups in bigger environments. In many cases a discrete workspace model is sufficient to represent desired features of the beings and reproduce phenomena observed in real systems, like in the well-recognized Nagel-Schreckenberg freeway traffic model [8]. Many other researchers use inspirations from cellular automata in simulation of different types of beings (swarms of bees [1] or groups of pedestrians [13]) because of simplicity, elegance and sufficient expressiveness.

The common challenge identified in the vast majority of the publications in the area is the performance of the simulations. The need for increasing the complexity of models, simulating larger populations of beings and getting results faster is visible in almost all of the considered approaches. In many cases the performance issues prevent further development, and therefore a lot of effort is being put into creating parallel versions of simulation algorithms and general-purpose simulation frameworks. Scalable communication mechanisms have been added to Repast [4], dedicated frameworks are being built (like Pandora [14]). The results show that by defining models dedicated for parallel execution, scalability can be achieved, like in [3], where almost linear scaling is demonstrated up to 432 processing units with the *FLAME on HPC* framework.

Efficient parallelization of this type of computational task is non-trivial. The algorithm executed in each time step of the simulation operates on a single data structure, which represents the environment state. Parallel access to the data structure requires complex synchronization protocols, which imply significant and non-scalable overhead. Therefore, in our solution presented in [2], we focused on removing the need for accessing the remote parts of the data structure. The modeling method assumes that the information is explicitly pushed to computing units that might need it for updating their state. Implemented *Xinuk* simulation platform proves the scalability of the approach, offering linear scalability up to 3456 cores of a supercomputer. In this paper we present important extensions to the modeling methods supported by the platform, which allow representing complex environments, not possible to model with a 2D grid of cells.

3 Signal-Based Model of Complex Environments

The scalable simulation algorithm, implemented by the Xinuk framework, is capable of simulating any 2D environment while providing a flexible approach to the interpretation of the grid. Cells might represent a terrain with qualities appropriate for the simulation, actors exclusively occupying a specific place in the grid or a group of actors amassed in one location. Each cell is adjacent to its eight neighbor cells and is able to interact with any of them (e.g. move its contents or a part thereof), if the logic defined by the simulation designer allows such an action. However, the simulations that cannot be represented in simple 2D environment are difficult, if not impossible, to properly model using the framework.

While some 3D layouts can be mapped to 2D grid utilizing simplifications or other compromises (e.g. modelling upward/downward slopes or stairs as special cells that modify movement speed or behavior of the agent in such a cell), most terrain configurations would greatly benefit from—or even require—more general solution. One example of such configuration, which will be the main focus of this work, is a multi-story building with staircases located in the center. Each floor can be represented as an independent part of the 2D grid, but the connections between the floors would have to overlap with them. The standard, 2D Moore neighborhood is not sufficient to correctly represent aforementioned environments. One solution would be to generalize the method to represent 3D grid of cells, with each cell having 26 neighbors. However, in the considered class of problems, this approach would result in significant waste of memory and computational resources, as only few of the cells would be important for the simulation results.

From this problem stems the idea of the abstraction of the cell neighborhood. The proposed version of the modeling method introduces a neighborhood mapping for each direction. Each cell can connect to: top, top-right, right, bottom-right, bottom, bottom-left, left and top-left. Given a grid of dimensions $H \times W$, this mapping can be declared as in Eq. 1:

$$
\begin{aligned}
X &: \{0, 1, ..., H\}, \\
Y &: \{0, 1, ..., W\}, \\
D &: \{T, TR, R, BR, B, BL, L, TL\}, \\
N &: (X \times Y) \times D \to (X \times Y) \cup None,
\end{aligned}
\tag{1}
$$

where $X \times Y$ is a set of all possible coordinates in initial grid, D is a set of mentioned directions and N is a function mapping coordinates and direction to another set of coordinates or $None$, representing the absence of the neighbor in given direction. Likewise, the signal has been updated to be stored as a similar map containing signal strength in given direction. As a result, the signal propagation algorithm required reformulation to make use of the new representation. Firstly, the idea of the function of adjacent direction AD of a direction was necessary, which is presented in Eq. 2:

$$
AD : D \to D^2,
$$

$$
AD(d) = \begin{cases}
\{TL, TR\} & if\, d = T, \\
\{T, R\} & if\, d = TR, \\
\{TR, BR\} & if\, d = R, \\
\{B, R\} & if\, d = BR, \\
\{BL, BR\} & if\, d = B, \\
\{B, L\} & if\, d = BL, \\
\{TL, BL\} & if\, d = L, \\
\{T, L\} & if\, d = TL
\end{cases}
\tag{2}
$$

With use of this function, and assuming the function S that returns the current signal in the cell at given coordinates in given direction (Eq. 3), the new signal propagation function SP can be described as in Eq. 4:

$$S : (X \times Y) \times D \to \mathbb{R}, \tag{3}$$

$$SP : (X \times Y) \times D \to \mathbb{R},$$
$$SP((x,y),d) =$$
$$SPF \cdot \begin{cases} S(N((x,y),d),d) & \text{if } d \in \{TR, BR, BL, TL\}, \\ \\ S(N((x,y),d),d)+ \\ \sum_{ad \in AD(d)} S(N((x,y),d),ad) & \text{if } d \in \{T, R, B, L\} \end{cases} \tag{4}$$

where $SPF \in [0,1]$ is a global suppression factor of the signal.

It is worth noting that, should the need arise, this mechanism can be extended to any collection of directions, as long as the signal propagation function is updated to properly represent distribution in all directions.

Introduction of the new neighbor resolution allows seamless adaptation of the previous approach: by default, all the neighbors of the cell are the adjacent cells. The neighbor mapping function for such a case is defined as in Eq. 5, with an exception of the grid borders, where neighbors are nonexistent (*None*, as in (1)):

$$N((x,y),d) = \begin{cases} (x-1,y) & \text{if } d = T, \\ (x-1,y+1) & \text{if } d = TR, \\ (x,y+1) & \text{if } d = R, \\ (x+1,y+1) & \text{if } d = BR, \\ (x+1,y) & \text{if } d = B, \\ (x+1,y-1) & \text{if } d = BL, \\ (x,y-1) & \text{if } d = L, \\ (x-1,y-1) & \text{if } d = TL \end{cases} \tag{5}$$

Additionally, in the step of the grid creation any neighbor relation can be replaced to represent non-grid connection between cells. While the concept of remote connections appears trivial, it is critical to consider the possibility that in the process of the grid division neighbors are distributed to the separate computational nodes. Such a situation is certain to occur on the line of the division. In our previous approach, we applied buffer zones mechanism as a solution. The new concept required this part of the framework to be redesigned, to allow more flexible cell transfer. As a result, new type of cells was introduced: *remote cells*. Each represents a cell that is not present in the part of grid processed in this worker and contains information regarding:

- the identifier of the worker responsible for processing the part of grid containing target cell,
- the coordinates of target cell,
- the contents of the cell awaiting the transfer to the target cell.

Following each simulation step, contents of all remote cells are sent to their respective workers, using the logic previously associated with the buffer zones.

The modification of the framework did not introduce any alterations in the overall complexity of the simulation process. Communication between processes was not altered and utilizes the same methods as the previous synchronization of the buffer zones. Creation of the grid containing large number of non-standard neighborhood relations does introduce additional cell contents that need to be transmitted to the target worker, however it is the minimal volume of data required for the simulation to be processed.

Summarizing, as a result of all the mentioned modifications, the simulation algorithm acquired the ability to model environments that are not possible to be represented on the 2D grid. Buffer zones mechanism has been abstracted to allow more flexible communication without assuming that the communication can only occur at the grid part borders.

The scalability of the framework has been preserved, since the amount of the data sent between workers remains unchanged for the same simulation model represented in the previous and the proposed approach. It is possible to define more complex communication schemes, however, the number of communication targets remains fixed and relatively low for each worker.

4   Building Evacuation Model

Signal-based methods can be used to simulate evacuations of people from buildings. In such a signal-based model, it is enough to place a signal sources in exits, so beings will move accordingly to egress routes created by signal emitted by the sources, leaving the building. A negative signal can be used to make beings stay away from potential threats, for instance fire or smoke. In the presented model, the repelling signal was used for representing the reluctance for creating large crowds when alternative routes are possible.

In *Xinuk* framework, there are two basic cell types:

- *Obstacle* - a cell that does not let signal through,
- *EmptyCell* - an empty cell traversable by a signal and accessible by any being.

In the proposed model, *Obstacle* type of cells was used to create walls. In addition, the following cells were added to create the evacuation model:

- *PersonCell* - representing a person that was to evacuate,
- *ExitCell* - representing an exit from a building,
- *TeleportationCell* - a remote cell, described in the previous section, that was moving a being to a destination cell,
- *EvacuationDirectionCell* - source of a static signal.

A being of *PersonCell* type was able to move to any accessible adjacent cell, that is an *EmptyCell*, an *ExitCell* or a *TeleportationCell*. Movements were possible in 8 directions if there were no walls or other people in the neighborhood, as shown in Fig. 1.

In the created model, two types of signals were used:

– static signal field - a snapshot of a signal propagated in a particular number of initial iterations, where cells of *EvacuationDirectionCell* type were the signal sources. Propagated signal created egress routes that were static during the simulation,
– dynamic signal field - signal emitted by moving *PersonCell* beings.

The static signal field can be compared to a floor field described in [11] or potential field [13].

Two different models of evacuating people behaviors were implemented and tested:

– *Moving variant* - always move if a movement is possible.
– *Standing variant* - move only when the best movement is possible.

In the *moving variant*, a person's destination was calculated as follows:

1. signals in directions of neighbor cells were calculated based on dynamic and static signals by summing both signals,
2. calculated signals were sorted in a descending order,
3. from sorted destinations, a first destination was chosen that was currently available (the cell was not occupied by any other person).

In the *standing variant*, a person did not move if the best direction was not available, preventing unnatural movements to directions further away from targeted exit. Thus the 3rd step from the *moving variant* algorithm was changed as follows:

3. from sorted destinations, a first destination was chosen. If the destination was not available, the being would not move.

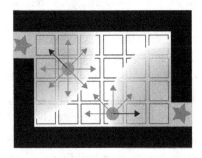

Fig. 1. Possible directions of beings. Beings marked as circles, signal sources marked as stars. Round gradient around signal sources symbolizes propagation of signal - the further from a signal source, the weaker the signal is

In a high congestion of beings trying to get to a particular location, conflicts are highly possible. One solution to this problem is to let two beings enter one cell, creating a *crowd*. A simpler solution, implemented in our model, is to check if a destination cell chosen by a being is empty both in current and in the next iteration. This way, all conflicts will be avoided.

Each floor of a building was mapped onto a 2D grid reflecting its shape and dimensions. To simulate a multi-story buildings using 2D grids, a special *TeleportationCell* was created. A being that entered the *TeleportationCell* at the end of a staircase on a floor *N*, was moved to a cell corresponding to the beginning of a staircase at floor *N − 1*.

5   Model Validation

To validate the evacuation model, a simulation of a real-life evacuation described in [7] was implemented. The evacuation drill took place in a 14-story tower connected to a 3-story low-rise structure. The highest floor of the tower was ignored in the research, thus in the implementation there is no *XII* floor in the tower, as shown on the building plan (Fig. 2). The highest floor of a low-rise structure was ignored as well and the data focused on the tower, which is a reason why the low-rise structure is shown as one floor on the building plan. Each floor was connected to two staircases, each 0,91m wide, allowing only a single line of pedestrians to be created. The staircase length was not mentioned in the paper.

Evacuation started earlier on 3 floors: *V*, *VI* and *VII*. After 3 min, there was a general alarm on the remaining floors. Pre-evacuation time curve was approximately linear (Fig. 6 in [7]). Evacuation rate was over one person per second for the first 75% of people in the building. Afterwards, the rate was slightly smaller because of discontinuation of use of one of exits (the reason not explained in the article). Results from the drill can be seen in Fig. 6.

Based on the above data, an implementation of evacuation model described in the previous section was created in *Xinuk* platform. Crucial parameters were set as follows:

- *gridSize = 264*; size of a grid. The whole grid was a square of 264 × 264 cells, each cell representing a square with side length of 0,91 m.
- *iterationsNumber = 1000*; number of iterations completed in a single simulation. 1 iteration was corresponding to 1 s.
- *evacuationDirectionInitialSignal = 1*; signal intensity that was emitted by *EvacuationDirectionCell*.
- *personInitialSignal = -0.036*; signal intensity that was emitted by *PersonCell*. In contrast to *evacuationDirectionInitialSignal*, the value was negative so people were repelling each other slightly.

In the simulation, 1 iteration corresponds to 1 s of an evacuation. Number of people on each floor was set accordingly to *Table 1* from the source paper. The

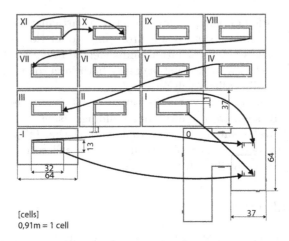

Fig. 2. Plan of the building described in [7] mapped onto a 2D grid. Arrows show transitions of people at the end of corridors - once a person reaches a corridor's end cell, it is moved to the corresponding corridor on the neighboring floor. For clarity, only a few arrows are drawn - remaining transitions were analogous.

number of people placed on low-rise structure's floor was equal to the number of all people in low-rise structure stated in the source paper.

The implemented simulation had three phases:

– *Phase 1* - 1st to 18th iteration - creating static signal field by placing signal sources in the cells corresponding to the exits and corridors' ends,
– *Phase 2* - 19th to 199th iteration - evacuation after the initial alarm, evacuating people from *V*, *VI* and *VII* floors,
– *Phase 3* - 200th to 1000th iteration - evacuation after the general alarm, evacuating people from the whole building.

To achieve linear pre-evacuation times, all of the people were not placed on a grid in the first iteration of the 2nd and 3rd phase of the simulation, but they were placed on a grid linearly – one person per iteration on each floor (relying on the Fig. 6 in [7]).

Validation of the model was based on visual analysis of simulation results and on the *evacuation curve* metric which is a dependency between the number of people evacuated from the building and time. Empirical data is shown in Fig. 6.

During the observations, two anomalies were visible:

1. **Problem 1**: Crowds of people next to the upper and lower entrance of staircases on each floor were consistently different—*lower* entrance seemed to generate larger crowds—and an evacuation using *upper* stairs tended to take longer (Fig. 3),
2. **Problem 2**: People in corridors that could not move forward but could go back, were moving back and forth waiting to go further or were choosing to go back to staircase entrances.

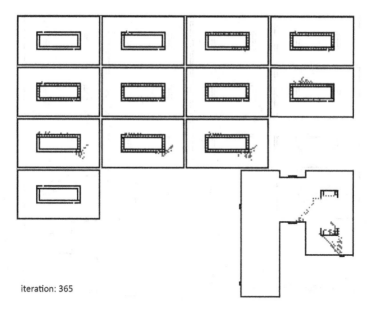

iteration: 365

Fig. 3. Arrangement of people in the building using a sequential order of cells updating (with moving variant). Iteration: 365. Link to the full video: https://bit.ly/2QO8lHA

Problem 1 was a result of a sequential order of updating cells. A person that was trying to enter an upper staircase was above the people in the staircase, thus the person was updated before the people in the corridor. This way, the person's movement was favored and congestion was visible in the corridors and there were no crowds close to the upper entrances to the upper staircases. Similarly, trying to enter a lower staircase, the movement of people already in the staircase was favored. A person trying to enter the staircase was updated after people in the staircase managed to occupy all empty spaces, preventing the person from entering the staircase. Figure 5 shows both of the situations. A simple solution to this problem was to update locations of people in a random order. Figure 4 shows the results of such an approach. Crowds are distributed equally close to both of staircase entrances and evacuation in both staircases progresses similarly.

Problem 2 was a result of the decision making algorithm. According to the algorithm, people would make any movement if it was possible, even a bad one, rather than stay in place. A solution was to use the *Standing variant*: choose only the most attractive destination or not to move at all.

The visual analysis of the final model behavior is satisfactory. The formation of crowds and selection of the exits resembled the expected phenomena. The people were eager to follow the shortest path towards the exits, while avoiding excessive crowding which might lead to trampling.

The four combinations of the model update algorithm, *moving/standing variant* and *sequential/random variant*, were executed 30 times. We used the

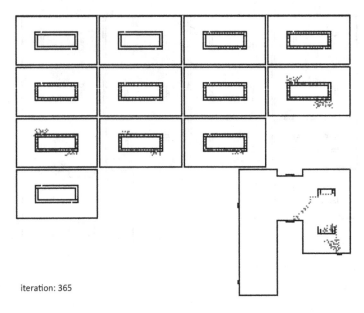

iteration: 365

Fig. 4. Arrangement of people in the building using a random order of cells updating (with moving variant). Iteration: 365. Link to the full video: https://bit.ly/2MX6SO8

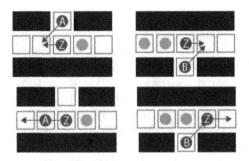

Fig. 5. Favoring movements of people that are upper in a grid in sequential way of updating cells. On the left - a situation when person A is trying to enter a staircase that is below. On the right - a situation when person B is trying to enter a staircase that is above

Prometheus supercomputer (part of PL-Grid infrastructure[1]), which allowed completion of all computation within several minutes. The resulting chart (Fig. 7) shows that this particular metrics is not influenced significantly by the selected variant. An average rate of people leaving the building is a little over 1 person per second, which matches the data on 6. After an evacuation of 75% of people, the rate did not change, as in the simulation people continued using two exits. On a source chart (Fig. 6) people have already reached exits in first seconds while

[1] http://www.plgrid.pl/en.

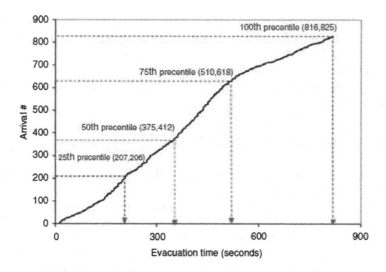

Fig. 6. Number of people evacuated in time, empirical results from [7]

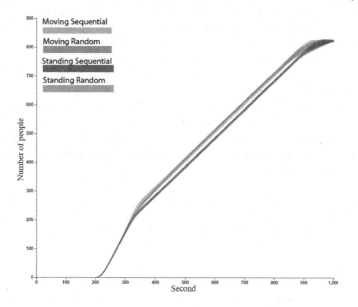

Fig. 7. Evacuation curves for four tested model update algorithms. The ribbons represent the median of 30 runs.

on the resulting chart (Fig. 7) the first person left the building after 200th second of evacuation. According to [7], people on the floors other than *V, VI* and *VII* did not evacuate before the general alarm. Thus, it is not clear why the chart shown in Fig. 6 suggests that some people have left the building in such a short time overcoming at least 5 floor long stairs.

The results of the experiments are not perfectly matching the empirical data from [7]. Nonetheless, taking into consideration the ambiguity, contradictions or lack of some details of the described evacuation, we conclude that the resulting model yielded realistic and consistent outcomes. In the context of validation of the presented method as a building block of similar environments, the results are satisfactory.

6 Conclusions and Further Work

In this paper we proposed an extension to the signal propagation modeling method and the Xinuk framework, addressing the limitations of the existing approach in the context of complex 3D environments. We introduced the alternative to the standard, grid-based neighborhood as a means to generalizing the idea of buffer zones present in the framework. The new mechanism greatly increased the flexibility of the method, while avoiding the massive growth of complexity that would result from switching to full 3D grid environment. At the same time, the scalability of the original solution remained unhindered, as the alteration did not increase the amount of data exchanged between computational nodes in the synchronization step. The evacuation scenario used as a demonstration of the new capabilities of the method provided results confirming that such a scenario can be accurately represented in the resulting framework.

As a followup of this work, we intend to further explore the possibilities arising from the flexible neighborhood declaration, especially the environments that would benefit from additional directions, e.g. layers of an anthill utilizing the vertical directions. Research dedicated to further reduction in the communication might yield interesting results as well, e.g. investigating a trade-off between accuracy and performance of the system while performing the synchronization after several simulation steps.

Acknowledgments. The research presented in this paper was supported by the Polish Ministry of Science and Higher Education funds assigned to AGH University of Science and Technology. The authors acknowledge using the PL-Grid infrastructure.

References

1. Becher, M.A., Grimm, V., Thorbek, P., Horn, J., Kennedy, P.J., Osborne, J.L.: BEEHAVE: a systems model of honeybee colony dynamics and foraging to explore multifactorial causes of colony failure. J. Appl. Ecol. **51**(2), 470–482 (2014)
2. Bujas, J., Dworak, D., Turek, W., Byrski, A.: High-performance computing framework with desynchronized information propagation for large-scale simulations. J. Comput. Sci. **32**, 70–86 (2019)
3. Coakley, S., Gheorghe, M., Holcombe, M., Chin, S., Worth, D., Greenough, C.: Exploitation of high performance computing in the flame agent-based simulation framework. In: 2012 IEEE 14th International Conference on High Performance Computing and Communication & 2012 IEEE 9th International Conference on Embedded Software and Systems, pp. 538–545. IEEE (2012)

4. Collier, N., North, M.: Repast HPC: a platform for large-scale agent-based model-ing. In: Large-Scale Computing, pp. 81–109 (2012)
5. Engelmann, C., Geist, A.: Super-scalable algorithms for computing on 100,000 pro-cessors. In: Sunderam, V.S., van Albada, G.D., Sloot, P.M.A., Dongarra, J.J. (eds.) ICCS 2005. LNCS, vol. 3514, pp. 313–321. Springer, Heidelberg (2005). https:// doi.org/10.1007/11428831_39
6. Gerlough, D.L.: Simulation of freeway traffic on a general-purpose discrete variable computer. Ph.D. thesis, University of California, Los Angeles (1955)
7. Gwynne, S., Boswell, D.: Pre-evacuation data collected from a mid-rise evacuation exercise. J. Fire. Prot. Eng. **19**(1), 5–29 (2009)
8. Nagel, K., Schreckenberg, M.: A cellular automaton model for freeway traffic. Jour-nal de physique I **2**(12), 2221–2229 (1992)
9. Renc, P., et al.: HPC large-scale pedestrian simulation based on proxemics rules. In: Wyrzykowski, R., Deelman, E., Dongarra, J., Karczewski, K. (eds.) PPAM 2019. LNCS, vol. 12044, pp. 489–499. Springer, Cham (2020). https://doi.org/10. 1007/978-3-030-43222-5_43
10. Sargent, R.G.: Verification and validation of simulation models. In: Proceedings of the 2010 Winter Simulation Conference, pp. 166–183. IEEE (2010)
11. Varasa, A., et al.: Cellular automaton model for evacuation process with obstacles. Physica A Stat. Mech. Appl. (PHYSICA A) **382**, 631–642 (2007)
12. van Wageningen-Kessels, F., van Lint, H., Vuik, K., Hoogendoorn, S.: Genealogy of traffic flow models. EURO J. Transp. Logistics **4**(4), 445–473 (2014). https:// doi.org/10.1007/s13676-014-0045-5
13. Wąs, J., Lubaś, R.: Towards realistic and effective agent-based models of crowd dynamics. Neurocomputing **146**, 199–209 (2014)
14. Wittek, P., Rubio-Campillo, X.: Scalable agent-based modelling with cloud HPC resources for social simulations. In: 4th IEEE International Conference on Cloud Computing Technology and Science Proceedings, pp. 355–362. IEEE (2012)
15. Zhan, B., Monekosso, D.N., Remagnino, P., Velastin, S.A., Xu, L.Q.: Crowd anal-ysis: a survey. Mach. Vis. Appl. **19**(5–6), 345–357 (2008)

Design of Loss Functions for Solving Inverse Problems Using Deep Learning

Jon Ander Rivera[1,2(✉)], David Pardo[1,2,3], and Elisabete Alberdi[1]

[1] University of the Basque Country (UPV/EHU), Leioa, Spain
riverajonander@gmail.com
[2] BCAM-Basque Center for Applied Mathematics, Bilbao, Spain
[3] IKERBASQUE, Basque Foundation for Science, Bilbao, Spain

Abstract. Solving inverse problems is a crucial task in several applications that strongly affect our daily lives, including multiple engineering fields, military operations, and/or energy production. There exist different methods for solving inverse problems, including gradient based methods, statistics based methods, and Deep Learning (DL) methods. In this work, we focus on the latest. Specifically, we study the design of proper loss functions for dealing with inverse problems using DL. To do this, we introduce a simple benchmark problem with known analytical solution. Then, we propose multiple loss functions and compare their performance when applied to our benchmark example problem. In addition, we analyze how to improve the approximation of the forward function by: (a) considering a Hermite-type interpolation loss function, and (b) reducing the number of samples for the forward training in the Encoder-Decoder method. Results indicate that a correct design of the loss function is crucial to obtain accurate inversion results.

Keywords: Deep learning · Inverse problems · Neural network

1 Introduction

Solving inverse problems [17] is of paramount importance to our society. It is essential in, among others, most areas of engineering (see, e.g., [3,5]), health (see, e.g. [1]), military operations (see, e.g., [4]) and energy production (see, e.g. [11]). In multiple applications, it is necessary to perform this inversion in real-time. This is the case, for example, of geosteering operations for enhanced hydrocarbon extraction [2,10].

Traditional methods for solving inverse problems include gradient based methods [13,14] and statistics based methods (e.g., Bayesian methods [16]). The main limitation of these kind of methods is that they lack an explicit construction of the pseudo-inverse operator. Instead, they only *evaluate* the inverse function for a given set of measurements. Thus, for each set of measurements, we need to perform a new inversion process. This may be time consuming.

Deep Learning (DL) seems to be a proper alternative to overcome the aforementioned problem. With DL methods, we explicitly build the pseudo-inverse

© Springer Nature Switzerland AG 2020
V. V. Krzhizhanovskaya et al. (Eds.): ICCS 2020, LNCS 12139, pp. 158–171, 2020.
https://doi.org/10.1007/978-3-030-50420-5_12

operator rather than only evaluating it. Recently, the interest on performing inversion using DL techniques has grown exponentially (see, e.g., [9,15,18,19]). However, the design of these methods is still somehow *ad hoc* and it is often difficult to encounter a comprehensive road map to construct robust Deep Neural Networks (DNNs) for solving inverse problems.

One major problem when designing DNNs is the error control. Several factors may lead to deficient results. Such factors include: poor loss function design, inadequate architecture, lack of convergence of the optimizer employed for training, and unsatisfactory database selection. Moreover, it is sometimes elusive to identify the specific cause of poor results. Even more, it is often difficult to asses the quality of the results and, in particular, determine if they can be improved.

In this work, we take a simple but enlightening approach to elucidate and design certain components of a DL algorithm when solving inverse problems. Our approach consists of selecting a simple inverse benchmark example with known analytical solution. By doing so, we are able to evaluate and quantify the effect of different DL design considerations on the inversion results. Specifically, we focus on analyzing a proper selection of the loss function and how it affects to the results. While more complex problems may face additional difficulties, those observed with the considered simple example are common to all inverse problems.

The remainder of this article is as follows. Section 2 describes our simple model inverse benchmark problem. Section 3 introduces several possible loss functions. Section 4 shows numerical results. Finally, Sect. 5 summarizes the main findings.

2   Simple Inverse Benchmark Problem

We consider a benchmark problem with known analytical solution. Let $\mathcal{F}$ be the forward function and $\mathcal{F}^\dagger$ the pseudo-inverse operator. We want our benchmark problem to have more than one solution since this is one of the typical features exhibited by inverse problems. For that, we need $\mathcal{F}$ to be non-injective. We select the non-injective function $y = \mathcal{F}(x) = x^2$, whose pseudo-inverse has two possible solutions: $x = \mathcal{F}^\dagger(y) = \pm\sqrt{y}$. (See Fig. 1). The objective is to design a NN that approximates one of the solutions of the inverse problem.

2.1   Database and Data Rescaling

We consider the domain $\Omega = [-33, 33]$. In there, we select a set of 1000 equidistant numbers. The corresponding dataset of input-output pairs $\{(x_i, \mathcal{F}(x_i))\}_{i=1}^{1000}$ is computed analytically.

In some cases, we perform a change of coordinates in our output dataset. Let's name $\mathcal{R}$ the linear mapping that goes from the output of the original dataset into the interval $[0,1]$. Instead of approximating function $\mathcal{F}$, our NN will approximate function $\mathcal{F}^\mathcal{R}$ given by

$$\mathcal{F}^\mathcal{R} := \mathcal{R} \circ \mathcal{F}. \tag{1}$$

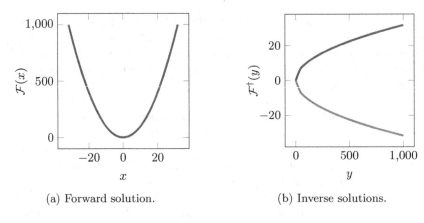

(a) Forward solution.   (b) Inverse solutions.

Fig. 1. Benchmark problem.

In the cases we perform no rescaling, we select $\mathcal{R} = \mathcal{I}$, where $\mathcal{I}$ is the identity mapping.

3   Loss Functions

We consider different loss functions. The objective here is to discern between adequate and poor loss functions for solving the proposed inverse benchmark problem.

We denote as $\mathcal{F}_\varphi$ and $\mathcal{F}_\theta^\dagger$ the NN approximations of the forward function and the pseudo-inverse operator, respectively. Weights φ and θ are the parameters to be trained (optimized) in the NN. Each value within the set of weights is a real number.

In a NN, we try to find the weights φ^* and θ^* that minimize a given loss function L. We express our problem mathematically as

$$(\varphi^*, \theta^*) = \arg \min_{\varphi, \theta} L(\varphi, \theta). \tag{2}$$

Loss Based on the Missfit of the Inverse Data: We first consider the traditional loss function:

$$L_1(\theta) = \left\| \mathcal{F}_\theta^{\dagger \mathcal{R}}(y) - x \right\|. \tag{3}$$

Theorem 1. *Solution of minimization problem* (2) *with the loss function given by Eq.* (3) *has analytical solution for our benchmark problem in both the l_1 norm*

$$\|x\|_1 = \sum_{i=1}^{n} |x_i| \tag{4}$$

and the l_2 norm

$$||x||_2 = \sum_{i=1}^{n}(x_i)^2. \tag{5}$$

These solutions are such that:

- *For l_1, $\mathcal{F}_{\theta_*}^{\dagger}{}^{\mathcal{R}}(y) = x$, where x is any value in the interval $[-\sqrt{y}, \sqrt{y}]$.*
- *For l_2, $\mathcal{F}_{\theta_*}^{\dagger}{}^{\mathcal{R}}(y) = 0$.*

Proof. We first focus on norm $||\cdot||_1$. We minimize the loss function:

$$\sum_{i \in I}|\mathcal{F}_{\theta}^{\dagger}{}^{\mathcal{R}}(y_i) - x_i|, \tag{6}$$

where $I = \{1, ..., N\}$ denotes the training dataset. For the exact pseudo-inverse operator $\mathcal{F}_{\theta}^{\dagger}{}^{\mathcal{R}}$, we can express each addend of (6) as follows:

$$|\mathcal{F}_{\theta}^{\dagger}{}^{\mathcal{R}}(y_i) - x_i| = \begin{cases} -2x_i, & \text{if } x_i \leq -\sqrt{y_i}, \\ 0, & \text{if } -\sqrt{y_i} \leq x_i \leq \sqrt{y_i}, \\ 2x_i, & \text{if } x_i \geq \sqrt{y_i}. \end{cases} \tag{7}$$

Taking the derivative of Eq. (6) with respect to x_i, we see in view of Eq. (7) that the loss function for the exact solution attains its minimum at every point $x_i \in [-\sqrt{y_i}, \sqrt{y_i}]$.

In the case of norm $||\cdot||_2$, for each value of y we want to minimize:

$$\sum_{i \in I}\left(\mathcal{F}_{\theta}^{\dagger}{}^{\mathcal{R}}(y_i) - x_i\right)^2. \tag{8}$$

Again, for the exact pseudo-inverse operator $\mathcal{F}_{\theta}^{\dagger}{}^{\mathcal{R}}$, we can express each addend of Eq. (8) as:

$$\left(-\sqrt{y_i} - x_i\right)^2 + \left(\sqrt{y_i} - x_i\right)^2. \tag{9}$$

Taking the derivative of Eq. (8) with respect to x_i and equaling it to zero, we obtain:

$$2x_i + \sqrt{y_i} - \sqrt{y_i} = 0 \Rightarrow x_i = \frac{\sqrt{y_i} - \sqrt{y_i}}{2} = 0. \tag{10}$$

Thus, the function is minimized when the approximated value is 0. $\square$

Observation: Problem of Theorem 1 has infinite solutions in the l_1 norm. In the l_2 norm, the solution is unique; however, it differs from the two desired exact inverse solutions.

Loss Based on the Missfit of the Effect of the Inverse Data: As seen with the previous loss function, it is inadequate to look at the misfit in the inverted space. Rather, it is desirable to search for an inverse solution such that after applying the forward operator, we recover our original input. Thus, we consider the following modified loss function, where $\mathcal{F}^{\mathcal{R}_1}$ corresponds to the analytic forward function:

$$L_2(\theta) = \left\| \mathcal{F}^{\mathcal{R}_1}(\mathcal{F}_\theta^{\mathcal{R}_2\dagger}(y)) - y \right\|. \tag{11}$$

Unfortunately, computation of $\mathcal{F}^{\mathcal{R}_1}$ required in L_2 involves either (a) implementing $\mathcal{F}^{\mathcal{R}_1}$ in a GPU, which may be challenging in more complex examples, or (b) calling $\mathcal{F}^{\mathcal{R}_1}$ as a CPU function multiple times during the training process. Both options may considerably slow down the training process up to the point of making it impractical.

Encoder-Decoder Loss: To overcome the computational problems associated with Eq. (11), we introduce an additional NN, named $\mathcal{F}_\varphi^{\mathcal{R}_1}$, to approximate the forward function. Then, we propose the following loss function:

$$L_3(\varphi, \theta) = \left\| \mathcal{F}_\varphi^{\mathcal{R}_1}(x) - y \right\| + \left\| \mathcal{F}_\varphi^{\mathcal{R}_1}(\mathcal{F}_\theta^{\dagger\mathcal{R}_2}(y)) - y \right\|. \tag{12}$$

Two NNs of this type that are being simultaneously trained are often referred to as Encoder-Decoder [6,12].

Two-Steps Loss: It is also possible to train $\mathcal{F}_\varphi^{\mathcal{R}_1}$ and $\mathcal{F}_\theta^{\dagger\mathcal{R}_2}$ separately. By doing so, we diminish the training cost. At the same time, it allows us to separate the analysis of both NNs, which may simplify the detection of specific errors in one of the networks. Our loss functions are:

$$L_{4.1}(\varphi) = \left\| \mathcal{F}_\varphi^{\mathcal{R}_1}(x) - y \right\| \tag{13}$$

and

$$L_{4.2}(\theta) = \left\| \mathcal{F}_{\varphi^*}^{\mathcal{R}_1}(\mathcal{F}_\theta^{\dagger\mathcal{R}_2}(y)) - y \right\|. \tag{14}$$

We first train $\mathcal{F}_\varphi^{\mathcal{R}_1}$ using $L_{4.1}$. Once $\mathcal{F}_\varphi^{\mathcal{R}_1}$ is fixed (with weights φ^*), we train $\mathcal{F}_\theta^{\dagger\mathcal{R}_2}$ using $L_{4.2}$.

4   Numerical Results

We consider two different NNs. The one approximating the forward function has 5 fully connected layers [8] with ReLU activation function [7]. The one approximating the inverse operator has 11 fully connected layers with ReLU activation function. ReLU activation function is defined as

$$\mathrm{ReLu}(x) = \begin{cases} x, \ x \geq 0 \\ 0, \ x < 0. \end{cases} \tag{15}$$

These NN architectures are "overkilling" for approximating the simple benchmark problem studied in this work. Moreover, we also obtain results for different NN architectures, leading to identical conclusions that we omit here for brevity.

4.1   Loss Function Analysis

Loss Based on the Missfit of the Inverse Data: We produce two models using norms l_1 and l_2, respectively. Figure 2 shows the expected disappointing results (see Theorem 1). The approximated NN values (green circles) are far from the true solution (blue line). From an engineering point of view, the recovered solution is worthless. The problem resides on the selection of the loss function.

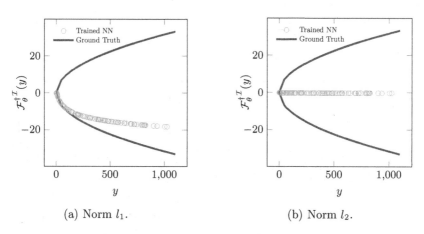

(a) Norm l_1.   (b) Norm l_2.

Fig. 2. Predicted ($\mathcal{F}_\theta^{\mathcal{I}\dagger}$, green circles) vs exact ($\mathcal{F}^{\mathcal{I}\dagger}$, blue line) inverse solutions evaluated over the testing dataset. (Color figure online)

Loss Based on the Missfit of the Effect of the Inverse Data: Figure 3 shows the real values of y (ground truth) vs their predicted pseudo-inverse values. The closer the predicted values are to the blue line, the better the result from the NN. We now observe an excellent match between the exact and approximated

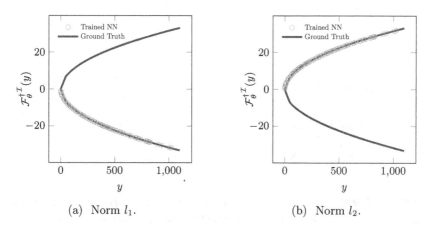

(a) Norm l_1.   (b) Norm l_2.

Fig. 3. Solution of the pseudo-inverse operator approximated by the NN. (Color figure online)

solutions. However, as mention in Sect. 3, this loss function entails essential limitations when considering complex problems.

Encoder-Decoder Loss: Figure 4 shows the results for norm l_1 and Fig. 5 for norm l_2. We again recover excellent results, without the limitations provided by loss function L_2. Coincidentally, different norms recover different solution branches of the inverse problem. Note that in this problem, it is possible to prove that the probability of recovering either of the solution branches is identical.

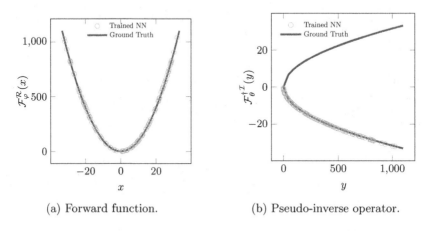

(a) Forward function.   (b) Pseudo-inverse operator.

Fig. 4. Exact vs NN solutions using loss function L_3 and norm l_1.

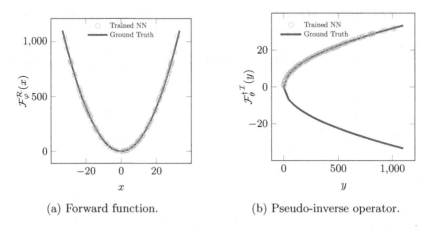

(a) Forward function.   (b) Pseudo-inverse operator.

Fig. 5. Exact vs NN solutions using loss function L_3 and norm l_2.

Two-Steps Loss: Figures 6 and 7 show the results for norms l_1 and l_2, respectively. The approximations of forward function and pseudo-inverse operator are accurate in both cases.

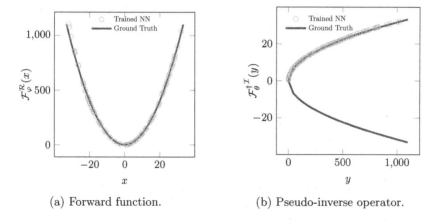

(a) Forward function.   (b) Pseudo-inverse operator.

Fig. 6. Exact vs NN solutions using loss functions $L_{4.1}$ and $L_{4.2}$ and norm l_1.

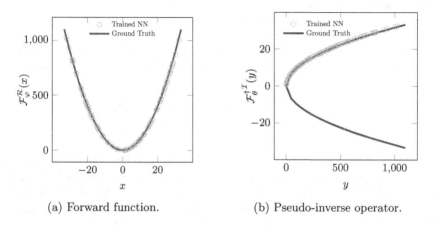

(a) Forward function.   (b) Pseudo-inverse operator.

Fig. 7. Exact vs NN solutions using loss functions $L_{4.1}$ and $L_{4.2}$ and norm l_2.

4.2   Hermite-Type Loss Functions

We now consider the two-steps loss function and we focus only on the forward function approximation given by Eq. (13). This is frequently the most time consuming part when solving an inverse problem with NNs. In this section, we analyze different strategies to work with a reduced dataset, which entails a dramatic reduction of the computational cost. We consider a dataset of three input-outputs pairs $(x, y) = \{(-33, 1089), (1, 1), (33, 1089)\}$.

Figure 8 shows the results for norms l_1 and l_2. Training data points are accurately approximated. Other points are poorly approximated.

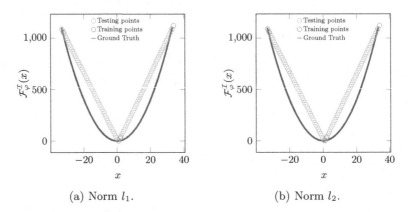

(a) Norm l_1.   (b) Norm l_2.

Fig. 8. Results of the NN that approximates the forward function. Red points correspond to the evaluation of the training dataset x and to $x+1$. Green points correspond to the evaluation of a testing dataset. (Color figure online)

To improve the approximation, we introduce another term to the loss function. We force the NN to approximate the derivatives at each training point. This new loss is:

$$\left\| \mathcal{F}_\varphi^{\mathcal{I}}(x) - y \right\| + \left\| \frac{\mathcal{F}_\varphi^{\mathcal{I}}(x + \epsilon) - \mathcal{F}_\varphi^{\mathcal{I}}(x)}{\epsilon} - \frac{\partial \mathcal{F}^{\mathcal{I}}(x)}{\partial x} \right\|. \tag{16}$$

From a numerical point of view, the term that approximates the first derivatives could be very useful. If we think about x as a parameter of a Partial Differential Equation (PDE), we can efficiently evaluate derivatives via the adjoint problem.

Figure 9 shows the results when we use norms l_1 and l_2 for the training. For this benchmark problem, we select $\epsilon = 1$. Thus, to approximate derivatives, we evaluate the NN at the points $x + 1$.

We observe that points nearby the training points are better approximated via Hermite interpolation, as expected. However, the entire approximation still lacks accuracy and exhibits undesired artifacts due to an insufficient number of training points. Thus, while the use of Hermite interpolation may be highly beneficial, especially in the context of certain PDE problems or when the derivatives are easily accessible, there is still a need to have a sufficiently dense database of sampling points. Figure 10 shows the evolution of the terms composing the loss function.

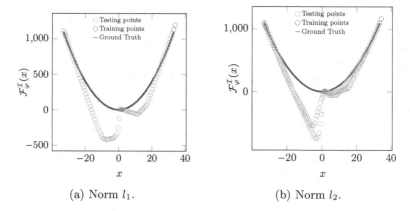

(a) Norm l_1.　　　　　　　　　　　(b) Norm l_2.

Fig. 9. Results of the NN that approximates the forward function. Red points correspond to the evaluation of the training dataset x and to $x+1$. Green points correspond to the evaluation of a testing dataset. (Color figure online)

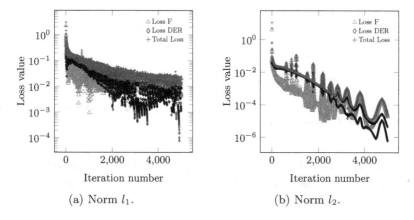

(a) Norm l_1.　　　　　　　　　　　(b) Norm l_2.

Fig. 10. Evolution of the loss value when we train the NN that approximates $\mathcal{F}_\varphi^\mathcal{I}$ using as loss Eq. (16). "Loss F" corresponds to the loss of the first term of Eq. (16). "Loss DER" corresponds to the loss of the second term of Eq. (16). "Total Loss" corresponds to the total value of Eq.(16).

4.3　Loss Function with a Reduced Number of Samples for the Forward Training

We now consider an Encoder-Decoder loss function, as described in Eq. (12). The objective is to minimize the number of samples employed to approximate the forward function since producing such database is often the most time-consuming part in a large class of inverse problems governed by PDEs.

We employ a dataset of three input-output pairs $\{(-33, 1089), (1, 1), (33, 1089)\}$ for the first term of Eq. (12) and a dataset of 1000 values of y obtained with an equidistant distribution on the interval $[0, 1089]$ for the second term of Eq. (12).

Figure 11 shows the results of the NNs trained with norm l_1. Results are disappointing. The forward function is far from the blue line (real forward function), specially nearby zero. The forward function leaves excessive freedom for the training of the inverse function. This allows the inverse function to be poorly approximated (with respect the to real inverse function).

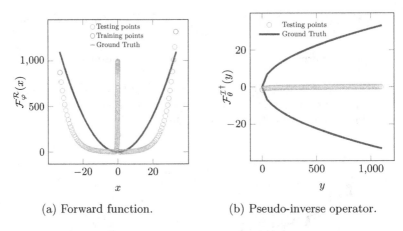

(a) Forward function.   (b) Pseudo-inverse operator.

Fig. 11. Exact vs NN solutions using loss function L_3, norm l_1, and a reduced number of samples for the forward evaluation.

In order to improve the results, we train the NNs adding a regularization term to Eq. (12). We add the following regularization term maximizing smoothness on $\mathcal{F}_\varphi^\mathcal{R}$:

$$L_{3.1}(\varphi, \theta) = ||\mathcal{F}_\varphi^\mathcal{R}(x) - y|| + ||\mathcal{F}_\varphi^\mathcal{R}(\mathcal{F}_\theta^{\mathcal{I}\dagger}(y)) - y|| + \left|\left|\frac{\mathcal{F}_\varphi^\mathcal{R}(x + \epsilon) - \mathcal{F}_\varphi^\mathcal{R}(x)}{\epsilon}\right|\right|. \quad (17)$$

We evaluate this regularization term over a dataset of 1000 samples obtained with an equidistant distribution on the interval $[-33, 33]$ and we select $\epsilon = 1$.

Figure 12 shows the results of the NN. Now, the forward function is better approximated around zero. Unfortunately, the approximation is still inaccurate, indicating the need for additional points on the approximation. Figure 13 shows the evolution of the terms composing the loss function. The loss values associated with the first and the second terms are minimized. The loss corresponding to the regularization term remains as the largest one.

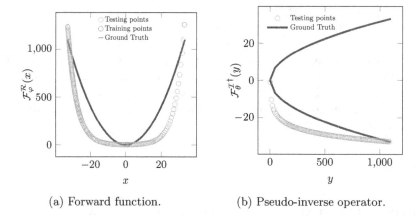

(a) Forward function.   (b) Pseudo-inverse operator.

Fig. 12. Exact vs NN solutions using loss function L_3, norm l_1, and a reduced number of samples for the forward evaluation.

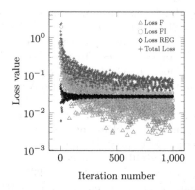

Iteration number

Fig. 13. Evolution of the loss value for Encoder-Decoder method trained with loss function $L_{3.1}$ and norm l_1. "Loss F" corresponds to the loss of the first term of Eq. (17). "Loss FI" corresponds to the loss of the second term of Eq. (17). "Loss REG" corresponds to the loss of the third term of Eq. (17). "Total Loss" corresponds to the total value of Eq. (17).

5 Conclusions

We analyze different loss functions for solving inverse problems. We demonstrate via a simple numerical benchmark problem that some traditional loss functions are inadequate. Moreover, we propose the use of an Encoder-Decoder loss function, which can also be divided into two loss functions with a one-way coupling. This enables to decompose the original DL problem into two simpler problems.

In addition, we propose to add a Hermite-type interpolation to the loss function when needed. This may be especially useful in problems governed by PDEs where the derivative is easily accessible via the adjoint operator. Results indicate that Hermite interpolation provides enhanced accuracy at the training points and in the surroundings. However, we still need a sufficient density of points in our database to obtain acceptable results.

Finally, we evaluate the performance of the Encoder-Decoder loss function with a reduced number of samples for the forward function approximation. We observe that the forward function leaves excessive freedom for the training of the inverse function. To partially alleviate that problem, we incorporate a regularization term. The corresponding results improve, but they still show the need for additional training samples.

References

1. Albanese, R.A.: Wave propagation inverse problems in medicine and environmental health. In: Chavent, G., Sacks, P., Papanicolaou, G., Symes, W.W. (eds.) Inverse Problems in Wave Propagation, pp. 1–11. Springer, New York (1997). https://doi.org/10.1007/978-1-4612-1878-4_1
2. Beer, R., et al.: Geosteering and/or reservoir characterization the prowess of new generation LWD tools. 51st Annual Logging Symposium Society of Petrophysicists and Well-Log Analysts (SPWLA) (2010)
3. Bonnet, M., Constantinescu, A.: Inverse problems in elasticity. Inverse Prob. **21**(2), R1–R50 (2005). https://doi.org/10.1088/0266-5611/21/2/r01
4. Broquetas, A., Palau, J., Jofre, L., Cardama, A.: Spherical wave near-field imaging and radar cross-section measurement. IEEE Trans. Antennas Propag. **46**(5), 730–735 (1998)
5. Burczyński, T., Beluch, W., Dugosz, A., Orantek, P., Nowakowski, M.: Evolutionary methods in inverse problems of engineering mechanics. In: Inverse Problems in Engineering Mechanics II, pp. 553–562. Elsevier Science Ltd., Oxford (2000). https://doi.org/10.1016/B978-008043693-7/50131-8. http://www.sciencedirect.com/science/article/pii/B9780080436937501318
6. Cho, K., et al.: Learning phrase representations using RNN encoder-decoder for statistical machine translation (2014)
7. Hara, K., Saito, D., Shouno, H.: Analysis of function of rectified linear unit used in deep learning. In: 2015 International Joint Conference on Neural Networks (IJCNN), pp. 1–8 (2015)
8. Huang, G., Liu, Z., van der Maaten, L., Weinberger, K.Q.: Densely connected convolutional networks. In: The IEEE Conference on Computer Vision and Pattern Recognition (CVPR), July 2017
9. Jin, Y., Wu, X., Chen, J., Huang, Y.: Using a physics-driven deep neural network to solve inverse problems for LWD azimuthal resistivity measurements, pp. 1–13, June 2019
10. Li, Q., Omeragic, D., Chou, L., Yang, L., Duong, K.: New directional electromagnetic tool for proactive geosteering and accurate formation evaluation while drilling (2005)
11. Liu, G., Zhou, B., Liao, S.: Inverting methods for thermal reservoir evaluation of enhanced geothermal system. Renew. Sustain. Energy Rev. **82**, 471–476 (2018). https://doi.org/10.1016/j.rser.2017.09.065. http://www.sciencedirect.com/science/article/pii/S1364032117313175
12. Mao, X.J., Shen, C., Yang, Y.B.: Image restoration using very deep convolutional encoder-decoder networks with symmetric skip connections (2016)

13. Neto, A.S., Soeiro, F.: Solution of implicitly formulated inverse heat transfer problems with hybrid methods. In: Computational Fluid and Solid Mechanics 2003, pp. 2369–2372. Elsevier Science Ltd., Oxford (2003). https://doi.org/10.1016/B978-008044046-0.50582-0

14. Oberai, A.A., Gokhale, N.H., Feijóo, G.R.F.: Solution of inverse problems in elasticity imaging using the adjoint method. Inverse Prob. **19**(2), 297–313 (2003). https://doi.org/10.1088/0266-5611/19/2/304

15. Puzyrev, V.: Deep learning electromagnetic inversion with convolutional neural networks. Geophys. J. Int. **218**, 817–832 (2019). https://doi.org/10.1093/gji/ggz204

16. Stuart, A.M.: Inverse problems: a Bayesian perspective. Acta Numerica **19**, 451–559 (2010). https://doi.org/10.1017/S0962492910000061

17. Tarantola, A.: Inverse Problem Theory and Methods for Model Parameter Estimation. Society for Industrial and Applied Mathematics, USA (2004)

18. Xu, Y., et al.: Schlumberger: Borehole resistivity measurement modeling using machine-learning techniques (2018)

19. Zhu, G., Gao, M., Kong, F., Li, K.: A fast inversion of induction logging data in anisotropic formation based on deep learning. IEEE Geosci. Remote Sens. Lett., 1–5 (2020). https://doi.org/10.1109/LGRS.2019.2961374

Asynchronous Actor-Based Approach to Multiobjective Hierarchical Strategy

Michał Idzik[(✉)] , Aleksander Byrski , Wojciech Turek,
and Marek Kisiel-Dorohinicki

AGH University of Science and Technology, Kraków, Poland
{miidzik,olekb,wojciech.turek,doroh}@agh.edu.pl

Abstract. Hierarchical Genetic Strategy (HGS) is a general-purpose optimization metaheuristic based on multi-deme evolutionary-like optimization, while demes are parts of adaptive dynamically changing tree. The paper focuses on adaptation of the classic HGS algorithm for multi-criteria optimization problems, coupling the HGS with Particle Swarm Optimization demes. The main contribution of the paper is showing the efficacy and efficiency of the actor-based implementation of this metaheuristic algorithm.

Keywords: Hierarchic genetic search · Metaheuristics · Actor model

1 Introduction

Keeping balance between exploration and exploitation is a crucial task of an expert, trying to solve complex problems with metaheuristics. A number of different approaches were made (like automatic adaptation of variation operators in evolution strategies [1] or multi-deme evolution model [2]).

About 20 years ago an interesting metaheuristics has been proposed by Schaefer and Kolodziej, namely Hierarchic Genetic Search [12], managing a whole tree of demes which can be dynamically constructed or removed, depending on the quality of their findings. Recently this metaheuristic was adapted to solving multi-criteria optimization problems by Idzik et al. [9], and coupled with Particle Swarm Optimization as working nodes (demes) instead of classic genetic algorithms. This new algorithm has proven to be efficient in many multi-criteria benchmark problems, however as it is quite normal in the case of metaheuristics, it is cursed with high computational complexity.

Nowadays HPC-related solutions (supercomputers, hybrid infrastructures etc.) are very common and easy to access, especially for scientists[1]. High-level

---

The research presented in this paper was financed by Polish National Science Centre PRELUDIUM project no. 2017/25/N/ST6/02841.

[1] E.g. Academic Computing Centre "Cyfronet" of AGH University of Science and Technology makes possible to utilize supercomputing facilities to all scientists working in Poland or cooperating with Polish scientists, free of charge.

V. V. Krzhizhanovskaya et al. (Eds.): ICCS 2020, LNCS 12139, pp. 172–185, 2020.
https://doi.org/10.1007/978-3-030-50420-5_13

programming languages, such as Scala/Akka or Erlang make the use of such facilities very easy, and significantly decrease steepness of the learning curve for people who want to try such solutions.

This paper deals with a concept of actor-based design and implementation of Multiobjective HGS aiming at showing that such an approach yields a very efficient and efficacious computing system. The next section discusses existing actor-based implementations of metaheuristics, then the basics of HGS are given, later the actor model of HGS is presented and the obtained experimental results are shown and discussed in detail. Finally the paper is concluded and future work plans are sketched out.

2   Actor-Based Implementation of Metaheuristics

Actor-based concurrency is a popular, easy to apply paradigm for paralleliza-tion and distribution of computing. As this paper is focused on parallelization of a complex metaheuristic algorithm using the actor-based approach, let us refer to existing similar high-level distributed and parallel implementations of metaheuristics.

Evolutionary multi-agent system is an interesting metaheuristic algorithm putting together the evolutionary and agent-based paradigms. Thus an agent becomes not only a driver for realization of certain computing task, but also a part of the computation, carrying the solution (genotype) and working towards improving its quality throughout the whole population. The agents undergo decentralized selection process based on non-renewable resources assigned to agents and the actions of reproduction and death. Such an approach (high decentralization of control) resulted in several distributed and parallel imple-mentations, while one utilizing actor-model and defining so-called "arenas" was particularly efficient and interesting.

Its first implementation was realized in Scala [8]. The arenas were designed as meeting places, where particular actors (agents) can go and do a relevant task. E.g. at the meeting arena the agents were able to compare the qualities of their solutions while at the reproduction arena they could produce offspring etc. Another very efficient implementation of EMAS based on actor model was realized using Erlang [18]. Those implementations were tested in supercomputing environment and yielded very promising results.

Another interesting high-level approach to parallelization of metaheuristics consists in using a high-level parallel patterns in order to be able to automatically parallelize and distribute certain parts of code [17]. This approach has also been widely tested on many available supercomputing facilities.

Finally, a high-level distributed implementation of Ant Colony Optimization type algorithms was realized using Scala and Akka [16]. The approach is based on distributing the pheromone table and assuring the updates of its state. Currently the research shows that it might be possible to accept certain delays or even lack of updates of the pheromone table, still maintaining the high quality and scalability of the computing.

Observing efficient outcomes of the above-mentioned approaches, we decided to apply actor-based concurrency model for implementing HGS.

3   Hierarchic Genetic Strategy

Another approach involving multiple evolving populations is Hierarchic Genetic Strategy (HGS) [12]. The algorithm dynamically creates subpopulations (*demes*) and lays them out in tree-like hierarchy (see Fig. 1). The search accuracy of a particular deme depends on its depth in the hierarchy. Nodes closer to the root perform more chaotic search to find promising areas. Each tree node is assigned with an internal evolutionary algorithm (*driver*, in single-objective processing it is the Simple Genetic Algorithm).

The process of HGS can be divided into several steps, called *metaepochs*. Each metaepoch consists of several epochs of a driver. Additionally, HGS-specific mechanics are applied. The most promising individuals of each node have a chance to become seeds of next-level *child nodes* (*sprouting* procedure). A child node runs with reduced variance settings, so that its population will mostly explore that region. To eliminate risk of redundant exploration of independently evolving demes, branch comparison procedure is performed. If the area is already explored by one of other children, then sprouting is cancelled and next candidate for sprouting is considered. Furthermore, the *branch reduction* compares and removes populations at the same level of the HGS tree that perform search in the common landscape region or in already explored regions.

Following very good results obtained for single-objective problems, the HGS metaheuristic has been adapted for multi-objective optimization tasks (MO-HGS [3]). This direction was further explored and improved [9] to create more generic solution, able to incorporate any multi-objective evolutionary algorithm (MOEA) as HGS driver.

The basic structure of this approach, denoted as Multiobjective Optimization Hierarchic Genetic Strategy with maturing (MO-mHGS) is similar to classical HGS, but the following features of the basic algorithm had to be adapted in order to tackle multi-objective problems:

- Flexibility of proposed model – different MOEA approaches may vary substantially. There are algorithms basing on Pareto dominance relation, algorithms built around quality indicator or algorithms decomposing multi-objective problem into subproblems to handle larger set of objectives. Moreover, some of them are generational approaches, others are steady-state. They all should be supported as MO-mHGS drivers.
- Evaluating quality of internal algorithm outcome – MOEA have specific set of quality indicators measuring distance from Pareto front, individuals spread, coverage of separate parts of a front, etc.
- Killing of the node based on lack of improvement of the solution in the last metaepoch – in MO-mHGS hypervolume metric is used as a factor for checking the spread of the population processed at the node.

Hierarchic strategies are also a subject of hybridization. They can be combined with local methods, e.g. local gradient-based search [6], and with clustering ([11] and [7], for HGS and MO-HGS, respectively).

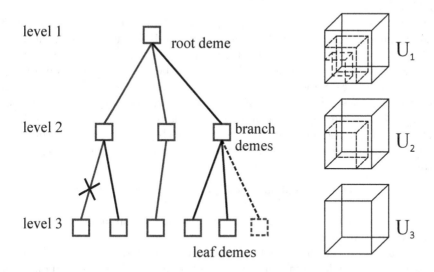

Fig. 1. Left panel: exemplary HGS tree with three levels. The left-most leaf is being reduced, the right-most leaf is being created. Right panel: three-dimensional genetic spaces in the real-number encoding, corresponding to levels in HGS tree.

4   Actor-Based HGS Models

In recent research we have shown [9] that hierarchical multi-deme model can be used in combination with any single-deme MOEA. We considered mainly variable evaluation accuracy as a crucial MO-mHGS feature. The results were promising and proved its applicability to real-world problems, where time of calculating a single fitness evaluation depends on the accuracy.

However, HGS model has additional properties we can take advantage of. Similarly to other multi-deme approaches such as Island Model [10], calculations can be naturally divided into several processes. These processes are able to run simultaneously, ensuring benefits of parallel execution.

4.1   Adapting HGS to Asynchronous Environment

In classic MO-mHGS initial population is evaluated with assigned MOEA *driver* (e.g. NSGAII). After reaching specified progress ratio between 2 last metaepochs, HGS may create additional sprouts – MOEA nodes with smaller populations created basing on the most promising individuals. On final stages of computing, MO-mHGS may consist of multiple sprouts connected in tree-like node structure. Each sprout can be treated as separate, independent unit. This specific calculation structure can be naturally transformed into actor-based model with parallel execution capabilities. Therefore, our goal was to adapt and optimize MO-mHGS to meet the actor model's requirements. It resulted in creating new improved model, Multiobjective Optimization Distributed HGS (denoted as MO-DHGS or DHGS).

The key problem was to identify types of actors and messages exchanged between them without loosing basic flow of MO-mHGS algorithm. Eventually, we defined two kinds of actors: Supervisor and Node. The actor system consist of one Supervisor and multiple Nodes that can be added dynamically during sprouting procedure. Supervisor should manage metaepochs and ensure parallel execution of Nodes is limited by the algorithm flow. Node is just an equivalent of single HGS sprout. At the beginning of simulation Supervisor creates single (root) node. Supervisor communicates with Nodes by passing messages that may contain additional data. There are several messages types used during single metaepoch:

- *GetState* – asking node about it current HGS state (alive/dead), maturity, etc;
- *StartMetaepoch* – triggering new epoch's calculation in each node that receive this message;
- *GetPopulation* – asking a node about current population. Some of this messages may be used on different stages of computing, but we focused on minimizing unnecessary data exchange due to its impact on the performance.

Apart from architectural changes, we also had to modify the control flow. Core MO-mHGS procedure was left intact, so a single DHGS step can be described as the same sequence of phases: 1) Run Metaepoch 2) Trim Not Progresssing Sprouts 3) Trim redundant sprouts 4) Release sprouts 5) Revive dead tree. However, each phase was adapted to the actor model. In case of phases 2–5 it required major changes, because these procedures require exchanging a lot of information between supervisor and nodes (or even between nodes). In classical approach, where whole memory was shared between all units, it could be easily achieved. In case of asynchronous model it often leads to messages overload or disturbing order of processed data. It may not only have negative impact on performance, but also violate the algorithm assumptions.

Algorithm 1 shows new, actor-based, supervised sprouting procedures. In classical approach, sprouting was simple, recursive procedure. Every node asked its old sprouts to perform sprouting and, at the end, released new sprouts if constraints (ξ – *sproutivness*, amount of children created after a metaepoch and ξ_{max} – maximum number of living children in a tree node) had been preserved. In actor-based approach there are two major problems that need to be addressed. First of all, in order to preserve sproutivness, node has to filter out alive sprouts. It can be achieved by sending *GetState* message to each sprout and wait for all responses at the beginning of procedure. There is guarantee that state can't be changed at this point of the algorithm so it suffices to do it once per sprouting procedure. The second issue is more difficult to handle. Comparing best individuals (Γ) to choose seed of new sprout's population requires gathering information

Algorithm 1. Supervisor Sprouting Task

1: **procedure** INITSPROUTING($hgsNodes$)
2:     $currentLevelStates \leftarrow \varnothing$
3:     $nextLevelStates \leftarrow \varnothing$
4:     $lvl \leftarrow FindLeafLevel(hgsNodes)$
5:     $i \leftarrow 0$
6:     $node \leftarrow hgsNodes_{lvl,i}$
7:     Send($node$, "$SproutingRequest$", $nextLevelStates$)
8: **end procedure**
9:
10: **procedure** RECEIVESPROUTINGANSWER($node, state, sproutStates$)
11:     $currentLevelStates \leftarrow currentLevelStates \cup \{state\}$
12:     $nextLevelStates \leftarrow currentLevelStates \cup sproutStates$
13:     $i \leftarrow i + 1$
14:     **if** $i \geq |hgsNodes_{lvl}|$ **then**
15:         $lvl \leftarrow lvl - 1$
16:         $i \leftarrow 0$
17:         $nextLevelStates \leftarrow currentLevelStates$
18:         $currentLevelStates \leftarrow \varnothing$
19:         **if** $lvl < 0$ **then**
20:             **return**   ▷ End of sprouting procedure
21:         **end if**
22:     **end if**
23:     $node \leftarrow hgsNodes_{lvl,i}$
24:     Send($node$, "$SproutingRequest$", $nextLevelStates$)
25: **end procedure**

---

from all next level nodes (not only sprouts of the current node). If we let each node to perform sprouting asynchronously, new sprouts will impact comparison procedure of other nodes. Depending of gathering time the results may vary – some nodes might have finished sprouting, others have not even started. This is why the whole sprouting procedure in actor-based version should be supervised. Our approach is to divide sprouting into 2 tasks. First task is conducted by the Supervisor. It runs sprouting sequentially traversing all nodes from a tree level, starting with leaves. Supervisor gathers information about sprout states that may be used in comparison procedure. After receiving sprouting summary from a node that finished its procedure, supervisor updates $nextLevelStates$ set with states of newly created sprouts. This set is later passed to another node starting sprouting. It also keeps updating $currentLevelStates$ that eventually becomes $nextLevelStates$ when supervisor goes to a lower tree level. That's how we ensure every node operates on the most current knowledge about HGS tree. Moreover, each node sends this information to supervisor only once so that we can minimize required communication.

Sprouting procedure is the most complex (and important) part of HGS. The remaining HGS procedures also required adjusting to the new model, but usually solutions were similar to described idea and other differences lie mainly in implementation details. Full implementation of DHGS can be found in Evogil project, our open-source evolutionary computing platform[2].

4.2   Evaluation Methodology

In order to measure hypothetical impact of parallel HGS nodes on overall MO-DHGS performance, we have measured the cost of calculations. Cost is expressed as number of fitness function evaluations. Each algorithm had the same cost constraints (*budget*). We assumed that in case of parallel execution a cost of single metaepoch can be reduced to cost of longest running node. In other words, for set of costs $\{c_1, c_2, ...c_n\}$ where c_i is a cost of metaepoch of a node i, the overall DHGS metaepoch cost can be expressed as $C_{DHGS} = max\{c_1, c_2, ...c_n\}$, while cost of sequential MO-mHGS is $C_{HGS} = \sum_{i=1}^{n} c_i$.

All runs of the system were performed on Evogil platform with the same algorithms' parameters as in our previous research [9]. This time we focused on comparing 2 specific algorithms that were combined with MO-mHGS and DHGS models: classical dominance-based approach NSGAII [4] and a hybrid of MOEA and particle swarm optimizer, OMOPSO [15]. In the later case, choice was dictated by promising results of previous research, where OMOPSO characteristics have proved to be well fit to HGS meta-model. Algorithms were evaluated on ZDT and CEC09 benchmark families.

Cost and error scaling on different tree levels were also adjusted as in our previous work, simulating real-world problems behaviour. Again, we chose to take into consideration only the best working set of parameters, thus cost modifiers were set to $\langle 0.1, 0.5, 1.0 \rangle$ and error variation levels to $\langle 10\%, 1\%, 0.1\% \rangle$. These parameters were configured for both MO-mHGS and MO-DHGS.

4.3   MO-DHGS Results

For each result set we have evaluated several quality indicators. In this paper we focus on three popular metrics: Average Hausdorff Distance (AHD)] [14] (combination of GD and IGD metrics), Hypervolume [5] and Spacing [13]. Figure 2 shows summary of results from end of the system run for first considered algorithm, NSGAII. It is clear that DHGS improves performance of single-deme NSGAII, but also almost always wins with classical MO-mHGS regardless of metric. In terms of distance from Pareto front and individuals distribution there are significant differences between HGS methods, up to 50% of AHD value (UF4).

If we take a closer look on UF4 problem case and distribution of metric value over consumed budget (Fig. 3), we can observe that DHGS achieves good results also on earlier stages of system run. However, DHGS converges much faster than MO-mHGS and gap between these two increases with time. It is related

---

[2] https://github.com/Soamid/evogil.

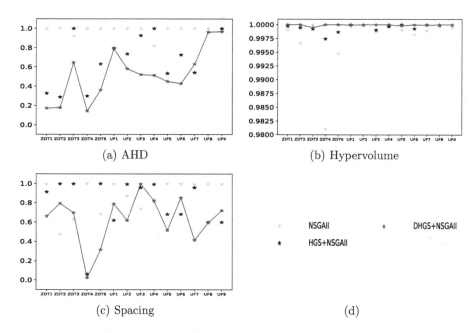

Fig. 2. Run summary (budget = 2500, NSGAII hybridization). Markers represent algorithms' normalized metric outcomes at final stage of system run. MO-DHGS outcomes are connected with lines.

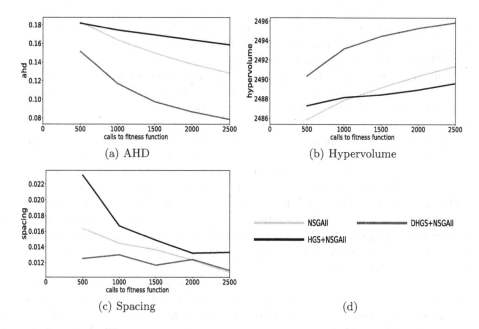

Fig. 3. Detailed results of NSGAII hybridization for all considered quality indicators tackling UF4 benchmark.

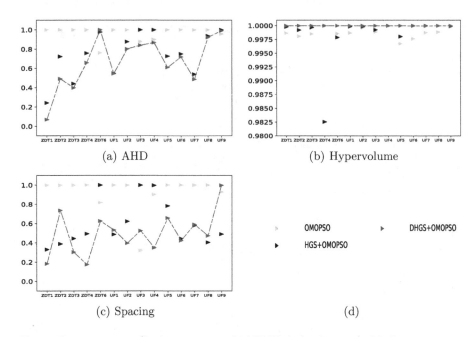

(a) AHD   (b) Hypervolume

(c) Spacing   (d)

Fig. 4. Run summary (budget = 2500, OMOPSO hybridization). Markers represent algorithms' normalized metric outcomes at final stage of system run. MO-DHGS outcomes are connected with lines.

to parallel architecture of DHGS: at the beginning it operates on single root node. Additional nodes (evaluated in parallel) are created later, after reaching satisfying progress ratio on lower level nodes.

Similar behaviour can be observed during OMOPSO computing. Summary of results across all benchmark problems (Fig. 4) lead to the same conclusions: DHGS outperforms other solutions in vast majority of cases. Detailed charts of representative example (UF5, Fig. 5) present how taking advantage of parallelism may eventually provide tremendous results. It is worth to note that in case of UF5 problem DHGS struggled to beat MO-mHGS in first half of the process. Later (after reaching promising outcomes in root node) it released sprouts, pushing final results farther and resulting in higher Hypervolume. It also found better individuals in terms of AHD after reaching about 60% of the total budget (even in comparison to final MO-mHGS results). This observation is important for our research, as MO-mHGS with OMOPSO driver was our best hybridization attempt in the previous research. This version of algorithm always reached optimal values very fast.

In order to summarize DHGS experiments, the outcomes we calculated score for each algorithm at two stages: after reaching budget equal to 1000 fitness evaluations and at the end of system run (2500 evaluations). Score was calculated basing on number of strong and weak wins for a given problem and budget step. We define weak win as a situation when specified algorithm is no worse than all other solutions with some arbitrary chosen confidence value. In our experiments we set the confidence to 0.5% for Hypervolume and 5% for all other metrics. Algorithm is marked as strong winner if there are no weak winners for the considered result set. At the end, each strong winner obtains 2 points and weak winner scores 1 point. Table 1 contains summarized values of strong and weak wins in all considered situations. Again, DHGS versions of algorithms gather best scores in all cases. In multiple examples DHGS score value is improved at the end of system run (with one notable exception of NSGAII spacing: DHGS score decreases from 24 to 14 while single-deme NSGAII gathers some points).

5   Evaluating DHGS in Asynchronous Environment

Results presented in previous section should be considered as hypothetical. They are based on several assumptions about possibility of fitness adjustment and cost-free parallel execution. Real world problems results will be determined by proper model configuration, but also by technical conditions. In order to test DHGS model we implemented realistic multi-process version of the algorithm. We used thespian and rxPy Python libraries to include DHGS as new Evogil platform component. Then we created new simulation mode: to measure realistic performance we applied time constraints instead of budget. During simulation we have measured time of each metaepoch and sampled current results every 2 s. Whole simulation had 20 s timeout.

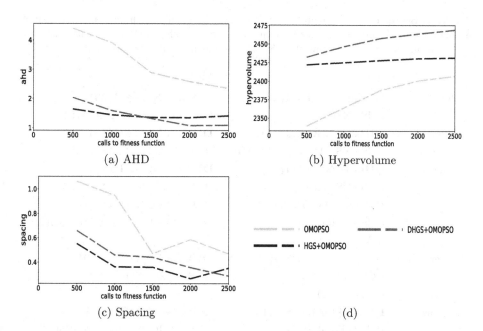

(a) AHD   (b) Hypervolume

(c) Spacing   (d)

Fig. 5. Detailed results of OMOPSO hybridization for all considered quality indicators tackling UF5 benchmark.

Note that all presented simulations were run without fitness evaluation adjusting – these are the same benchmark problems that were used in previous simulations, but they evaluate fitness the same way on every HGS node, regardless of tree level. That's why in time-bound tests even MO-mHGS ends up with worse results than single-deme algorithm.

All experiments were conducted on Windows 10 with Intel Core i9-9900K (3.6 GHz with 8 cores and 16 threads) and 16 GB RAM.

As predicted, results summary of realistic OMOPSO hybridization (Fig. 6 look different than its simulated version. DHGS still outperforms MO-mHGS and single-deme algorithm in most situations, but it happens less frequently. Especially in terms of spacing indicator final results do not seem to have one dominating solution. On the other hand, basic AHD metric looks much better: usually asynchronous DHGS beats at least one of its competitors and almost always improves synchronous MO-mHGS results.

In fact, there are to separate sources of the problem here. First of all, the lack of aforementioned fitness evaluation adjusting that can be implemented in real-world problems, but it is not applicable to popular benchmark problems. That's why DHGS (and MO-mHGS) can't beat NSGAII performance even though both were clearly better in simulated version of our experiment. Moreover, original reviving procedure does not allow us to remove dead nodes (or reuse their processes for creating new nodes), as they can be bring back to life later. Therefore, if the algorithm releases many sprouts in specific problem case, there will be also

Table 1. Scores of NSGAII and OMOPSO system in all benchmark problems. For each metric and budget stage, all wining methods are shown. Values in parentheses represent methods' scores. If method is not present in a cell, its score is 0.

	1000	2500
ahd	**DHGS+NSGAII(22)**, HGS+NSGAII(6)	**DHGS+NSGAII(23)**, HGS+NSGAII(5), NSGAII(2)
hypervolume	**DHGS+NSGAII(15)**, HGS+NSGAII(13), NSGAII(2)	**DHGS+NSGAII(18)**, HGS+NSGAII(9), NSGAII(5)
spacing	**DHGS+NSGAII(24)**, HGS+NSGAII(4)	**DHGS+NSGAII(14)**, HGS+NSGAII(7), NSGAII(7)
ahd	**DHGS+OMOPSO(15)**, HGS+OMOPSO(13), OMOPSO(1)	**DHGS+OMOPSO(22)**, OMOPSO(4), HGS+OMOPSO(2)
hypervolume	**DHGS+OMOPSO(17)**, HGS+OMOPSO(11), OMOPSO(2)	**DHGS+OMOPSO(19)**, HGS+OMOPSO(9), OMOPSO(3) .
spacing	**DHGS+OMOPSO(14)**, HGS+OMOPSO(12), OMOPSO(2)	**DHGS+OMOPSO(17)**, HGS+OMOPSO(9), OMOPSO(2)

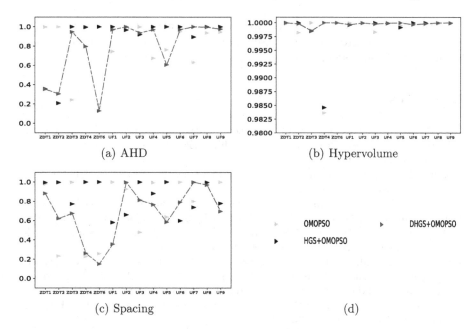

Fig. 6. Realistic simulation summary (time=20s, OMOPSO hybridization). Markers represent algorithms' normalized metric outcomes at final stage of simulation. MO-DHGS outcomes are connected with lines.

many processes. Obviously, it has negative impact on performance of whole simulation. This interesting observation leaves space for further research of adjusting number of processing units or improving HGS reviving procedure.

6   Conclusion

Striving toward construction of efficient metaheuristics can help in reasonable utilization of nowadays very popular supercomputing facilities. Moreover, focusing on high-level programming languages makes possible for the user to focus on the algorithm itself, relying on the technology supporting the development process.

In this paper we have shown, that actor-based DHGS model can be a natural improvement of MO-mHGS. Tree-like structure of HGS concept gave us opportunity to create generic, multi-deme asynchronous algorithm able to outperform its synchronous version and converge faster on later stages of the system run.

We have shown that the proposed solution is sensitive to environment conditions and implementation details nodes which leaves room for further investigation.

Nevertheless, in combination with fitness evaluation adjusting mechanism, DHGS is a powerful tool for solving real-world multi-objective problems.

In the future we will focus on experimenting scalability in broader range, which seems to be promising because of the actual actor model used for synchronization of the systems work.

References

1. Bäck, T., Schwefel, H.P.: Evolutionary computation: an overview. In: Fukuda, T., Furuhashi, T. (eds.) Proceedings of the Third IEEE Conference on Evolutionary Computation. IEEE Press (1996)
2. Cantú-Paz, E.: Efficient and Accurate Parallel Genetic Algorithms. Kluwer Academic Publishers, Norwell (2000)
3. Ciepiela, E., Kocot, J., Siwik, L., Dreżewski, R.: Hierarchical approach to evolutionary multi-objective optimization. In: Bubak, M., van Albada, G.D., Dongarra, J., Sloot, P.M.A. (eds.) ICCS 2008. LNCS, vol. 5103, pp. 740–749. Springer, Heidelberg (2008). https://doi.org/10.1007/978-3-540-69389-5_82
4. Deb, K., Agrawal, S., Pratap, A., Meyarivan, T.: A fast Elitist non-dominated sorting genetic algorithm for multi-objective optimization: NSGA-II. In: Schoenauer, M., Deb, K., Rudolph, G., Yao, X., Lutton, E., Merelo, J.J., Schwefel, H.-P. (eds.) PPSN 2000. LNCS, vol. 1917, pp. 849–858. Springer, Heidelberg (2000). https://doi.org/10.1007/3-540-45356-3_83
5. Fonseca, C.M., Paquete, L., López-Ibánez, M.: An improved dimension-sweep algorithm for the hypervolume indicator. In: IEEE Congress on Evolutionary Computation, 2006. CEC 2006, pp. 1157–1163. IEEE (2006)
6. Gajda-Zagorska, E., Smolka, M., Schaefer, R., Pardo, D., Álvarez Aramberri, J.: Multi-objective hierarchic memetic solver for inverse parametric problems. Procedia Comput. Sci. **51**, 974–983 (2015)
7. Gajda-Zagórska, E.: Multiobjective evolutionary strategy for finding neighbourhoods of pareto-optimal solutions. In: Esparcia-Alcázar, A.I. (ed.) EvoApplications 2013. LNCS, vol. 7835, pp. 112–121. Springer, Heidelberg (2013). https://doi.org/10.1007/978-3-642-37192-9_12
8. Krzywicki, D., Turek, W., Byrski, A., Kisiel-Dorohinicki, M.: Massively concurrent agent-based evolutionary computing. J. Comput. Science **11**, 153–162 (2015). https://doi.org/10.1016/j.jocs.2015.07.003
9. Lazarz, R., Idzik, M., Gadek, K., Gajda-Zagorska, E.: Hierarchic genetic strategy with maturing as a generic tool for multiobjective optimization. J. Comput. Sci. **17**, 249–260 (2016)
10. Martin, W.N., Lienig, J., Cohoon, J.P.: Island (migration) models: evolutionary algorithms based on punctuated equilibria. In: Handbook of Evolutionary Computation, vol. 6, no. 3 (1997)
11. Schaefer, R., Adamska, K., Telega, H.: Clustered genetic search in continuous landscape exploration. Eng. Appl. Artif. Intell. **17**(4), 407–416 (2004)
12. Schaefer, R., Kolodziej, J.: Genetic search reinforced by the population hierarchy. Found. Genet. Algorithms **7**, 383–401 (2002)
13. Schott, J.R.: Fault tolerant design using single and multicriteria genetic algorithm optimization. Tech. rep, DTIC Document (1995)
14. Schütze, O., Esquivel, X., Lara, A., Coello, C.A.C.: Using the averaged hausdorff distance as a performance measure in evolutionary multiobjective optimization. IEEE Trans. Evol. Comput. **16**(4), 504–522 (2012)

15. Sierra, M., Coello Coello, C.: Improving PSO-based multi-objective optimization using crowding, mutation and ϵ-dominance. In: Coello Coello, C., Hernández Aguirre, A., Zitzler, E. (eds.) Evolutionary Multi-Criterion Optimization. Lecture Notes in Computer Science, vol. 3410, pp. 505–519. Springer, Berlin Heidelberg (2005). https://doi.org/10.1007/978-3-540-31880-4_35

16. Starzec, M., Starzec, G., Byrski, A., Turek, W.: Distributed ant colony optimization based on actor model. Parallel Comput. **90** (2019). https://doi.org/10.1016/j.parco.2019.102573

17. Stypka, J., et al.: The missing link! A new skeleton for evolutionary multi-agent systems in erlang. Int. J. Parallel Prog. **46**(1), 4–22 (2018). https://doi.org/10.1007/s10766-017-0503-4

18. Turek, W., et al.: Highly scalable erlang framework for agent-based metaheuristic computing. J. Comput. Science **17**, 234–248 (2016). https://doi.org/10.1016/j.jocs.2016.03.003

MeshingNet: A New Mesh Generation Method Based on Deep Learning

Zheyan Zhang$^{(\boxtimes)}$ ⓘ, Yongxing Wang ⓘ, Peter K. Jimack ⓘ, and He Wang ⓘ

School of Computing, University of Leeds, Leeds LS2 9JT, UK
sczz@leeds.ac.uk

Abstract. We introduce a novel approach to automatic unstructured mesh generation using machine learning to predict an optimal finite element mesh for a previously unseen problem. The framework that we have developed is based around training an artificial neural network (ANN) to guide standard mesh generation software, based upon a prediction of the required local mesh density throughout the domain. We describe the training regime that is proposed, based upon the use of *a posteriori* error estimation, and discuss the topologies of the ANNs that we have considered. We then illustrate performance using two standard test problems, a single elliptic partial differential equation (PDE) and a system of PDEs associated with linear elasticity. We demonstrate the effective generation of high quality meshes for arbitrary polygonal geometries and a range of material parameters, using a variety of user-selected error norms.

Keywords: Mesh generation · Error equidistribution · Machine learning · Artificial neural networks

1 Introduction

Mesh generation is a critical step in the numerical solution of a wide range of problems arising in computational science. The use of unstructured meshes is especially common in domains such as computational fluid dynamics (CFD) and computational mechanics, but also arises in the application of finite element (FE) and finite volume (FV) methods for estimating the solutions of general partial differential equations (PDEs): especially on domains with complex geometries [19,20]. The quality of the FE/FV solution depends critically on the nature of the underlying mesh. For an ideal mesh the error in the FE solution (we focus on the FE method (FEM) in this paper) will be distributed equally across the elements of the mesh, implying the need for larger elements where the local error density is small and smaller elements where the local error density is large. This "equidistribution" property tends to ensure that a prescribed global error tolerance (i.e. an acceptable error between the (unknown) true solution and the computed FE solution) can be obtained with the fewest number of elements in the mesh [9,22]. This is a desirable feature since the computational work generally grows superlinearly with the number of elements (though, in some special cases, this can be linear [18]).

© Springer Nature Switzerland AG 2020
V. V. Krzhizhanovskaya et al. (Eds.): ICCS 2020, LNCS 12139, pp. 186–198, 2020.
https://doi.org/10.1007/978-3-030-50420-5_14

The conventional approach to obtain high quality meshes involves multiple passes, where a solution is initially computed on a relatively coarse uniform mesh and then a post-processing step, known as *a posteriori* error estimation, is undertaken [2,4,24]. This typically involves solving many auxiliary problems (e.g. one per element or per patch of elements) in order to estimate the local error in the initial solution [11]. These local errors can be combined to form an overall (global) error estimate but they can also be used to determine where the local mesh density most needs to be increased (mesh refinement), and by how much, and where the local mesh density may be safely decreased (mesh coarsening), and by how much. A new mesh is then generated based upon this *a posteriori* error estimate and a new FE solution is computed on this mesh. A further *a posteriori* error estimate may be computed on this new mesh to determine whether the solution is satisfactory (i.e. has an error less than the prescribed tolerance) or if further mesh refinement is required.

A necessary requirement for efficient *a posteriori* error estimators is that they should be relatively cheap to compute (whilst still, of course, providing reliable information about the error in a computed solution). For example, the approaches of [2–4] each solve a supplementary local problem on each finite element in order to estimate the 2-norm of the local error from the local residual. Alternatively, recovery-based error estimators use local "superconvergence" properties of finite elements to estimate the energy norm of the local error based purely on a locally recovered gradient: for example the so-called ZZ estimator of [24]. Here, the difference between the original gradient and a patch-wise recovered gradient indicates the local error. In the context of linear elasticity problems, the elasticity energy density of a computed solution is evaluated at each element and the recovered energy density value at each vertex is defined to be the average of its adjacent elements. The local error is then proportional to the difference between the recovered piece-wise linear energy density and the original piece-wise constant values.

In this paper we exploit a data-driven method to improve the efficiency of non-uniform mesh generation compared with existing approaches. The core of non-uniform mesh generation is to find an appropriate mesh density distribution in space. Rather than utilizing expensive error estimators at each step, we compute and save high quality mesh density distributions obtained by FEM, followed by accurate error estimation, as a pre-processing step. If a model can successfully learn from the data, it no longer needs an FE solution and error estimator to predict a good mesh density distribution, but instead can reply on learning from a set of similar problems for prediction. Artificial Neural Networks (ANNs) are mathematical models that use a network of "neurons" with activation functions to mimic biological neural networks. Even a simple ANN can approximate continuous functions on compact subsets of R^n [8]. An ANN is composed of a large number of free parameters which define the network that connects the "neurons". These trainable parameters are generally not explainable. However, with them ANNs can approximate the mapping between inputs and outputs. A training loss function reflects how well the predicted output of an ANN performs for a given input (i.e. measured by the difference between the ANN's prediction and the ground truth). Furthermore, an ANN can be trained by gradient decent methods because this loss function is generally differentiable with respect to the network parameters. In recent years, With the developments of parallel hardware, larger/deeper neural networks (DNNs) have been proven to supersede existing methods on various high-level tasks such as object recognition [15]. Within computational science, DNNs have also been explored to solve ordinary differential equations (ODEs) and PDEs under both supervised [6,16] and unsupervised [12] settings.

In the work reported here we propose a DNN model, MeshingNet, to learn from the *a posteriori* error estimate on an initial (coarse) uniform mesh and predict non-uniform mesh density for refinement, without the need for (or computational expense of) solving an FE system or computing an *a posteriori* error estimate. MeshingNet is trained using an accurate error estimation strategy which can be expensive but is only computed offline. Hence, the mesh generation process itself is extremely fast since it is able to make immediate use of standard, highly-tuned, software (in our case [19]) to produce a high quality mesh at the first attempt (at similar cost to generating a uniform mesh with that number of elements). Note that it is not our intention in this work to use deep learning to solve the PDEs directly (as in [16,21] for example): instead we simply aim to provide a standard FE solver with a high quality mesh, because we can provide more reliable predictions in this way, based upon the observation that a greater variation in the quality of predictions can be tolerated for the mesh than for the solution itself. For example, in an extreme worst case where the DNN predicts a constant output for all inputs, the result would be a uniform mesh (which is tolerable) however such a poor output would be completely unacceptable in the case where the network is used to predict the solution itself.

Formally, we propose, what is to the best of our knowledge, the first DNN-based predictor of *a posteriori* error, that can: (i) efficiently generate non-uniform meshes with a desired speed; (ii) seamlessly work with existing mesh generators, and; (iii) generalize to different geometric domains with various governing PDEs, boundary conditions (BCs) and parameters.

The remainder of this paper is structured as follows. In the next section we describe our proposed deep learning algorithm. This is not the first time that researchers have attempted to apply ANNs to mesh generation. However, previous attempts have been for quite specific problems [5,17] and have therefore been able to assume substantially more *a priori* knowledge than our approach. Consequently, the novelty in Sect. 2 comes through both the generality of the approach used in formulating the problem (i.e. generality of the inputs and outputs of the DNN) as well as the network itself. In Sect. 3, we demonstrate and assess the performance of our approach on two standard elliptic PDE problems. These tests allow us to account for variations in the PDE system, the domain geometry, the BCs, the physical problem parameters and the desired error norm when considering the efficacy of our approach. Finally, in Sect. 4 we discuss our plans to further develop and apply this deep learning approach.

2   Proposed Method

2.1   Overview

We consider a standard setting where the FEM is employed. Given a geometry and a mesh generator, a low density uniform mesh (LDUM) can be easily computed, then refined non-uniformly based on the *a posteriori* error distribution, for better accuracy. Since this iterative meshing process is very time-consuming, we propose a DNN-based supervised learning method to accelerate it.

Given a family of governing PDEs and material parameters, we assume that there is a mapping

$$F : \Gamma, B, M, X \rightarrow A(X) \tag{1}$$

that can be learned, where Γ is a collection of domain geometries, B is a set of BCs, M is a set of PDE parameters (e.g. material properties), $x \in X$ is a location in the domain and $A(X)$ is the target area upper bound distribution over the whole domain. To represent an interior location, we use Mean Value Coordinates [10] because they provide translational and rotational invariance with respect to the boundary. Given Γ, B, M and X, we aim to predict $A(X)$ quickly. The mapping F is highly non-linear and is therefore learned by our model, MeshingNet. Under the supervised learning scheme, we first build up our training data set by computing high-accuracy solutions (HASs) on high-density uniform meshes (HDUMs) using a standard FE solver. The same computation is also done on LDUMs to obtain lower accuracy solutions (LAS). Then an *a posteriori* error distribution $E(X)$ is computed based upon interpolation between these solutions. According to $E(X)$, we compute $A(X)$ for refinement. The training data is enriched by combining different geometries with different parameters and BCs. Next, MeshingNet is trained, taking as input the geometry, BCs and material properties, with the predicted local area upper bound $A(X)$ as output. After training, MeshingNet is used to predict $A(X)$ over a new geometry with a LDUM. The final mesh is generated either by refining the LDUM non-uniformly or using it to generate a completely new mesh (e.g. using the method in [19]), guided by the predicted target local area upper bound. Figure 1 illustrates the whole workflow of our mesh generation system.

The approach that we propose has a number of components that are not fixed and may be selected by the user. MeshingNet is agnostic about both the mesh generator and the particular FE/FV solver that are used. It is designed to work with existing methods. Furthermore, the *a posteriori* error can be computed using any user-defined norm (in this paper we consider L1 and energy norms respectively in our two validation tests). Some specific examples of governing equations, geometries, boundary conditions and material parameters are illustrated in the evaluation section. Prior to this however we provide some additional details of the components used in our paper.

2.2   Component Details

Mesh Generation. All meshes (LDUM, HDUM and the refined mesh) are created using the software *Triangle* [19] which conducts Delaunay triangulations.

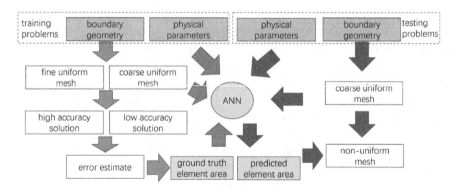

Fig. 1. A high-level diagram of MeshingNet workflow showing the ANN alongside the training regime (left) and testing regime (right). Grey, orange and blue arrows represent data generation, training and testing processes. (Color figure online)

Triangle reads a planar graph, representing the boundary of a two-dimensional polygonal domain as its input and the user may define a single maximum element area across the entire polygon when a uniform mesh is desired. To refine a mesh non-uniformly the user specifies, within each coarse element, what the element area upper bound should be after refinement. *Triangle* ensures a smooth spatial variation of the element size. The refinement is not nested since, although it does not eliminate the pre-existing vertices of the original mesh, it does break the original edges.

Mesh Refinement via Error Estimation. Broadly speaking, the finer the mesh is, the closer the FE result is expected to be to the ground truth. We regard the HAS as the ground truth by ensuring that HDUMs are always significantly finer than LDUMs. Linear interpolation is used to project the LAS to the fine mesh to obtain LAS*, where LAS* has the same dimension as HAS. An error estimate approximates the error E by comparing LAS* and HAS in the selected norm, on the HDUM, and this is then projected back to the original LDUM. The target area upper bound for the refined mesh within each LDUM element is then defined to be inversely correlated with E. In this paper, we select $K/E(x_i)^\alpha$ as the area upper bound for element number i of the LDUM (to be refined), where x_i is the center of the ith element, K and α determines the overall target element number and, in the examples given here, we always choose $\alpha = 1$. By varying K appropriately it is possible to adjust the refined mesh to reach a target total number of elements.

MeshingNet Model and Training. As outlined in Subsect. 2.1, for a given PDE system, MeshingNet approximates the target local element area upper bound $A(x)$ at a given point within a polygonal domain based upon inputs which include: the coordinates of the polygon's vertices, key parameters of the PDE and the mean value coordinates of the specified point (mean value coordinates, [10], parameterizes a location within a 2D polygon as a convex combination of polygon vertices). Two types of DNNs are considered: a fully connected network (FCN) and two residual networks (ResNets). The dimensions of our FCN layers are X-32-64-128-128-64-32-8-1 (where X represents the dimension of the input and is problem-specific) and each hidden layer uses rectified linear units (ReLU) as the activation function. To further improve and accelerate training, two ResNets are also experimented with to enhance FCN (Fig. 2). ResNet1 enhances FCN by adding a connection from the first hidden layer to the output of the last one. Note that residual connections can help to resolve the vanishing gradient problem in deep networks and improve training [23]. ResNet2 enhances FCN by adding multiple residual connections. This is inspired by recent densely connected convolutional networks [13] which shows superior data-fitting capacity with a relatively small number of parameters. The training data set samples over geometries, BCs and parameter values. Each geometry with fixed BCs and parameters uniquely defines a problem. In a problem, each LDUM element centroid (represented by its mean value coordinates) and its target $A(x)$ forms an input-output training pair. We randomly generated 3800 problems of which 3000 are used for training and 800 for testing (because each LDUM contains approximately 1000 elements, there are over 3 million training pairs). We then use stochastic gradient descent, with a batch size 128, to optimize the network. We use mean square error as the loss function and *Adam* [14] as the optimizer. The implementation is done using *Keras* [7] on *Tensorflow* [1] and the training is conducted on a single *NVIDIA Tesla* K40c graphics card.

Guiding Mesh Generation via MeshingNet. After training, the network is able to predict the target distribution $A(x)$ on a previously unseen polygonal domain. Given a problem, a LDUM is first generated; next, MeshingNet predicts the local target area upper bound, $A(x)_i$, at the centre, x, of the ith element; *Triangle* then refines the LDUM to generate the non-uniform mesh based upon $KA(x)_i$ within each element of the LDUM (to be refined). Optionally, this last step may be repeated with an adjusted value of K to ensure that the refined mesh has a desired total number of elements (this allows an automated approximation to "the best possible mesh with X elements").

Error Norms. Error estimation provides a means of quantifying both the local and the global error in an initial FE solution. However the precise magnitude and distribution of the error depends on the choice of the norm used to compute the difference between the LAS and the HAS. Different norms lead to different non-uniform meshes. Consequently, for any given PDE system, a single norm should be selected in order to determine A from Γ, B and M (Equation (1)). The appropriate choice of norm is a matter for the user to decide, similar to the choice of specific *a posteriori* error estimate in the conventional adaptive approach. In the following section two different norms are considered for the purposes of illustration.

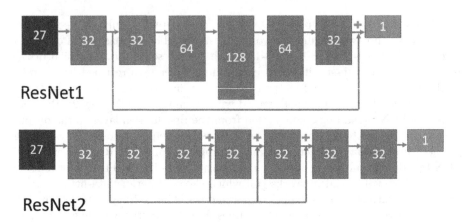

Fig. 2. Our two residual networks, illustrated with 27 input parameters. ResNet1 is a modification of the FCN with the output of the first hidden layer added to the output of the last hidden layer. ResNet2 has all hidden layers of the same dimension and the output of the first hidden layer is added to outputs of the three front hidden layers.

3   Validation Results

We now assess the performance of MeshingNet through two computational examples: a single PDE, for which we consider only the effect of the domain geometry on the optimal mesh; and a system of PDEs, for which we consider the influence of geometry, BCs and material parameters on the optimal mesh.

3.1   2D Poisson's Equation

We solve Poisson's equation ($\nabla^2 u + 1 = 0$) on a simply connected polygon Ω with boundary $\partial\Omega$ (on which $u = 0$). The polygons in our data set are all octagons, generated randomly, subject to constraints on the polar angle between consecutive vertices and on the radius being between 100 to 200 (so the polygons are size bounded). The L1 norm relative error estimate is

$$E = \left| \frac{u^{LAS} - u^{HAS}}{u^{HAS}} \right|. \tag{2}$$

As expected, in Fig. 3 (which shows a typical test geometry), the mesh generated by MeshingNet is dense where the error for the LAS is high and coarse where it is low. Figure 4 quantifies the improvement of MeshingNet relative to an uniform mesh (Fig. 3 right) by showing, for the entire test data set, the error distributions of the computed FE solutions (relative to the ground truth solutions) in each case.

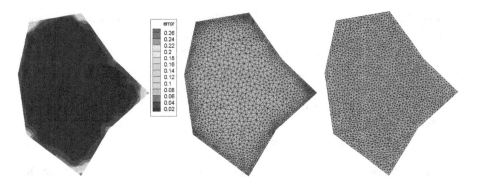

Fig. 3. For Poisson's equation: the L1 error distribution for the LAS (left); the 4000 elements mesh generated by *Triangle* under the guidance of MeshingNet (middle); and the uniform mesh with 4000 elements generated by *Triangle* (right).

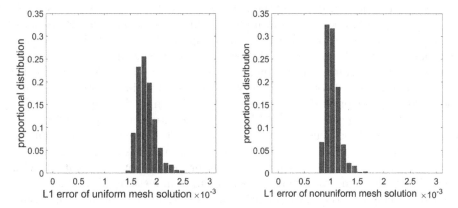

Fig. 4. For Poission's problem, L1 error distribution on uniform meshes (4000 elements) and MeshingNet meshes (4000 elements). Each bar shows the proportion of test data meshes whose FE solution error is in the range of the bar: the uniform meshes give FE errors between 0.0015 to 0.0025, whilst the MeshingNet meshes give FE errors between 0.0007 and 0.0015.

3.2   2D Linear Elasticity

We solve 2D plane stress problems on a set of *different* polygons (6–8 edges). Each polygon edge is associated with one of three possible BCs. BC1: zero displacement; BC2: uniformly distributed pressure or traction (with random amplitude up to 1000); and BC3: unconstrained. For different geometries, we number the vertexes and edges anti-clockwise with the first vertex always on the positive x-axis, without loss of generality. To get a combinations of BCs, we always apply BC1 on the first edge, BC2 on the fourth and fifth edges and BC3 on the rest. We also allow different (homogeneous) material properties: density up to a value of 1 and Poisson's ratio between 0 to 0.48. The error approximation uses energy norm

$$E = (\epsilon^L - \epsilon^H) : (\sigma^L - \sigma^H) \tag{3}$$

where ϵ^L and ϵ^H are strains of LAS* and HAS, σ^L and σ^H are stresses of LAS* and HAS. This is the "natural norm" for this problem since the PDEs are the Euler-Lagrange equations for the minimization of the following energy functional:

$$Ep = \int \frac{1}{2}\epsilon : \sigma - F \cdot u d\Omega - \int \sigma \cdot u d\Gamma \tag{4}$$

where F is the body force and u is the displacement. Due to the linearity of the problem, the relative accuracy of two FE solutions may be determined equivalently by which has the lower error in the energy norm or which has the lower total potential energy (we exploit this in our validation below).

There are 27 dimensions in MeshingNet's input: 16 for the polygon vertices, 8 for the mean value coordinates of the target point, and 1 each for the traction BC magnitude, density and Poisson's ratio. We train FCN, Resnet1 and Resnet2 for 50 epochs, each taking 142, 134 and 141 min respectively. Figure 5 shows that the training processes all converge, with ResNet training typically converging faster than FCN. After training, predicting the target A (on all LDUM elements) for one problem takes 0.046 s on average, which is over 300 times faster than using the *a posteriori* error method that generates the training data set.

Figure 6 shows a comparison of FE results computed on MeshingNet meshes, uniform meshes of the same number of elements (4000 elements) and non-uniform meshes (also of the same number of elements) computed based upon local refinement following ZZ error estimation. The former meshes have FE solutions with potential energy significantly lower than the uniform mesh and ZZ refined mesh (and much closer to the HAS potential energy). Figure 7 illustrates some typical meshes obtained using MeshingNet: the non-uniform meshes correspond to the error distributions in the LAS. Though not shown here due to space constraints, we also find that the traction-to-density ratio impacts the non-uniform mesh most significantly. Overall, this example shows that MeshingNet can generate high quality meshes that not only account for geometry but also the given material properties and BCs.

Finally we compare different DNN models, using the average potential energy on the testing data set. The baseline is from the HDUMs, whose FE solutions have a mean energy of -7.7293, followed by the meshes from ResNet2 (mean energy -7.6816), ResNet1 (-7.6815), and FCN (-7.6813). The lower the better. These are all far superior to the uniform meshes of the same size (4000 elements), which yield FE solutions with a mean energy of -7.6030. Note that ResNet not only shortens the training time over FCN but, on average, produces better solutions.

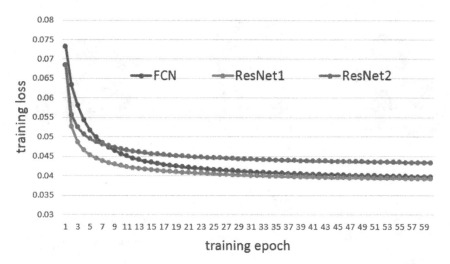

Fig. 5. L2 training loss on elasticity training dataset during 60 training epochs. Three curves representing FCN (blue), ResNet1 (red) and ResNet2 (green) converge individually. (Color figure online)

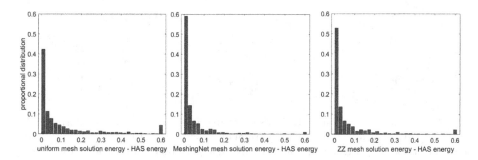

Fig. 6. Potential energy comparison of solving 2D elasticity test problems on 4000 element meshes: uniform element size (left); MeshingNet (using FCN) meshes (centre); and ZZ refined meshes (right). We use the HAS energy as our baseline: the MeshingNet mesh solutions have energies that are significantly closer to the HAS energies than the ZZ mesh solutions and uniform mesh solutions (since a greater proportion of results are distributed near zero). The rightmost bar represents the proportion of all tests where the energy difference is no smaller than 0.6.

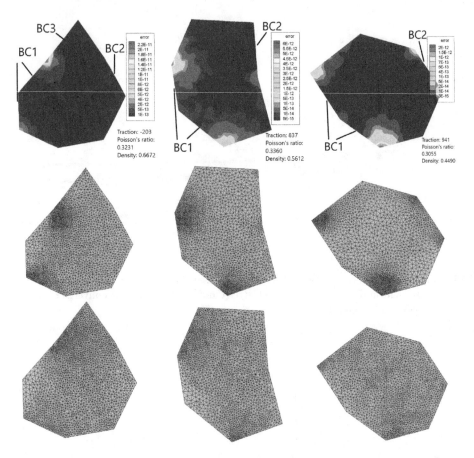

Fig. 7. FE error (top) relative to HAS on coarse uniform mesh, non-uniform mesh guided by MeshingNet (middle) and non-uniform mesh refined by ZZ (bottom). The left geometry is an octagon and other two are heptagons (defined by placing their final vertex at the centre of edge 7).

4   Discussion

In this paper, we have proposed a new non-uniform mesh generation method based on DNN. The approach is designed for general PDEs with a range of geometries, BCs and problem parameters. We have implemented a two-dimensional prototype and validated it on two test problems: Poisson's equation and linear elasticity. These tests have shown the potential of the technique to successfully learn the impact of domain geometries, BCs and material properties on the optimal finite element mesh. Quantitatively, meshes generated by MeshingNet are shown to be more accurate than uniform meshes and non-uniform ZZ meshes of the number of elements. Most significantly, MeshingNet avoids the expense of *a posteriori* error estimation whilst still predicting these errors efficiently. Even though generating the training data set is expensive, it is offline and is thus acceptable in practice.

The meshes generated via MeshingNet may be used in a variety of ways. If our goal is to obtain a high quality mesh with a desired number of elements then the approach described in this paper provides a cost-effective means of achieving this. If however the goal is to produce a solution with an estimated *a posteriori* error that is smaller than a desired tolerance, then the generated mesh may not meet this criterion. This may be addressed either by regenerating the mesh based upon a higher target number of elements in the final mesh, or through the use of a traditional *a posteriori* estimate on the computed solution in order to guide further mesh refinement. In the latter case we can view MeshingNet as a means to obtaining an improved initial mesh within a traditional mesh adaptivity workflow.

In future, we plan to generalize MeshingNet onto more general problems: 3D geometries, more complex PDE systems and BCs (e.g. Navier-Stokes and fluid-structure interactions), and time-dependent cases. For 3D problems, *Tetgen* [20] is able to do a similar refinement process to what *Triangle* does in 2D in this paper, which will enable us to directly apply MeshingNet to 3D problems. Furthermore, it would be desirable to develop an interface to enable the mesh generator to read geometries from standard computer aided design software. For complex three-dimensional problems it also seems unlikely that the accurate, but very expensive, approach to training the error estimator that is used in this paper will always be computationally viable (despite its excellent performance). In such cases we may replace this estimator with a more traditional method such as [2, 4, 24], on relatively fine uniform grids for training purposes.

Finally, we note that the ANNs used in this initial investigation are relatively simple in their structure. In the future the use of new DNN models, such as Convolutional/Graph Neural Networks, should be considered. These may be appropriate for problems in three dimensions or with larger data sets, such as arising from more general geometries and boundary conditions.

References

1. Abadi, M., et al.: TensorFlow: Large-scale machine learning on heterogeneous systems (2015). http://tensorflow.org/, software available from tensorflow.org
2. Ainsworth, M., Oden, J.T.: A posteriori error estimation in finite element analysis. Comput. Methods Appl. Mech. Eng. **142**(1–2), 1–88 (1997)
3. Apel, T., Grosman, S., Jimack, P.K., Meyer, A.: A new methodology for anisotropic mesh refinement based upon error gradients. Appl. Numer. Math. **50**(3–4), 329–341 (2004)
4. Bank, R.E., Weiser, A.: Some a posteriori error estimators for elliptic partial differential equations. Math. Comput. **44**(170), 283–301 (1985)
5. Chedid, R., Najjar, N.: Automatic finite-element mesh generation using artificial neural networks-Part I: Prediction of mesh density. IEEE Trans. Magn. **32**(5), 5173–5178 (1996)

6. Chen, T.Q., Rubanova, Y., Bettencourt, J., Duvenaud, D.K.: Neural ordinary differential equations. In: Advances in Neural Information Processing Systems, pp. 6571–6583 (2018)
7. Chollet, F., et al.: Keras (2015). https://keras.io
8. Csáji, B.C.: Approximation with artificial neural networks. Faculty Sci. Etvs Lornd Univ. Hungary **24**, 7–48 (2001)
9. Dörfler, W.: A convergent adaptive algorithm for poisson's equation. SIAM J. Numer. Anal. **33**(3), 1106–1124 (1996)
10. Floater, M.S.: Mean value coordinates. Comput. Aided Geom. Des. **20**(1), 19–27 (2003)
11. Grätsch, T., Bathe, K.J.: A posteriori error estimation techniques in practical finite element analysis. Comput. Struct. **83**(4–5), 235–265 (2005)
12. Han, J., Jentzen, A., Weinan, E.: Solving high-dimensional partial differential equations using deep learning. Proc. Natl. Acad. Sci. **115**(34), 8505–8510 (2018)
13. Huang, G., et al.: Densely connected convolutional networks. In: 2017 IEEE Conference on Computer Vision and Pattern Recognition (CVPR), pp. 2261–2269, July 2017. https://doi.org/10.1109/CVPR.2017.243
14. Kingma, D.P., Ba, J.: Adam: A method for stochastic optimization. arXiv preprint arXiv:1412.6980 (2014)
15. Krizhevsky, A., Sutskever, I., Hinton, G.E.: Imagenet classification with deep convolutional neural networks. In: Advances in Neural Information Processing Systems, pp. 1097–1105 (2012)
16. Long, Z., Lu, Y., Ma, X., Dong, B.: PDE-net: Learning pdes from data. arXiv preprint arXiv:1710.09668 (2017)
17. Lowther, D., Dyck, D.: A density driven mesh generator guided by a neural network. IEEE Trans. Magn. **29**(2), 1927–1930 (1993)
18. Notay, Y.: An aggregation-based algebraic multigrid method. Electr. Trans. Numer. Anal. **37**(6), 123–146 (2010)
19. Shewchuk, J.R.: Delaunay refinement algorithms for triangular mesh generation. Comput. Geom. **22**(1–3), 21–74 (2002)
20. Si, H.: TetGen, a Delaunay-based quality tetrahedral mesh generator. ACM Trans. Math. Softw. (TOMS) **41**(2), 11 (2015)
21. Sirignano, J., Spiliopoulos, K.: DGM: a deep learning algorithm for solving partial differential equations. J. Comput. Phys. **375**, 1339–1364 (2018)
22. Stevenson, R.: Optimality of a standard adaptive finite element method. Found. Comput. Math. **7**(2), 245–269 (2007)
23. Veit, A., Wilber, M.J., Belongie, S.: Residual networks behave like ensembles of relatively shallow networks. In: Advances in Neural Information Processing Systems, pp. 550–558 (2016)
24. Zienkiewicz, O., Zhu, J.: Adaptivity and mesh generation. Int. J. Numer. Methods Eng. **32**(4), 783–810 (1991)

A Block Preconditioner for Scalable Large Scale Finite Element Incompressible Flow Simulations

Damian Goik and Krzysztof Banaś[(✉)]

AGH University of Science and Technology,
al. A. Mickiewicza 30, 30-059 Kraków, Poland
pobanas@cyf-kr.edu.pl

Abstract. We present a block preconditioner, based on the algebraic multigrid method, for solving systems of linear equations, that arise in incompressible flow simulations performed by the stabilized finite element method. We select a set of adjustable parameters for the preconditioner and show how to tune the parameters in order to obtain fast convergence of the standard GMRES solver in which the preconditioner is employed. Additionally, we show some details of the parallel implementation of the preconditioner and the achieved scalability of the solver in large scale parallel incompressible flow simulations.

Keywords: Finite element method · Navier-stokes equations · Solvers of linear equations · Block preconditioning · Algebraic multigrid

1 Introduction

Stabilized finite elements are one of the popular techniques for solving Navier-Stokes equations of incompressible flows [5]. We are interested in the strategy, in which the finite element method is applied for space discretization, with some form of implicit discretization in time, either for transient problems or for pseudo-transient continuation employed to achieve steady-state [11]. The resulting non-linear systems are usually solved either by some form of Newton iterations or Picard, fixed-point, iterations [7] – we select the latter technique in our numerical examples.

In each of the considered scenarios there is a sequence of systems of linear equations to be solved. Due to the incompressibility condition, in the form of the requirement for divergence free velocity field, the systems are ill conditioned and the standard Krylov subspace methods with typical preconditioners (like ILU(k) – incomplete factorization algorithms) become inefficient [16,18].

The work was realized as a part of fundamental research financed by the Ministry of Science and Higher Education, grant no. 16.16.110.663.

V. V. Krzhizhanovskaya et al. (Eds.): ICCS 2020, LNCS 12139, pp. 199–211, 2020.
https://doi.org/10.1007/978-3-030-50420-5_15

We aim at developing a solver for large scale parallel incompressible flow simulations. Therefore we renounce the direct solvers, due to their super-linear complexity and poor parallel scalability [12,14]. We try to find a scalable and sufficiently strong preconditioner, in order to guarantee the convergence of the standard restarted GMRES method [17].

We consider block preconditioners [15,16] that split the linear systems into two parts: the first related to velocity components, with time derivative terms and better convergence properties, and the second related to the pressure, i.e. the incompressibility condition.

For the latter part we use an algorithm based on the algebraic multigrid, as the black-box version of the multigrid method [19], the only method that properly takes into account the infinite speed of propagation of pressure changes throughout the domain [7].

Several versions of algebraic multigrid (AMG) have been proposed for dealing with the pressure related part of the system that arise in block preconditioners for the Navier-Stokes equations [8]. The extensive studies in [6] showed the strong deterioration of convergence properties of the solvers for the increasing Reynolds and CFL numbers in the case of using smoothed aggregation multigrid. In the recent article [20] classical AMG is compared with smooth aggregation AMG for a large scale transient problem.

In the current paper we propose a block preconditioner based on the algorithm outlined in [18], combined with the classical AMG algorithm by Stuben [19]. We investigate the optimization options available for the standard AMG and show how to obtain, based on the proper selection of techniques and parameters, good convergence properties together with the scalability for the whole solver in large scale stationary incompressible flow simulations.

2   Problem Statement

We solve the Navier-Stokes equations of incompressible fluid flow, formulated for the unknown fluid velocity $u(x, t)$ and pressure $p(x, t)$ that satisfy:

$$\rho \left(\frac{\partial u}{\partial t} + (u \cdot \nabla)u - \nu \nabla^2 u \right) + \nabla p = f$$

$$\nabla \cdot u = 0 \tag{1}$$

together with boundary conditions:

$$u = \hat{u}_0 \quad \text{on } \Gamma_D$$

$$(\nu \nabla u)n - pn = g \quad \text{on } \Gamma_N$$

where ν and ρ denote kinematic viscosity and density of fluid respectively, and f is a source term that includes gravity forces (the system is considered in the dimensional form). The vector fields $\hat{u}_0$ and g are given on two disjoint parts of the boundary of the 3D computational domain Ω, Γ_D for velocities and Γ_N for stresses, respectively.

We discretize the Navier-Stokes equations using the spaces of continuous, piecewise linear polynomials. For velocity and pressure unknowns we consider the spaces V_u^h and V_p^h (vector valued for velocities), with functions that satisfy Dirichlet boundary conditions, while for test functions we apply the spaces V_w^h and V_r^h, with zero values on the Dirichlet parts of the boundary. We use SUPG stabilized finite element formulation [9] for space discretization that can be written (in index notation with the summation convention for repeated indices):

Find approximate functions $u^h \in V_u^h$ and $p^h \in V_p^h$ such that the following statement:

$$\int_\Omega \rho \frac{\partial u_j^h}{\partial t} w_j^h d\Omega + \int_\Omega \rho u_{j,l}^h u_l^h w_j^h d\Omega + \int_\Omega \rho \nu u_{j,l}^h w_{j,l}^h d\Omega - \int_\Omega p^h w_{j,j}^h d\Omega$$

$$- \int_\Omega u_{j,j}^h r^h d\Omega + \sum_e \int_{\Omega_e} R_j^{NS}(u^h, p^h) \tau_{jl} R_l^{NS}(w^h, r^h) d\Omega$$

$$+ \sum_e \int_{\Omega_e} u_{j,l}^h \delta w_{j,l}^h d\Omega = \int_\Omega f_j w_j^h d\Omega - \int_{\Gamma_N} g_j w_j^h d\Gamma$$

holds for every test function $w^h \in V_w^h$ and $r^h \in V_r^h$.

Above, $R_j^{NS}(u^h, p^h)$ and $R_l^{NS}(w^h, r^h)$ denote residuals of the Navier-Stokes equations computed for respective arguments, while τ_{jl} and δ are coefficients of SUPG stabilization [9].

Since in the current paper we are mainly interested in stationary problems and pseudo-transient continuation technique, we use the implicit Euler time discretization, due to its stability. When applied to (2), it leads to a non-linear problem for each time step. We use Picard's (simple) iterations for solving non-linear problems that finally lead to a series of linear problems.

The structure of each original linear system consists of 4×4 blocks for every finite element node in the mesh (three velocity components and pressure). For the purpose of applying block preconditioning the system is rearranged. In the vector of unknowns, first, all velocity components at all nodes are placed (we will denote that part of u by $\mathbf{u_v}$), followed by the pressure degrees of freedom (denoted by $\mathbf{u_p}$). With this approach the system of equations can be written as:

$$\begin{pmatrix} \mathbf{D_{vv}} & \mathbf{D_{vp}} \\ \mathbf{D_{pv}} & \mathbf{D_{pp}} \end{pmatrix} \cdot \begin{pmatrix} \mathbf{u_v} \\ \mathbf{u_p} \end{pmatrix} = \begin{pmatrix} \bar{\mathbf{b}}_\mathbf{w} \\ \mathbf{b_q} \end{pmatrix} \tag{2}$$

For classical mixed formulations without stabilization terms, the part $\mathbf{D_{vp}}$ is just the transpose of $\mathbf{D_{pv}}$, while the part $\mathbf{D_{pp}}$ vanishes. For the stabilized formulation additional terms appear in $\mathbf{D_{vp}}$, $\mathbf{D_{pv}}$ and $\mathbf{D_{pp}}$, while $\mathbf{D_{vv}}$ keeps its diagonally dominant form, due to the discretized time derivative term. The matrix $\mathbf{D_{vv}}$ depends additionally on the solution at the previous Picard's iteration, while the right hand side $\bar{\mathbf{b}}_w$ depends on the solution at the previous time step.

We solve the system (2) using the restarted GMRES method with left preconditioning [17]. At each GMRES iteration the two most time consuming steps are the multiplication of the residual vector (for the whole system) by the system matrix and then the application of the preconditioner. Formally, the preconditioner is represented as a matrix, that tries to approximate the inverse of the system matrix (the better the approximation, the faster the GMRES convergence) and the action of the preconditioner is represented as matrix-vector multiplication. In practice, the preconditioner matrix is usually not formed, instead, an algorithm is applied for the input vector, that is equivalent to a linear operator. For block preconditioners, the algorithm becomes complex, with several matrices involved, and iterative methods used for approximating inverses.

3   A Multigrid Based Block Preconditioner for Linear Equations

Following the approach in SIMPLE methods for solving Navier-Stokes equations [15], we observe that the inverse of the system matrix in Eq. 2 can be decomposed into the following product:

$$\begin{pmatrix} \mathbf{D_{vv}} & \mathbf{D_{vp}} \\ \mathbf{D_{pv}} & \mathbf{D_{pp}} \end{pmatrix}^{-1} = \\ \begin{pmatrix} \mathbf{I} & -\mathbf{D_{vv}^{-1}D_{vp}} \\ 0 & \mathbf{I} \end{pmatrix} \times \begin{pmatrix} \mathbf{D_{vv}^{-1}} & 0 \\ 0 & \mathbf{S^{-1}} \end{pmatrix} \times \begin{pmatrix} \mathbf{I} & 0 \\ -\mathbf{D_{pv}D_{vv}^{-1}} & \mathbf{I} \end{pmatrix}$$

where $\mathbf{S}$ is the Schur complement for $\mathbf{D_{pp}}$

$$\mathbf{S} = \mathbf{D_{pp}} - \mathbf{D_{pv}D_{vv}^{-1}D_{vp}}$$

The action of the inverse of the system matrix on a vector (with the parts related to velocity components and pressure denoted by $\mathbf{z_v}$ and $\mathbf{z_p}$ respectively) can be written as:

$$\begin{pmatrix} \mathbf{D_{vv}} & \mathbf{D_{vp}} \\ \mathbf{D_{pv}} & \mathbf{D_{pp}} \end{pmatrix}^{-1} \begin{pmatrix} \mathbf{z_v} \\ \mathbf{z_p} \end{pmatrix} = \begin{pmatrix} \mathbf{D_{vv}^{-1} \left(z_v - D_{vp}\bar{z}_p \right)} \\ \mathbf{\bar{z}_p} \end{pmatrix}$$

with

$$\bar{\mathbf{z}}_\mathbf{p} = \mathbf{S^{-1} \left(z_p - D_{pv}D_{vv}^{-1}z_v \right)}$$

The above formulae would correspond to the application of the perfect preconditioner (being the exact inverse of the system matrix), that would guarantee the convergence of GMRES in a single iteration [17]. However, the construction of the presented exact form does not satisfy the requirement for the preconditioner to be relatively cheap, hence, some appoximations have to be performed.

We consider the approximation based on the SIMPLEC (Semi-Implicit Pressure Linked Equation Corrected) algorithm, where the action of the preconditioner is split into three steps [8]:

1. solve approximately: $\mathbf{D_{vv}\tilde{z}_v = z_v}$
2. solve approximately: $\mathbf{\tilde{S}\hat{z}_p = z_p - D_{pv}\tilde{z}_v}$
3. substitute: $\mathbf{\hat{z}_v = \tilde{z}_v - \tilde{D}_{vv}^{-1}D_{vp}\hat{z}_p}$

where any vector with parts $\mathbf{z_v}$ and $\mathbf{z_p}$ is used as an input, and the output is stored in $\hat{\mathbf{z}}_\mathbf{v}$ and $\hat{\mathbf{z}}_\mathbf{p}$ (with an intermediate vector $\tilde{\mathbf{z}}_\mathbf{v}$). Above, $\tilde{\mathbf{S}}$ is an approximation to the original Schur complement matrix $\mathbf{S}$, that changes the original block $\mathbf{D}_{\mathbf{vv}}^{-1}$ to some approximation. The approximation to $\mathbf{D}_{\mathbf{vv}}^{-1}$ is also used in step 3 of the algorithm (denoted their by $\hat{\mathbf{D}}_{\mathbf{vv}}^{-1}$), although these two approximations to $\mathbf{D}_{\mathbf{vv}}^{-1}$ can be different.

The presented above three-step algorithm is used as the preconditioner in our GMRES solver, with the input parts $\mathbf{z_v}$ and $\mathbf{z_p}$ provided by the product of the system matrix and a suitable GMRES residual vector.

There are several factors influencing the quality of the preconditioner, and in consequence the convergence and scalability of the solver. The first factor is the quality of the approximation to $\mathbf{D}_{\mathbf{vv}}^{-1}$ in Step 1 of the procedure. In our implementation we solve approximately the system, simply by employing some number of Gauss-Seidel iterations to $\mathbf{D_{vv}}$. The convergence is sufficient to decrease the residual fast, partially due to the diagonal dominance of time derivative terms in $\mathbf{D_{vv}}$.

The second factor is the accuracy of obtaining $\hat{\mathbf{z}}_\mathbf{p}$ in Step 2. It is influenced by the solution procedure for the system of equations, as well as the choice of the approximation to $\mathbf{D}_{\mathbf{vv}}^{-1}$ in the approximate Schur complement matrix $\tilde{\mathbf{S}}$.

Usually the approximation to $\mathbf{D}_{\mathbf{vv}}^{-1}$ in $\tilde{\mathbf{S}}$ uses the diagonal form, with the matrix having inverted diagonal entries of $\mathbf{D_{vv}}$ as the simplest choice (the original SIMPLE algorithm). In the SIMPLEC approach, adopted in our implementation, the diagonal matrix approximating $\mathbf{D}_{\mathbf{vv}}^{-1}$ have at each diagonal position the inverse of the sum of the absolute values of the entries in the corresponding row of $\mathbf{D_{vv}}$. We use the same approximation to $\mathbf{D}_{\mathbf{vv}}^{-1}$ in Step 3 of the SIMPLEC algorithm.

Given the approximation to $\tilde{\mathbf{S}}$, the most important, from the computational point of view, is the choice of the solution procedure for the associated subsystem. We want to apply an iterative solver, since direct solvers become infeasible for large scale 3D problems. However, the system is difficult to solve due to its ill-conditioning (related to infinite speed of propagation of pressure changes in incompressible flows). Usually a multilevel solver is required in order to guarantee good convergence rates, independent of the mesh size (classical Krylov solvers with ILU preconditioning are not scalable with that respect). In the case of the stabilized methods, the addition of non-zero $\mathbf{D_{pp}}$ deteriorates the convergence by a large factor [18]. This is because without the stabilization the system has positive eigenvalues exclusively and the GMRES worst case convergence is different for matrices having eigenvalues of both signs [17].

We select a classical algebraic multigrid method, AMG [19], for the approximate solution of the system in Step 2 of the SIMPLEC algorithm. Gauss-Seidel iterations are used for smoothing at each level of the solver, with the coarser systems obtained using the Galerkin projection. The system at the last level is solved exactly using a direct solver. For the approximate solution we employ a single V-cycle of the AMG algorithm. Standard restriction and prolongation AMG operators [19] are used for projecting the solution between levels.

The key to the efficiency of the solver lies in the procedure for creating coarser levels, i.e. the selection of the degrees of freedom from the finer level that are retained at the coarser level. The construction of the hierarchy of system levels is done during the levels set-up phase of the solver, performed once per system solution and followed by some number of V-cycle iterations. Usually one can select less levels that would lead to slower convergence but a faster single V-cycle iteration or more levels, with faster convergence and slower individual V-cycle iterations.

The cost of the set-up phase depends on the number of levels and a particular algorithm used for coarse level creation. We use a classical approach, that is based on partitioning, at each level, the degrees of freedom into two sets: interpolatory (retained at the coarser level) and non-interpolatory (removed form the system). In order to partition the DOFs, the notions of dependence and importance are introduced [19]. The importance is a measure of how many other rows are influenced by the solution for a given row. The relation opposite to the influence is called dependence.

We adopt the following formula for finding the set S_i of DOFs influencing a given DOF [10] (with the system matrix entries denoted by a_{ij}):

$$S_i \equiv \{j \neq i : -a_{ij} \geq \alpha \max_{k \neq i}(-a_{ik})\}$$

with the parameter $\alpha \in (0,1)$ specifying the threshold for the strength of influence, that determines the inclusion of a DOF into the set S_i (we call α the strength threshold). An important observation is that only a single row of a matrix itself defines what influences a particular DOF, making this definition easy to use when the matrix is distributed on different computational nodes.

The DOFs are selected in the order of the number of DOFs that a given DOF influences. After selection of a DOF all DOFs that are influenced by this DOF are moved to the set of noninterpolatory degrees of freedom and the algorithm continues.

In our solution we chose to run this algorithm on every computational node separately, thus there is a possibility of dependencies between two DOFs on the boundaries of the subdomains. In our approach we let the owner of a DOF to assign it either to the set of interpolatory or noninterpolatory degrees of freedom and then broadcast this decision to all adjacent subdomains. Because of that procedure, the actual selections for different numbers of subdomains are not the same. This can lead to different convergence properties, that eventually make the numerical scalability of the solver problem dependent.

After the partition of DOFs, an interpolation matrix is created with each DOF having it's row and each interpolatory DOF having it's corresponding column. The rows associated with the interpolatory DOFs have just a single entry equal to one in a column related to this DOF. The rest of the rows is filled by the classic direct interpolation rules [19]. The interpolation matrix is used for the restriction and prolongation operations, as well as for the creation of the coarse systems using the Galerkin projection.

4   Parallel Implementation

The whole solution procedure is implemented within the finite element framework ModFEM [13] with the PETSC library [1] employed for linear algebra operations. ModFEM is a general purpose finite element software framework, with modular structure [2], that uses special problem dependent modules to create codes for different application domains [4]. In our setting the generic ModFEM modules are used to manage computational grids (in particular to perform domain decomposition for parallel execution) and to calculate element stiffness matrices and load vectors for the particular incompressible flow problem formulation that we employ.

We created a special ModFEM module for solving systems of linear equations that implements the algorithm described in the paper. The module receives local element matrices and vectors for the system of linear equations during Navier-Stokes simulations, assembles them to the global system matrix and right hand side vector, creates the other necessary preconditioner matrices, in particular the AMG levels structure, and performs preconditioned residual computations for the GMRES solver.

The module is built around matrix and vector data structures provided by the PETSC library. The PETSC linear algebra data structures and operations (including the sparse matrix-matrix product) serve as building blocks for the preconditioner responsibilities. Apart from basic matrix and vector operations, the only PETSC algorithm utilized during the set-up and solution phases is parallel matrix successive over-relaxation which is adapted to serve as Gauss-Seidel smoother. The parallel successive over-relaxation executes a configurable number of local iterations for each subdomain and a configurable number of global block Jacobi iterations, where blocks correspond to subdomain matrices. Such hybrid algorithm, frequently used in parallel iterative solvers, results in lower convergence rates than the Gauss-Seidel method at the global level, but has much lower cost and provides good scalability [3].

During the parallel solution procedures, communication steps are required only for global vector operations (norm, scalar product) and the exchange of data during Gauss-Seidel/Jacobi iterations. The scheme for exchanging data for ghost nodes is created by the generic ModFEM domain decomposition module and passed to the special module.

In our implementation we finally have the following set of control parameters to achieve the best GMRES convergence when using the developed preconditioner (in terms of CPU time for reducing relative error by a specified factor):

– the number of Gauss-Seidel iterations in step 1 of the SIMPLE procedure
– the form of $\mathbf{D}_{\mathbf{vv}}^{-1}$ in the construction of the preconditioner for step 2
– the method and accuracy of solving the system in step 2
  • the number of pre-smooth steps at each level
  • the number of post-smooth steps at each level
  • the number of outer, additive Schwarz (block Jacobi) iterations and the number of inner, multiplicative Schwarz (block Gauss-Seidel) iterations for each smoothing step

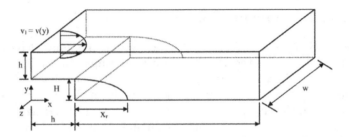

Fig. 1. 3D backward facing step - problem definition.

- the number of AMG levels and the size of the last level, for which a direct solver is used
- strength threshold which affects how many DOFs are dropped on subsequent levels

5   Numerical Example

As a numerical example for testing the performance of the developed solver we take a well known stationary flow problem – backward facing step in its 3D form (Fig. 1). All boundary conditions are assumed as no-slip with zero velocity, except the inflow boundary with parabolic (with respect to y dimension) inflow velocity. The parameters are chosen in such a way, that the Reynolds number of the flow is equal to 800, which makes the finite element as well as the linear solver convergence difficult, but still remains in laminar regime. The step height, H, and the inflow height, h, are both assumed equal to 0.5.

The actual computational domain with the dimensions $1 \times 1 \times 20$ is triangulated with an initial (G0) mesh having 38 000 elements and 23 221 nodes (Fig. 2). For this mesh a stationary solution is obtained and then the mesh is uniformly refined to produce a new, generation 1 (G1), mesh with 304 000 elements and 168 441 nodes (673 764 degrees of freedom in the solved linear system). For this mesh the solution procedure is continued until the convergence and the same procedure is repeated for the next meshes. The uniform refinements produce the meshes:

- generation 2 (G2): 2 432 000 elements, 1 280 881 nodes, 5 123 524 DOFs
- generation 3 (G3): 19 456 000 elements, 9 985 761 nodes, 39 943 044 DOFs

The numerical experiments were performed on different numbers of nodes from the Prometheus system at Cyfronet AGH computing centre. Each node has two 12-core Intel Xeon E5-2680v3 CPUs (2.5GHZ) and 128GB DRAM and runs under Centos7 Linux version.

We present the performance measurements for a single typical time step and one non-linear iteration during the simulation. For the purpose of our tests we performed the GMRES solver iterations with high accuracy, stopping the process when the relative residual dropped by the factor of 10^9.

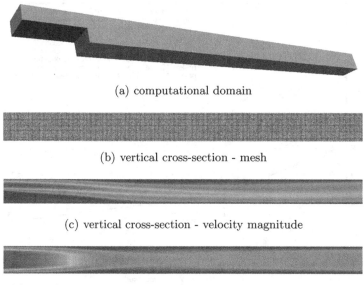

(a) computational domain

(b) vertical cross-section - mesh

(c) vertical cross-section - velocity magnitude

(d) horizontal cross-section at height H+0.5h - velocity magnitude

Fig. 2. 3D backward facing step problem – computational domain, G0 mesh and Re=800 solution contours for a part of the computational domain.

During the tests we tried to establish the best set of control parameters to be used for the multigrid phase of solving the system with approximated Schur complement. Fig. 3 presents the comparison of execution times for a single iteration of our solver on mesh G2, obtained with different sets of parameters. The symbols used for different lines on the plot contain encoded parameter values, in such a way that the symbol nr-lev-a-pre-b-post-c-in-d-out-e-alpha-f indicates the configuration with: a levels, b presmoothing steps, c post-smoothing steps, d inner, Gauss-Seidel iterations within single multigrid smoothing step, e outer, block Jacobi (additive Schwarz) iterations within single multirid smoothing step and the value of strength threshold α equal to f (the value 0 for the last parameter indicates a very small coefficient, in our experiments equal to 0.0001).

Similar results were obtained for executions on other mesh sizes, thus some general guidelines regarding the solver tuning can be deduced:

- additional AMG levels always speed up the convergence (less GMRES iterations required) and execution time, hence the number of levels should be such that additional level would not reduce the number of rows on the last level
- more aggressive coarsening tends to produce better level structure in terms of memory consumption and iterations execution time
- there is not much to be gained by manipulating local to global iterations ratio
- time spent in GMRES should be prioritized over time spent in the preconditioner i.e. adding more iterations does not always improve the convergence

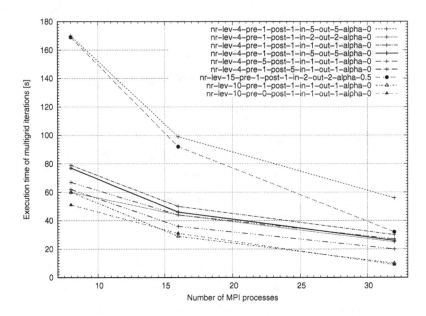

Fig. 3. 3D backward facing step problem – execution time on the Prometheus cluster for the iteration phase of the multigrid algorithm for different combinations of parameters (description of line symbols in the text)

The best configuration in our experiments had the most aggressive coarsening possible, no limit imposed on the number of AMG levels and just one post-smoothing iteration with one global and local iteration, without pre-smoothing.

For this best configuration we show in Table 1 the comparison of the total linear system solution time on a single cluster node for our developed solver and a high performance direct solver (in this case the PARDISO from the Intel MKL library). The execution of the ModFEM code was done in MPI only mode, while the direct PARDISO solver was running in OpenMP only mode.

Table 1. 3D backward facing step problem – execution time (in seconds) for solving the system of linear equations (673 764 DOFs) using two solvers: the direct PARDISO solver from the Intel MKL library and the GMRES solver with the developed block preconditioner based on AMG

	Number of cores				
Solver	1	2	4	8	16
PARDISO	176.46	109.09	60.34	39.17	25.19
GMRES+AMG	115.51	56.61	29.93	16.87	10.39

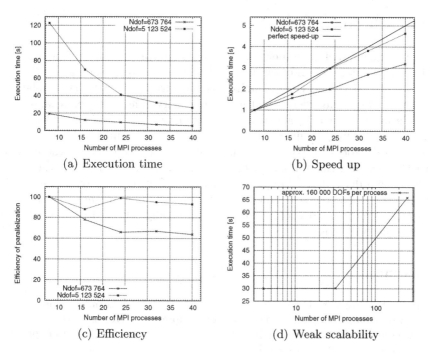

(a) Execution time

(b) Speed up

(c) Efficiency

(d) Weak scalability

Fig. 4. 3D backward facing step problem – parallel performance metrics for two problem sizes (Ndof equal to 673 764 and 5 123 524): execution time, speed up and parallel efficiency for the growing number of cluster processors plus weak scalability results as the execution time for 4, 32 and 256 processes with approx. 160 000 DOFs per process (subdomain)

In order to assess the computational scalability of the developed solver we performed a series of tests for the best algorithmic version. Figure 4 presents the parallel performance characteristics obtained for two problem sizes (the number of DOFs equal to 673 764 and 5 123 524) and parallel runs on different number of processors in the Prometheus cluster. The metrics include execution time, standard parallel speed up and efficiency, as well as the results for weak scalability study. The latter was performed for approximately 160 000 DOFs per subdomain and three numbers of cluster nodes, with the last system, solved using 256 processes, having 39 943 048 DOFs.

The weak scalability study was also used to assess the numerical scalability of the algorithm. The results indicated that the convergence of the solver has not deteriorated for subsequent, refined meshes. The overall GMRES convergence rates were equal to 0.24 for G1 mesh, 0.14 for G2 mesh and 0.19 for G3 mesh. This strong convergence was obtained with the same time step length for all meshes and, hence, the growing CFL number for refined meshes. The different numbers of subdomains for each case influenced the convergence results as well. We plan to investigate these issues in forthcoming papers.

6 Conclusions

We have shown that a proper design and tuning of parameters for algebraic multigrid method, used in block preconditioners employed to accelerate the convergence of GMRES method in finite element incompressible flow simulations, can lead to fast convergence and good scalability. The tests for a large scale example problem of 3D backward facing step produced solution times in the range of one minute for linear systems with approx. 40 million degrees of freedom and 256 processes (cores). For this size the standard direct methods or the GMRES method with standard ILU preconditioners cannot produce the results in the same order of time. We plan further investigations to show the strategies for optimal parameter selection depending on the CFL number in time discretization, that should lead to methods specifically adjusted to transient and steady-state problems.

References

1. Balay, S., Gropp, W.D., McInnes, L.C., Smith, B.F.: Efficient management of parallelism in object oriented numerical software libraries. In: Arge, E., Bruaset, A.M., Langtangen, H.P. (eds.) Modern Software Tools in Scientific Computing, pp. 163–202. Birkhäuser Press, Basel (1997)
2. Banaś, K.: A modular design for parallel adaptive finite element computational kernels. In: Bubak, M., van Albada, G.D., Sloot, P.M.A., Dongarra, J. (eds.) ICCS 2004. LNCS, vol. 3037, pp. 155–162. Springer, Heidelberg (2004). https://doi.org/10.1007/978-3-540-24687-9_20
3. Banaś, K.: Scalability analysis for a multigrid linear equations solver. In: Wyrzykowski, R., Dongarra, J., Karczewski, K., Wasniewski, J. (eds.) PPAM 2007. LNCS, vol. 4967, pp. 1265–1274. Springer, Heidelberg (2008). https://doi.org/10.1007/978-3-540-68111-3_134
4. Banaś, K., Chłoń, K., Cybułka, P., Michalik, K., Płaszewski, P., Siwek, A.: Adaptive finite element modelling of welding processes. In: Bubak, M., Kitowski, J., Wiatr, K. (eds.) eScience on Distributed Computing Infrastructure. LNCS, vol. 8500, pp. 391–406. Springer, Cham (2014). https://doi.org/10.1007/978-3-319-10894-0_28
5. Brooks, A., Hughes, T.: Streamline upwind/Petrov-Galerkin formulations for convection dominated flows with the particular emphasis on the incompressible Navier-Stokes equations. Comput. Methods Appl. Mech. Eng. **32**, 199–259 (1982)
6. Cyr, E.C., Shadid, J.N., Tuminaro, R.S.: Stabilization and scalable block preconditioning for the Navier-Stokes equations. J. Comput. Physics **231**(2), 345–363 (2012). https://doi.org/10.1016/j.jcp.2011.09.001
7. Elman, H., Silvester, D., Wathen, A.: Finite Elements and Fast Iterative Solvers with Applications in Incompressible Fluid Dynamics. Oxford University Press, Oxford (2005)
8. Elman, H., Howle, V., Shadid, J., Shuttleworth, R., Tuminaro, R.: A taxonomy and comparison of parallel block multi-level preconditioners for the incompressible Navier-Stokes equations. J. Comput. Phys. **227**(3), 1790–1808 (2008). https://doi.org/10.1016/j.jcp.2007.09.026

9. Franca, L., Frey, S.: Stabilized finite element methods II: the incompressible Navier-Stokes equations. Comput. Methods Appl. Mech. Eng. **99**, 209–233 (1992)
10. Henson, V.E., Yang, U.M.: Boomeramg: a parallel algebraic multigrid solver and preconditioner. Appl. Numer. Math. **41**(1), 155–177 (2002). https://doi.org/10.1016/S0168-9274(01)00115-5
11. Kelley, C., Keyes, D.: Convergence analysis of pseudo-transient continuation. SIAM J. Numer. Anal. **35**, 508–523 (1998)
12. Koric, S., Lu, Q., Guleryuz, E.: Evaluation of massively parallel linear sparse solvers on unstructured finite element meshes. Comput. Struct. **141**, 19–25 (2014). https://doi.org/10.1016/j.compstruc.2014.05.009
13. Michalik, K., Banaś, K., Płaszewski, P., Cybułka, P.: ModFEM - a computational framework for parallel adaptive finite element simulations. Comput. Methods Mater. Sci. **13**(1), 3–8 (2013)
14. Pardo, D., Paszynski, M., Collier, N., Alvarez, J., Dalcin, L., Calo, V.M.: A survey on direct solvers for galerkin methods. SeMA J. **57**, 107–134 (2012). https://doi.org/10.1007/BF03322602
15. Patankar, S.V.: Numerical Heat Transfer and Fluid Flow. Series on Computational Methods in Mechanics and Thermal Science, Hemisphere Publishing Corporation (CRC Press, Taylor & Francis Group) (1980)
16. Pernice, M., Tocci, M.: A multigrid-preconditioned Newton-Krylov method for the incompressible Navier-Stokes equations. Ind. Appl. Math. **23**, 398–418 (2001). https://doi.org/10.1137/S1064827500372250
17. Saad, Y.: Iterative Methods for Sparse Linear Systems. PWS Publishing, Boston (1996)
18. Segal, A., Rehman, M., Vuik, C.: Preconditioners for incompressible Navier-Stokes solvers. Numer. Math. Theory Methods Appl. **3** (2010). https://doi.org/10.4208/nmtma.2010.33.1
19. Stüben, K.: A review of algebraic multigrid. J. Comput. Appl. Math. **128**(1), 281–309 (2001). https://doi.org/10.1016/S0377-0427(00)00516-1
20. Thomas, S.J., Ananthan, S., Yellapantula, S., Hu, J.J., Lawson, M., Sprague, M.A.: A comparison of classical and aggregation-based algebraic multigrid preconditioners for high-fidelity simulation of wind turbine incompressible flows. SIAM J. Sci. Comput. **41**(5), S196–S219 (2019). https://doi.org/10.1137/18M1179018

Integrating Agent-Based Modelling with Copula Theory: Preliminary Insights and Open Problems

Peter Fratrič[1,3(✉)], Giovanni Sileno[1], Tom van Engers[1,2], and Sander Klous[1]

[1] Informatics Institute, University of Amsterdam, Amsterdam, The Netherlands
p.fratric@uva.nl
[2] Leibniz Institute, TNO/University of Amsterdam, Amsterdam, The Netherlands
[3] Nikhef – National Institute for Subatomic Physics, Amsterdam, The Netherlands

Abstract. The paper sketches and elaborates on a framework integrating agent-based modelling with advanced quantitative probabilistic methods based on copula theory. The motivation for such a framework is illustrated on a artificial market functioning with canonical asset pricing models, showing that dependencies specified by copulas can enrich agent-based models to capture both micro-macro effects (e.g. herding behaviour) and macro-level dependencies (e.g. asset price dependencies). In doing that, the paper highlights the theoretical challenges and extensions that would complete and improve the proposal as a tool for risk analysis.

Keywords: Agent-based modelling · Copula theory · Dependencies · Risk

1 Introduction

Complex systems like markets are known to exhibit properties and phenomenal patterns at different levels (e.g. trader decision-making at micro-level and average asset price at macro-level, individual defaults and contagion of defaults, etc.). In general, such stratifications are irreducible: descriptions at the micro-level cannot fully reproduce phenomena observed at macro-level, plausibly because additional variables are failed to be captured or cannot be so. Yet, anomalies of behaviour at macro-level are typically originated by the accumulation and/or structuration of divergences of behaviour occurring at micro-level, (see e.g. [1] for trade elasticities). Therefore, at least in principle, it should be possible to use micro-divergences as a means to evaluate and possibly calibrate the macro-level model. One of the crucial aspects for such an exercise would be to map which features of the micro-level models impact (and do not impact) the macro-level model. From a conceptual (better *explainability*) and a computational (better *tractability*) point of view, such mapping would enable a practical decomposition of the elements at stake, thus facilitating parameter calibration and estimation from data. Moreover, supposing these parameters to be adequately extracted, one could put the system in stress conditions and see what kind of systematic

V. V. Krzhizhanovskaya et al. (Eds.): ICCS 2020, LNCS 12139, pp. 212–225, 2020.
https://doi.org/10.1007/978-3-030-50420-5_16

response would be entailed by the identified dependence structure. The overall approach could provide an additional analytical tool for *systematic risk*.

With the purpose of studying and potentially providing a possible solution to these requirements, we are currently working on establishing a theoretical framework integrating agent-based modelling (ABM) with advanced quantitative probabilistic methods based on *copula theory*. The intuition behind this choice is the possibility to connect the causal, agentive dependencies captured by agent-based models with the structural dependencies statistically captured by copulas, in order to facilitate the micro-macro mappings, as well as the extraction of dependencies observable at macro-level.

To the best of our knowledge, even if many research efforts targeting hybrid qualitative-quantitative methods exist in the computational science and artificial intelligence literature, the methodological connection of ABM with copula theory is still an underexplored topic. A large-scale agent-based model of trader agents incorporating serial dependence analysis, copula theory and co-evolutionary artificial market, allowing traders to change their behaviour during crisis periods, has been developed in [2]; the authors rely on copula to capture cross-market linkages on macro-level. A similar approach is taken in [4]. Examples of risk analysis in network-based setting can be found for instance in [5,6], in which the mechanisms of defaults and default contagion are separated from other dependencies observed in the market. In [3], copula is used to model low-level dependencies of natural hazards with agent-based models, in order to study their impact at macro-level. In the present paper, we will use copula to model dependencies among agents at micro-level, and we will propose a method to combine aggregated micro-correlations at market scale.

The paper is structured as follows. Section 2 provides some background: it elaborates on the combined need of agent-based modeling and of quantitative methods, illustrating the challenges on a running example based on canonical trader models for asset pricing, and gives a short presentation on copula theory. Section 3 reports on the simulation of one specific hand-crafted instanciation of copula producing a relevant result from the running example, and will elaborate on extensions and theoretical challenges that remain to be solved for the proposal to be operable. A note on future developments ends the paper.

2   Agent-Based Modeling and Copula Theory

In financial modelling, when statistical models are constructed from time series data, it is common practice to separately estimate serial dependencies and cross-sectional dependencies. The standard approach to capture serial dependence (also referred to as *autocorrelation*) is to use autoregressive models [10]. If the time series exhibit *volatility clustering*–i.e. large changes tend to be followed by large changes, of either sign, small changes tend to be followed by small changes— then it is typical to use the *generalized autoregressive conditional heteroskedasticity* model (GARCH) [11], or one of its variants [12]. Once the GARCH model is estimated, the cross-sectional dependence analysis can be performed on the residuals. Unfortunately, autoregressive models provide only little information useful for interpretation; this is no surprise since these models are purely quantitative, and suffer problems similar to all data-driven methods. As low interpretability goes along with limited possibility of performing counterfactual or *what-if* reasoning (see e.g. the discussion in [9]), such models are weakly justifiable in policy-making contexts, as for instance in establishing sound risk balancing measures to be held by economic actors.

2.1   Agent-Based Modelling

An alternative approach is given by *agent-based modelling* (ABM), through which several interacting heterogeneous agents can be used to replicate patterns in data (see e.g. [2]). Typically, agent models are manually specified from known or plausible templates of behaviour. To a certain extent, their parameters can be set or refined by means of some statistical methods. Model validation is then performed by comparing the model execution results against some expected theoretical outcome or observational data. These models can be also used to discover potential states of the system not yet observed [23], thus becoming a powerful tool for policy-making. Independently from their construction, agent models specify, at least qualitatively, both serial dependencies (for the functional dependence between actions) and cross-sectional dependencies (for the topological relationships between components), and are explainable in nature. However, they do not complete the full picture of the social system, as the focus of the designers of agent-based models is typically on the construction of its micro-level components. Nevertheless, to elaborate on the connection between micro-level and macro-level components of a social system we still need to start from capturing *behavioural* variables associated to micro-level system components, assuming them to have the strongest effect at micro-level (otherwise the micro-level would be mostly *determined* by the macro-level).

Running Example: Asset Market. We will then consider a paradigmatic scenario for ABM: an asset market, in which traders concurrently sell, buy or hold their assets. Our running example is based on canonical asset pricing models, proceeding along [17].

Fundamental Value. The target asset has a publicly available fundamental value given by a random walk process:

$$F_t = F_{t-1} + \eta \tag{1}$$

where η is a normally distributed random variable with mean zero and standard deviation σ_η.

Market-Maker Agent. At the end of each trading day, a market-maker agent sets the price at which a trader agent can buy or sell the asset according to a simple rule:

$$p_{t+1} = p_t + \Delta p_t \tag{2}$$

where:

$$\Delta p_t = a(1 + D(t) - S(t)) + \delta$$

The variable $D(t)$ denotes the number of buy orders at time t, $S(t)$ denotes number of sell orders at time t and δ is a normally distributed random variable with zero mean and constant standard deviation σ_δ. The positive coefficient a can be interpreted as the speed of price adjustment.

Fundamental Traders. Fundamental traders operate with the assumption that the price of an asset eventually returns to its fundamental value. Therefore, for them it is rational to sell if the value of an asset is above its fundamental value and buy if the value of an asset is below its fundamental value. Their price expectation can be written as:

$$E_t^{fund}[p_{t+1}] = p_t + x_{fund}(F_t - p_t) + \alpha \tag{3}$$

where α is a normally distributed random variable with mean zero and standard deviation σ_α. x_{fund} can be interpreted as the strength of a *mean-reverting belief* (i.e. the belief that the average price will return to the fundamental value).

Technical Traders. In contrast, technical traders, also referred to as *chartists*, decide on the basis of past trend in data. They will buy if the value of an asset is on the rise, because they expect this rise to continue and sell if the value is on decline. Their expectation can be written as:

$$E_t^{tech}[p_{t+1}] = p_t + x_{tech}(p_t - p_{t-1}) + \beta \tag{4}$$

where β is a normally distributed random variable with mean zero and standard deviation σ_β. x_{tech} can be interpreted as a strength of reaction to the trend.

Relevant Scenario: Herding Behaviour. Because they are intrinsic characteristics of each agent, x_{fund} and x_{tech} can be seen as capturing the behavioural variables we intended to focus on at the beginning of this section.

Now, if for all traders x_{fund} or x_{tech} happen to be realized with unexpectedly large values at the same time, the effect of α and β will be diminished, and this will result in higher (lower) expected value of the asset price and then in the consequent decision of traders to buy (sell). Purchases (sales) in turn will lead the market-maker agent to set the price higher (lower) at the next time step, thus reinforcing the previous pattern and triggering *herding behaviour*. Such chain of events are known to occur in markets, resulting in periods of rapid increase of asset prices followed by periods of dramatic falls. Note however that this scenario is not directly described by the agent-based models, but is entailed as a possible consequence of specific classes of instantiations.

2.2 Combining Agent-Based Models with Probability

Herding behaviour is recognized to be a destabilizing factor in markets, although extreme time-varying volatility is usually both a cause and an effect of its occurrence. In the general case, factors contributing to herding behaviour are: the situation on the global market, the situation in specific market sectors, policies implemented by policy makers, etc. All these factors are somehow processed by each human trader. However, because such mental reasoning is only partially similar amongst the agents, and often includes non-deterministic components (including the uncertainty related to the observational input), it is unlikely that it can be specified by predefined, deterministic rules. For these reasons, probabilistic models are a suitable candidate tool to recover the impossibility to go beyond a certain level of model depth, in particular to capture the mechanisms behind behavioural variables as x_{fund} and x_{tech}. In the following, we will therefore consider two normally distributed random variables X_{fund} and X_{tech} realizing them.[1] This means that the traders will perceive the price difference in parenthesis of Eqs. (3) and (4) differently, attributing to it different importance at each time step.

Looking at Eqs. (3) and (4) we can see that the essence of an agent's decision making lies in balancing his *decision rule* (e.g. for the fundamental trader $x_{fund}(F_t - p_t)$) with the uncertainty about the asset price (e.g. α). If for instance the strength of the mean-reversing belief x_{fund} happens to be low (in probabilistic terms, it would be a value from the lower tail), then the uncertainty α will dominate the trader's decision. In contrast, if x_{fund} happens to be very high (i.e. from the upper tail), then the trader will be less uncertain and trader's decision to buy or sell will be determined by $(F_t - p_t)$. Similar considerations apply to technical traders.

Assuming that behavioural random variables are normally distributed, obtaining values from the upper tail is rather unlikely and, even if some agent's behavioural variable is high, it will not influence the asset price very much since the asset price is influenced collectively. However, if all traders have strong beliefs about the rise or fall of the price of the asset, then the price will change dramatically. The dependence of the price on a *collective* increase in certainty cannot be directly modeled by the standard toolkit of probability, and motivates the use of copulas.

2.3 Copulas

Copula theory is a sub-field of probability theory dealing with describing dependencies holding between random variables. Application-wise, *copulas* are well established tools for quantitative analysis in many domains, as e.g. economic time series [15] and hydrological data [16].

---

[1] We are not aware of behavioural research justifying this assumption of normality, but for both variables it seems plausible that deviations larger than twice the standard deviation from the mean will be rather improbable.

Consider a random variable $\mathbf{U} = (U_1, ..., U_d)$ described as a d-dimensional vector. If all components of $\mathbf{U}$ are independent, we can compute its joint probability distribution function as $F_{\mathbf{U}}(u_1, .., u_d) = F_{U_1}(u_1) \cdot ... \cdot F_{U_d}(u_d)$, i.e. the product of marginal distributions. In case of dependence among components we need some function that specifies this dependence. The concept of *copula* is essentially presented as a specific class of such functions, specifically defined on uniform marginals [13]:

Definition 1. $C : [0,1]^d \rightarrow [0,1]$ *is a d-dimensional copula if C is a joint cumulative distribution function of a d-dimensional random vector on the unit cube $[0,1]^d$ with uniform marginals.*

To obtain a uniform marginal distribution from any random variable we can perform a *probability integral transformation* $u_i = F_i(x_i)$, where F_i is the marginal distribution function of the random variable X_i. In practice, when we estimate the copula from data we estimate the *marginals* and *copula* components separately. We can then introduce the most important theorem of copula theory:

Theorem 1. *Let F be a distribution function with marginal distribution functions $F_1, ..., F_d$. There exists a copula C such that for all $(x_1, ..., x_d) \in [-\infty, \infty]^d$,*

$$F(x_1, ..., x_d) = C(F_1(x_1), ..., F_d(x_d)) \tag{5}$$

If $F_1, ..., F_d$ are continuous this copula is unique.

If we consider the partial derivatives of Eq. (5) we obtain the density function in the form:

$$f(x_1, ..., x_d) = c(F_1(x_1), ..., F_d(x_d)) \prod_{i=1}^{d} f_i(x_i). \tag{6}$$

where c is density of the copula and f_i is marginal density of random variable X_i. The reason why copulas gained popularity is that the cumulative distribution function F_i contains all the information about the marginal, while the copula contains the information about the structure of dependence, enabling a principled decomposition for estimations.

Correspondingly to the high variety of dependence structures observed in the real world, there exists many parametric families of copulas, specializing for specific types of dependence. The most interesting type for economic applications is *tail dependence*. For example, if there is nothing unusual happening on the market and time series revolve around its mean value, then the time series might seem only weakly correlated. However, co-movements far away from mean value tend to be correlated much more strongly. In probabilistic terms, the copula describing such dependence between random variables has strong tail dependence. Tail dependence does not have to be always symmetrical. Certain types of copulas have strong upper tail dependence and weaker lower tail dependence. In simple terms this means that there is higher probability of observing random variables together realized in the upper quantile of their distribution than in the lower quantile. One of the parametric copulas having such properties is the *Joe copula* [13], illustrated in Fig. 1.

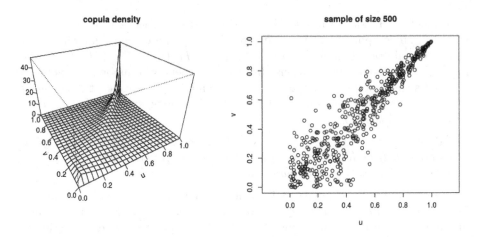

Fig. 1. Density and scatter plot of a bivariate Joe copula, with parameter set to 8. (picture taken from http://pilot.52north.org/shinyApps/copulatheque/copulas/)

3 Tracking Dependencies

This section aims to show how copulas can be effectively used to enrich stochastic agent-based models with additional dependencies, relative to micro- and macro-levels. In particular, we will focus on the phenomenon of dependence of the asset price to a collective increase in strength of belief associated to herding behaviour scenarios, observed in Sect. 2.2. To model the balance between certainty and uncertainty of each trader with respect to current price of the asset, we need to set the marginal distribution functions of X_{fund} and X_{tech} to have a mean value and standard deviation such that if herding behaviour occurs then the uncertainty parameters α and β will play essentially no role, but if herding behaviour is not occurring, then α and β will stop traders from massive buying or selling. Therefore the parameters for X_{fund}, X_{tech} and α, β are not entirely arbitrary.

3.1 Market Simulation

To illustrate the influence of dependence structures of behavioural random variables we will compare simulations of the market with *independent* behavioural random variables to simulations of the market whose random variables have a *dependence* structure defined by Joe copula. It is important to make clear that this exercise does not have any empirical claim: it is meant just to show that copula can be used for a probabilistic characterization of the social behaviour of the system.

Consider a group of $N_{total} = 1000$ trader agents consisting of $N_{fund} = 300$ fundamental traders and $N_{tech} = 700$ technical traders (this ration roughly reflects the situation of a real market). Let us denote with **X** the vector collecting all behavioural random variables of the traders. Table 1 reports all parameters

Table 1. Table of parameters used for the simulations

Parameter	Definition	Value
N	Number of traders	1000
N_{fund}	Number of fundamental traders	300
N_{tech}	Number of technical traders	700
T	Time steps (not including initial value)	500
Sim	Number of simulations	100
a	Price adjustment	$0.2 * 10^{-2}$
σ_δ	S.d. of random factor in price process	0.005
α	Uncertainty of fundamental trader	$N(0,1)$
β	Uncertainty of technical trader	$N(0,1)$
x_{fund}	Behavioural r.v. fundamental trader	$N(0.6, 0.1)$
x_{char}	Behavioural r.v. technical trader	$N(0.8, 0.4)$

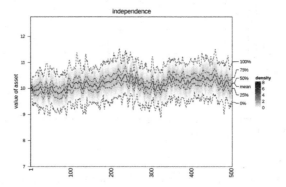

Fig. 2. 100 simulations when behavioural variables are independent.

we have used for our simulation. The sequence of fundamental values $\{F_t\}_{t=1}^T$ is generated by Eq. (1) at the beginning and remains the same for all simulations.

Independence Scenario. At first, we will assume that the (normally distributed) behavioural random variable assigned to each trader is independent of other normally distributed behavioural variables. This means that the probability density function capturing all behavioural variables of market can be written as a simple product of marginal density functions:

$$f_{\mathbf{X}}(\mathbf{x}) = \prod_{i=1}^{N_{fund}} f_{X_{fund}}(x_{fund}) \prod_{i=1}^{N_{tech}} f_{X_{tech}}(x_{tech}) \qquad (7)$$

We have considered $T = 502$ time steps with initialization $p_1 = 10$ and $p_2 = 10.02574$. Figure 2 illustrates the output of 100 simulations of the market with a probability density function given by the Eq. (7). The simulations

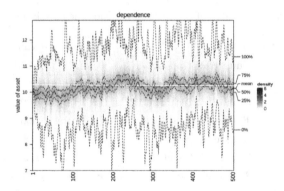

Fig. 3. 100 simulations when behavioural random variables have a dependence structure of defined by Joe copula.

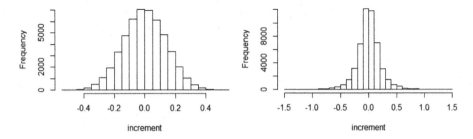

Fig. 4. Histogram of increments for an independence structure (left) and dependence structure defined by Joe copula (right).

on average follow the fundamental price of an asset. The marginal distribution of the increments Δp_t taking realizations of the generated time series (Fig. 2) clearly follows a normal distribution (left of Fig. 4).

Dependence Scenario. Let us keep same number of agents and exactly the same parameters, but this time consider a dependence structure between the behavioural variables described by a Joe copula with parameter equal to 8. As shown in Fig. 1, this copula has strong right upper tail dependence. The Joe copula is an Archimedean copula, which means it attributes an univariate generator, hence drawing samples from this copula is not time consuming, even for large dimensions. In our case, each sample will be a N_{total} dimensional vector $\mathbf{U}$ with components u_i from the unit interval. For each agent, a quantile transformation will be made $x_{A,i} = Q_{X_{A,i}}(u_i)$ by the quantile function $Q_{X_{A,i}}$ of i-the agent $A = \{fund, tech\}$ to obtain the realization from the agent's density function. Here the i-th agent A's behavioural random variable is again distributed following what specified in Table 1.

Running 100 market simulations, the time-series we obtain this time are much more unstable (Fig. 3). This is due to the structural change in the marginal distribution function of Δp_t, which has now much fatter tails. The fatter tails

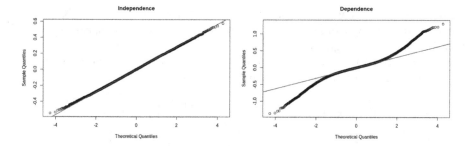

Fig. 5. Normal QQ-plots comparing distribution functions of increments Δp_t to normal distribution of behavioural variables with an independence structure (left) and with a dependence structure defined by Joe copula (right).

can be seen on the right histogram on Fig. 4, and on the comparison of both histograms by normal QQ-plots in Fig. 5. We see that for independence the increments follow a normal distribution very closely, but for dependence defined by Joe copula the tails of marginal distribution deviate greatly and approximates a normal distribution only around the mean.

3.2   Directions for Model Aggregation

For applications in finance it would be desirable to extend this model to consider broader contexts: e.g. a market with many assets, whose prices in general may exhibit mutual dependencies. A way to obtain this extension is to introduce adequate aggregators, following for instance what was suggested in [14, 3.11.3]. In order to apply this method, however, we need to make explicit a few assumptions.

Step-Wise Computation Assumption. The increase or the decrease of the price of the asset Δp_t is obtained via the simulation of the agent-based market model. As it can be seen in formulas (3) and (4), the increment Δp_t at time t depends on the realization of random variables $\mathbf{X}$ at time t with agents observing previous market values p_{t-1} and p_{t-2}. This means that Δp_t is also a continuous random variable and its probability density function should be written as $f_{\Delta p_t | p_{t-1}, p_{t-2}, \mathbf{X}}$. Note that this function can be entirely different at each time t depending on what values of p_{t-1}, p_{t-2} the agents observe. Since the time steps are discrete, then the density functions forms a sequence $\{f_{\Delta p_t | p_{t-1}, p_{t-2}, \mathbf{X}}\}_{t=t_0}^{t=T}$. In this paper, for simplicity, we will not describe the full dynamics of this sequence; we will focus only on one density function in one time step, assuming therefore that the computation can be performed step-wise. By fixing time, p_{t-1}, p_{t-2} are also fixed (they have already occurred), so we can omit them and write just $f_{\Delta p_t | \mathbf{X}}$.

Generalizing to Multiple Assets. Consider m non-overlapping groups of absolutely continuous random variables $\mathbf{X}_1, ..., \mathbf{X}_m$, where each group consists of behavioural random variables, or, to make our interpretation more general, predictor variables that determine the value of an asset. Each group $\mathbf{X}_g$ forms a random vector and has a scalar *aggregation random variable* $V_g = h_g(\mathbf{X}_g)$. This means that each value of an asset is determined by a mechanism specified by the function h_g, which might be an agent-based model similar to the one explored in the previous section, but this time each group of predictor random variables will have its own distribution function. We can write then:

$$f_{\mathbf{X}_1,...,\mathbf{X}_m}(\mathbf{x}_1,...,\mathbf{x}_m) = \frac{f_{\mathbf{V}}(v_1,...,v_m)}{\prod_{i=1}^{m} f_{V_i}(v_i)} \prod_{i=1}^{m} f_{\mathbf{X}_i}(\mathbf{x}_i) \tag{8}$$

where f denotes the marginal (joint) probability density function of the corresponding variable (variables) written as a subscript. The validity of (8) relies on two assumptions: (a) conditional independence of the groups given aggregation variables $V_1,..,V_m$ and (b) conditional distribution function of the group $\mathbf{X}_g$ conditioned on $V_1,...,V_m$ is the same as the conditional distribution of $\mathbf{X}_g$ conditioned on V_g (for a 2-dimensional proof see [14]). These assumptions are in principle not problematic in our application because we are assuming that all interactions on micro-level of the agents are sufficiently well captured by the distribution of aggregation variables. Hence formula (8) should be viewed as a crucial means for *simplification*, because it enables a principled decomposition.

Expressing the density function of $\mathbf{V}$ in (8) using formula (6) as copula, we obtain:

$$f_{\mathbf{X}_1,...,\mathbf{X}_m}(\mathbf{x}_1,...,\mathbf{x}_m) = c_{\mathbf{V}}(F_{V_1}(v_1),...,F_{V_m}(v_m)) \prod_{i=1}^{m} f_{\mathbf{X}_i}(\mathbf{x}_i) \tag{9}$$

This formula provides us a way to integrate in the same model mechanisms associated to the different assets in the market by means of a copula at aggregate level. In other words, by this formula, it is possible to calculate the probability of rare events, and therefore estimate systematic risk, based on the dependencies of aggregation variables and on the knowledge of micro-behaviour specified by group density functions of the agent-based models. The marginal distribution functions $F_{V_i}(v_i)$ can be estimated either from real world data (e.g. asset price time series), or from simulations. Note that whether we estimate from real world data or an agent-based market model should not matter in principle, since, if well constructed, the agent-based model should generate the same distribution of Δp as the distribution estimated from real world data. The density function $f_{\mathbf{X}_g}(\mathbf{x}_g)$ of an individual random vector $\mathbf{X}_g$ can be defined as we did in our simulation study. However, to bring this approach into practice, three problems remain be investigated:

- **estimation** of copula. We need to consider possible structural time dependencies and serial dependencies in the individual aggregation variables. Additionally, the agents might change their behavioural script (e.g. traders might pass from technical to fundamental at certain thresholds conditions).
- **high dimensionality** of Eqs. (8) and (9). If we consider N predictor variables for each group $g = 1,...,m$ we will end up with a $N \cdot m$-dimensional density function.

- **interpolation** of function h_g. Calculating high-dimensional integrals that occur for instance in formula (8), with function h_g being implicitly computed by the simulation of an agent-based market model, is clearly intractable.

For the first problem, we observe that time dependence with respect to copulas is still an active area of research. Most estimation methods do not allow for serial dependence of random variables. One approach to solve this is to filter the serial dependence by an autoregressive model as described at the beginning of Sect. 2. Another approach is to consider a dynamic copula, similarly as in [14,20,21]. A very interesting related work is presented in [22], where ARMA-GARCH and ARMA-EGARCH are used to filter serial dependence, but the regime switching copula is considered on the basis of two-states Markov models. Using an agent-based model instead of (or integrated with) a Markov model would be a very interesting research direction, because the change of regime would also have a qualitative interpretation.

For the second problem, although in our example we have used 1000 agents, in general this might be not necessary, considering that ABMs might be not as heterogeneous, and aggregators might work with intermediate layers between micro and macro-levels.

For the third problem, a better approach would be to interpolate the ABM simulation by some function with closed form. In future works, we are going to evaluate the use of neural networks (NNs), which means creating a model of our agent-based model, that is, a *meta-model*. The general concept of meta-models is a well-established design pattern [18] and the usage of NNs for such purposes dates back to [19]. In our example the basic idea would be to obtain samples from the distribution $f_{\mathbf{X}_g}(\mathbf{x}_g)$ as input, the results of an ABM simulation v_g as output, and then feed both input and output to train a dedicated NN, to be used at runtime. This would be done for each group g. The biggest advantage of this approach, if applicable in our case, is that we will have both a quick way to evaluate a function approximating h_g, but we will also have the interpretative power of the agent-based market model, resulting in an overall powerful modelling architecture.

4 Conclusions and Future Developments

Agent-based models are a natural means to integrate expert (typically qualitative) knowledge, and directly support the interpretability of computational analysis. However, both the calibration on real-data and the model exploration phases cannot be conducted by symbolic means only. The paper sketched a framework integrating agent-based models with advanced quantitative probabilistic methods based on copula theory, which comes with a series of data-driven tools for dealing with dependencies. The framework has been illustrated with canonical asset pricing models, exploring dependencies at micro- and macro-levels, showing that it is indeed possible to capture quantitatively social characteristic of the systems. This also provided us with a novel view on market destabilization, usually explained in terms of strategy switching [24,25]. Second, the paper formally sketched a principled model decomposition, based on theoretical contributions presented in the literature.

The ultimate goal of integrating agent-based models, advanced statistical methods (and possibly neural networks) is to obtain an unified model for risk evaluation, crucially centered around Eq. (9). Clearly, additional theoretical challenges for such a result remains to be investigated, amongst which: (a) probabilistic models other than copulas to be related to the agent's decision mechanism, (b) structural changes of dependence structures, (c) potential causal mechanisms on the aggregation variables and related concepts as time dependencies (memory effects, hysteresis, etc.) and latency of responses. These directions, together with the development of a prototype testing the applicability of the approach, set our future research agenda.

Acknowledgments. The authors would like to thank Drona Kandhai for comments and discussions on preliminary versions of the paper. This work has been partly funded by the Marie Sklodowska-Curie ITN Horizon 2020-funded project INSIGHTS (call H2020-MSCA-ITN-2017, grant agreement n.765710).

References

1. Bas, M., Mayer, T., Thoenig, M.: From micro to macro: demand, supply, and heterogeneity in the trade elasticity. J. Int. Econ. **108**(C), 1–19 (2017)
2. Serguieva, A., Liu, F., Paresh, D.: Financial contagion: a propagation simulation mechanism. SSRN: Risk Management Research, vol. 2441964 (2013)
3. Poledna, S., et al.: When does a disaster become a systemic event? estimating indirect economic losses from natural disasters. arXiv Preprint. https://arxiv.org/abs/1801.09740 (2018)
4. Hochrainer-Stigler, S., Balkovič, J., Silm, K., Timonina-Farkas, A.: Large scale extreme risk assessment using copulas: an application to drought events under climate change for Austria. Comput. Manag. Sci. **16**(4), 651–669 (2018). https://doi.org/10.1007/s10287-018-0339-4
5. Hochrainer-Stigler, S., et al.: Integrating systemic risk and risk analysis using copulas. Int. J. Disaster Risk Sci. **9**(4), 561–567 (2018). https://doi.org/10.1007/s13753-018-0198-1
6. Anagnostou, I., Sourabh, S., Kandhai, D.: Incorporating contagion in portfolio credit risk models using network theory. Complexity **2018** (2018). Article ID 6076173
7. Brilliantova A., Pletenev A., Doronina L., Hosseini H.: An agent-based model of an endangered population of the Arctic fox from Mednyi Island. In: The AI for Wildlife Conservation (AIWC) Workshop at IJCAI 2018 (2018)
8. Sedki, K., de Beaufort, L.B.: Cognitive maps and Bayesian networks for knowledge representation and reasoning. In: 24th International Conference on Tools with Artificial Intelligence, pp. 1035–1040, Greece (2012)
9. Pearl, J.: Causal inference in statistics: an overview. Statist. Surv. **3**, 96–146 (2009)
10. Brockwell, P.J., Davis, R.A.: Introduction to Time Series and Forecasting. Springer Texts in Statistics. Springer, New York (1996). https://doi.org/10.1007/b97391

11. Engle, R.F.: Autoregressive conditional heteroscedasticity with estimates of the variance of United Kingdom inflation. Econometrica (1982)
12. Bollerslev, T., Russell, J., Watson, M.: Volatility and Time Series Econometrics: Essays in Honor of Robert Engle, 1st edn, pp. 137–163. Oxford University Press, Oxford (2010)
13. Nelsen, R.B.: An Introduction to Copulas. Springer, New York (1999). https://doi.org/10.1007/978-1-4757-3076-0
14. Joe, H.: Dependence Modeling with Copulas. CRC Press Taylor & Francis Group, Boca Raton (2015)
15. Patton, A.J.: A review of copula models for economic time series. J. Multivar. Anal. **110**, 4–18 (2012)
16. Chen, L., Guo, S.: Copulas and Its Application in Hydrology and Water Resources. SW. Springer, Singapore (2019). https://doi.org/10.1007/978-981-13-0574-0
17. Zeeman, E.C.: On the unstable behaviour of stock exchanges. J. Math. Econ. **1**(1), 39–49 (1974)
18. Simpson, T., Poplinski, J., Koch, P., et al.: Eng. Comput. **17**(2) (2001)
19. Pierreval, H.: A metamodeling approach based on neural networks. Int. J. Comput. Simul. **6**(3), 365 (1996)
20. Fermanian, J.D., Wegkamp, M.H.: Time-dependent copulas. J. Multivar. Anal. **110**, 19–29 (2012)
21. Peng, Y., Ng, W.L.: Analysing financial contagion and asymmetric market dependence with volatility indices via copulas. Ann. Finance **8**, 49–74 (2012)
22. Fink, H., Klimova, Y., Czado, C., Stöber, J.: Regime switching vine copula models for global equity and volatility indices. Econometrics **5**(1), 1–38 (2017). MDPI
23. Epstein, J.M.: Why model? J. Artif. Soc. Soc. Simul. **11**(4), 12 (2008)
24. Westerhoff, F.A.: Simple agent-based financial market model: direct interactions and comparisons of trading profits. Nonlinear Dynamics in Economics, Finance and Social Sciences: Essays in Honour of John Barkley Rosser Jr. (2009)
25. Hessary, Y.K., Hadzikadic, M.: An agent-based simulation of stock market to analyze the influence of trader characteristics on financial market phenomena. J. Phys. Conf. Series **1039**(1) (2016)
26. Gu, Z.: Circlize implements and enhances circular visualization in R. Bioinformatics **30**(19), 2811–2812 (2014)

Computational Complexity
of Hierarchically Adapted Meshes

Marcin Skotniczny[✉]

Department of Computer Science, Faculty of Computer Science,
Electronics and Telecommunications, AGH University of Science and Technology,
Al. Adama Mickiewicza 30, 30-059 Kraków, Poland
mskotn@agh.edu.pl

Abstract. We show that for meshes hierarchically adapted towards singularities there exists an order of variable elimination for direct solvers that will result in time complexity not worse than $\mathcal{O}(\max(N, N^{3\frac{q-1}{q}}))$, where N is the number of nodes and q is the dimensionality of the singularity. In particular, we show that this formula does not change depending on the spatial dimensionality of the mesh. We also show the relationship between the time complexity and the Kolmogorov dimension of the singularity.

Keywords: Computational complexity · Direct solvers ·
h-adaptation · Hierarchical grids · Kolmogorov dimension

1 Introduction

Computational complexity, especially time complexity, is one of the most fundamental concepts of theoretical computer science, first defined in 1965 by Hartmanis and Stearns [1]. The time complexity of direct solvers [2,3] for certain classes of meshes, especially regular meshes, is well known. The problem of finding the optimal order of elimination of unknowns for the direct solver, in general, is indeed NP-complete [7], however there are several heuristical algorithms analyzing the sparsity pattern of the resulting matrix [8–12].

In particular, for three-dimensional uniformly refined grids, the computational cost is of the order of $\mathcal{O}(N^2)$ [4,5]. For three-dimensional grids adapted towards a point, edge, and face, the time complexities are $\mathcal{O}(N), \mathcal{O}(N)$, and $\mathcal{O}(N^{1.5})$, respectively [6]. These estimates assume a prescribed order of elimination of variables [13,14].

Similarly, for two dimensions, the time complexity for uniform grids is known to be $\mathcal{O}(N^{1.5})$, and for the grids refined towards a point or edge it is $\mathcal{O}(N)$ [15]. These estimates assume a prescribed order of elimination of variables [16]. The orderings resulting in such linear or quasi-linear computational costs can also be used as preconditioners for iterative solvers [17].

For all others, there is no known general formula or method of calculation of the computational complexity. It is neither hard to imagine that such a formula

V. V. Krzhizhanovskaya et al. (Eds.): ICCS 2020, LNCS 12139, pp. 226–239, 2020.
https://doi.org/10.1007/978-3-030-50420-5_17

or method might not be discovered any time soon, nor that such formula or method would be simple enough to be applicable in real-world problems. Thus in this paper, we focus on a specific class of problems: hierarchically adapted meshes refined towards singularities. This paper generalizes results discussed in [6] into an arbitrary spatial dimension and arbitrary type of singularity. We, however, do not take into the account the polynomial order of approximation p, and we assume that this is a constant in our formula.

Singularities in the simulated mesh can lead, depending on the stop condition, to an unlimited number of refinements. Thus, usually the refinement algorithm will be capped by a certain refinement level that is common to the whole mesh. In this paper, we analyze how the computational cost grows when the refinement level is increased. The analysis is done only for a very specific class of meshes, however the conclusions extend to a much wider spectrum of problems.

This publication is structured as follows. First, in Sect. 2 we define the problem approached by this paper. Secondly, in Sect. 3 we show a method to calculate an upper bound for the computational cost of direct solvers for sparse matrices. Then, in Sect. 4 we calculate time complexity for meshes meeting a certain set of criteria. Section 5 contains the analysis of properties of typical h-adaptive meshes and the final Sect. 6 concludes the proof by showing how those properties fit in with the earlier calculations.

2 Problem Definition

In this paper we analyze finite element method meshes that are hierarchically adapted towards some singularity – by *singularity* we mean a subset of the space over which the finite element method never converges. The existence of a singularity might cause the mesh to grow infinitely large, so a limit of refinement rounds is necessary – we will call it the *refinement level* of the mesh. As the refinement level of the mesh grows, the number of variables will grow as well and so will the computational cost. For instance, if the singularity is defined over a two-dimensional h-adaptive grid as shown in Fig. 1, each refinement of an element will increase the number of elements by three and (assuming that $p = 2$ over the whole grid).

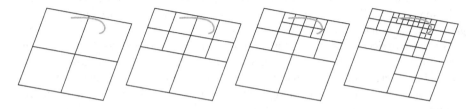

Fig. 1. Example of hierarchical adaptation towards a singularity (in green). Here, apart from the refinement of elements overlapping the singularity, the 1-irregularity rule is applied (see Sect. 5 for definition). (Color figure online)

When referring to a mesh hierarchically refined towards a singularity S up to a refinement level R we will have in mind a mesh that has been through R refinement rounds from its initial state. During each refinement round, all elements that have an overlap with the singularity S of at least one point will undergo a division into smaller elements, then, if required by given type of the finite element method, some extra elements might also get divided to keep some mesh constraints. We generally assume that the hierarchical refinements are done uniformly for all dimensions of the grid, however the findings from this paper should also extend to other types of refinements.

In this paper we analyze the relationship between the computational cost of the solver as a direct function of the number of variables and state it using big $\mathcal{O}$ notation. By definition of the big $\mathcal{O}$ notation, when we denote that the time complexity of the problem is $T(N) = \mathcal{O}(f(N))$, it means that for some large enough N_0 there exists a positive constant C such as the computational cost is no larger than $C \cdot f(N)$ for all $N \geq N_0$. In this paper, we are considering how the computational cost increases as the number of variables N grows. As N is a function of the refinement level R, we can alternatively state the definition of time complexity as follows:

$$T(N) = \mathcal{O}(f(N)) \Longleftrightarrow \exists R_0 \geq 0, C > 0 : \forall R \geq R_0 : |T(N(R))| \leq C \cdot f(N(R))$$
$$(1)$$

As the polynomial order p is constant in the analyzed meshes, the number of variables grows linearly proportional with the growth of the number of all elements of the mesh. Thus, in almost all formulas for time complexity, we can use the number of elements and the number of variables interchangeably.

3   Time Complexity of Sparse Matrix Direct Solvers

In the general case, the time complexity of solving a system of equations for a finite element method application in exact numbers (that is using a direct solver) will be the complexity of general case of Gaussian elimination – that is a pessimistic $\mathcal{O}(N^3)$, where N is the number of variables. However, on sparse matrices the complexity of the factorization can be lowered if a proper row elimination order is used – in some cases even a linear time complexity can be achieved. The matrices corresponding to hierarchically adapted meshes are not only sparse but will also have a highly regular structure that corresponds to the geometrical structure of the mesh.

In the matrices constructed for finite element method, we assign exactly one row and column (with the same indexes) to each variable. A non-zero value on an intersection of a row and column is only allowed if the basis functions corresponding to the variables assigned to that row and column have overlapping supports.

If a proper implementation of the elimination algorithm is used, when eliminating a row we will only need to do a number of subtractions over the matrix that is equal to the number of non-zero elements in that row multiplied by the number of non-zero values in the column of the leading coefficient in the remaining rows. This way, as long as we are able to keep the sparsity of the matrix high during the run of the algorithm, the total computation cost will be lower. To achieve that we can reorder the matrix rows and columns.

In this paper we will analyze the speed improvements gained from the ordering created through recursive section of the mesh and traversal of generated elimination tree. The tree will be built according to the following algorithm, starting with the whole mesh with all the variables as the input:

1. If there is only one variable, create an elimination tree node with it and return it.
2. Else:
   (a) Divide the elements in the mesh into two or more continuous submeshes using some division strategy. The division strategy should be a function that assigns each element to one and exactly one resulting submesh.
   (b) For each submesh, assign to it all the variables, which have their support contained in the elements of that submesh.
   (c) Create an elimination tree node with all the variables that have not been assigned in the previous step. Those variables correspond to basis functions with support spread over two or more of the submeshes.
   (d) For each submesh that has been assigned at least one variable, recursively run the same algorithm, using that submesh and the assigned variables as the new input. Any submeshes without any variables assigned can be skipped.
   (e) Create and return the tree created by taking the elimination tree node created in step 2c as its root and connecting the roots of the subtrees returned from the recursive calls in step 2d as its children.

An example of such process has been shown on Fig. 2.

From the generated tree we will create an elimination order by listing all variables in nodes visited in post-order direction. An example of an elimination tree with the order of elimination can be seen on Fig. 3.

The created tree has an important property that if two variables have non-zero values on the intersection of their columns and rows, then one of them will either be listed in the elimination tree node that is an ancestor of elimination tree node containing the other variable, or they can be both listed in the same node.

The computational cost of elimination can be analyzed as a sum of costs of elimination of rows for each tree node. Let us make the following set of observations:

1. A non-zero element in the initial matrix happens when the two basis functions corresponding to that row and column have overlapping supports (see the example in the left side of Fig. 4). Let us denote the graph generated from the initial matrix (if we consider it to be an adjacency matrix) as an *overlap graph*. Two variables cannot be neighbors in overlap graph unless the supports of their basis functions overlap.

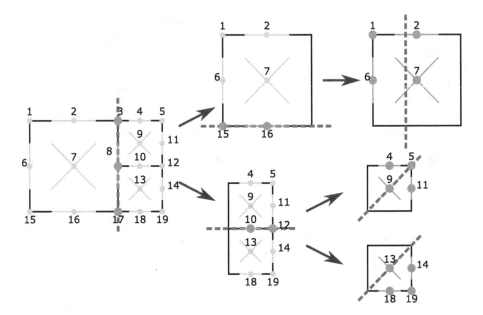

Fig. 2. Example of the generation of an elimination tree for h-adaptive mesh of $p = 2$. Large green nodes are assigned to the new node created during each recursive call. In the first recursion, a vertical cut through the vertices 3 and 17 is made. Vertex basis functions 3 and 17 and edge basis function 8 have supports spreading over both sides of the cut. The remaining sets of variables are then divided into the two submeshes created and the process is repeated until all variables are assigned. The cuts shown on the example are not necessarily optimal. (Color figure online)

2. When a row is eliminated, the new non-zero elements will be created on the intersection of columns and rows that had non-zero values in the eliminated row or the column of the lead coefficient (see the right side of Fig. 4). If we analyze the matrix as a graph, then elimination of the row corresponding to a graph node will potentially produce edges between all nodes that were neighbors of the node being removed.

3. If at any given time during the forward elimination step a non-zero element exists on the intersection of a row and column corresponding to two grid nodes, then either those two nodes are neighbors in the overlap graph, or that there exists a path between those two nodes in the overlap graph that traverses only graph nodes corresponding to rows eliminated already.

4. All variables corresponding to the neighboring nodes of the graph node of the variable x in the overlap graph will be either:

    (a) listed in one of the elimination tree nodes that are descendants of the elimination tree node listing the variable x – and those variables are eliminated already by the time this variable is eliminated, or

    (b) listed in the same elimination tree node as the variable x, or

(c) having support of the corresponding basis function intersected by the boundary of the submesh of the elimination tree node containing the variable x – those graph nodes will be listed in one of the ancestors of the elimination tree node listing the variable x.

Thus, in the overlap graph there are no edges between nodes that belong to two different elimination tree nodes that are not in an ancestor-descendant relationship. At the same time, any path that connects a pair of non-neighboring nodes in the overlap graph will have to go through at least one graph node corresponding to a variable that is listed in a common ancestor of the elimination tree nodes containing the variables from that pair of nodes.

We can infer from Observations 3 and 4 that by the time a row (or analogously column) is eliminated, the non-zero values can exist only on intersections with other variables eliminated together in the same tree node and variables corresponding to basis functions intersected by the boundary of the subspace of the tree node. During traversal of one tree node we eliminate a variables from b variables with rows or columns potentially modified, where $b - a$ is the number of variables on the boundary of that tree node. Thus, the cost of elimination of variables of a single tree node is equal to the cost of eliminating a variables from a matrix of size b – let us denote that by $C_r(a, b) = \mathcal{O}(ab^2)$.

4   Quasi-optimal Elimination Trees for Dimensionality q

A method of elimination tree generation will generate *quasi-optimal elimination trees for dimensionality q*, if each elimination tree generated:

(a) is a K-nary tree, where $K \geq 2$ is shared by all trees generated (a tree is K-nary if each node has up to K children; a binary tree has $K = 2$.);
(b) has height not larger than some $H = \lceil \log_K N \rceil + H_0$, where H_0 is a constant for the method (for brevity, let us define the height of the tree H as the longest path from a leaf to the root plus 1; a tree of single root node will have height $H = 1$). This also means that $\mathcal{O}(N) = \mathcal{O}(K^H)$;

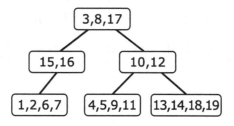

Fig. 3. Elimination tree generated in the process shown on Fig. 2. Induced ordering is: $1, 2, 6, 7$, then $15, 16$, then $4, 5, 9, 11$, then $13, 14, 18, 19$, then $10, 12$, then $3, 8, 17$.

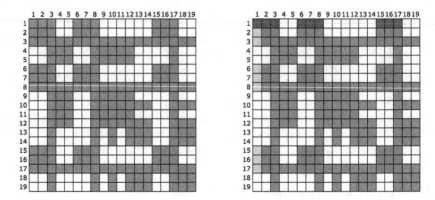

Fig. 4. On the left, the matrix generated for the mesh from Fig. 2 – potentially non-zero elements are marked in grey. On the right, the first elimination of a row (variable 1) is shown; all the elements changed during that elimination are marked: blue elements are modified with a potentially non-zero value, yellow elements are zeroed. (Color figure online)

(c) its nodes will not have more than $J \cdot \max(h, K^{\frac{q-1}{q}h})$ variables assigned to each of them; J is some constant shared by each tree generated and h is the height of the whole tree minus the distance of that node from the root (i.e. $h = H$ for the root node, $h = H - 1$ for its children, etc. For brevity, later in this paper we will call such defined h as the *height of tree node*, slightly modifying the usual definition of the height of a node). This limit is equivalent to $J \cdot h$ for $q \leq 1$ and $J \cdot K^{\frac{q-1}{q}h}$ for $q > 1$);

(d) for each node, the supports of the variables assigned to that tree node will overlap with no more than $J \cdot \max(h, K^{h\frac{q-1}{q}})$ supports of the variables assigned to tree nodes that are the ancestors of that tree node.

For a quasi-optimal elimination tree, the cost of eliminating of all the variables belonging to a tree node S_h with height h will be no more than:

$$T = \begin{cases} C_r(J \cdot h, 2J \cdot h) = O(4J^3 h^3) & \text{if } q \leq 1 \\ C_r(J \cdot (K^{h\frac{q-1}{q}}), 2J \cdot (K^{h\frac{q-1}{q}})) = \mathcal{O}(4J^3 \cdot (K^{3h\frac{q-1}{q}})) & \text{if } q > 1 \end{cases} \quad (2)$$

This means that for $q \leq 1$, the computational cost of a solver will be no more than:

$$T(N) = \sum_{h=1}^{H} \mathcal{O}(K^{H-h}) \cdot \mathcal{O}(4(J^3 h^3))$$

// *There are no more than K^{H-h} nodes of height h.*

$$= \mathcal{O}(4J^3 K^H \sum_{h=1}^{H} (K^{-h} h^3))$$

$$= \mathcal{O}(4J^3 K^H (\frac{1^3}{K^1} + \frac{2^3}{K^2} + \frac{3^3}{K^3} + \cdots + \frac{(H-1)^3}{K^{H-1}} + \frac{H^3}{K^H}))$$

// *This summation converges to some $C \le 26$ ($C = 26$ for $K = 2$).*

$$= \mathcal{O}(4J^3 K^H C)$$

// *Eliminating the constants.*

$$= \mathcal{O}(K^H) = \mathcal{O}(K^{\log_K(N)} K^{H_0}) = \mathcal{O}(N \cdot K^{H_0}) = \mathcal{O}(N) \qquad (3)$$

For $q > 1$, the computational cost of the solver will be no more than:

$$T(N) = \sum_{h=1}^{H} \mathcal{O}(K^{H-h}) \cdot \mathcal{O}(4J^3 (K^{3h\frac{q-1}{q}})) = \mathcal{O}(4J^3 \sum_{h=1}^{H} K^{H-h+3h\frac{q-1}{q}})$$

$$= \mathcal{O}(4J^3 K^H \sum_{h=1}^{H} K^{(2-\frac{3}{q})h}) = \mathcal{O}(4J^3 K^H \sum_{h=1}^{H} (K^{2-\frac{3}{q}})^h)$$

$$(4)$$

Then, depending on the value of q, this will become:

If $1 < q < \dfrac{3}{2}$:
$$T(N) = \mathcal{O}(4J^3 K^H K^2 \frac{1 - K^{H(2-\frac{3}{q})}}{K^{\frac{3}{q}} - K^2})$$
$$= \mathcal{O}(4J^3 K^{H+2}) = \mathcal{O}(K^H) = \mathcal{O}(N) \qquad (5)$$

If $q = \dfrac{3}{2}$:
$$T(N) = \mathcal{O}(4J^3 K^H H) = \mathcal{O}(K^H H)$$
$$= \mathcal{O}(K^{\log_K N} \log_K N) = \mathcal{O}(N \log N) \qquad (6)$$

If $q > \dfrac{3}{2}$:
$$T(N) = \mathcal{O}(4J^3 K^H K^2 \frac{K^{H(2-\frac{3}{q})} - 1}{K^2 - K^{\frac{3}{q}}})$$
$$= \mathcal{O}(4J^3 K^H K^2 K^{H(2-\frac{3}{q})})$$
$$= \mathcal{O}(K^{H+H(2-\frac{3}{q})}) = \mathcal{O}((K^H)^{3-\frac{3}{q}})$$
$$= \mathcal{O}(N^{3-\frac{3}{q}}) = \mathcal{O}(N^{3\frac{q-1}{q}}) \qquad (7)$$

Let us denote a singularity of a shape with Kolmogorov dimension q as q-dimensional singularity. For example, point singularity is 0-dimensional, linear singularity (that is, a singularity in a shape of a curve or line segment) is 1-dimensional, surface singulararity (in shape of a finite surface) is 2-dimensional, etc. A union of finite number of non-fractal singularities will have the Kolmogorov dimension equal to the highest of dimensions of the components, so, for example, a singularity consisting of a finite number of points will also be 0-dimensional.

With the calculation above, if we can show that for a sequence of meshes of consecutive refinement levels over a q-dimensional singularity there exists a method that constructs quasi-optimal elimination trees for dimensionality q, we will prove that there exists a solver algorithm with time complexity defined by the following formula:

$$T(N) = \begin{cases} \mathcal{O}(N) & \text{if } q < \frac{3}{2} \\ \mathcal{O}(N \log N) & \text{if } q = \frac{3}{2} \\ \mathcal{O}(N^{3\frac{q-1}{q}}) & \text{if } q > \frac{3}{2} \end{cases} \tag{8}$$

5 h-adaptive Meshes

In this section we will show how the proof framework from the preceding Sect. 4 applies to real meshes by focusing on h-adaptive meshes. Those meshes have basis functions defined over geometrical features of its elements: vertices and segments in 1D, vertices, edges and interiors in 2D, vertices, edges, faces and interiors in 3D, vertices, edges, faces, hyperfaces and interiors in 4D, etc. Basis functions in 2D with their supports have been shown on Fig. 5. For illustrative purposes the h-adaptation analyzed here has only uniform refinements into 2^D elements with length of the edge 2 times smaller, where D is the dimensionality of the mesh – the observations will also hold for refinements into any K^D elements for any small natural K.

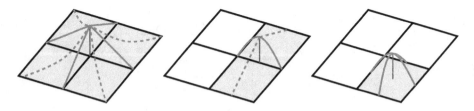

Fig. 5. Basis functions of 2-dimensional h-adaptive mesh with $p = 2$ and their support – based on vertex, edge and interior respectively.

It is worth noting that, during the mesh adaptation, if the basis functions are created naively, one basis function of lower refinement level can possibly have a support that completely contains the support of another basis function of the same type but higher level. In such instance, the function of lower refinement level will be modified by subtracting the other function multiplied by such factor that it cancels out part of the support of the lower level function, as illustrated in Fig. 6 – let us refer to this procedure as *indentation* of a basis function. This modification doesn't change the mathematical properties of the basis, but will significantly reduce the number of non-zero elements in the generated matrix.

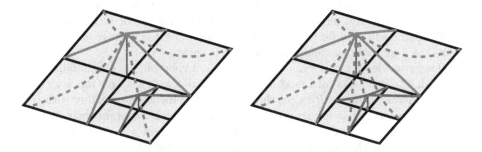

Fig. 6. Basis function indentation.

In addition to the above, a *1-irregularity rule* will be enforced: two elements sharing any point cannot differ in refinement level by more than 1 (a variant of 1-irregularity rule constraining elements sharing an edge instead of a vertex is also sometimes used, but the vertex version is easier to analyze). When splitting an element in the refinement process, any larger elements sharing a vertex should be split as well. This procedure will reduce the number of different shapes of basis functions created during the indentation, which will make the implementation easier. At the same time, if not for this rule, a single basis function could have overlap with an arbitrarily large number of other functions what would disrupt the sparsity of the generated matrix, as illustrated in Fig. 7.

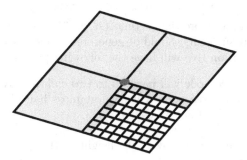

Fig. 7. If 1-irregularity rule is not met, one basis function can overlap any number of others.

It is important to note that during a refinement round an element will be split only if it overlaps a part of singularity or one of its neighbors overlaps a part of singularity, the distance between an element and the closest point of singularity cannot be larger than the length of the size of the element multiplied by a small constant. This has several important implications:

(a) The number of new elements created with each refinement level is $\mathcal{O}(2^{Rq})$ as refinement level $R \to \infty$, where q is the Kolmogorov dimension (also known as Minkowski dimension or box-counting dimension) of the singularity (or $\mathcal{O}(K^{Rq})$ if the refinements are into K^D elements). This also means that as $R \to \infty$, the total number of elements is $\mathcal{O}(R)$ if $q = 0$ or $\mathcal{O}(2^{Rq})$ if $q > 0$. The same will hold for the number of variables, as it is linearly proportional to the number of elements (as long as p is constant).

(b) If we consider a *cut* (i.e. a surface, volume, hypervolume, etc. dividing the mesh into two parts) of a mesh with a singularity with a simple topology (i.e. not fractal), then, as $R \to \infty$, the number of elements intersecting with the cut (assuming that elements are open sets, i.e. don't contain their boundaries) will grow as $\mathcal{O}(2^{Rs})$ (or $\mathcal{O}(R)$ for $s = 0$), where s is the Kolmogorov dimension of the intersection of that cut and the singularity. Intuitively, this observation should also extend to singularities that are well-behaved fractals.

6   Time Complexity of Hierarchical Meshes Based on Singularities

It is easy to see that for topologically simple sets (i.e. non-fractals) of integer Kolmogorov dimension q it is possible to divide them by a cut such that the Kolmogorov dimension of the intersection will be no more than $q-1$. For example, if we consider a singularity in a shape of line segment embedded in 3-dimensional space, any plane that is not parallel to the segment will have intersection of at most a single point. By recursively selecting such a cut that divides elements into two roughly equal parts (that is, as $R \to \infty$, the proportion of the number of elements in the two parts will approach 1:1), we can generate an elimination tree with cuts having no more than $\mathcal{O}(\max(R, (2^{Rq})^{\frac{q-1}{q}}))$ elements (finding the exact method of how such cuts should be generated is out of scope of this paper). The resulting elimination tree will have the following properties:

1. Each elimination tree node will have up to two children ($K = 2$).
2. As each subtree of a tree node will have at most half of the variables, the height of the tree H will not exceed $\lceil log_2 N \rceil$.
3. Depending on the value of q:
   (a) If $q > 1$, as the R grows by 1, the height of the given subtree grows by about q, but the number of variables that will be cut through grows by about 2^{q-1}. Thus the number of variables in a root of a subtree of height h will be no more than $J \cdot 2^{\frac{q-1}{q}h}$.
   (b) If $q \leq 1$, each tree node will have at most $J \cdot h$ variables – the number of elements of each refinement level that are being cut through such cut is limited by a constant.

4. Variables from a tree node can only overlap with the variables on the boundary of the subspace corresponding to that tree node. The amount of variables of the boundary should not exceed then $J \cdot 2^{\frac{q-1}{q}h}$ if $q > 1$ or $J \cdot h$ if $q \leq 1$.

Thus, h-adaptive meshes with q-dimensional singularity are *quasi-optimal* as defined in Sect. 4. To generalize, the same will hold true for any hierarchical meshes that fulfill the following constraints:

(a) each basis function can be assigned to an element (let us call it an *origin* of the function) such that its support will not extend further than some constant radius from that element, measured in the number of elements to be traversed (for example, in case of h-adaptation, supports do not extend further than 2 elements from its origin).

(b) the elements will not have more than B basis functions with supports overlapping them, where B is some small constant specific for the type of mesh; in particular, one element can be the origin of at most B basis functions. For example, h-adaptive mesh in 2D with $p = 2$ will have $B = 9$, as each element can have at most 4 vertex basis functions, 4 edge basis functions and 1 interior basis function.

(c) and no basis functions will overlap more than C elements, where C is another some small constant specific for the type of mesh. For example, $C = 8$ in h-adaptive mesh in 2D with $p = 2$ as long as the 1-irregularity rule is observed, as a basis function defined over a vertex and indented twice can overlap 8 elements.

Those constraints are also met by meshes with T-splines or with meshes in which elements have shapes of triangles, which means that well formed hierarchically adapted meshes can be solved by direct solvers with time complexity of $\mathcal{O}(\max(N, N^{3\frac{q-1}{q}}))$, where N is the number of nodes and q is the dimensionality of the singularity.This means that for point, edge and face singularities, the time complexity will be $\mathcal{O}(N)$, $\mathcal{O}(N)$ and $\mathcal{O}(N^{1.5})$ respectively, which corresponds to the current results in the literature for both two and three dimensional meshes [6,15]. For higher dimensional singularities, the resulting time complexity is $\mathcal{O}(N^2)$, $\mathcal{O}(N^{2.25})$, $\mathcal{O}(N^2.4)$ and $\mathcal{O}(N^2.5)$ for singularities of dimensionality 3, 4, 5 and 6 respectively.

An important observation here is that the time complexity does not change when the dimensionality of the mesh changes, as long as the dimensionality of the singularity stays the same.

7   Conclusions and Future Work

In this paper we have shown that for meshes hierarchically adapted towards singularities there exists an order of variable elimination that results in computational complexity of direct solvers not worse than $\mathcal{O}(\max(N, N^{3\frac{D-1}{D}}))$, where N is the number of nodes and q is the dimensionality of the singularity. This formula does not depend on the spatial dimensionality of the mesh. We have also shown the relationship between the time complexity and the Kolmogorov dimension of the singularity.

Additionally, we claim the following conjecture:

Conjecture 1. For any set of points S with Kolmogorov dimension $q \geq 1$ defined in Euclidean space with dimension D and any point in that space p there exists a cut of the space that divides the set S into two parts of equal size, for which the intersection of that cut and the set S has Kolmogorov dimension of $q - 1$ or less. *Parts of equal size* for $q < D$ are to be meant intuivitely: as the size of covering boxes (as defined in the definition of Kolmogorov dimension) decreases to 0, the difference between the number of boxes in both parts should decrease to 0.

It remains to be verified if the Conjecture 1 is true for *well-behaved fractals* and what kinds of fractals can be thought of as *well behaved*. If so, then meshes of the kinds that fulfill the constraints set in previous Sect. 6 built on a signularities in the shape of well-behaved fractals of non-integer Kolmogorov dimension q can be also solved with time complexity stated in Eq. 8. The proof of the conjecture is left for future work.

We can however illustrate the principle by the example of Sierpinski's triangle – a fractal of Kolmogorov dimension of $\frac{\log 3}{\log 2} = 1.58496250072116\ldots$. To build an elimination tree, we will divide the space into three roughly equal parts as shown in Fig. 8. The Kolmogorov dimension of the boundary of such division is 0, which is less than $\frac{\log 3}{\log 2} - 1$.

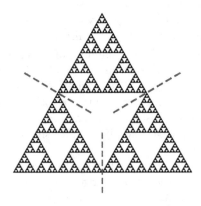

Fig. 8. Division of the space for Sierpinski triangle.

As the refinement level R grows, the number of elements grows as $\mathcal{O}(2^{\frac{\log 3}{\log 2} R}) = \mathcal{O}(3^R)$ and as the elimination tree s ternary, the partition tree will have height of $\log_3 3^R + H_0 = \log_K N + H_0$. In addition, as the Kolmogorov dimension of the boundary of each elimination tree node is 0, there are at most $\mathcal{O}(\log h)$ variables in each elimination tree node and on the overlap with nodes of ancestors is also limited by the same number. As this number is less than $\mathcal{O}(3^{h\frac{q-1}{q}})$, both conditions of *quasi-optimal mesh with q-dimensional singularity* are met, which means that it is possible to solve system build on a singularity of the shape of Sierpinski triangle in time not worse than $\mathcal{O}(N^{3\frac{\log 3/\log 2-1}{\log 3/\log 2}}) = \mathcal{O}(N^{1.10721...})$.

References

1. Hartmanis, J., Stearns, R.: On the computational complexity of algorithms. Trans. Am. Math. Soc. **117**, 285–306 (1965)
2. Duff, I.S., Reid, J.K.: The multifrontal solution of indefinite sparse symmetric linear systems. ACM Trans. Math. Softw. **9**, 302–325 (1983)
3. Duff, I.S., Reid, J.K.: The multifrontal solution of unsymmetric sets of linear systems. SIAM J. Sci. Stat. Comput. **5**, 633–641 (1984)
4. Liu, J.W.H.: The multifrontal method for sparse matrix solution: theory and practice. SIAM Rev. **34**, 82–109 (1992)
5. Calo, V.M., Collier, N., Pardo, D., Paszyński, M.: Computational complexity and memory usage for multi-frontal direct solvers used in p finite element analysis. Procedia Comput. Sci. **4**, 1854–1861 (2011)
6. Paszyński, M., Calo, V.M., Pardo, D.: A direct solver with reutilization of previously-computed LU factorizations for h-adaptive finite element grids with point singularities. Comput. Math. Appl. **65**(8), 1140–1151 (2013)
7. Yannakakis, M.: Computing the minimum fill-in is NP-complete. SIAM J. Algebraic Discrete Methods **2**, 77–79 (1981)
8. Karypis, G., Kumar, V.: A fast and high quality multilevel scheme for partitioning irregular graphs. SIAM J. Scientiffic Comput. **20**(1), 359–392 (1998)
9. Heggernes, P., Eisenstat, S.C., Kumfert, G., Pothen, A.: The computational complexity of the minimum degree algorithm. ICASE Report No. 2001–42 (2001)
10. Schulze, J.: Towards a tighter coupling of bottom-up and top-down sparse matrix ordering methods. BIT **41**(4), 800 (2001)
11. Amestoy, P.R., Davis, T.A., Du, I.S.: An approximate minimum degree ordering algorithm. SIAM J. Matrix Anal. Appl. **17**(4), 886–905 (1996)
12. Flake, G.W., Tarjan, R.E., Tsioutsiouliklis, K.: Graph clustering and minimum cut trees. Internet Math. **1**, 385–408 (2003)
13. Paszyńska, A.: Volume and neighbors algorithm for finding elimination trees for three dimensional h-adaptive grids. Comput. Math. Appl. **68**(10), 1467–1478 (2014)
14. Skotniczny, M., Paszyński, M., Paszyńska, A.: Bisection weighted by element size ordering algorithm for multi-frontal solver executed over 3D h-refined grids. Comput. Methods Mater. Sci. **16**(1), 54–61 (2016)
15. Paszyńska, A., et al.: Quasi-optimal elimination trees for 2D grids with singularities. Scientific Program. Article ID **303024**, 1–18 (2015)
16. AbouEisha, H., Calo, V.M., Jopek, K., Moshkov, M., Paszyńska, A., Paszyński, M.: Bisections-weighted-by-element-size-and-order algorithm to optimize direct solver performance on 3D hp-adaptive grids. In: Shi, Y., et al. (eds.) ICCS 2018. LNCS, vol. 10861, pp. 760–772. Springer, Cham (2018). https://doi.org/10.1007/978-3-319-93701-4_60
17. Paszyńska, A., et al.: Telescopic hybrid fast solver for 3D elliptic problems with point singularities. Procedia Comput. Sci. **51**, 2744–2748 (2015)

A Novel Bio-inspired Hybrid Metaheuristic for Unsolicited Bulk Email Detection

Tushaar Gangavarapu[1,2]([⊠])(iD) and C. D. Jaidhar[2]

[1] Automated Quality Assurance (AQuA) Machine Learning Research, Content Experience and Quality Algorithms, Amazon.com, Inc., Bellevue, USA
tusgan@amazon.com

[2] Department of Information Technology, National Institute of Technology Karnataka, Mangalore, India

Abstract. With the recent influx of technology, Unsolicited Bulk Emails (UBEs) have become a potential problem, leaving computer users and organizations at the risk of brand, data, and financial loss. In this paper, we present a novel bio-inspired hybrid parallel optimization algorithm (Cuckoo-Firefly-GR), which combines Genetic Replacement (GR) of low fitness individuals with a hybrid of Cuckoo Search (CS) and Firefly (FA) optimizations. Cuckoo-Firefly-GR not only employs the random walk in CS, but also uses mechanisms in FA to generate and select fitter individuals. The content- and behavior-based features of emails used in the existing works, along with Doc2Vec features of the email body are employed to extract the syntactic and semantic information in the emails. By establishing an optimal balance between intensification and diversification, and reaching global optimization using two metaheuristics, we argue that the proposed algorithm significantly improves the performance of UBE detection, by selecting the most discriminative feature subspace. This study presents significant observations from the extensive evaluations on UBE corpora of $3,844$ emails, that underline the efficiency and superiority of our proposed Cuckoo-Firefly-GR over the base optimizations (Cuckoo-GR and Firefly-GR), dense autoencoders, recurrent neural autoencoders, and several state-of-the-art methods. Furthermore, the instructive feature subset obtained using the proposed Cuckoo-Firefly-GR, when classified using a dense neural model, achieved an accuracy of 99%.

Keywords: Evolutionary computing · Feature selection · Internet security · Metaheuristics · Natural language processing · Phishing · Spam

1 Introduction

In recent years, due to the increased ease of communication via emails, Unsolicited Bulk Emails (UBEs) have become a common problem. UBEs can be

© Springer Nature Switzerland AG 2020
V. V. Krzhizhanovskaya et al. (Eds.): ICCS 2020, LNCS 12139, pp. 240–254, 2020.
https://doi.org/10.1007/978-3-030-50420-5_18

majorly divided into two related categories, i.e., spam and phishing emails. Spam emails constitute the category of bulk emails sent without users' consent, primarily with the intent of marketing (e.g., diet supplements, unlicensed medicines, etc.). Phishing is a more severe type of semantic attack aimed at deceiving users into providing sensitive information such as bank details, account numbers, and others. According to the internet security threat report [15], 55% of the emails constituted spam in 2017 (2% more than in 2015–16). Gartner study in the United States showed that approximately 109 million adults received phishing email attacks which resulted in an average loss of $1,244 per victim.

Evidently, spam and phishing rates are proliferating and the effects of such UBEs include theft of user identities, intellectual properties, degradation of mailing efficiency and recipient's productivity. Automatically detecting such UBEs has become a prominent area of research and hence has drawn a variety of considerations from researchers including behavior-based [16,20] and content-based [5] anti-UBE methods. In spite of the continuous efforts to avoid UBEs, attackers continuously change their strategies of executing spam and phishing attacks which makes it crucial to develop UBE detection methods with high performance. Furthermore, most of the UBEs are very similar to ham emails, making it extremely challenging to curb UBE attacks purely based on the email content. Moreover, most of the current email filtering approaches are static and can be defeated by modifying the link strings and email content. In this study, we mine forty content- and behavior-based features of emails from the existing literature along with 200 Doc2Vec features of the email body content, to extract the syntactic and semantic information embedded in the emails.

Many attempts to detect and classify UBEs have been made [7]. These methods include white and blacklisting, content- and network-based filtering, client-side toolbars, server-side filters, firewalls, and user awareness [19]. Usually, the data contained in the emails is very complex and multi-dimensional, resulting in higher time and space complexity, and low classifier performance [3]. Thus the cost of computation can be reduced while increasing the classification performance with a discriminative and informative feature subset. Dimensionality reduction to aid the classification of UBEs can be performed either by feature selection or feature extraction. This study focuses on a bio-inspired hybrid feature selection approach to discriminate between the email types.

A metaheuristic aims at generating or selecting a low-level heuristic which might provide a better solution than classical approaches to an optimization problem. The success of metaheuristics can be attributed to the nature of swarm intelligence algorithms being flexible and versatile, in the sense that they mimic the best features in nature [8]. Recently, Cuckoo Search (CS) [18] and Firefly Algorithm (FA) [17] have gained popularity from many researchers—their performance proved to be more efficient in solving global optimization problems than other metaheuristics [6,14]. CS was inspired by the obligate brood parasitism of certain species of the cuckoo bird by laying their eggs in the nests of other species birds, in combination with their Lévy flight behavior to search for food. FA was inspired by the observations of the flashing patterns and practices of fireflies who attract their partners using intensity of the emitted light.

This paper presents a novel hybrid bio-inspired metaheuristic called the Cuckoo-Firefly-GR to select the most discriminative and informative feature subset from a set of both content-based and behavior-based features needed to classify UBEs. The hybrid metaheuristic combines the evolutionary natures of CS and FA using the concepts of random walk in CS and mechanisms of FA to generate or select fitter individuals. Furthermore, the hybrid metaheuristic is combined with a Genetic Replacement (GR) of low fitness individuals. The novelty of our hybrid metaheuristic lies in the way that the abandoned nests in CS and low fitness fireflies in FA are genetically replaced, and in the way that combines the complementary strengths and advantages of Cuckoo-GR and Firefly-GR. To the best of our knowledge, the existing literature that combines CS or FA, and the Genetic Algorithm (GA) modifies the non-abandoned nests using GA. Moreover, the previously available methods that combine CS and FA do not use GR. The goal of the proposed hybrid metaheuristic is to produce more accurate feature subset and globally optimize the task of feature selection, by reducing the average number of fitness evaluations, in turn, improving the selection performance. The experimental results emphasize the superiority of the proposed hybrid metaheuristic over the base optimizations (Cuckoo-GR and Firefly-GR), dense autoencoders, and Long Short Term Memory (LSTM) autoencoders. The key contributions of this paper can be summarized as:

- Leveraging vector space content modeling, to extract the syntactic and semantic relationships between the textual features of the email body.
- Design of a hybrid metaheuristic based on the evolutionary natures of CS and FA that uses GR to establish an optimal balance between intensification and diversification in the population.
- Evaluating the effectiveness of the proposed hybrid metaheuristic (Cuckoo-Firefly-GR) in the classification of UBEs. Our results indicate the efficacy of the proposed metaheuristic over various state-of-the-art methods.

The rest of this paper is organized as follows: Sect. 2 reviews the relevant aspects of CS, FA, and GA algorithms. Section 3 elucidates the approach followed to extract content-based and behavior-based features, and presents the proposed metaheuristic. Results of the proposed metaheuristic on ham, spam, and phishing corpora are presented and discussed in Sect. 4. Finally, Sect. 5 concludes this study and presents future research directions.

2   Cuckoo Search, Firefly, and Genetic Algorithms

In this section, we review the relevant aspects of CS, FA, and GA used in designing the hybrid metaheuristic (Cuckoo-Firefly-GR).

2.1   Cuckoo Search Algorithm

Yang and Deb [18] developed CS, a metaheuristic search algorithm to efficiently solve the global optimization problems. Existing body of literature reports that

CS is far more efficient than many other metaheuristic search approaches including particle swarm optimization and GA. CS has since been successfully applied to various fields including biomedical engineering, antenna design, power systems, and microwave applications [14].

CS is primarily inspired by the Lévy flight behavior of the cuckoo birds while searching for food, along with their aggressive reproduction strategy. Generally, cuckoo birds do not build their nests; instead, they lay their eggs in communal nests so that surrogate parents unwittingly raise the cuckoo brood. Moreover, cuckoo birds may remove host bird's eggs to increase the hatching probability of their eggs. The host bird can build a new nest at a new location or throw away the cuckoo eggs if it finds that the eggs are not its own. The following three idealized rules are employed in the design of CS using the breeding analogy:

- Each cuckoo lays one egg at a time and deposits the egg in a randomly chosen nest among the available communal nests.
- A certain number of best nests with high-quality eggs will be carried on to the subsequent generations, thus ensuring that good solutions are preserved.
- There are a fixed number of nests, and the probability of a host bird discovering a cuckoo egg is $p_a \in [0, 1]$. When the host bird encounters a cuckoo egg, it can either build a new nest at a new location or throw away the egg.

In an optimization problem, every egg in a communal nest represents a possible solution, and a cuckoo egg constitutes a new solution candidate $x^{(t+1)} = (x_1^{(t+1)}, x_2^{(t+1)}, \ldots, x_d^{(t+1)})^{\mathrm{T}}$. For each iteration, a cuckoo egg is randomly selected to generate new solutions. This random search can be executed efficiently using a Lévy flight. The Lévy flights are a type of random walks where the step lengths follow a specific heavy-tailed probabilistic distribution and the step directions are random and isotropic. Since Lévy flights have infinite mean and variance, some solutions will be closer to the current best solutions and others will be placed away from the best solutions, thus enabling a broader search space exploration. For a given cuckoo, say c, the Lévy flight on the current solution $x_c^{(t)}$ generates the new solution parameter $x_c^{(t+1)}$ which is computed as $x_c^{(t)} + \alpha \oplus \mathrm{L\acute{e}vy}(\lambda)$, where $\oplus$ indicates entry-wise multiplication and α is a positive constant, scaled using the dimensions of the search space. The value of α determines how far a particle can move by a random walk, in a fixed number of cycles. The computation of $x_c^{(t+1)}$ is a Markov chain, which is a stochastic equation for a random walk and the new solution is only reliant on: the current solution $(x_c^{(t)})$ and the probability of transition. The probability of transition is modulated using the Lévy distribution as: $\mathrm{L\acute{e}vy}(\lambda) \sim u = t^{-\lambda}$, where λ defines the decay of the probability density function with t. For most cases, $\alpha = 0.01$ and $1 < \lambda \le 3$ [18]. This study employs CS with $N = 25$ nests, $\alpha = 0.01$, $\lambda = 2.5$, and $p_a = 0.4$, for a maximum of 50 cycles. In nature, many insects and animals often follow the properties of Lévy flights while searching food in a random or quasi-random manner [14]. Using CS via Lévy flights helps to explore the search space more effectively when compared to algorithms using standard Gaussian process, by avoiding the problem of being trapped around local optima.

2.2 Firefly Algorithm

Yang developed FA [17] based on the flashing patterns and social behaviors of fireflies. Fireflies produce luminescent flashes by process of bioluminescence, as a signal system to attract mating partners and potential prey. The rhythmic flashing, rate of flashing, and the amount of time form part of the signaling system that brings both sexes together. Owing to the effectiveness of FA in solving global optimization problems, it has been applied to various fields including stock forecasting, structure design, and production scheduling [6]. The existing literature corroborates that, although CS has outperformed FA in multimodal optimization, FA is better at generating optimum or near-optimum value in limited time [2]. The following three idealized rules are employed in the development of FA using the firefly signaling analogy:

- All the fireflies are unisexual, and every individual firefly will be attracted to every other firefly regardless of their sex.
- The attractiveness is proportional to the brightness, and they both decrease as the distance increases. Thus, for any two flashing fireflies, the less bright firefly will move towards the brighter one. Also, a firefly will move randomly, if there is no brighter firefly.
- The light intensity or brightness of a firefly is associated with the landscape of the objective function to be optimized.

There are two main concerns in FA: the variation of the light intensity and formulation of the attractiveness among fireflies. For simplicity, it can be assumed that the attractiveness is determined by the light intensity or brightness of a firefly, which is in turn dependent on the objective function ($f(x)$). For a maximization problem, the light intensity $I_i(x)$ of a particular firefly, say i, at a location, say x, can be chosen such that $I_i(x) \propto f(x), \forall i$. However, the attractiveness (β) is relative and is judged by other fireflies, thus varies with the distance between the fireflies (r). Since the brightness decreases with an increase in the distance and the flashing light is also absorbed in the media, attractiveness must be modeled using both these parameters. In the simplest form, the brightness $I(r)$ with I' source brightness follows the inverse square law, $I(r) = I'/r^2$ However, in a given medium with γ light absorption coefficient, $I(r)$ varies monotonically and exponentially as: $I(r) = I' \cdot exp(-\gamma r)$. Since the attractiveness of a firefly is proportional to the brightness seen by the adjacent fireflies, β can be computed as: $\beta = \beta' \cdot exp(-\gamma r^2)$, where β' is the attractiveness when $r = 0$. The distance between any two fireflies, say i and j at positions x_i and x_j, can be computed as the Cartesian between them, i.e., $r_{i,j} = ||x_i - x_j||_2$ or the l_2-norm. The movement of a less bright firefly, say i, towards a brighter (more attractive) firefly, say j, is determined using $x_i = x_i + \beta' \cdot exp(-\gamma r_{i,j}^2) \cdot (x_j - x_i) + \alpha \cdot \mathcal{G}_i$, where the second term accounts for attraction while the third term is randomization with a vector of random variables $\mathcal{G}_i$ drawn from a Gaussian distribution. In most cases, $\gamma = 1$, $\beta' = 1$, and $\alpha \in [0, 1]$. This study employs FA with $N = 25$ fireflies, $\gamma = 1$, $\alpha \in [0.1, 1]$, and $\beta' = 1$, for a maximum of 50 cycles.

2.3   Genetic Algorithm

GA is a classical optimization algorithm inspired by Darwin's theory of evolution, natural selection process, and various genetic operators such as crossover and mutation. GA has been successfully applied to various fields, including job-shop scheduling, rule set prediction, feature selection, and others.

To solve an optimization problem, GA constructs a random population of individual solutions. Subsequently, the fitness function, i.e., the objective function to be optimized measures the quality of an individual in the current population. Bio-inspired operators including selection, crossover, and mutation aid in the conversion of one generation to the next generation. First, the reproduction or selection procedure such as roulette-wheel, truncation, or tournament, is executed to select a potential set of parents based on their fitness scores. The crossover and mutation follow the selection procedure, aiding in the construction of the subsequent generation. Computationally, the variables are denoted as genes in a chromosome and are codified using a string of bits. These parent bit sequences are selected randomly based on their fitness scores, to produce the next generation of individuals. Several crossovers are performed with a crossover probability p_c, to replace the population and replicate the randomness involved in an evolutionary process. Finally, certain bits in the newly formed individuals are swept and changed with a small mutation probability p_m. The quality of the new generation of chromosomes is assessed using the fitness function. This procedure is repeated until the fitness values of the generated solutions converge.

3   Proposed Methodology

First, we present the procedure utilized to extract features from emails, followed by the mathematical formulation of the optimization problem. Then, we describe the proposed hybrid metaheuristic employed to facilitate selection of an informative feature subspace. Finally, we discuss the Multi-Layer Perceptron (MLP) model used in the UBE detection. The employed pipeline is presented in Fig. 1.

3.1   Extraction of Content-Based and Behavior-Based Features

The features of emails considered in this study are internal to emails, rather than those from external sources like domain registry information or search engine information. Existing literature has shown that internal features form a comparatively more informative feature subset, as most of the external data such as DNS information changes regularly [16]. This study considers forty potential content- and behavior-based features used in the existing literature along with 200 Doc2Vec features extracted purely from the message content.

Every email consists of two parts: email header and email body. The email header consists of general identification information including subject, date, from, to, and the route information followed by the email transfer agent. Forty significant features extracted from the email messages can be roughly categorized

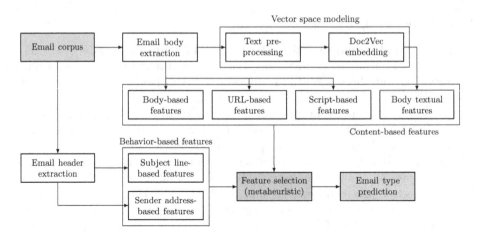

Fig. 1. Pipeline used to efficiently predict the email type using a hybrid of content-based and behavior-based email features.

into five major categories: subject line-based features, sender address-based features, body-based features, URL-based features, and script-based features.

The subject line-based features include boolean flags to check if the email is a reply or forwarded email, presence of words like *verify*, *debit*, or *bank*, along with numeric features including word count, character count, and subject richness. The richness of a given text is computed as the ratio between number of words and number of characters. The sender address is mined for features including word count and character count, in addition to boolean checks aimed at detecting if the sender's domain, email's modal domain, and the reply-to domain are different. The subject line-based features and sender address-based features model the behavioral aspects of an email.

The body-based features include boolean flags indicating the presence or absence of HTML tags, forms, words like *suspension*, and phrases like *verify your account*, along with numeric attributes such as word count, character count, distinct word count, function word (e.g., *account*, *identity*, etc.) count, and body richness. The features extracted from the URLs in the email body include continuous attributes like total IP address count, total URL count, internal and external link count, image link count, total domain count, port count, and highest period count. In addition to these numeric features, certain boolean flags are also used to check for the presence of @ symbol in URLs, words such as *login*, *click*, *here*, or *update* in the link text, links mapping to a non-modal domain, URLs accessing ports other than 80, and IP addresses rather than domain names. Finally, the script-based features include boolean flags to check for the presence of scripts, JavaScript, and non-modal external JavaScript forms, and to check if the script overwrites the status bar in the email client. Furthermore, the number of on-click events are also measured as a part of the script-based features. The body-based features, URL-based features, and script-based features constitute the content-based aspects of an email.

In addition to these forty potential features of an email, Doc2Vec embedding of the email message content was performed to capture the syntax and semantics of the UBE's content. Various preprocessing steps including tokenization, stopword removal, and stemming/lemmatization were performed to achieve text normalization. First, punctuation marks, URLs, multiple spaces, and special characters were removed. Then, the resultant text is split into several smaller tokens during text tokenization. Most frequently occurring words (stopwords) were removed using the NLTK English stopword corpus. Furthermore, character casing and space trimming were performed to achieve normalized data representations. Finally, stemming was performed to achieve suffix stripping, followed by lemmatization, to convert the stripped words into their base forms. To avoid overfitting and lower the computational complexity, the words occurring in less than ten emails were removed. The preprocessed text is then modeled using Doc2Vec, to derive optimal data representations.

Doc2Vec or Paragraph Vectors (PVs) efficiently learn the term representations in a data-driven manner [11]. Doc2Vec embeddings aim at capturing the intuition that semantically similar email contents will have near-identical vector representations (e.g., *debit* and *bank*). Doc2Vec is essentially a neural network with a shallow hidden layer that facilitates the model to learn the distributed representations of paragraphs to provide a content-related assessment. The implementations available in Python Gensim package were used to extract the Doc2Vec style features. This study employs a PV distributed memory variant of the Doc2Vec, with a dimension size of 200 (trained for 25 epochs) due to its ability to preserve the word order in the email content. In this study, we used a combination of these 200 features along with forty potential features, resulting in a total of 240 features.

3.2 Problem Formulation

Feature selection aims at projecting a set of points from the $m-$dimensional (original) space to a smaller $k-$dimensional space with a substantial reduction of noise and minimal loss of information. When such a projection specifies a subset of the available dimensions, the feature selection problem has a combinatorial nature. Let $\mathcal{E}$ denote the set of n emails, indexed by e. Each email has a set of m features that define the email, $\Pi^{(e)} = \{\mathcal{F}_f^{(e)}\}_{f=1}^m$, with each feature ($\mathcal{F}$), indexed by f. Our ultimate goal is to learn a function (g) that estimates the probability of an email belonging to one of the email types (ham, spam, or phishing), given its features: $g(\Pi^{(e)}) \approx Pr(\text{email type} \mid \Pi^{(e)})$. However, the complexity and high dimensionality of $\Pi^{(e)}$ often make it challenging to train and generalize a classifier. Furthermore, the nature of $\mathcal{E}$ makes it cost and time intensive to learn g as a mapping from emails to probabilities. Thus we need a transformation (T) from $\Pi^{(e)}$ into a machine processable and easier-to-use form, $T : \Pi^{(e)} \longrightarrow \mathbb{R}^k$. Usually, $k \ll m$ is used to achieve dimensionality reduction. Now, the transformed data, $\pi^{(e)} = T(\Pi^{(e)})$, $\pi^{(e)} \in \mathbb{R}^k$ is used generalize g and make the problem more tractable: $g(\pi^{(e)}) = g(T(\Pi^{(e)})) \approx Pr(\text{email type} \mid \Pi^{(e)})$. Now, the problem is decomposed into two steps: estimating T using a feature selection approach and estimating g using $\{(\pi^{(e)}, \text{email type}^{(e)})\}_{e \in \mathcal{E}_{\text{train}}}$.

The first step is formulated as determining a feature subset from the given m−features resulting in the best discriminating capabilities. Thus, this step involves an optimization problem in the search space of all the possible feature subsets whose criterion function is the accuracy (Φ_c) obtained with a classifier (c). This study poses the selection of an optimal subset of features as an optimization problem such that: $\arg\max_{\mathcal{S}} \Phi_c(\mathcal{S}, \{\pi^{(e)}\}_{e \in \mathcal{E}_{\text{train}}}, \{\pi^{(e)}\}_{e \in \mathcal{E}_{\text{test}}})$, where $\mathcal{S}$ denotes the feature subset and $\pi^{(e)} = \{\mathcal{F}_f^{(e)}\}_{f=1}^{|\mathcal{S}|}$ is obtained by projecting $\Pi^{(e)}$ onto the subspace spanned by features in $\mathcal{S}$. In optimization problems, often a coding scheme is needed to transform the selected subset of features into a string form. In this study, every individual solution is represented as a binary string of m−bits where the fth bit corresponds to f^{th} feature. We use a bijection as a coding scheme, i.e., a feature is only included if its corresponding bit value is set. The classification accuracy is used to compute the fitness values of each solution in the population. This study uses a random forest classifier with 100 classification and regression trees of maximum depth 2, as the fitness function, evaluated as 3-fold cross-validation on the training set. Furthermore, this study uses a classifier that can learn and generalize from the informative feature subset obtained from transformation T, as the function g.

3.3   Genetic Replacement of Low Fitness Individuals

Intensification aims at measuring the ability to exploit the local neighborhood of the existing solutions needed to improve them. Diversification relates to the ability to explore the global search space to generate diverse solutions. An optimal balance between intensification and diversification is crucial in attaining better performance using global heuristic search methods [12].

Both CS and FA algorithms provide intensification and diversification. The diversification in CS and FA is rendered by the random initialization of population, intensity and attractiveness (in FA), and Lévy flights (in CS). The proposed hybrid metaheuristic manages the abandoned nests more effectively by replacing a part of them genetically, using bio-inspired operators including crossover and mutation. Such replacement aids in the effective convergence of solutions within a limited number of cycles by reducing the total number of iterations [14]. We define p_g as the probability of genetic replacement such that: $N_a = p_a \times N$; $N_g = p_g \times N_a$, where N is the population size and p_a is the probability of abandonment. The individual solutions for genetic replacement are randomly selected from the set of abandoned solutions. The N_g solutions are generated using the parents selected from the set of better fitness individuals using a roulette-wheel selection. The genetic replacement of abandoned solutions enhances the balance between diversification and intensification via genetic mutation. In this study, we use $p_a = 0.4$, $p_g = 0.6$, and $p_m = 0.025$.

3.4 Proposed Cuckoo-Firefly-GR Hybrid Metaheuristic

Both CS and FA have their advantages and are successful in solving a wide range of optimization problems. These metaheuristics are further diversified using the concepts of genetic algorithm, aimed at replacing low fitness individuals. We propose a hybrid of these two metaheuristic optimizations called the Cuckoo-Firefly-GR, to facilitate the use of random walks and Lévy flights in CS in addition to the mechanisms of attractiveness and intensity in FA.

Algorithm 1: Proposed Cuckoo-Firefly-GR hybrid metaheuristic.

1: Divide the population into two diverse groups, say P_1 and P_2 each of size N
2: Random initialization of the populations P_1 and P_2
3: Evaluate the fitness of each individual solution
4: **while** *(t < max generation)* **do**
5:    | do *in parallel*
6:    | | Perform CS with GR (using p_a and p_g) on P_1
7:    | | Perform FA with GR (using p_a and p_g) on P_2
8:    | Rank solutions and find the current global best solution among P_1 and P_2
9:    | Mix P_1 and P_2 and randomly shuffle the entire population
10:   | Regroup the entire population into equisized diverse groups, P_1 and P_2
11:   | Evaluate the fitness of each individual solution
12: Postprocess the results

Existing literature indicates that FA could automatically subdivide the entire population into subgroups based on attractiveness via light intensity variations [6]. Such highlights of FA, when combined with the reproductive behavior of CS, and enhanced with GR, provide exploration and diversification needed to obtain optimal solutions with faster convergence. We employ a parallel hybridization, wherein the location information of the current best solution is mixed and re-grouped, instead of re-finding new positions using a random walk. Such parallelization ensures the search in the optimal location of the previous cycle, rather than having to re-search randomly.

The pseudocode in Algorithm 1 summarizes the procedure followed in the Cuckoo-Firefly-GR parallel hybridization. This study employs a population size (N) of 25, for a maximum of 25 cycles.

3.5 Classification of Unsolicited Bulk Emails

This study uses an MLP model to facilitate the UBE classification. MLP offers several advantages including fault tolerance, adaptive learning, and parallelism. Above all, MLP learns distributed data representations, enabling generalization to new combinations of values of features, beyond those seen while training.

MLP is a feed-forward neural network with an input layer, one or more non-linear hidden layers, and one prediction layer. The first layer takes the optimal email features ($\mathcal{I}$) as the input, and the output of each layer is used to activate the input in the subsequent layer. The transformation at each layer l is given as: $\mathcal{I}^{(l+1)} = f^{(l)}(\mathcal{I}^{(l)}) = g^{(l)}(W^{(l)} \cdot \mathcal{I}^{(l)} + b^{(l)})$, where $W^{(l)}$ and $b^{(l)}$ are the weight matrix and bias at layer l, and $g^{(l)}$ is a non-linear activation function such as ReLU, logistic sigmoid, or tanh function.

MLP uses the backpropagation algorithm to compute the gradient of the loss function needed to learn an optimal set of biases and weights. In this study, we aim at optimizing the cross-entropy prediction loss using an MLP model with one hidden layer of 75 nodes with a ReLU activation function.

Table 1. Statistics of the datasets used in UBE classification.

Dataset	Components	#Samples	#Classes
$D_{h,s}$	H and S	3,051	2
$D_{h,p}$	H and P	3,344	2
$D_{h,s,p}$	H, S, and P	3,844	3

H: Ham, S: Spam, and P: Phishing.

4   Experimental Validation and Discussion

This section describes the datasets used, the conducted experiments, and the obtained results concerning the classification of emails. Finally, we benchmark against the existing state-of-the-art methods in the field of UBE classification.

4.1   Raw Email Corpus

The raw email corpus used in this study consists of 3,844 emails comprising of 2,551 ham emails (66%), 793 phishing emails (21%), and 500 spam emails (13%), obtained from [16]. From these emails, three datasets were created. The first dataset was used to investigate the effectiveness of the proposed approach in spam classification: the dataset combined ham and spam emails from the corpus. The second dataset combined ham and phishing emails and aimed at testing the feature selection by the hybrid metaheuristic. To account for the fact that real-world email system could simultaneously receive ham, spam, and phishing emails, we created the third dataset comprising of all the emails. The statistics of these datasets are tabulated in Table 1.

Table 2. Comparison of the performance (%) of the hybrid metaheuristic over various feature selection approaches.

Approach	Metric	Dataset		
		$D_{h,s}$	$D_{h,p}$	$D_{h,s,p}$
Cuckoo-Firefly-GR	ACC	**99.78**	**99.40**	98.74
	MCC	**99.20**	**98.35**	**98.10**
Cuckoo-GR	ACC	98.47	97.11	**98.79**
	MCC	94.34	91.97	97.57
Firefly-GR	ACC	97.37	97.61	92.36
	MCC	92.73	93.36	94.16
Deep Autoencoders (4 Layers)	ACC	89.85	95.62	83.54
	MCC	57.82	87.65	65.92
Deep Autoencoders (8 Layers)	ACC	87.77	95.82	82.06
	MCC	47.96	88.23	62.65
LSTM Autoencoders (Compression = 0.4)	ACC	89.08	92.33	77.30
	MCC	53.83	78.35	50.90
LSTM Autoencoders (Compression = 0.8)	ACC	88.43	67.83	77.90
	MCC	50.42	39.61	51.39

4.2 Results and Discussion

The experiments were performed using a server running Ubuntu OS with two cores of Intel Xeon processors, 8 GB RAM, and one NVIDIA Tesla C-2050 GPU. All the algorithms were implemented in Python 2.7. To effectively test the performance of the proposed metaheuristic, we divide the entire email corpus using 70 − 30 train-to-test split percentage. To facilitate exhaustive benchmarking of the proposed approach, we utilize Accuracy (ACC) and MCC scores as the evaluation metrics. Furthermore, we compare performance of the proposed Cuckoo-Firefly-GR over the base optimization approaches (Cuckoo-GR and Firefly-GR), dense autoencoders (two variants: four encoding layers, a compression factor of 0.4, and trained for 25 epochs and eight encoding layers, a compression factor of 0.8, trained for 50 epochs), and LSTM autoencoders (two variants: one encoding layer, a compression factor of 0.4, trained for 25 epochs and one encoding layer, a compression factor of 0.8, trained for 50 epochs). Table 2 compares the performance of various dimensionality reduction approaches. We observe that the feature subset obtained using hybrid metaheuristic, when classified using an MLP model outperforms various other feature selection approaches.

From the statistics of the email corpus and Table 1, it can be observed that the dataset is class imbalanced. MCC score facilitates a balanced score even in an imbalanced scenario by considering true and false positives and negatives [4]. From Table 2, it can be remarked that the feature subset obtained using the proposed Cuckoo-Firefly-GR results in comparatively higher MCC scores (closer to $+1$), indicating the superior predictive capability of the proposed approach. Furthermore, the better and fitter solutions obtained using the proposed meta-heuristic can be attributed to the enhancement of balance between diversification and exploration, resulting in superior performance in accuracy.

Our results also signify the impact of capturing syntactic and semantic email content features. Table 3 compares the proposed approach with various state-of-the-art methods. Early works [1,5] used email content-based features to aid in the UBE classification. The results presented by Chandrasekaran *et al.* [5] show that a better classification is obtained with a larger number of features. Abu-Nimeh *et al.* [1] showed that an MLP model outperforms other machine learning classifiers. Several recent works [10,13,16,20] establish the need for both content-based and behavior-based features in phishing email classification. Zareapoor and Seeja [19] establish the need for mining email body content and using effective dimensionality reduction techniques to classify the email types effectively. This work employs a bio-inspired metaheuristic to mimic the best features in nature including natural selection, reproduction, and social behavior. The metaheuristic optimally selects a discriminative feature subset needed for the classifier to learn and generalize. Using an MLP model, we obtained an overall accuracy and MCC score of approximately 99% for all three UBE datasets. The large heterogeneity

Table 3. Comparison of the performance of the hybrid metaheuristic over various state-of-the-art approaches.

Work	#Features	Approach	Dataset	Accuracy (%)
Toolan and Carthy [16]	22	Content- and behavior- based features	Phishing: 6, 458	Set a: 97.00
			Non-phishing: 4, 202	Set b: 84.00
				Set c: 79.00
Zhang *et al.* [20]	7	Behavior-based features	Host-based: 2, 328	Train: 95.80
				Test: 99.60
Ma *et al.* [13]	7	Content- and behavior- based features	Phishing: 46, 525 (7%)	99.00
			Non-phishing: 613, 048	
Hamid and Abawajy [9]	7	Content- and behavior-based features	Set a (H and P): 1, 645	Set a: 96.00
			Set b (H and P): 2, 495	Set b: 92.00
			Set c (H and P): 4, 594	Set c: 92.00
Zareapoor and Seeja [19]	Varying (10 − 2, 000)	Content- and behavior- based features with dimensionality reduction	Phishing: 1, 000	LSA^a: 96.80
			Non-phishing: 1, 700	IG^b: 94.20
				PCA^c: 96.40
				χ^2: 94.50
This work	Set a: 164	Content- and behavior- based features with dimensionality reduction	Set a (H and S): 3, 051	Set a: 99.78
	Set b: 167		Set b (H and P): 3, 344	Set b: 99.40
	Set c: 172		Set c (H, S, and P): 3, 844	Set c: 98.79

^{a} *Latent Semantic Analysis,* ^{b} *Information Gain,* ^{c} *Principal Component Analysis*; H: *Ham,* S: *Spam,* and P: *Phishing.*

in the datasets used in various existing studies makes it challenging to compare their results with the proposed approach, despite which, our results are robust and comparable in terms of the overall accuracies and MCC scores.

5   Summary

UBE classification is a challenging problem due to the dynamic nature of UBE attacks. In this paper, we used forty prominent content- and behavior-based features used in the existing literature, in addition to the Doc2Vec features of email content. Doc2Vec modeling captures the syntactic and semantic textual features in the email content. We also designed Cuckoo-Firefly-GR, a hybrid metaheuristic to obtain a discriminative and informative feature subset crucial to UBE classification. The proposed Cuckoo-Firefly-GR combines the concepts of random walks in CS with the mechanisms of FA such as attractiveness and intensity to enhance the optimal balance between diversification and exploration. Moreover, the hybrid metaheuristic incorporates the evolutionary strategies of GA including selection, crossover, and mutation, to genetically replace the lower fitness individuals resulting in faster convergence and superior performance.

The proposed algorithm has been extensively tested using a corpus of $3,844$ emails to evaluate its efficacy. We underlined the superiority in the performance of the proposed Cuckoo-Firefly-GR over various feature selection methods, including base optimizations (Cuckoo-GR and Firefly-GR), dense autoencoders, and LSTM autoencoders. We also presented a comparative analysis of the proposed method over various state-of-the-art methods. The discriminative feature subset obtained by the proposed hybrid metaheuristic, when classified using an MLP model resulted in the overall performance of 99%. Our results revealed the impact of using effective email content modeling strategies along with efficient feature selection approaches in the classification of email types. In the future, we aim at extending the proposed approach to model the graphical features popularly used to exploit content- and behavior-based anti-UBE mechanisms.

References

1. Abu-Nimeh, S., Nappa, D., Wang, X., Nair, S.: A comparison of machine learning techniques for phishing detection. In: Proceedings of the Anti-Phishing Working Groups 2nd Annual eCrime Researchers Summit, pp. 60–69. ACM (2007)
2. Arora, S., Singh, S.: A conceptual comparison of firefly algorithm, bat algorithm and cuckoo search. In: 2013 International Conference on Control, Computing, Communication and Materials (ICCCCM), pp. 1–4. IEEE (2013)
3. BİRİCİK, G., Diri, B., SÖNMEZ, A.C.: Abstract feature extraction for text classification. Turk. J. Electr. Eng. Comput. Sci. 20(Sup. 1), 1137–1159 (2012)
4. Boughorbel, S., Jarray, F., El-Anbari, M.: Optimal classifier for imbalanced data using matthews correlation coefficient metric. PloS one 12(6), e0177678 (2017)
5. Chandrasekaran, M., Narayanan, K., Upadhyaya, S.: Phishing email detection based on structural properties. In: NYS Cyber Security Conference, pp. 2–8 (2006)

6. Elkhechafi, M., Hachimi, H., Elkettani, Y.: A new hybrid cuckoo search and firefly optimization. Monte Carlo Methods Appl. **24**(1), 71–77 (2018)
7. Gangavarapu, T., Jaidhar, C.D., Chanduka, B.: Applicability of machine learning in spam and phishing email filtering: review and approaches. Artif. Intell. Rev. 1–63 (2020). https://doi.org/10.1007/s10462-020-09814-9
8. Gangavarapu, T., Patil, N.: A novel filter-wrapper hybrid greedy ensemble approach optimized using the genetic algorithm to reduce the dimensionality of high-dimensional biomedical datasets. Appl. Soft Comput. **81**, 105538 (2019)
9. A. Hamid, I.R., Abawajy, J.: Hybrid feature selection for phishing email detection. In: Xiang, Y., Cuzzocrea, A., Hobbs, M., Zhou, W. (eds.) ICA3PP 2011. LNCS, vol. 7017, pp. 266–275. Springer, Heidelberg (2011). https://doi.org/10.1007/978-3-642-24669-2_26
10. Hamid, I.R.A., Abawajy, J.H.: An approach for profiling phishing activities. Comput. Secur. **45**, 27–41 (2014)
11. Le, Q., Mikolov, T.: Distributed representations of sentences and documents. In: International Conference on Machine Learning, pp. 1188–1196 (2014)
12. Lozano, M., García-Martínez, C.: Hybrid metaheuristics with evolutionary algorithms specializing in intensification and diversification: overview and progress report. Comput. Oper. Res. **37**(3), 481–497 (2010)
13. Ma, L., Yearwood, J., Watters, P.: Establishing phishing provenance using orthographic features. In: eCrime Researchers Summit, 2009, eCRIME 2009. IEEE (2009)
14. de Oliveira, V.Y., de Oliveira, R.M., Affonso, C.M.: Cuckoo search approach enhanced with genetic replacement of abandoned nests applied to optimal allocation of distributed generation units. IET Gener. Transm. Distrib. **12**(13), 3353–3362 (2018)
15. Symantec: Internet security threat report. Technical Report (March 2018)
16. Toolan, F., Carthy, J.: Feature selection for spam and phishing detection. In: 2010 eCrime Researchers Summit, pp. 1–12. IEEE (2010)
17. Yang, X.-S.: Firefly algorithms for multimodal optimization. In: Watanabe, O., Zeugmann, T. (eds.) SAGA 2009. LNCS, vol. 5792, pp. 169–178. Springer, Heidelberg (2009). https://doi.org/10.1007/978-3-642-04944-6_14
18. Yang, X.S., Deb, S.: Cuckoo search via lévy flights. In: 2009 World Congress on Nature & Biologically Inspired Computing (NaBIC), pp. 210–214. IEEE (2009)
19. Zareapoor, M., Seeja, K.: Feature extraction or feature selection for text classification: a case study on phishing email detection. Int. J. Inf. Eng. Electron. Bus. **7**(2), 60 (2015)
20. Zhang, J., Du, Z.H., Liu, W.: A behavior-based detection approach to mass-mailing host. In: 2007 International Conference on Machine Learning and Cybernetics, vol. 4, pp. 2140–2144. IEEE (2007)

Applications of Computational Methods in Artificial Intelligence and Machine Learning

Link Prediction by Analyzing Temporal Behavior of Vertices

Kalyani Selvarajah[1][(✉)] (iD), Ziad Kobti[1][(✉)] (iD), and Mehdi Kargar[2][(✉)] (iD)

[1] School of Computer Science, University of Windsor, Windsor, ON, Canada
{selva111,kobti}@uwindsor.ca
[2] Ted Rogers School of Management, Ryerson University, Toronto, ON, Canada
kargar@ryerson.ca

Abstract. Complexity and dynamics are challenging properties of real-world social networks. Link prediction in dynamic social networks is an essential problem in social network analysis. Although different methods have been proposed to enhance the performance of link prediction, these methods need significant improvement in accuracy. In this study, we focus on the temporal behavior of social networks to predict potential future interactions. We examine the evolving pattern of vertices of a given network $\mathcal{G}$ over time. We introduce a time-varying score function to evaluate the activeness of vertices that uses the number of new interactions and the number of frequent interactions with existing connections. To consider the impact of timestamps of the interactions, the score function engages a time difference of the current time and the time of the interaction occurred. Many existing studies ignored the weight of the link in the given network $\mathcal{G}$, which brings the time-varied details of the links. We consider two additional objective functions in our model: a weighted shortest distance between any two nodes and a weighted common neighbor index. We used Multi-Layer Perceptron (MLP), a deep learning architecture as a classifier to predict the link formation in the future and define our model as a binary classification problem. To evaluate our model, we train and test with six real-world dynamic networks and compare it with state-of-the-art methods as well as classic methods. The results confirm that our proposed method outperforms most of the state-of-the-art methods.

Keywords: Link prediction · Multilayer perceptron · Dynamic networks · Social network analysis

1 Introduction

Social Networks (SN) can be used to model a comprehensive range of real-life phenomena and examine the world around us. It ranges from online social interaction, including Facebook, Twitter, and LinkedIn to human interactions such as co-authorship, healthcare, and terrorist networks. Social networks analysis is the study of such networks to discover common structural patterns and explains

© Springer Nature Switzerland AG 2020
V. V. Krzhizhanovskaya et al. (Eds.): ICCS 2020, LNCS 12139, pp. 257–271, 2020.
https://doi.org/10.1007/978-3-030-50420-5_19

their emergence through computational models of network formation. The complexity and dynamics are essential properties of real-world social networks. Since these networks evolve quickly over time through the appearance or disappearance of new links and nodes, the connection becomes stronger and weaker, and underlying network structure changes with time. Therefore it has become high challenges for researchers in order to examine various research issues in social network analysis such as classification of a node, detecting the communities, formation of teams, and predicting links between nodes.

Understanding the mechanism of how the networks change over time is a crucial problem that is still not well understood [13]. Significant efforts have been made to explain the evolution of networks during the past decades [5]. However, such researches are yet to achieve the desired results, leaving the door open for further advances in the field. Throughout the last decades, analyzing temporal networks has received much attention among researchers as it has enormous applications in different disciplines such as co-authorship [2], the recommendation of friends [30] and website links. Recently, dynamic link predictions have been approached by various mechanisms and achieved promising results. However, the features of networks vary from each other, and the existing studies are not efficient to represent the importance of nodes and links. The objective of this paper is to address these issues and examines the dynamic nature of social networks.

Link prediction problem needs to be solved by determining the potentialities of the appearance or disappearance of links between all node pairs of the given network [28]. However, for example, in a collaborative network, if two experts (*i.e.*, vertices) collaborate once on any project, their link remains permanent, although any one of them stops interacting with the others further. Therefore, their link become weak in the future, while experts who have frequent interactions their link become strong. In this regard, we observe the behavior of individuals in the collaborative network when they need to decide which new collaborations might prove fruitful in addition to existing connections. Before anyone connects with others in the network, they usually examine several factors, including whether the people are active throughout the past or not, and are they working on similar projects that they have skills. Therefore, our study takes these factors into account and propose a new model, LATB, to predict the links which occur with others in the future. In this paper, we propose a model for link prediction problem on dynamic social networks and make the following major contributions.

1. Active individuals in social networks are popular among both existing members and new members who like to join the network. They believe that active individuals, for instance, in the co-authorship network, always update their research with current trends as well as being open to new ideas. So, to evaluate the activeness of any member, we consider two factors on the temporal network. (a) The score for constructing new connections (b) The score for the increased number of interactions with existing connections (How much the existing link becomes strong). We introduce a new score function to incor-

porate the impact of the timestamps and the gap between the current time and the time of the interaction occurred. Besides this, we introduce a probability function based on the activeness score of a pair of nodes to decide the likelihood of occurring a new link.

2. The smaller distance between any two individuals is higher the chances of future interaction. We incorporate the weighted shortest distance in LATB. In addition to this, we include another objective function, the weighted common neighbor index, which incorporates the time to evaluate the changes of strength of the neighbors' relationship.

3. In LATB, we used Multi-Layer Perceptron (MLP) as a classifier to predict the link formation in the future and defined our model as a binary classification problem.

The remaining of the paper is organized as follows. Section 2 describes related works. Section 3 specifies the problem definitions. Section 4 presents the experimental setup and the corresponding results. Finally, Sect. 5 concludes the research idea of this paper with directions for future work.

2   Related Works

Link prediction problem on the static network examines a single snapshot of a network structure at time t as an input, and then predicts possible unobserved links at time $t'(t \leq t')$ [13]. On the other hand, link prediction in dynamic networks investigates the evolution of networks over time as a sequence of snapshots and then predicts new links in the future. This section presents an overview of the link prediction problems on social networks. Several methods have been proposed to deal with the link prediction problem on the temporal network systems during the past decade.

The researchers designed a lot of topology-based similarity metrics for link prediction such as Common Neighbors (CN) [16], Adamic-Adar Coefficient (AA) [1], and Katz (KZ) [9]. Since the weights of links are rarely taken into account, many researchers modified those metrics in order to adopt the dynamic features of the social networks. The authors [15] examine the link prediction based on connection weight score structural properties of a given network. Zhu et al. [32] proposed a weighted mutual information model which is to estimate the effect of network structures on the connection likelihood by considering the benefits of both structural properties and link weights.

Potgieter et al. [21] showed that temporal metrics are valuable features in terms of accuracy. Tylenda et al. [25] proposed a graph-based link prediction algorithm and integrated it with temporal information and extended the local probabilistic model to involve time awareness. Yao et al. [31] used time-decay to manage the weight of the links and modified the common neighbor index to includes nodes in 2-hop. The authors [10] presented a time frame based unsupervised link prediction method for directed and weighted networks and derived a score for potential links in a time-weighted manner.

Tong W et al. [29] examined the concepts of the temporal trend of nodes by considering the changes of the degree over time using the structural perturbation method. Munasinge et al. [14] studied the impact of a relationship between timestamps of interactions and strength of the link for the future.

Xiaoyi Li et al. [12] proposed a deep learning method, conditional temporal restricted Boltzmann machine, which adopted a combination of feature engineering and CNN to predicts links. Recently, Goyal et al. [6] proposed DynGEM, which uses the recent advances in deep learning methods, autoencoders for graph embeddings to handle growing, dynamic graphs and for link prediction. Wang et al. [27] examined relational deep learning to jointly model high-dimensional node attributes and link structures with layers of latent variables and proposed generalized variational inference algorithm for learning the variables and predicting the links.

3   A Model for Dynamic Link Prediction

3.1   Problem Definition

A dynamic network is evolving over time and can be considered as a sequence of network snapshots within a time interval. The size of the network can occasionally shrink or expand as the network evolves. In this work, we focus on undirected weighted graphs.

Given a series of snapshots $\{\mathcal{G}_1, \mathcal{G}_2, \ldots, \mathcal{G}_{t-1}\}$ of an evolving graph $\mathcal{G}_T = \langle \mathcal{V}, \mathcal{E}_T \rangle$, where the edge $e = (u, v) \in \mathcal{E}_{t'}$ represents a link between $u \in \mathcal{V}_{t'}$ and $v \in \mathcal{V}_{t'}$ at a particular time t'. The dynamic link prediction approaches attempt to predict the likelihoods of links in the next time step $\mathcal{G}_t$. The list of graphs $\{\mathcal{G}_1, \mathcal{G}_2, \ldots, \mathcal{G}_{t-1}\}$ corresponding to a list of symmetric adjacency matrices $\{A_1, A_2, \ldots, A_{t-1}\}$. The adjacency matrix A_T of $\mathcal{G}_T$ is a $\mathcal{N} \times \mathcal{N}$ matrix where each element $A_T(i, j)$ takes 1 if the nodes $v_i \in \mathcal{V}$ and $v_j \in \mathcal{V}$, are connected at least once within time period T and takes 0 if they are not. Given a sequence of $(t-1)$ snapshots $\{A_1, A_2, \ldots, A_{t-1}\}$, the goal is to predict the adjacency matrix A_t at future time t.

3.2   Node Activeness

The idea of node activeness is highly related to the temporal behaviors of nodes. We can determine the active nodes through the analysis of the time-varying historical information of the nodes. To decide the activeness of the nodes, we can examine how they interact with others (nodes) throughout the timeframe. With any temporal network involving humans, we believe that the following factors are highly relevant to decide the activeness of nodes.

New Connections: A node can remain inactive or active. It can be decided based on how often a node made the new interactions throughout the time frame. Let us consider any two members A and B of a given dynamic network

$\mathcal{G}$ with the same number of new connections in the past; A might be connected at an early stage and not creating any new connection later (can be named as sleeping node), while node B formed most of his or her connection at the later stage (active node). The node B would attract more new people and have a high probability of generating new connections in the near future. We believe that considering this behavior of the node is significant in predicting the new links.

Definition 1 *(Score for New Connections). Given a network $\mathcal{G}_T\langle\mathcal{V}, \mathcal{E}_T\rangle$ and a set of nodes $A = \langle a_1, a_2, \dots, a_n\rangle$ at time t_m. Let's say the time windows to estimate the new connections are $|t_m^1 - t_m^2|, |t_m^2 - t_m^3| \dots |t_m^{n-1} - t_m^n|$, where $t_m^1, t_m^2 \dots t_m^n$ are consecutive time stamps. The score of building new connection at time t_m^n of a member or node of the network a_k can be defined as:*

$$SN(a_k) = \sum_{i=1}^{t_m^n} \frac{\mathcal{NC}_{a_k}^{t_i}}{|t_m^n - t_i| + 1} \tag{1}$$

where $\mathcal{NC}_{a_k}^{t_i}$ is the number of new connection made by the node a_k at time t_i. The term $|t_m^n - t_i|$ is the timestamp between current time and the selected time that the number of new connections has been made by the node a_k. If the difference between t_m^n and t_i is high, the value of $\mathcal{NC}_{a_k}^{t_i}$ over $|t_m^n - t_i| + 1$ become smaller although a node made more number of new connections at early time ($\mathcal{NC}_{a_k}^{t_i}$ is large). However, this is opposite for the nodes which made new connection recently. The timestamp has an addition of one to avoid the denominator become infinite when the two timestamps are equal.

The Frequency of Interaction: In addition to the new interactions, a node can continuously associate with the existing connected nodes. It is another way to decide the activeness of the node. Let us consider two nodes A and B with the same number of existing connections in the past; A might have several interactions at an early stage and not interacting with them later, while B is having frequent interactions with existing connections throughout the time frame. The later one (B) would attract more new nodes and have a high probability of generating new connections in the near future.

Definition 2 *(Score for Frequent Interactions). The score of frequent collaborations with existing connection at time t_m^n of a node a_k can be defined as:*

$$SE(a_k) = \sum_{i=1}^{t_m^n} \frac{\mathcal{EC}_{a_k}^{t_i}}{|t_m^n - t_i| + 1} \tag{2}$$

where $\mathcal{EC}_{a_k}^{t_i}$ is the number of frequent collaboration with existing connection by a node a_k at time t_i.

For instance, let's consider a network of authors (nodes) at certain year (2012) and the number new connection which both A and C made through last seven years (till 2019) can be shown as Fig. 1.

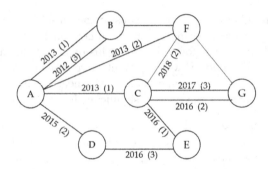

Fig. 1. Interaction with new and existing nodes throughout the time

$$\mathcal{SN}(A) = \sum_{i=2012}^{2019} \frac{\mathcal{NC}_A^i}{|2019 - i| + 1}$$

$$= \frac{3}{|2019 - 2012| + 1} + \frac{4}{|2019 - 2013| + 1} + \frac{2}{|2019 - 2015| + 1}$$

$$= 1.382$$

Similarly, the value of $\mathcal{SN}(C)$ can be evaluated to 2.89 as the information given in Fig. 1. Although both A and C made the equal number of new connections ($= 9$) in the past seven years, the score has a huge difference. The reason is C built new connections recently than A. This shows that nodes, which generate new connections recently have more attractive than others. Likewise, we can evaluate the score value of active nodes in terms of frequent interaction with existing connections.

Since both building new interactions and frequent interactions with existing connections have an influence on deciding the active node, the combination of these scores is a proper way to determine the score of the active node as given in Eq. 3.

$$\mathcal{SA}(a_k) = \lambda \sum_{i=1}^{t_m^n} \frac{\mathcal{NC}_{a_k}^{t_i}}{|t_m^n - t_i| + 1} + (1 - \lambda) \sum_{i=1}^{t_m^n} \frac{\mathcal{EC}_{a_k}^{t_i}}{|t_m^n - t_i| + 1} \tag{3}$$

where λ is a tradeoff value between the score for building new connections and expanding existing connections. Algorithm 1 describe the step by step process of calculating node activeness score for all nodes and store it in a lookup table.

At this point, every node is assigned by a score based on their activity on dynamic networks. However, to maintain the range of values between 0 and 1, we normalize each score. Inspired by configuration model, the probability $\mathcal{P}_{a_i,a_j}$ of a link exists between any two nodes a_i and a_j would be proportional to $\mathcal{SA}(a_i).\mathcal{SA}(a_j)$. Since the probability should be between 0 and 1, we drive the

---

Algorithm 1. Node Activeness

Input: Current Time stamp t_c, Snapshots of a given network G_T, List of vertices L_V, Trade off value λ

Output: Activeness Score lookup table $\mathcal{SA}$

1: $\mathcal{SA} \leftarrow \{\}$
2: $T[\,] \leftarrow \{t_1, t_2, \ldots t_k\}$ time steps
3: **for all** $n \in L_V$ **do**
4:    $Ex \leftarrow 0$ Existing Connection Score
5:    $Nw \leftarrow 0$ New Connection Score
6:    $nb \leftarrow \Gamma[G_{t_0}(n)]$ collaborated nodes at time $t = 0$
7:    **for all** t_i in T **do**
8:       $nb_i[\,] \leftarrow \Gamma[G_{t_i}(n)]$
9:       **for all** $m \in nb$ **do**
10:          **if** $m \in nb_0$ **then**
11:             $Ex \leftarrow Ex + \frac{w_{G_{t_i}}(m,n)}{t_c - t_i + 1}$
12:          **else**
13:             $Nw \leftarrow Nw + \frac{w_{G_{t_i}}(m,n)}{t_c - t_i + 1}$
14:          **end if**
15:       **end for**
16:       $nb \leftarrow nb \cup nb_i$
17:    **end for**
18:    $P \leftarrow \lambda.Ex + (1 - \lambda)Nw$
19:    $\mathcal{SA} \leftarrow \{n\colon P\}$
20: **end for**
21: **return** $\mathcal{SA}$

---

equation for this value by multiplying the reciprocal of the total activeness of the nodes in the networks as given in Eq. 5.

$$\mathcal{P}_{a_i,a_j} = \frac{\mathcal{SA}(a_i)\mathcal{SA}(a_j)}{\sum_{k=1}^{n} \mathcal{SA}(a_k)} \qquad (4)$$

where $\mathcal{P}_{a_i,a_j}$ is the probability of existing link between node a_i and a_j, $\mathcal{SA}(a_i)$ and $\mathcal{SA}(a_j)$ are the popularity scores of nodes a_i and a_j respectively and n is the total number of nodes in the networks.

We name our proposed method LATB (Link prediction by Analyzing Temporal Behaviour of vertices), because it highlights the behaviors of vertices.

3.3 Similarity Metrics

In the past, the majority of researches [13,19] have examined the accuracy of several heuristics for link prediction such as Adamic Adar, Preferential Attachment, and Jaccard Coefficient. In this research, we consider two modified forms of heuristics: weighted shortest path distance and weighted common neighbors.

Weighted Shortest Path Distance: In the undirected social network G, if some nodes have past interaction, their associated nodes in G are connected by an edge. If many levels of past interactions between two nodes are taken into account, then the input graph G is weighted. In this case, the smaller the edge weight between two nodes, the two nodes had more interactions in the past and have higher chances of interactions in the future. The distance between two nodes a_i and a_j, specified as $dist(a_i, a_j)$, is equal to the sum of the weights on the shortest path between them in the input graph G. If a_i and a_j are not connected in graph G, i.e., there is no path between a_i and a_j in G, the distance between them is set to ∞.

$$SD_{a_i,a_j} = wdist(a_i, a_j) \tag{5}$$

Weighted Common Neighbors: The Common Neighbors (CN) is the most widely used index in link prediction and evidence to the network transitivity property. It counts the number of common neighbors between node pair a_i and a_j. Newman et al. [16] has estimated this quantity in the context of collaboration networks. The probability that a_i and a_j collaborate in the future can be written as 6.

$$CN_{a_i,a_j} = |\Gamma(a_i) \cup \Gamma(a_j)| \tag{6}$$

where $\Gamma(a_i)$ and $\Gamma(a_j)$ consists of number of neighbors of the node a_i and a_j in $\mathcal{G}$ respectively.

As mentioned in the above definition, the common neighbors only consider the binary relations between nodes and ignore the time-varying nature and number of link occurrences. We adopt the time-varied weights into the common neighbors, which can give better predictions [8].

$$CN_{a_i,a_j}^{tw} = \sum_{|\Gamma(a_i) \cup \Gamma(a_j)|} \mathcal{W}^t(a_i, a_k) + \mathcal{W}^t(a_k, a_j) \tag{7}$$

where $\mathcal{W}^t(a_i, a_k) = \mathcal{W}(a_i, a_k) - \beta(t' - t)$, $\mathcal{W}(a_i, a_k)$ is original weight at time t, β is an attenuation factor and t' is the time considered for prediction.

3.4 Multilayer Perceptron (MLP) Framework

We treat the link prediction problem as a binary classification problem. We used MLP as a classifier. In this regards, we generate a dataset for all existing links in a last time step of a given dynamic network $\mathcal{G}_T$, for the positive link class, where for any two vertices i and j, the link between i and j, $ij \in E_t$. We generate negative link class, where any two vertices i and j, $ij \notin E_t$ by using downsampling technique to avoid imbalance problem. We assign the binary cross-entropy as loss function, which can be written as:

$$BCE = -y.log(p) - (1 - y)log(1 - p) \tag{8}$$

where y is binary indicator (0 or 1), p is predicted probability.

Finally, we build and train a neural network for link prediction. MLP is one of the most common and a variant of the original Perceptron model [22]. Here, we only briefly discuss the components of an MLP since this paper is not about MLP innovations. A typical MLP system can be built with layers of neurons, as shown in Fig. 2. Each neuron in a layer calculates the sum of its inputs (x) that are carried through an activation function (f). The output (O) from the network can be written as:

$$\mathcal{O}_{jk} = F_k \left(\sum_{i=1}^{N_{k-1}} w_{ijk} x_{i(k-1)} + \beta_{jk} \right) \tag{9}$$

where $\mathcal{O}_{jk}$ is the neuron j^{th} output at k^{th} layer and β_{jk} is bias weight for neuron j in layer k, respectively.

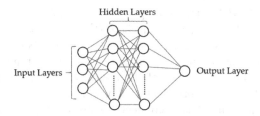

Fig. 2. MLP neural network

4   Experimental Results

We conduct extensive experiments to test our model with five real-world dynamic networks and use AUC (Area Under Curve) as evaluation metrics.

4.1   Dataset

We use six real world dynamic networks (Table 1): **Enron** corpus [11] and **Radoslaw** [23] are email communication networks. Each node specifies an employee and link represents email conversation among employees. Enron has the details from 6 January 1998 until 4 February 2004 while Randoslaw is from January 2nd 2010 to September 30th 2010. **Contact** [4] is data from wireless devices carried by people. Every node is people, and a link established when they contacted. The contact list represents the active contacts during 20-s intervals of the data collection. **College Messages** [17] have private messages sent on an online social network at the University among college people. **EU-core** [18] is an email data from a large European research institution. Link is the communication between members from 4 different departments. **Mathoverflow** [18] has the interactions on the stack exchange web site Math Overflow.

In the beginning, we sort the dataset in ascending order of time, and then we process a sequence of snapshots for each dataset at a fixed interval. We split every dynamic networks into five time frames G_1, G_2, G_3, G_4, G_5. We evaluate each scoring value at the last snapshot.

Table 1. The statistical information of each real-world dynamic networks.

Dataset	$|V|$	$|E_T|$	Time span in days
Enron	151	50571	165
Radoslaw	167	82900	272
Contact	274	28200	4
CollegeMessages	1899	59835	193
EU-core	986	332334	803
Mathoverflow	24818	506550	2350

4.2 Experimental Setup

We implement our model in Python3, processed the dataset on IBM cluster, the specification of POWER8 52 processor 256 GB of RAM. We trained and tested our model in an Nvidia GTX 1050Ti, 4 GB GPU with 768 CUDA cores.

In the weighted common neighbor index, we set the attenuation factor β to 0.001. To evaluate the active score of the nodes, we assign tradeoff factor λ to 0.5, because we believe that both the score for new connections and the score for frequent interaction with existing connections are equally important to decide the activeness of a person.

In the MLP, the first layer has four neurons with the ReLu activation function. We use two hidden layers of 32 neurons. The output layer contains a single neuron with the Sigmoid activation function. We train the neural network for 100 epochs. We use 80% training set, 10% validation set, and 10% testing set. We repeat the above process for ten times and find the average AUC.

4.3 Baseline Methods

We use various methods as the baselines, including classical methods such as Common Neighbors (CN), Jaccard coefficient (JC), Adamic Adar (AA) and Preferential attachment (PA), and network embedding methods such as node2vec, LINE, DeepWalk, and SDNE. The brief introduction of these methods is listed as follow:

– Common Neighbors (CN) [16]: It is one of the most common measurements used in link prediction problem. Having a large number of the common neighbors easily create a link.
– Jaccard Coefficient (JC): It is a normalized form of the CN index.
– Adamic-Adar Coefficient (AA) [1]: It evaluates the impotency of a node when having less number of neighbors when predicting links.
– Preferential Attachment (PA) [3]: It generates the belief that nodes with large number of neighbors are more likely to form more in the future.
– node2vec [7]: It is a node embedding method, which learns nodes representation of network by preserving higher-order proximity between nodes. It used a higher probability of node occurrence in a fixed-length random walk.

- LINE [24]: It used an objective function to preserves the first-order and second-order neighborhoods to learn node representations, most similar to node2vec. It is useful to apply for large-scale network embedding.
- DeepWalk [20]: It used random walk model to learn vertex representations. This embedding can be used to predict link existence.
- SDNE [26]: It used both the first-order and second-order proximities together in an autoencoder based deep model to generate the vertex representations.

Regarding the implementations, we evaluated the Link prediction problem from the original code by the authors for node2vec[1], LINE[2], DeepWalk[3] and SDNE[4].

Table 2. Experimental results based on AUC by comparing to classic and embedding methods.

Dataset	CN	JC	AA	PA	node2vec	LINE	DeepWalk	SDNE	LATB
Enron	0.8106	0.8751	0.8970	0.8442	0.7596	0.5042	0.7190	**0.9437**	0.9302
Radoslaw	0.8417	0.8307	0.9028	0.8753	0.7417	0.6153	0.7342	0.8709	**0.9457**
Contact	0.8457	0.9141	0.9142	0.9027	0.8741	0.7360	0.8451	0.9376	**0.9906**
CollegeMessages	0.5742	0.5774	0.5843	0.5901	0.7049	0.4905	0.7506	0.7806	**0.9576**
EU-core	0.9227	0.9302	0.9341	0.7553	0.8602	0.6587	0.8201	0.9574	**0.9626**
Mathoverflow	0.7774	0.7692	0.7430	0.7783	0.7478	0.6587	0.7456	0.9574	**0.9968**

4.4 Results

Our model achieves a significant improvement compared to other methods in various dynamic networks except the dataset Enron, which is better in SDNE. The standard network embedding methods, node2vec and LINE, perform the worst than other methods in terms of AUC. LATB performs significantly better in Contact and Eu-Core networks above 96% while the classic and embedding methods achieve below 92%. Moreover, LATB has an AUC higher than 93% among all tested dynamic networks. We can conclude that LATB can perform well in both very sparse (Enron) and dense (Radoslaw) networks. Tables 2 and Fig. 3 present the results comparison of other methods with LATB.

---

[1] https://github.com/aditya-grover/node2vec.
[2] https://github.com/tangjianpku/LINE.
[3] https://github.com/phanein/deepwalk.
[4] https://github.com/xiaohan2012/sdne-keras.

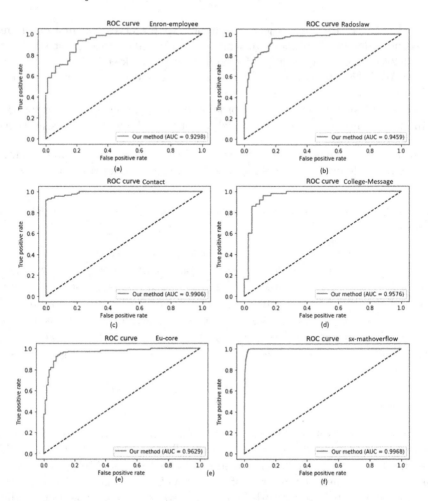

Fig. 3. Experiments on ROC curve. (a) ROC curve on College Message. (b) ROC curve on Enron-employee. (c) ROC curve on Radoslaw. (d) ROC curve on Contact. (e) ROC curve on Eu-Core.

5   Conclusions

In this paper, we propose a model for link prediction in dynamic networks by analyzing temporal behaviors of vertices, named LATB. To model the evolving pattern of each vertex, we propose a new scoring method, which can engage the historical changes of vertices. To further address LATB, temporal changes of vertices are analyzed in two ways to measure the activeness of a node: how often a vertex interacts with existing connected nodes - to measure the strength of the relationship with its neighbors, and how fast and often collaborate with new nodes. Because both measures have a strong influence on deciding the node activeness, we introduce a probability function based on the activeness of nodes to evaluate the chances of being connected in the future. We also use two other weighted indexes: shortest distance and common neighbors, to incorporate the time-varying nature and number of link occurrences in neighbor nodes. In LATB, MLP is used as a classifier, which results in the status of link existence. Empirically, we compare LATB with classical methods and traditional embedding methods in five different real-world dynamic networks. Overall our model, LATB, achieves significant improvements and reaches above 93% of AUC.

Acknowledgment. This research work was supported by International Business Machines (IBM); experiments were conducted on a high performance IBM Power System S822LC Linux Server.

References

1. Adamic, L.A., Adar, E.: Friends and neighbors on the web. Soc. Netw. **25**(3), 211–230 (2003)
2. Aydın, S.: Link prediction models for recommendation in academic collaboration network of Turkey. Ph.D. thesis (2017)
3. Barabâsi, A.L., Jeong, H., Néda, Z., Ravasz, E., Schubert, A., Vicsek, T.: Evolution of the social network of scientific collaborations. Phys. A Stat. Mech. Appl. **311**(3–4), 590–614 (2002)
4. Chaintreau, A., Hui, P., Crowcroft, J., Diot, C., Gass, R., Scott, J.: Impact of human mobility on opportunistic forwarding algorithms. IEEE Trans. Mob. Comput. **6**, 606–620 (2007)
5. Dorogovtsev, S.N., Mendes, J.F.: Evolution of networks. Adv. Phys. **51**(4), 1079–1187 (2002)
6. Goyal, P., Kamra, N., He, X., Liu, Y.: Dyngem: deep embedding method for dynamic graphs. arXiv preprint arXiv:1805.11273 (2018)
7. Grover, A., Leskovec, J.: node2vec: scalable feature learning for networks. In: Proceedings of the 22nd ACM SIGKDD International Conference on Knowledge Discovery and Data Mining, pp. 855–864. ACM (2016)
8. Huang, S., Tang, Y., Tang, F., Li, J.: Link prediction based on time-varied weight in co-authorship network. In: Proceedings of the 2014 IEEE 18th International Conference on Computer Supported Cooperative Work in Design (CSCWD), pp. 706–709. IEEE (2014)
9. Katz, L.: A new status index derived from sociometric analysis. Psychometrika **18**(1), 39–43 (1953)
10. Kaya, M., Jawed, M., Bütün, E., Alhajj, R.: Unsupervised link prediction based on time frames in weighted–directed citation networks. In: Missaoui, R., Abdessalem, T., Latapy, M. (eds.) Trends in Social Network Analysis. LNSN, pp. 189–205. Springer, Cham (2017). https://doi.org/10.1007/978-3-319-53420-6_8
11. Klimt, B., Yang, Y.: The enron corpus: a new dataset for email classification research. In: Boulicaut, J.-F., Esposito, F., Giannotti, F., Pedreschi, D. (eds.) ECML 2004. LNCS (LNAI), vol. 3201, pp. 217–226. Springer, Heidelberg (2004). https://doi.org/10.1007/978-3-540-30115-8_22
12. Li, X., Du, N., Li, H., Li, K., Gao, J., Zhang, A.: A deep learning approach to link prediction in dynamic networks. In: Proceedings of the 2014 SIAM International Conference on Data Mining, pp. 289–297. SIAM (2014)

13. Liben-Nowell, D., Kleinberg, J.: The link-prediction problem for social networks. J. Am. Soc. Inf. Sci. Technol. **58**(7), 1019–1031 (2007)
14. Munasinghe, L., Ichise, R.: Time aware index for link prediction in social networks. In: Cuzzocrea, A., Dayal, U. (eds.) DaWaK 2011. LNCS, vol. 6862, pp. 342–353. Springer, Heidelberg (2011). https://doi.org/10.1007/978-3-642-23544-3_26
15. Murata, T., Moriyasu, S.: Link prediction of social networks based on weighted proximity measures. In: Proceedings of the IEEE/WIC/ACM International Conference on Web Intelligence, pp. 85–88. IEEE Computer Society (2007)
16. Newman, M.E.: Clustering and preferential attachment in growing networks. Phys. Rev. E **64**(2), 025102 (2001)
17. Panzarasa, P., Opsahl, T., Carley, K.M.: Patterns and dynamics of users' behavior and interaction: network analysis of an online community. J. Am. Soc. Inf. Sci. Technol. **60**(5), 911–932 (2009)
18. Paranjape, A., Benson, A.R., Leskovec, J.: Motifs in temporal networks. In: Proceedings of the Tenth ACM International Conference on Web Search and Data Mining, pp. 601–610. ACM (2017)
19. Pavlov, M., Ichise, R.: Finding experts by link prediction in co-authorship networks. FEWS **290**, 42–55 (2007)
20. Perozzi, B., Al-Rfou, R., Skiena, S.: Deepwalk: online learning of social representations. In: Proceedings of the 20th ACM SIGKDD International Conference on Knowledge Discovery and Data Mining, pp. 701–710. ACM (2014)
21. Potgieter, A., April, K.A., Cooke, R.J., Osunmakinde, I.O.: Temporality in link prediction: understanding social complexity. Emergence Complex. Organ. (E: CO) **11**(1), 69–83 (2009)
22. Rosenblatt, F.: The perceptron: a theory of statistical separability in cognitive systems. United States Department of Commerce (1958)
23. Rossi, R., Ahmed, N.: The network data repository with interactive graph analytics and visualization. In: Twenty-Ninth AAAI Conference on Artificial Intelligence (2015)
24. Tang, J., Qu, M., Wang, M., Zhang, M., Yan, J., Mei, Q.: Line: large-scale information network embedding. In: Proceedings of the 24th International Conference on World Wide Web, pp. 1067–1077 (2015). International World Wide Web Conferences Steering Committee
25. Tylenda, T., Angelova, R., Bedathur, S.: Towards time-aware link prediction in evolving social networks. In: Proceedings of the 3rd Workshop on Social Network Mining and Analysis, p. 9. ACM (2009)
26. Wang, D., Cui, P., Zhu, W.: Structural deep network embedding. In: Proceedings of the 22nd ACM SIGKDD International Conference on Knowledge Discovery and Data Mining, pp. 1225–1234. ACM (2016)
27. Wang, H., Shi, X., Yeung, D.Y.: Relational deep learning: a deep latent variable model for link prediction. In: Thirty-First AAAI Conference on Artificial Intelligence (2017)
28. Wang, P., Xu, B., Wu, Y., Zhou, X.: Sci. China Inf. Sci. **58**(1), 1–38 (2014). https://doi.org/10.1007/s11432-014-5237-y
29. Wang, T., He, X.S., Zhou, M.Y., Fu, Z.Q.: Link prediction in evolving networks based on popularity of nodes. Sci. Rep. **7**(1), 7147 (2017)
30. Yang, S.H., Smola, A.J., Long, B., Zha, H., Chang, Y.: Friend or frenemy?: predicting signed ties in social networks. In: Proceedings of the 35th International ACM SIGIR Conference on Research and Development in Information Retrieval, pp. 555–564. ACM (2012)

31. Yao, L., Wang, L., Pan, L., Yao, K.: Link prediction based on common-neighbors for dynamic social network. Procedia Comput. Sci. **83**, 82–89 (2016)
32. Zhu, B., Xia, Y.: Link prediction in weighted networks: a weighted mutual information model. PloS one **11**(2), e0148265 (2016)

Detecting the Most Insightful Parts of Documents Using a Regularized Attention-Based Model

Kourosh Modarresi[✉]

AI Metrics, Sunnyvale, CA, USA
kouroshm@alumni.stanford.edu

Abstract. Every individual text or document is generated for specific purpose(s). Sometime, the text is deployed to convey a specific message about an event or a product. Other occasions, it may be communicating a scientific breakthrough, development or new model and so on. Given any specific objective, the creators and the users of documents may like to know which part(s) of the documents are more influential in conveying their specific messages or achieving their objectives. Understanding which parts of a document has more impact on the viewer's perception would allow the content creators to design more effective content. Detecting the more impactful parts of a content would help content users, such as advertisers, to concentrate their efforts more on those parts of the content and thus to avoid spending resources on the rest of the document. This work uses a regularized attention-based method to detect the most influential part(s) of any given document or text. The model uses an encoder-decoder architecture based on attention-based decoder with regularization applied to the corresponding weights.

Keywords: Artificial neural networks · Natural Language Processing · Sparse loss function · Regularization · Transformer

1 Motivation of This Work

1.1 Review

The main purpose of NLP (Natural Language Processing) and NLU (Natural Language Understanding) is to understand the language. More specifically, they are focused on not just to see the context of text but also to see how human uses the language in daily life. Thus, among other ways of utilizing this, we could provide an optimal online experience addressing needs of users' digital experience. Language processing and understanding is much more complex than many other applications in machine learning such as image classification as NLP and NLU involve deeper context analysis than other machine learning applications. This paper is written as a short paper and focuses on explaining only the parts that are contribution of this paper to the state-of-the art. Thus, this paper does not describe the state-of-the-art works in details and uses those works [2, 4, 5, 8, 53, 60, 66, 70, 74, 84] to build its model as a modification and extension of the state of

© Springer Nature Switzerland AG 2020
V. V. Krzhizhanovskaya et al. (Eds.): ICCS 2020, LNCS 12139, pp. 272–281, 2020.
https://doi.org/10.1007/978-3-030-50420-5_20

the art. Therefore, a comprehensive set of reference works have been added for anyone interested in learning more details of the previous state of the art research [3, 5, 10, 17, 33, 48, 49, 61–63, 67–73, 76, 77, 90, 91, 93].

1.2 Attention Based Model

Deep Learning has become a main model in natural language processing applications [6, 7, 11, 22, 38, 55, 64, 71, 75, 78–81, 85, 88, 94]. Among deep learning models, often RNN-based models like LSTM and GRU have been deployed for text analysis [9, 13, 16, 23, 32, 39–42, 50, 51, 58, 59]. Though, modified version of RNN like LSTM and GRU have been improvement over RNN (recurrent neural networks) in dealing with vanishing gradients and long-term memory loss, still they suffer from many deficiencies. As a specific example, a RNN-based encoder-decoder architecture uses the encoded vector (feature vector), computed at the end of encoder, as the input to the decoder and uses this vector as a compressed representation of all the data and information from encoder (input). This ignores the possibility of looking at all previous sequences of the encoder and thus suffers from information bottleneck leading to low precision, especially for texts of medium or long sequences. To address this problem, global attention-based model [2, 5] where each of the encoder sequence uses all of the encoder sequences. Figure 1 shows an attention-based model.

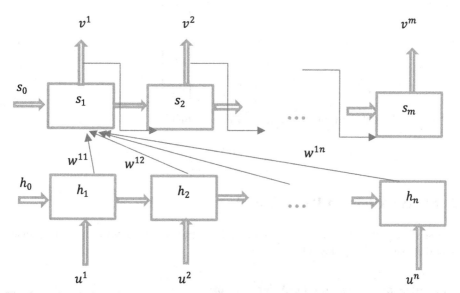

Fig. 1. A description of attention-based encoder-decoder architecture. The attention weights for one of the decoder sequences (the first decoder sequence) are displayed.

Where $i = 1{:}n$ is the encoder sequences and, $t = 1{:}m$ represents the decoder sequences. Each of the encoder states looks into the data from all the encoder sequences with specific attention measured by the weights. Each weight, w^{ti}, indicates the attention decoder network t pays for the encoder network i. These weights are dependent on the previous decoder and output states and present encoder state as shown in Fig. 2.

Given the complexity of these dependencies, a neural network model is used to compute these weights. Two layers (1024) of fully connected layers and ReLU activation function is used.

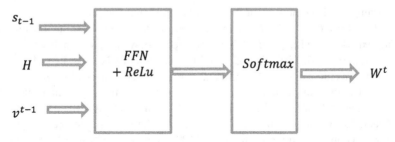

Fig. 2. The computation model of weights using fully-connected networks and SoftMax layer.

Where H is the state of the encoder networks, s_{t-1} is the previous state of the decoder and v^{t-1} is the previous decoder output. Also, W^t is the weights of the encoder state t.

$$W^t = \begin{bmatrix} W^{1t} \\ W^{2t} \\ \cdot \\ \cdot \\ \cdot \\ W^{nt} \end{bmatrix}, \quad H = \begin{bmatrix} h_i \\ h_2 \\ \cdot \\ \cdot \\ \cdot \\ h_n \end{bmatrix}, \tag{1}$$

Since W^t are the output from softmax function, then,

$$\sum_{i=1}^{n} w^{it} = 1 \tag{2}$$

2 Sparse Attention-Based Model

This section overviews of the contribution of this paper and explains the extension made over the state-of-the-art model.

2.1 Imposing Sparsity on the Weight Vectors

A major point of attention for many texts related analysis is to determine which part(s) of the input text has had more impact in determining the output. he length of input text could be very long combining of potentially hundreds and thousands of words or sequences, i.e., n could be very large number. Thus, there are many weights (w^{ti}) in determining any part of output v^t, and also since many of these weights are correlated, it's difficult to determine the significance of any input sequence in computing any output sequence v^t. To make these dependencies clearer and to recognize the most significant input sequences for any output sequence, we apply a zero-norm penalty to make the corresponding weight vector to become a sparse vector. To achieve the desired sparsity, zero-norm (L_0) is applied to make any corresponding W^t vector very sparse as the penalty leads to minimization of the number of non-zero entries in W^t. The process is implemented by imposing the constraint of,

$$\|W^t\|_0 \leq \vartheta \tag{3}$$

Since L_0 is computationally intractable, we could use surrogate norms such as L_1 norm or Euclidean norm, L_2. To impose sparsity, the L_1 norm, LASSO [8, 14, 15, 18, 21] is used in this work,

$$\|W^t\|_1 \leq \vartheta \tag{4}$$

Or,

$$\beta \|W^t\|_1 \tag{5}$$

As the penalty function to enforce sparsity on the weight vectors.

This penalty, $\beta \|W^t\|_1$, is the first extension to the attention model [2, 5]. Here, β is the regularization parameter which is set as a hyperparameter where its value is set before learning. Higher constraint leads to higher sparsity with higher added regularization biased error and lower values of the regularization parameter leads to lower sparsity and lesser regularization bias.

2.2 Embedding Loss Penalty

The main goal of this work is to find out which parts of encoder sequences are most critical in determining and computing any output. The output could be a word, a sentence or any other subsequence. The goal is critical especially in application such as machine translation, image captioning, sentiment analysis, topic modeling and predictive modeling such as time series analysis and prediction.

To add another layer of regularization, this work imposes embedding error penalty to the objective function (usually, cross entropy). This added penalty also helps to address the "coverage problem" (the phenomenon of often observed dropping or frequently repeating words - - or any other subsequence - - by the network). The embedding regularization is,

$$\alpha \|Embedding\ Error\|_2 \tag{6}$$

Input to any model has to be a number and hence the raw input of words or text sequence needs to be transformed to continuous numbers. This is done by using one-hot encoding of the words and then using embedding as shown in Fig. 3.

Where $\tilde{u}^i$ is the raw input text, $\widetilde{u}^i$ is the one-hot encoding representation of the raw input and u^i is the embedding of the i-th input or sequence. Also, α is the regularization parameter.

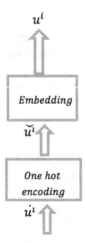

Fig. 3. The process of representation of input words by one-hot encoding and embedding.

The idea of embedding is based on that embedding should preserve word similarities, i.e., the words that are synonyms before embedding, should remain synonyms after embedding. Using this concept of embedding, the scaled embedding error is,

$$L(U) = \sum_{i=1}^{n} \sum_{j=1}^{n} \left(L\left(\breve{u}^i, \breve{u}^j \right) - L\left(u^i, u^j \right) \right)^2 \tag{7}$$

Or, after scaling the embedding error,

$$L(U) = \frac{1}{2n} \sum_{i=1}^{n} \sum_{j=1}^{n} \left(L\left(\breve{u}^i, \breve{u}^j \right) - L\left(u^i, u^j \right) \right)^2 \tag{8}$$

Which could be re-written, using a regularization parameter (α), as,

$$L(U) = \alpha \left(\frac{1}{2n} \sum_{i,j=1}^{n} \left(L\left(\breve{u}^i, \breve{u}^j \right) - L\left(u^i, u^j \right) \right)^2 \right) \tag{9}$$

Where L is the measure or metric of similarity of words representations. Here, for all similarity measures, both Euclidean norm and cosine similarity (dissimilarity) have been used. In this work, the embedding error using the Euclidean norm is used,

$$L(U) = \alpha \left(\frac{1}{2n} \sum_{i,j=1}^{n} \left(L_2\left(\breve{u}^i, \breve{u}^j \right) - L_2\left(u^i, u^j \right) \right)^2 \right) \tag{10}$$

Alternatively, we could include the embedding error of the output sequence in Eq. (10). When the input sequence (or the dictionary) is too long, to prevent high computational complexity of computing similarity of each specific word with all other words, we choose a random (uniform) sample of the input sequences to compute the embedding error. The regularization parameter, α, is computed using cross validation [26–31]. Alternatively, adaptive regularization parameters [82, 83] could be used.

2.3 Results and Experiments

This model was applied on Wikipedia datasets for English-German translation (one-way translation) with 1000 sentences. The idea was to determine which specific input word (in English) is the most important one for the corresponding German translation. The results were often an almost diagonal weight matrix, with few non-zero off diagonal entries, indicating the significance of the corresponding word(s) in the original language (English). Since the model is an unsupervised approach, it's hard to evaluate its performance without using domain knowledge. The next step in this work would be to develop a unified and interpretable metric for automatic testing and evaluation of the model without using any domain knowledge and also to apply the model to other applications such as sentiment analysis.

References

1. Anger, G., Gorenflo, R., Jochum, H., Moritz, H., Webers, W. (eds.): Inverse Problems: Principles and Applications in Geophysics, Technology, and Medicine. Akademic Verlag, Berlin (1993)
2. Vaswani, A., et al.: Polosukhin: attention is all you need. arXiv:1706.03762 (2017)
3. Axelrod, A., He, X., Gao, J.: Domain adaptation via pseudo in-domain data selection. In: Proceedings of the ACL Conference on Empirical Methods in Natural Language Processing (EMNLP), pap. 355–362. Association for Computational Linguistics (2011)
4. Ba, J.L., Mnih, V., Kavukcuoglu, K.: Multiple object recognition with visual attention. arXiv: 1412.7755, December 2014
5. Bahdanau, D., Cho, K., Bengio, Y.: Neural machine translation by jointly learning to align and translate. arXiv:1409.0473, September 2014
6. Baldi, P., Sadowski, P.: The dropout learning algorithm. Artif. Intell. **210**, 78–122 (2014)
7. Bastien, F., et al.: Theano: new features and speed improvements. In: Deep Learning an Unsupervised Feature Learning NIPS 2012 Workshop (2012)
8. Becker, S., Bobin, J., Candès, E.J.: NESTA, a fast and accurate first-order method for sparse recovery. SIAM J. Imaging Sci. **4**(1), 1–39 (2009)
9. Bengio, Y., Simard, P., Frasconi, P.: Learning long-term de-pendencies with gradient descent is difficult. IEEE Trans. Neural Networks **5**(2), 157–166 (1994)
10. Bengio, Y., Ducharme, R., Vincent, P., Janvin, C.: A neural probabilistic language model. J. Mach. Learn. Res. **3**, 1137–1155 (2003)
11. Bergstra, J., et al.: Theano: a CPU and GPU math expression compiler. In: Proceedings of the Python for Scientific Computing Conference (SciPy). Oral Presentation (2010)
12. Boulanger-Lewandowski, N., Bengio, Y., Vincent, P.: Audio chord recognition with recurrent neural networks. In: ISMIR (2013)
13. Cai, J.-F., Candès, E.J., Shen, Z.: A singular value thresholding algorithm for matrix completion. SIAM J. Optim. **20**(4), 1956–1982 (2008)
14. Candès, E.J., Recht, B.: Exact matrix completion via convex optimization. Found. Comput. Math. **9**, 717–772 (2008)
15. Candès, E.J.: Compressive sampling. In: Proceedings of the International Congress of Mathematicians, Madrid, Spain (2006)
16. Cheng, J., Dong, L., Lapata, M.: Long short-term memory-networks for machine reading. arXiv preprint arXiv:1601.06733 (2016)
17. Cho, K., et al.: Learning phrase representations using RNN encoder-decoder for statistical machine translation. arXiv preprint arXiv:1406.1078. 2, 12, 13, 14 (2014)

18. D'Aspremont, A., El Ghaoui, L., Jordan, M.I., Lanckriet, G.R.G.: A direct formulation for sparse PCA using semidefinite programming. SIAM Rev. **49**(3), 434–448 (2007)
19. Davies, A.R., Hassan, M.F.: Optimality in the regularization of ill-posed inverse problems. In: Sabatier, P.C. (ed.) Inverse Problems: An Interdisciplinary Study. Academic Press, London (1987)
20. DeMoor, B., Golub, G.H.: The restricted singular value decomposition: properties and applications. SIAM J. Matrix Anal. Appl. **12**(3), 401–425 (1991)
21. Donoho, D.L., Tanner, J.: Sparse nonnegative solutions of underdetermined linear equations by linear programming. Proc. Natl. Acad. Sci. **102**(27), 9446–9451 (2005)
22. Dozat, T., Qi, P., Manning, C.D.: Stanford's graph-based neural dependency parser at the CoNLL 2017 shared task. In: CoNLL 2017 Shared Task: Multilingual Parsing from Raw Text to Universal Dependencies, pp. 20–30 (2017)
23. Dyer, C., Kuncoro, A., Ballesteros, M., Smith, N.A.: Recurrent neural network grammars. In: Proceedings of NAACL (2016)
24. Efron, B., Hastie, T., Johnstone, I., Tibshirani, R.: Least angle regression. Ann. Stat. **32**, 407–499 (2004)
25. Elden, L.: Algorithms for the regularization of ill-conditioned least squares problems. BIT **17**, 134–145 (1977)
26. Elden, L.: A note on the computation of the generalized cross-validation function for ill-conditioned least squares problems. BIT **24**, 467–472 (1984)
27. Engl, H.W., Hanke, M., Neubauer, A.: Regularization methods for the stable solution of inverse problems. Surv. Math. Ind. **3**, 71–143 (1993)
28. Engl, H.W., Hanke, M., Neubauer, A.: Regularization of Inverse Problems. Kluwer, Dordrecht (1996)
29. Engl, H.W., Kunisch, K., Neubauer, A.: Convergence rates for Tikhonov regularisation of non-linear ill-posed problems. Inverse Prob. **5**, 523–540 (1998)
30. Engl, H.W., Groetsch, C.W. (eds.): Inverse and Ill-Posed Problems. Academic Press, London (1987)
31. Gander, W.: On the linear least squares problem with a quadratic Constraint. Technical report STAN-CS-78-697, Stanford University (1978)
32. Gers, F., Schraudolph, N., Schmidhuber, J.: Learning precise timing with LSTM recurrent networks. J. Mach. Learn. Res. **3**, 115–143 (2002)
33. Goldwater, S., Jurafsky, D., Manning, C.D.: Which words are hard to recognize? Prosodic, lexical, and disfluency factors that increase speech recognition error rates. Speech Commun. **52**, 181–200 (2010)
34. Golub, G.H., Van Loan, C.F.: Matrix Computations. Computer Assisted Mechanics and Engineering Sciences, 4th edn. Johns Hopkins University Press, Baltimore (2013)
35. Golub, G.H., Van Loan, C.F.: An analysis of the total least squares problem. SIAM J. Numer. Anal. **17**, 883–893 (1980)
36. Golub, G.H., Kahan, W.: Calculating the singular values and pseudo-inverse of a matrix. SIAM J. Numer. Anal. Ser. B **2**, 205–224 (1965)
37. Golub, G.H., Heath, M., Wahba, G.: Generalized cross-validation as a method for choosing a good ridge parameter. Technometrics **21**, 215–223 (1979)
38. Goodfellow, I., Warde-Farley, D., Mirza, M., Courville, A., Bengio, Y.: Maxout networks. In: Proceedings of The 30th International Conference on Machine Learning, pp. 1319–1327 (2013)
39. Graves, A.: Sequence transduction with recurrent neural networks. In: Proceedings of the 29th International Conference on Machine Learning, ICML (2012)
40. Graves, A., Fern'andez, S., Schmidhuber, J.: Bidirectional LSTM networks for improved phoneme classification and recognition. In: Proceedings of the 2005 International Conference on Artificial Neural Networks, Warsaw, Poland (2005)

41. Graves, A., Schmidhuber, J.: Framewise phoneme classification with bidirectional LSTM and other neural network architectures. Neural Netw. **18**, 602–610 (2005)
42. Graves, A.: Generating sequences with recurrent neural networks. arXiv:1308.0850 [cs.NE] (2013)
43. Hastie, T., Tibshirani, R., Friedman, J.: The Elements of Statistical Learning; Data miNing, Inference and Prediction. Springer, New York (2001). https://doi.org/10.1007/978-0-387-216 06-5
44. Hastie, T.J., Tibshirani, R.: Handwritten digit recognition via deformable prototypes. Technical report, AT&T Bell Laboratories (1994)
45. Hastie, T., et al.: 'Gene Shaving' as a method for identifying distinct sets of genes with similar expression patterns. Genome Biol. **1**, 1–21 (2000)
46. Hastie, T., Mazumder, R.: Matrix Completion via Iterative Soft-Thresholded SVD (2015)
47. Hastie, T., Tibshirani, R., Narasimhan, B., Chu, G.: Package 'impute'. CRAN (2017)
48. Hermann, K., Blunsom, P.: Multilingual distributed representations without word alignment. In: Proceedings of the Second International Conference on Learning Representations, ICLR (2014)
49. Hirschberg, J., Manning, C.D.: Advances in natural language processing. Science **349**(6), 261–266 (2015)
50. Hochreiter, S., Schmidhuber, J.: Long short-term memory. Neural Comput. **9**(8), 1735–1780 (1997)
51. Hochreiter, S., Bengio, Y., Frasconi, P., Schmidhuber, J.: Gradient flow in recurrent nets: the difficulty of learning long-term dependencies (2001)
52. Hofmann, B.: Regularization for Applied Inverse and Ill-Posed problems. Teubner, Stuttgart (1986)
53. Hudson, D.A., Manning, C.D.: Compositional attention networks for machine reasoning. In: International Conference on Learning Representations, ICLR (2018)
54. Jeffers, J.: Two case studies in the application of principal component. Appl. Stat. **16**, 225–236 (1967)
55. Jolliffe, I.: Principal Component Analysis. Springer, New York (1986). https://doi.org/10. 1007/978-1-4757-1904-8
56. Jolliffe, I.T.: Rotation of principal components: choice of normalization constraints. J. Appl. Stat. **22**, 29–35 (1995)
57. Jolliffe, I.T., Trendafilov, N.T., Uddin, M.: A modified principal component technique based on the LASSO. J. Comput. Graph. Stat. **12**(3), 531–547 (2003)
58. Kalchbrenner, N., Blunsom, P.: Recurrent continuous translation models. In: Proceedings of the ACL Conference on Empirical Methods in Natural Language Processing (EMNLP), pp 1700–1709. Association for Computational Linguistics (2013)
59. Koehn, P.: Statistical Machine Translation. Cambridge University Press, New York (2010)
60. Kim, Y., Denton, C., Hoang, L., Rush, A.M.: Structured attention networks. In: International Conference on Learning Representations (2017)
61. Koehn, P., Och, F.J., Marcu, D.: Statistical phrase-based translation. In: Proceedings of the 2003 Conference of the North American Chapter of the Association for Computational Linguistics on Human Language Technology, NAACL 2003, vol. 1, pp. 48–54. Association for Computational Linguistics, Stroudsburg (2003)
62. Cho, K., van Merrienboer, B., Gulcehre, C., Bougares, F., Schwenk, H., Bengio, Y.: Learning phrase representations using RNN encoder-decoder for statistical machine translation. CoRR, abs/1406.1078 (2014)
63. Lafferty, J., McCallum, A., Pereira, F.: Conditional random fields: Probabilistic models for segmenting and labeling sequence data. In: Proceedings 18th International Conference on Machine Learning, pp. 282–289. Morgan Kaufmann, San Francisco (2001)

64. LeCun, Y., Bottou, L., Orr, Genevieve B., Müller, K.-R.: Efficient BackProp. In: Orr, Genevieve B., Müller, K.-R. (eds.) Neural Networks: Tricks of the Trade. LNCS, vol. 1524, pp. 9–50. Springer, Heidelberg (1998). https://doi.org/10.1007/3-540-49430-8_2

65. Lin, Z., et al.: A structured self-attentive sentence embedding. arXiv preprint arXiv:1703. 03130 (2017)

66. Luong, H.P., Manning, C.D.: Effective Approaches to attention-based neural machine translation. In: EMNLP (2015)

67. Bill, M., Grenager, T., de Marneffe, M.-C., Cer, D., Manning, C.D.: Learning to recognize features of valid textual entailments. In: Proceedings of the Human Language Technology Conference of the North American Chapter of the Association for Computational Linguistics (HLT-NAACL 2006), pp. 41–48 (2006)

68. Bill, M., Manning, C.D.: Natural logic for textual inference. In: ACL-PASCAL Workshop on Textual Entailment and Paraphrasing, pp. 193–200 (2007)

69. Manning, C.D.: Ergativity. In: Brown, K. (ed.) Encyclopedia of Language & Linguistics, vol. 4, 2nd edn, pp. 210–217. Elsevier, Oxford (2006)

70. Manning, C.D., Raghavan, P., Schütze, H.: Introduction to information retrieval. Computer Science (2008)

71. Manning, C.D., Surdeanu, M., Bauer, J., Finkel, J.R., Bethard, S., McClosky, D.: The Stanford CoreNLP Natural Language Processing Toolkit. Computer Science, ACL (2014)

72. Manning, C.D.: Computational linguistics and deep learning. Comput. Linguist. **41**(4), 701–707 (2015)

73. McFarland, D.A., Ramage, D., Chuang, J., Heer, J., Manning, C.D., Jurafsky, D.: Differentiating language usage through topic models. Poetics **41**(6), 607–625 (2013)

74. Luong, M.-T., Pham, H., Manning, C.D.: Effective approaches to attention based neural machine translation. arXiv preprint arXiv:1508.04025 (2015)

75. Modarresi, K.: Application of DNN for Modern Data with two Examples: Recommender Systems & User Recognition. Deep Learning Summit, San Francisco, CA, 25–26 January 2018

76. Modarresi, K., Munir, A.: Standardization of featureless variables for machine learning models using natural language processing. In: Shi, Y., Fu, H., Tian, Y., Krzhizhanovskaya, V.V., Lees, M.H., Dongarra, J., Sloot, P.M.A. (eds.) ICCS 2018. LNCS, vol. 10861, pp. 234–246. Springer, Cham (2018). https://doi.org/10.1007/978-3-319-93701-4_18

77. Modarresi, K., Munir, A.: Generalized variable conversion using k-means clustering and web scraping. In: Shi, Y., Fu, H., Tian, Y., Krzhizhanovskaya, V.V., Lees, M.H., Dongarra, J., Sloot, P.M.A. (eds.) ICCS 2018. LNCS, vol. 10861, pp. 247–258. Springer, Cham (2018). https://doi.org/10.1007/978-3-319-93701-4_19

78. Modarresi, K., Diner, J.: An efficient deep learning model for recommender systems. In: Shi, Y., Fu, H., Tian, Y., Krzhizhanovskaya, V.V., Lees, M.H., Dongarra, J., Sloot, P.M.A. (eds.) ICCS 2018. LNCS, vol. 10861, pp. 221–233. Springer, Cham (2018). https://doi.org/10.1007/978-3-319-93701-4_17

79. Modarresi, K.: Effectiveness of Representation Learning for the Analysis of Human Behavior. American Mathematical Society, San Francisco State University, San Francisco, CA, 27 October 2018

80. Modarresi, K., Diner, J.: An evaluation metric for content providing models, recommendation systems, and online campaigns. In: Rodrigues, J.M.F., Cardoso, P.S., Monteiro, J., Lam, R., Krzhizhanovskaya, V.V., Lees, Michael H., Dongarra, J.J., Sloot, P.M.A. (eds.) ICCS 2019. LNCS, vol. 11537, pp. 550–563. Springer, Cham (2019). https://doi.org/10.1007/978-3-030-22741-8_39

81. Modarresi, K.: Combined Loss Function for Deep Convolutional Neural Networks. American Mathematical Society, University of California, Riverside, Riverside, CA, 9–10 November 2019

82. Modarresi, K., Golub, G.H.: A Randomized Algorithm for the Selection of Regularization Parameter. Inverse Problem Symposium, Michigan State University, MI, 11–12 June 2007
83. Modarresi:, K.: A local regularization method using multiple regularization levels. Ph.D. thesis, Stanford University, Stanford, CA (2007)
84. Parikh, A., Täckström, O., Das, D., Uszkoreit, J.. A decomposable attention model. In: Empirical Methods in Natural Language Processing (2016)
85. Pascanu, R., Mikolov, T., Bengio, Y.: On the difficulty of training recurrent neural networks. In: ICML (2013)
86. Pascanu, R., Mikolov, T., Bengio, Y.: On the difficulty of training recurrent neural networks. In: Proceedings of the 30th International Conference on Machine Learning, ICML (2013)
87. Pascanu, R., Gulcehre, C., Cho, K., Bengio, Y.: How to construct deep recurrent neural networks. In: Proceedings of the Second International Conference on Learning Representations, ICLR (2014)
88. Schraudolph, N.N.: Fast curvature matrix-vector products for second-order gradient descent. Neural Comp. **14**, 1723–1738 (2002)
89. Schuster, M., Paliwal, K.K.: Bidirectional recurrent neural networks. IEEE Trans. Signal Process. **45**, 2673–2681 (1997)
90. Schwenk, H.: Continuous space translation models for phrase-based statistical machine translation. In: Kay, M., Boitet, C. (eds.) Proceedings of the 24th International Conference on Computational Linguistics (COLIN), pp. 1071–1080. Indian Institute of Technology Bombay (2012)
91. Schwenk, H., Dchelotte, D., Gauvain, J.-L.: Continuous space language models for statistical machine translation. In: Proceedings of the COLING/ACL on Main Conference Poster Sessions, pp. 723–730. Association for Computational Linguistics (2006)
92. Sutskever, I., Vinyals, O., Le, Q.: Sequence to sequence learning with neural networks. In: Advances in Neural Information Processing Systems, NIPS (2014)
93. Wu, Y., et al.: Google's neural machine translation system: bridging the gap between human and machine translation. arXiv preprint arXiv:1609.08144 (2016)
94. Zeiler, M.D.: ADADELTA: an adaptive learning rate method. arXiv:1212.5701 (2012)

Challenge Collapsar (CC) Attack Traffic Detection Based on Packet Field Differentiated Preprocessing and Deep Neural Network

Xiaolin Liu[1,3], Shuhao Li[1,2(✉)], Yongzheng Zhang[1,2,3], Xiaochun Yun[4], and Jia Li[4]

[1] Institute of Information Engineering, Chinese Academy of Sciences, Beijing, China
liuxiaolin191@mails.ucas.ac.cn, {lishuhao,zhangyongzhen}@iie.ac.cn
[2] Key Laboratory of Network Assessment Technology, Chinese Academy of Sciences, Beijing, China
[3] School of Cyber Security, University of Chinese Academy of Sciences, Beijing, China
[4] National Computer Network Emergency Response Technical Team/Coordination Center of China, Beijing, China
{yunxiaochun,lijia}@cert.org.cn

Abstract. Distributed Denial of Service (DDoS) attack is one of the top cyber threats. As a kind of application layer DDoS attack, Challenge Collapsar (CC) attack has become a real headache for defenders. However, there are many researches on DDoS attack, but few on CC attack. The related works on CC attack employ rule-based and machine learning-based models, and just validate their models on the outdated public datasets. These works appear to lag behind once the attack pattern changes. In this paper, we present a model based on packet Field Differentiated Preprocessing and Deep neural network (FDPD) to address this problem. Besides, we collected a fresh dataset which contains 7.92 million packets from real network traffic to train and validate FDPD model. The experimental results show that the *accuracy* of this model reaches 98.55%, the F_1 value reaches 98.59%, which is 3% higher than the previous models (SVM and Random Forest-based detection model), and the training speed is increased by 17 times in the same environment. It proved that the proposed model can help defenders improve the efficiency of detecting CC attack.

Keywords: Malicious traffic detection · CC attack · Packet Field Differentiated Preprocessing · Deep neural network

Supported by the National Key Research and Development Program of China (Grant No.2016YFB0801502), and the National Natural Science Foundation of China (Grant No.U1736218).

1   Introduction

With the development of the Internet and the advancement of technology, Distributed Denial of Service (DDoS) attack has become more and more serious. Challenge Collapsar (CC) attack is a type of DDoS attack that sends forged HTTP requests to some target web server frequently. These requests often require complicated time-consuming caculations or database operations, in order to exhaust the resource of the target web server. Because the HTTP request packets of CC attack are standard and sometimes their IPs are true, it's difficult to defend. According to reports, in February 2016, hackers launched a large-volume CC attack against XBOX , one of the world's largest online game-playing platforms, causing a 24-h impact on the business [5]. Similar incidents are happening endlessly, causing serious effects.

At present, the research on CC attack detection has the following limitations:

(1) **High feature extraction dependency.** Most of the previous detection models are based on specific rules of CC attack. For example, Moore et al. [13] found that the attack packet size and time interval had a certain regularity when the attack occurred. So they use size and time to detect. It requires a lot of statistical calculations and can't be automatically updated to cope with variant attacks.

(2) **Long model training time.** Current detection models are mostly machine learning model. We have implemented representative machine learning models to train the data, including Support Vector Machine (SVM) and Random Forest [7,17,23]. Experiments show that in the face of large-traffic dataset, these models need a long training period, which is severely delayed in practice. In order to show the contrast, we divide the massive data into several small datasets, and the results show that previous model is 17 times slower than our model.

(3) **Lack of real and fresh dataset.** Most previous works are based on public datasets [4,7,8,19], which is not new enough to cope with changing attack patterns. Public datasets are often obtained by experts through empirical and statistical analysis. Therefore, in the face of new attack variants, the samples of the public datasets appear to be backward, making it difficult to meet the ever-changing attack detection requirements in the real environment.

The contributions of this paper are as follows:

(1) **We propose a packet field differentiated preprocessing algorithm.** Considering the specific meanings of the fields in the packet, we divide all fields into four types (misleading fields, useless fields, discrete fields, nondiscrete fields) and process them differently (drop, one-hot encoding, ASCII encoding). In traditional models, it is often necessary to extract the feature artificially, which depends on expert experience severely. Compared with this, our model greatly saves labor costs.

(2) **We present a CC attack detection model based on packet Field Differentiated Preprocessing and Deep neural network (FDPD).** Our model leverages its powerful self-learning capabilities to learn the implicit feature in traffic. The experimental results show that the accuracy reaches 98.55%, the recall is 98.41%, the precision is 98.76%, and the comprehensive evaluation index F_1 reaches 98.59%. Compared with the traditional SVM and Random Forest-based malicious traffic detection models, the comprehensive evaluation index value F_1 of our proposed model is increased by 3%, and the processing speed is increased by 17 times.

(3) **We collect a dataset of 7.92 million packets.** This dataset contains two parts. One is the CC attack packets from the CC attack script program based on HTTP protocol. The other is non-CC attack packtes from the backbone network traffic. And the data is labeled based on the system and manual sampling check. Compared with the public dataset, our data is more representative, which covers more CC attack ways.

The remainder of this paper is structured as follows. Section 2 discusses related work. Section 3 introduces traffic detection model of challenge collapsar attacks. The training and optimization process, experimental results and analysis will be showed in Sect. 4. Section 5 discusses the limitation of the proposed model. Finally, Sect. 6 concludes this paper.

2   Related Work

As an application layer attack based on HTTP protocol, CC attack is more subtle than traditional flood-based DDoS attack [16], posing an increasing challenge for Web service applications. Because of the low cost of attack and the disruptive impact, HTTP protocol packet is the primary target of GET flood attacks. As new attackers emerge, identifying these attacks is puzzling. Besides, due to the continuous upgrade of existing attack tools and the increasing network bandwidth [2], the cost of the attacker's continuous connection request to the website is getting lower and lower, and the Web server is more and more vulnerable. Sree et al. [18] suggested that the exponential growth of the usage of Internet led to cyberattacks. Among various attacks, the HTTP GET flood attack is one of the main threats to Internet services because it exhausts resources and services in the application layer. Because the attack request pattern is similar to a legitimate client, it is difficult to distinguish them.

In addition, HTTP DDoS attacks in cloud computing and SDN fields have intensified. Aborujilah et al. [1] mentioned that HTTP attack is one of the key attacks on cloud-based Web servers. Lin et al. [10] found that 43% of the three major network attacks are HTTP attack, and 27% of organizations face daily or weekly HTTP attack.

2.1   Rule-Based Detection

At present, many detection models rely on rules based on attack characteristics. Bin Xiao et al. [3] used the feature of time delay, obtained the feature distribution of the time internal by maximum likelihood estimation, and linked it to the self-organizing map neural network. And the abnormal detection is performed with the set of the detection threshold. Miao Tan et al. [20] extracted key information from different protocols packets, and proposed firefly group optimization algorithm by combining pattern search and boundary variation.

An Wang et al. [21] proposed a new dataset, and found three rules by analyzing this new dataset: 1. Location-based analysis shows most attackers come from active Botnet; 2. From the perspective of the target being attacked, multiple attacks on the same target also show a strong attack rule, so the start time of the next attack from some botnet families can be accurately predicted; 3. Different Botnets have a similar trend of initiating DDoS attacks against the same victim at the same time or in turn. Similarly, CC attack can be defended against by painting a picture of attacker, but it doesn't work when attackers change their attack way.

2.2 Machine Learning-Based Detection

In the field of CC attack detection research, machine learning models are widely used, including the original models and improvement models. Liu et al. [24] used improved neighbor propagation clustering algorithm to pre-classify the attackers' behavior with a small amount of prior knowledge, then merged and eliminated by Silhouette index, and timely made cluster perform re-cluster.

Xie et al. [22] established the Hidden Semi-Markov model and calculated the distance by the Euclidean distance formula. If the distance exceeds the limit value, they think that an attack has occurred. Kshira et al. [14] used the learning automaton (LA) model to implement DDoS defense based on the number of SYN requests and the actual number of TCP connections, and defended DDoS attacks on the hardware level to reduce the damage caused by the attack behavior. Sudip et al. [11] used learning automaton model to protect the network from DDoS attacks based on the existing optimized link state routing protocol.

2.3 Deep Learning-Based Detection

DDoS attacks often cause network delays. Xiao et al. [3] measures the delay from one network port to another, and use the maximum likelihood model to estimate the characteristic distribution of the traffic delay. They use a neural network of self-organizing maps to set thresholds based on learning results for anomaly detection. In order to solve the HTTP malicious traffic detection problem, Li et al. [9] proposed a model which combined convolutional neural network (CNN) and multilayer perceptron (MLP) based on the combination of raw data and empirical feature engineering. Their experiments showed good results. This proves the effectiveness of deep learning again. It's known that deep learning has excellent self-learning ability. Therefore, we employ deep learning to better cope with CC attack variants.

3   Modeling Methodology

Our FDPD model consists of two parts, one is a preprocessing algorithm and the other is a deep neural network, as is shown in Fig. 1.

3.1   Packet Field Differentiated Preprocessing Algorithm

CC attack is an interactive application layer attack based on the HTTP/HTTPs protocol. The potential features of this attack are often contained at the flow level, so the collected underlying TCP/IP data packets need to be restored and spliced, including CC attack flows and other flows. There are several factors that need to be considered. Firstly, the conditions that the packets can be regarded as a flow, here we introduce the flow period and the number of packets as the restoration and splice conditions. Secondly, the role of different fields in the packet should be considered while learning the CC attack behavior laws, here we classify the fields into four categories, a) misleading field, these fields can interfere with the model and have an adverse effect on model training; b) useless field, such as the fields that always stay same; c) discrete field; d) non-discrete field. Therefore, we propose a preprocessing algorithm for packet fields, as shown in Algorithm 1. Before explaining the algorithm, three functions that appear in the algorithm are illustrated.

$$f_1(protocolField) = \begin{cases} -1 & Misleading field \\ 0 & Constant field \\ 1 & Discrete field \\ 2 & Non-discrete field \end{cases} \tag{1}$$

$$f_2(protocolField) = f_{one-hot}(protocolField) \tag{2}$$

$$f_3(protocolField) = f_{ASCII}(protocolField) \tag{3}$$

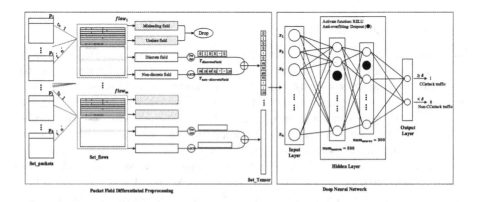

Fig. 1. The structure of FDPD model

In the algorithm, as for $p_j \in Set\_packets$, we firstly splice them into multiple flows by $quadruple = (sourceip, destinationip, sourceport, destinationport)$, and then we get $flow = \sum_{j=1}^{k} p_j$, each flow has the same quadruple. As a result, we get many flows, the set of flows is written as $Set\_flows = \sum_{i=1}^{m} flow_i$.

Considering the flow period and the number of packets, we define the time interval threshold α and the number of packets threshold β, where α is the time period constraint of a single flow, corresponding to the input variable $t\_flowPeriod$ in the algorithm, and β is the constraint of the number of packets in a single flow, corresponding to the input variable $n\_packetNumber$ in the algorithm. The p_j in each $flow_i$ is trimmed, and the original $flow_i$ is reorganized and updated. The updated $flow_i$ includes new header and payload. The structure is shown in Fig. 2. The header contains the TCP/IP packet header fields (such as "ip", "port", "reserved", "flags", etc.), and the payload contains the HTTP packet header fields and all other fields. Combining the specific meanings of the fields in the packet, we divide all fields into four types and process different types of fields differently.

(1) **Misleading field.** Misleading field mainly refers to "IP", "url", "host". For this type of fields, we adopt the strategy of dropping. Because the training sample is limited, and the value of this type of fields are fixed, but in fact, when the CC attack is launched, the attacker will constantly change the attack node and the attack target. The values of these fields are uncertain,

---

Algorithm 1. Packet Field Differentiated Preprocessing algorithm of FDPD

Input: Set_packets, t_flowPeriod, n_packetNumber
Output: Set_Tensor, Set_lengthOfTensor

1: *Initialization* $t' = t_0 + t\_flowPeriod, n' = n\_packetNumber$
2: Set_flows = Set_packets.groupby(quadruple) #classify packets to flows
3: **for** each $flow_i \in Set\_flows$ **do**
4:     **for** each $p_j \in flow_i$ **do**
5:         **if** $p_j.time < t'$ and $j < n'$ **then**
6:             $flow_i.header = merge1(flow_i.header, p_j.header)$ #merge the different part
7:             $flow_i.payload = merge2(flow_i.payload, p_j.payload)$ #merge all the part
8: **for** each $flow_i \in Set\_flows$ **do**
9:     $l_{payload}, T_{payload} = f_3(flow_i.payload)$
10:     **if** $f_1(flow_i.header.protocolField) == 1$ **then**
11:         $l_{discreteField}, T_{discreteField} = f_2(flow_i.header.protocolField)$
12:     **if** $f_1(flow_i.header.protocolField) == 2$ **then**
13:         $l_{nondiscreteField}, T_{nondiscreteField} = f_3(flow_i.header.protocolField)$
14:     $Tensor = concat(T_{payload}, T_{discreteField}, T_{nondiscreteField})$
15:     $lengthOfTensor = l_{payload} + l_{discreteField} + l_{nondiscreteField}$
16:     Set_Tensor.add(Tensor)
17:     Set_lengthOfTensor.add(lengthOfTensor)
    return Set_Tensor, Set_lengthOfTensor

so judging whether the traffic has CC aggressiveness through such fields has a very large error. Dropping these fields can prevent the overfitting problem of deep learning; By the way, this fields are sensitive, and removing them can protect privacy.

(2) **Useless field.** Useless field mainly refers to "reserved". We also adopt the strategy of dropping for such fields. Because such fields stay unchanged in the packet, they are not distinguishable. Although it may be helpful for other detections, such as hidden channel discovery, it does not make any sense to determine whether the traffic has CC aggressiveness. So retaining such fields will increase the cost of model training, and it may even reduce the accuracy of the model because it learns the features of other kinds of attack behaviors.

(3) **Discrete field.** Discrete field mainly refers to "flags" and "HTTP version". The values of these fields are discrete. For such fields, we use one-hot encoding. Because deep learning often uses distance when classifying, the one-hot encoding method will make the distance calculation between features more reasonable and protect the meaning of the original fields. For example, there are six values in "flags" field: 'URG' 'ACK' 'PSH' 'RST' 'SYN' 'FIN'. After one-hot encoding, the distance between two values is 1. Generally, the distance is used to measure the similarity between them, and the classification is based on the distance. By this way, they are obviously divided into the same category, it's in line with the facts; if ASCII encoding is used, the distance between values will be unreasonable.

(4) **Non-discrete field.** For Non-discrete field, such as "Payload" , use ASCII encoding to get the value from 0 to 127.

Finally, the processed fields are spliced to obtain a *Tensor*, which is used as the input of the deep neural network, as described in Sect. 3.2.

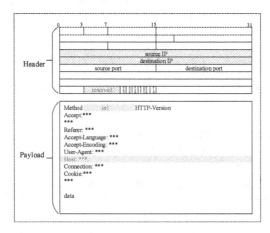

Fig. 2. Updated $flow_i$, containing the new header and payload

3.2   Deep Neural Network Structure

In addition to the necessary data preprocessing, rule-based or machine learning-based malicious traffic detection methods also require manual extraction of attack features. Most rule-based models need to summarize the representative rules of CC attack behavior through many statistical calculations. Then they use the rules directly or combine machine learning to classify the traffic, but once the attacker changes the CC attack mode slightly, these models will not work.

In the data preprocessing as described in Sect. 3.1, we get input data $Tensor$ that fully covers the packet information. After that, we didn't spend a lot of manual labor to perform feature extraction on the remaining useful fields. Instead, we automated this work. We set up a deep neural network after data preprocessing, and use its powerful self-learning capabilities to mine the hidden features in the traffic, let it automatically learn the difference between CC attack traffic and non-CC attack traffic at the packet level, which helps further improve the efficiency of CC Attack traffic detection.

Our deep neural network is a fully connected network with a four-layer structure, containing the input layer, two hidden layers, and the output layer, uses ReLU as the activation function in the hidden layers, uses Logistic Regression as the classifier in the output layer, and uses the cross entropy as the loss function to update the weight and bias values.

The number of input layer neurons depends on the dimention of $Tensor$, its value is related to three factors: $t\_flowPeriod$, $n\_packetNumber$ and *the length of packet*. The first two factors are the limiting conditions of data preprocessing, and the latter factor is needed because our deep neural network model can only process fixed-length inputs, so the length of the input data must be uniformly standardized. The determination of these three values will be discussed in the experimental section. As far hidden layers, the most important thing is the number of neurons in the hidden layers. We finally determined the numbers to be 500 and 300 respectively, and the detail is also discussed in the experimental section. The number of output layer neurons is the number of classifications, and our issue is a two-category problem. So there are two kinds of results: CC attack traffic or Non-CC attack traffic. There is a threshold δ, if the output value is not less than δ, we identify it as CC attack traffic and Non-CC attack traffic conversely.

The deep neural network learns the numerical input data, calculates the loss function, and updates the parameters through the back propagation. In order to prevent overfitting problem, we adopt Dropout strategy in all the hidden layers.

4   Experiments and Analysis

The basic experimental environment: CPU frequency is 2061 MHZ; CPU core is 64; RAM is 64 GB; GPU is GeForce GTX TITAN and the number is 8.

4.1 Dataset

The data source of this paper: a) CC attack traffic comes from the attack scripts based on HTTP protocol; b) Non-CC attack traffic comes from backbone network traffic. We annotate the dataset through system detection and manual verification. The dataset contains more than 1.8 million CC attack traffic data and more than 6.12 million non-CC attack traffic data. We share the dataset on Github[1] .

Table 1. Attacked domain name

Name	Types	Domain name
Baidu	Search	https://www.baidu.com/
Eastern Military Network	Military	http://mil.eastday.com/
Jingdong Mall	Shopping	https://www.jd.com/
Shanghai Pudong Development Bank	Bank	https://www.spdb.com.cn/
Starting point novel	Read	https://www.qidian.com
Blood war song	Online games	https://mir2.youxi.com/
Straight flush	Stock	https://www.10jqka.com.cn/
Wangyi cloud music	Music	https://music.163.com/
Ctrip	Travel	https://www.ctrip.com/
YouKu	Video	https://www.youku.com/
58City	Life	https://bj.58.com/
Tencent sports	Sports	https://sports.qq.com/
Today's headlines	News	https://www.toutiao.com/

The CC attack traffic data is generated by running CC attack scripts. These scripts covers a lot of CC attack ways, and the scripts are placed on Github[2]. In order to ensure that CC attack traffic will not cause substantial damage to the attack target, we build a network filter, by which the CC attack packets will be intercepted locally and stored as samples of the dataset. On the Ubuntu system, run the compiled CC attack scripts to launch attacks against 13 different kinds of domain names and open our network filter to capture the generated traffic. The domain name to be attacked are as shown in Table 1. And then more than 1.8 million CC attack packets are obtained through our filter.

Non-CC attack traffic data comes from communication among multiple servers and hosts. It is obtained in the backbone network, and captured through the whitelist. After capturing, more than 6.12 million packets are obtained, which basically covers the network communication among various web services.

---

[1] https://github.com/xiaolini/Sample_Dataset_afterprocess.

[2] https://github.com/xiaolini/Script.

4.2   Training and Validation

The evaluation indexes are as follows:

$$Accuracy = \frac{TP + TN}{TP + FN + TN + FP} \tag{4}$$

$$Precision = \frac{TP}{TP + FP} \tag{5}$$

$$Recall = \frac{TP}{TP + FN} \tag{6}$$

$$F_\beta = \frac{(1 + \beta^2) * Precision * Recall}{(\beta^2) * Precision + Recall}, \beta = 1 \tag{7}$$

where TP (True Positive): CC traffic is correctly identified as CC traffic; FP (False Positive): Non-CC attack traffic is incorrectly identified as CC traffic; FN (False Negative): CC traffic is incorrectly identified as Non-CC attack traffic; TN (True Negative): Non-CC attack traffic is correctly identified as Non-CC attack traffic.

We divided the dataset into two parts: Training set (60%) and Test set (40%). In order to determine specific model parameters, we further divided the training set into another two parts: Estimation set and Validation set, of which the estimation set accounts for 2/3, the validation set accounts for 1/3.

In the preprocessing phase, we first need to determine the time threshold $t\_flowPeriod$ and the number threshold of packets in a flow $n\_packetNumber$. In addition, the model can only adapt to fixed-length inputs, so we need to determine a fixed value. CC attackers often try to exhaust the victim's resources, the attack traffic is often short and frequent, so we set $t\_flowPeriod$ to 0.5s. $n\_packetNumber$ and *the length of packet* is determined by experiments that shown in Fig. 3. Figure 3(a) shows the percentage of the flow with less packets than $n(n = 13, 14, ..., 21)$. We can see 63% flows contains less than 18 packets, and the statistical average value is 17.95. Therefore, we determined

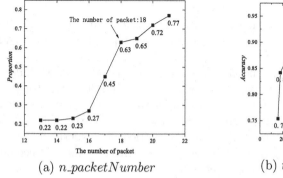

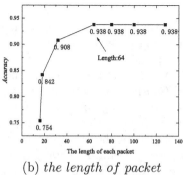

(a) $n\_packetNumber$ (b) *the length of packet*

Fig. 3. The parameter determination in the data preprocessing

$n\_packetNumber = 18$, so that it doesn't cause excessive abandonment and filling during data processing. Figure 3(b) is about the accuracy, we determined *the length of packet* is 64(bytes).

According to $t\_flowPeriod$, $n\_packetNumber$ and *the length of packet*, we drop the redundant packets, fill the insufficient packets with −1, and remove the excessive part. Then, we get updated flow with new header and payload. We drop the misleading fields and useless fields of the new flow, use one-hot encoding for discrete fields and use ASCII encoding for non-discrete fields. After that, we convert all the useful data into numeric data, and finally obtain 1152-dimension tensors as the input of DNN.

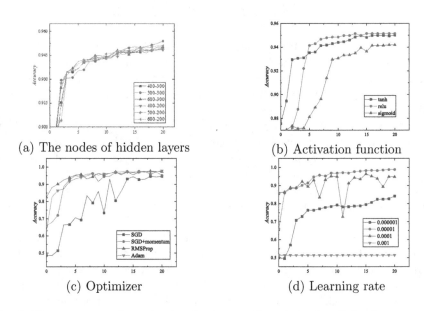

(a) The nodes of hidden layers   (b) Activation function

(c) Optimizer   (d) Learning rate

Fig. 4. The parameter determination in the model optimization ('x' axis represents 'iterations')

Our DNN is a fully connected network with a four-layer structure, containing the input layer, two hidden layers, and the output layer. Optimize DNN through four steps in Fig. 4.

1) We use two hidden layers and change the number of nodes in each hidden layer. Considering the final accuracy, we determine 500 and 300 (Fig. 4(a)). 2) We try to use *sigmoid*, *tanh* and *ReLU* as the activation function of hidden layer. It's obvious that *ReLU* has a great acceleration effect on the convergence compared with the other two (Fig. 4(b)). 3) We carried out experiments with *SGD*, *Momentum*, *RMSprop* and *Adam* respectively. As the gradient becomes sparser, *Adam* converges faster than the other three and the accuracy remains at a high level. So we choose the *Adam* optimizer to update the model parameters

(Fig. 4(c)). 4) We control the learning rate at 0.00001. Because the training process is more stable and the final accuracy rate is higher (Fig. 4(d)).

Through above steps, we obtain an optimal model. Finally, the model is trained in batches, the minimum batch is 20 flows every time, with a total of 20 iterations. The testing results are shown in Fig. 5, the accuracy increases from 75.45% to 98.55%, the recall reaches 98.41%, the precision reaches 98.76%, and the comprehensive evaluation index F_1 reaches 98.59%. Obviously, our model can accurately and quickly detect CC attack.

4.3 Robustness Analysis

In order to evaluate the robustness of the model, we adjust the ratio of attack traffic to non-CC attack traffic in training set to 1:1, 1:2, 1:3, 1:4, 1:8 and 1:11. For this, the number of CC-attack packets is controlled at 360,000 and the number of non-CC attack packets is changed several times. Besides, the test set contains 180,000 CC-attack packets and 720,000 non-CC attack packets.

The test results in different experimental environments are shown in Fig. 6. Through the experimental results, it can be found that the test results of the model under different environments are good. The accuracy is around 98%, indicating that the model has a good classification effect; the recall approaches 99%, indicating that the FN value is small, the model underreporting rate is low; the accuracy is about 95%, which means that the FC value is small and the model false positive rate is low; the F_1 value is about 97%, and the model comprehensive evaluation index is high, which can meet the needs of different practical application environments. In summary, the model has strong capability to detect attack traffic and can cope with more complex actual network environments.

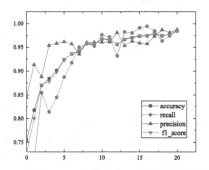
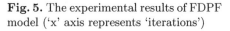

Fig. 5. The experimental results of FDPF model ('x' axis represents 'iterations')

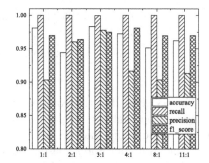

Fig. 6. The experimental results in different training set ratios

4.4 Comparison with Related Methods

In order to objectively evaluate the detection efficiency of our model, we choose classic and widely used related detection methods for comparative experiments.

Many good research, such as [7,17,23], used SVM classifier or random forest-based model for DDoS attack detection. Their experimental results show that these model has satisfying effect for their research problem. Therefore, we adopt SVM and random forest as comparison, and carry out the training and test under the same dataset used in this paper.

From the perspective of algorithm time complexity, the time complexity of the model used in this paper is $O(e * H * (K + d) * N_{sample})$ [6], where e is the number of network model training cycles; H is the number of hidden layers; K is the number of output layer nodes; d is the number of input layer nodes. The time complexity of Random Forest is $O(N_{sample} * N_{feature} * logN_{sample})$ [12], where N_{sample} is the number of samples, $N_{feature}$ is the dimension of features. The time complexity of SVM is $O(N_{Hessian}^2) + O(N_{sample}) + O(N_{sample} * N_{feature} * N_{Hessian})$ [15], where $N_{Hessian}$ is the number of matrix rows after feature matrixing, equals to N_{sample}, $O(N_{sample} * N_{feature} * N_{Hessian})$ is the time complexity of an iterative process, $O(N_{sample})$ is the time complexity of storing intermediate results, $O(N_{Hessian}^2)$ is the time complexity of storing feature matrix. Through this comparison, it can be found that when the dataset is large, the time complexity of the Random Forest and SVM algorithms will increase exponentially, and the training cost of the model will increase greatly; in the case of the same amount of data, SVM needs to spend more time to store data matrices and intermediate results.

Table 2. Evaluation index and speed comparison

	SVM-RBF	Random forest	FDPD model
Accuracy	95.00%	94.89%	**98.55%**
Recall	88.89%	93.00%	**98.42%**
Presicion	96.43%	95.00%	**98.76%**
F_1	94.12%	94.00%	**98.59%**
Speed	7.96 Mb/s	5.78 Mb/s	**137.37 Mb/s**

The SVM model, the Random Forest model, and our proposed model are trained on our dataset. Experimental comparison results are shown in Table 2, and it can be found that the accuracy of the proposed model is 3% higher than SVM and Random Forest-based malicious traffic detection model. Apart from this, the test speed of our model is nearly 17 times faster than the related machine learning model.

5 Discussion

Due to the changing attack patterns and limited traffic collection methods, the traffic contained in the dataset for experiments can't cover all the CC attack variants, so the trained model may have a unsatisfactory effect when detecting unknown CC attack traffic. However, if the training data is sufficient enough, the model proposed in this paper can detect more attack traffic efficiently, because it has good data-preprocessing ability and self-learning ability. The FDPD model can adequately learn the features implied in the traffic and use them for classification. Furthermore, the reason that we don't use other neural networks, such as CNN, RNN, is to improve the efficiency of training and detection while ensuring the effect. Since the dimension of the input after preprocessing is not very high and our model shows good result, it is not necessary to spend extra designing and training costs to build overly complex models.

6   Conclusion and Future Work

In order to cope with CC attack, we proposed a packet Field Differentiated Preprocessing and Deep neural network (FDPD) model to detect CC attack traffic. And we collected a fresh dataset that contains 7.92 million packets to validate our model. In the design of FDPD model, efficient strategies are adopted to prevent overfitting problem of deep learning. In the preprocessing phase, we combine the specific meaning of each field of the packet to drop misleading field such as 'IP' and useless fields such as 'reserved'. In the model training phase, a dropout strategy is adopted for each hidden layer. The experimental results show that the accuracy of FDPD model reaches 98%. In the same training environment, our model was compared with the SVM-based and Random Forest-based traffic detection models. The accuracy is increased by 3%, and the speed is increased by 17 times.

In the future, we will improve in the following aspects: 1) Train a more effective attack traffic detection model; 2) Enrich our dataset.

References

1. Abdulaziz, A., Shahrulniza, M.: Cloud-based DDoS http attack detection using covariance matrix approach. J. Comput. Netw. Commun. **2017**(38), 1–8 (2017)
2. Adi, E., Baig, Z., Hingston, P.: Stealthy denial of service (DoS) attack modelling and detection for http/2 services. J. Netw. Comput. Appl. **91**, S1084804517301637 (2017)
3. Xiao, B., Chen, W., He, Y., Sha, E.M.: An active detecting method against SYN flooding attack. In: International Conference on Parallel & Distributed Systems (2005)
4. Cheng, R., Xu, R., Tang, X., Sheng, V.S., Cai, C.: An abnormal network flow feature sequence prediction approach for DDoS attacks detection in big data environment. Comput. Mater. Continua **55**(1), 095–095 (2018)
5. Douglas, D., Santanna, J.J., Schmidt, R.D.O., Granville, L.Z., Pras, A.: Booters: can anything justify distributed denial-of-service (DDoS) attacks for hire? J. Inf. Commun. Ethics Soc. **15**(1), 90–104 (2017)
6. Alpaydin, E.: Introduction to Machine Learning. MIT press, Cambridge (2009)
7. Idhammad, M., Afdel, K., Belouch, M.: Detection system of HTTP DDoS attacks in a cloud environment based on information theoretic entropy and random forest. In: Security and Communication Networks (2018)

8. Kumar, D., Rao, C.G.: Leveraging big data analytics for real-time DDoS attacks detection in SDN. Int. J. Res. Eng. Appl. Manag. IJREAM **04**(02), 677–684 (2018)

9. Li, J., Yun, X., Li, S., Zhang, Y., Xie, J., Fang, F.: A HTTP malicious traffic detection method based on hybrid structure deep neural network. J. Commun. **40**(01), 28–37 (2019)

10. Lin, Y.H., Kuo, J.J., Yang, D.N., Chen, W.T.: A cost-effective shuffling-based defense against HTTP DDoS attacks with sdn/nfv. In: IEEE International Conference on Communications, pp. 1–7 (2017)

11. Misra, S., Krishna, P.V., Abraham, K.I., Sasikumar, N., Fredun, S.: An adaptive learning routing protocol for the prevention of distributed denial of service attacks in wireless mesh networks. Comput. Math. Appl. **60**(2), 294–306 (2010)

12. Idhammad, M., Afdel, K., Belouch, M.: Detection system of HTTP DDoS attacks in a cloud environment based on information theoretic entropy and random forest. Secur. Commun. Netw. **2018**(1263123), 1–13 (2018)

13. Moore, D., Voelker, G.M., Savage, S.: Inferring internet denial-of-service attack. In: Conference on Usenix Security Symposium (2001)

14. Sahoo, K.S., Tiwary, M., Sahoo, S., Nambiar, R., Sahoo, B., Dash, R.: A learning automata-based DDoS attack defense mechanism in software defined networks. In: Proceedings of the 24th Annual International Conference on Mobile Computing and Networking, pp. 795–797 (2018)

15. Sahu, S.K., Jena, S.K.: A multiclass SVM classification approach for intrusion detection. In: International Conference on Distributed Computing & Internet Technology (2016)

16. Singh, K., Singh, P., Kumar, K.: Application layer HTTP-get flood DDoS attacks: research landscape and challenges. Comput. Secur. **65**, 344–372 (2017)

17. Singh, K., Singh, P., Kumar, K.: User behavior analytics-based classification of application layer HTTP-GET flood attacks. J. Netw. Comput. Appl. **112**, 97–114 (2018)

18. Sree, T.R., Bhanu, S.M.S.: Hadm: detection of HTTP get flooding attacks by using analytical hierarchical process and dempster-shafer theory with mapreduce. Secur. Commun. Netw. **9**(17), 4341–4357 (2016)

19. Su, L., Yao, Y., Li, N., Liu, J., Lu, Z., Liu, B.: Hierarchical clustering based network traffic data reduction for improving suspicious flow detection. In: IEEE International Conference on Trust, Security and Privacy in Computing and Communications, pp. 744–753 (2018)

20. Tan, M.: Research and implementation of DDoS attack detection based on machine learning in distributed environment. Ph.D. thesis (2018)

21. Wang, A., Mohaisen, A., Chang, W., Chen, S.: Delving into internet DDoS attacks by botnets: Characterization and analysis. In: IEEE/IFIP International Conference on Dependable Systems & Networks, vol. 26, pp. 2843–2855 (2015)

22. Xie, Y., Yu, S.Z.: A large-scale hidden semi-markov model for anomaly detection on user browsing behaviors. IEEE/ACM Trans. Netw. **17**(1), 54–65 (2009)

23. Ye, J., Cheng, X., Zhu, J., Feng, L., Song, L.: A DDoS attack detection method based on SVM in software defined network. In: Security and Communication Networks (2018)

24. Liu, Z., Zhang, B.: Self-learning application layer DDoS detection method based on improved AP clustering algorithm (2018)

Deep Low-Density Separation for Semi-supervised Classification

Michael C. Burkhart[1]([✉]) [iD] and Kyle Shan[2] [iD]

[1] Adobe Inc., San José, USA
mburkhar@adobe.com
[2] Stanford University, Stanford, USA
kylecshan@gmail.com

Abstract. Given a small set of labeled data and a large set of unlabeled data, semi-supervised learning (SSL) attempts to leverage the location of the unlabeled datapoints in order to create a better classifier
than could be obtained from supervised methods applied to the labeled
training set alone. Effective SSL imposes structural assumptions on the
data, e.g. that neighbors are more likely to share a classification or that
the decision boundary lies in an area of low density. For complex and
high-dimensional data, neural networks can learn feature embeddings to
which traditional SSL methods can then be applied in what we call hybrid
methods.

Previously-developed hybrid methods iterate between refining a latent
representation and performing graph-based SSL on this representation. In
this paper, we introduce a novel hybrid method that instead applies low-
density separation to the embedded features. We describe it in detail
and discuss why low-density separation may better suited for SSL on
neural network-based embeddings than graph-based algorithms. We validate our method using in-house customer survey data and compare it
to other state-of-the-art learning methods. Our approach effectively classifies thousands of unlabeled users from a relatively small number of
hand-classified examples.

Keywords: Semi-supervised learning · Low-density separation · Deep
learning · User classification from survey data

1 Background

In this section, we describe the problem of semi-supervised learning (SSL) from
a mathematical perspective. We then outline some of the current approaches to
solve this problem, emphasizing those relevant to our current work.

Work completed while K.S. interned at Adobe.

© Springer Nature Switzerland AG 2020
V. V. Krzhizhanovskaya et al. (Eds.): ICCS 2020, LNCS 12139, pp. 297–311, 2020.
https://doi.org/10.1007/978-3-030-50420-5_22

1.1   Problem Description

Consider a small labeled training set $\mathcal{D}_0 = \{(x_1, y_1), (x_2, y_2), \dots, (x_\ell, y_\ell)\}$ of vector-valued features $x_i \in \mathbb{R}^d$ and discrete-valued labels $y_i \in \{1, \dots, c\}$, for $1 \leq i \leq \ell$. Suppose we have a large set $\mathcal{D}_1 = \{x_{\ell+1}, x_{\ell+2}, \dots, x_{\ell+u}\}$ of unlabeled features to which we would like to assign labels. One could perform supervised learning on the labeled dataset $\mathcal{D}_0$ to obtain a general classifier and then apply this classifier to $\mathcal{D}_1$. However, this approach ignores any information about the distribution of the feature-points contained in $\mathcal{D}_1$. In contrast, SSL attempts to leverage this additional information in order to either inductively train a generalized classifier on the feature space or transductively assign labels only to the feature-points in $\mathcal{D}_1$.

Effective SSL methods impose additional assumptions about the structure of the feature-data (i.e., $\{x : (x, y) \in \mathcal{D}_0\} \cup \mathcal{D}_1$); for example, that features sharing the same label are clustered, that the decision boundary separating differently labeled features is smooth, or that the features lie on a lower dimensional manifold within $\mathbb{R}^d$. In practice, semi-supervised methods that leverage data from $\mathcal{D}_1$ can achieve much better performance than supervised methods that use $\mathcal{D}_0$ alone. See Fig. 1 for a visualization. We describe both graph-based and low-density separation methods along with neural network-based approaches, as these are most closely related to our work. For a full survey, see [7,37,50].

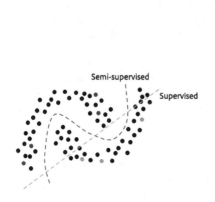

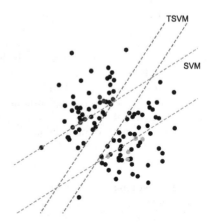

Fig. 1. A schematic for semi-supervised classification. The grey line corresponds to a decision boundary obtained from a generic supervised classifier (incorporating information only from the labeled blue and orange points); the red line corresponds to a boundary from a generic semi-supervised method seeking a low-density decision boundary. (Color figure online)

Fig. 2. A schematic for TSVM segmentation. The grey lines correspond to maximum margin separation for labeled data using a standard SVM; the red lines correspond to additionally penalizing unlabeled points that lie in the margin. In this example, the data is perfectly separable in two dimensions, but this need not always be true. (Color figure online)

1.2   Graph-Based Methods

Graph-based methods calculate the pairwise similarities between labeled and unlabeled feature-points and allow labeled feature-points to pass labels to their unlabeled neighbors. For example, label propagation [51] forms a $(\ell + u) \times (\ell + u)$ dimensional transition matrix T with transition probabilities proportional to similarities (kernelized distances) between feature-points and an $(\ell + u) \times c$ dimensional matrix of class probabilities, and (after potentially smoothing this matrix) iteratively sets $Y \leftarrow TY$, row-normalizes the probability vectors, and resets the rows of probability vectors corresponding to the already-labeled elements of $\mathcal{D}_0$. Label spreading [48] follows a similar approach but normalizes its weight matrix and allows for a (typically hand-tuned) clamping parameter that assigns a level of uncertainty to the labels in $\mathcal{D}_0$. There are many variations to the graph-based approach, including those that use graph min-cuts [4] and Markov random walks [40].

1.3   Low-Density Separation

Low-density separation methods attempt to find a decision boundary that best separates one class of labeled data from the other. The quintessential example is the transductive support vector machine (TSVM: [1,6,8,15,21,30]), a semi-supervised maximum-margin classifier of which there have been numerous variations. As compared to the standard SVM (cf., e.g., [2,32]), the TSVM additionally penalizes unlabeled points that lie close to the decision boundary. In particular, for a binary classification problem with labels $y_i \in \{-1, 1\}$, it seeks parameters w, b that minimize the non-convex objective function

$$J(w, b) = \frac{1}{2}\|w\|^2 + C\sum_{i=1}^{l} H(y_i \cdot f_{w,b}(x_i)) + C^* \sum_{i=l+1}^{u} H(|f_{w,b}(x_i)|), \quad (1)$$

where $f_{w,b} : \mathbb{R}^d \to \mathbb{R}$ is the linear decision function $f_{w,b}(x) = w \cdot x + b$, and $H(x) = \max(0, 1 - x)$ is the hinge loss function. The hyperparameters C and C^* control the relative influence of the labeled and unlabeled data, respectively. Note that the third term, corresponding to a loss function for the unlabeled data, is non-convex, providing a challenge to optimization. See Fig. 2 for a visualization of how the TSVM is intended to work and Ding et al. [13] for a survey of semi-supervised SVM's. Other methods for low-density separation include the more general entropy minimization approach [17], along with information regularization [39] and a Gaussian process-based approach [27].

1.4   Neural Network-Based Embeddings

Both the graph-based and low-density separation approaches to SSL rely on the geometry of the feature-space providing a reasonable approximation to the true underlying characteristics of the users or objects of interest. As datasets become increasingly complex and high-dimensional, Euclidean distance between feature vectors may not prove to be the best proxy for user or item similarity. As the Gaussian kernel is a monotonic function of Euclidean distance, kernelized methods such as label propagation and label spreading also suffer from this criticism. While kernel learning approaches pose one potential solution [9,49], neural network-based embeddings have become increasingly popular in recent years. Variational autoencoders (VAE's: [24]) and generative adversarial nets (GAN's: [12,29]) have both been successfully used for SSL. However, optimizing the parameters for these types of networks can require expert hand-tuning and/or prohibitive computational expense [35,53]. Additionally, most research in the area concentrates on computer vision problems, and it is not clear how readily the architectures and techniques developed for image classification translate to other domains of interest.

1.5   Hybrid Methods

Recently, Iscen et al. introduced a neural embedding-based method to generate features on which to perform label propagation [19]. They train a neural network-based classifier on the supervised dataset and then embed all featurepoints into an intermediate representation space. They then iterate between performing label propagation in this feature space and continuing to train their neural network classifier using weighted predictions from label propagation (see also [52]). As these procedures are similar in spirit to ours, we next outline our method in the next section and provide more details as part of a comparison in Subsect. 2.4.

2   Deep Low-Density Separation Algorithm

In this section, we provide a general overview of our algorithm for deep lowdensity separation and then delve into some of the details. We characterize our general process as follows:

1. We first learn a neural network embedding $f : \mathbb{R}^d \to \mathbb{R}^m$ for our featuredata optimized to differentiate between class labels. We define a network $g : \mathbb{R}^m \to \mathbb{P}^c$ (initialized as the initial layers from an autoencoder for the feature-data), where $\mathbb{P}^c$ is the space of c-dimensional probability vectors, and optimize $g \circ f$ on our labeled dataset $\mathcal{D}_0$, where we one-hot encode the categories corresponding to each y_i.
2. We map all of the feature-points through this deep embedding and then implement one-vs.-rest TSVM's for each class on this embedded data to learn classpropensities for each unlabeled data point. We augment our training data with the x_i from $\mathcal{D}_1$ paired with the propensities returned by this method and continue to train $g \circ f$ on this new dataset for a few epochs.
3. Our neural network f now provides an even better embedding for differentiating between classes. We repeat step 2 for a few iterations in order for the better embedding to improve TSVM separation, which upon further training yields an even better embedding, and so forth, etc.

Data: labeled dataset $\mathcal{D}_0$ and unlabeled dataset $\mathcal{D}_1$
Result: probabilistic predictions for the labels in $\mathcal{D}_1$
Initialize a deep neural network $f_\theta : \mathbb{R}^d \to \mathbb{R}^m$ with trainable parameters θ;
Initialize a neural network $g_\psi : \mathbb{R}^m \to \mathbb{P}^c$ with trainable parameters ψ;
Obtain θ_0, ψ_0 by minimizing cross entropy between $h(y_i)$ and $g_\psi(f_\theta(x_i))$ for $(x_i, y_i) \in \mathcal{D}_0$, where h is the encoding defined in (2);
for $t = 1, \ldots, T$ **do**

    Compute $\tilde{D}_0 = \{(f_{\theta_{t-1}}(x), y) : (x, y) \in \mathcal{D}_0\}$ and $\tilde{D}_1 = \{f_{\theta_{t-1}}(x) : x \in \mathcal{D}_1\}$;

    Perform one-vs.-rest TSVM training on $\tilde{D}_0$ and $\tilde{D}_1$ to obtain predicted probabilities $\hat{p}_i, i = \ell + 1, \ldots, \ell + u$ that the x_i in $\mathcal{D}_1$ lie in each class and then set $\check{D}_1 = \{(x_i, \hat{p}_i)\}$;

    Obtain θ_t, ψ_t by continuing to optimize $g_\psi \circ f_\theta$, using $\mathcal{D}_0 \cup \check{D}_1$;

end
return $g_{\psi_T}(f_{\theta_T}(x_i))$ or an exponential moving average of the probabilistic predictions $g_{\psi_t}(f_{\theta_t}(x_i))$ for $x_i \in \mathcal{D}_0$.

Algorithm 1: The Deep Segmentation Algorithm

This is our basic methodology, summarized as pseudo-code in Algorithm 1 and visually in Fig. 3. Upon completion, it returns a neural network $g \circ f$ that maps feature-values to class/label propensities that can easily be applied to $\mathcal{D}_1$ and solve our problem of interest. In practice, we find that taking an exponentially decaying moving average of the returned probabilities as the algorithm progresses provides a slightly improved estimate. At each iteration of the algorithm, we reinitialize the labels for the unlabeled points and allow the semi-supervised TSVM to make inferences using the new embedding of the feature-data alone. In this way, it is possible to recover from mistakes in labeling that occurred in previous iterations of the algorithm.

2.1 Details: Neural Network Training

In our instantiation, the neural network $f : \mathbb{R}^d \to \mathbb{R}^m$ has two layers, the first of size 128 and the second of size 32, both with hyperbolic tangent activation. In between these two layers, we apply batch normalization [18] followed by dropout at a rate of 0.5 during model training to prevent overfitting [38]. The neural network $g : \mathbb{R}^m \to \mathbb{P}^c$ consists of a single layer with 5 units and softmax activation. We let θ (resp. ψ) denote the trainable parameters for f (resp. g) and sometimes use the notation f_θ and g_ψ to stress the dependence of the neural networks on these trainable parameters. Neural network parameters receive Glorot normal initialization [16]. The network weights for f and g receive Tikhonov-regularization [43,44], which decreases as one progresses through the network.

We form our underlying target distribution by one-hot encoding the labels y_i and slightly smoothing these labels. We define $h : \{1, \ldots, c\} \to \mathbb{P}^c$ by its components $1 \le j \le c$ as

$$h(y)_j = \begin{cases} 1 - c \cdot \epsilon, & \text{if } y = j, \\ \epsilon, & \text{otherwise} \end{cases} \tag{2}$$

where we set $\epsilon = 10^{-3}$ to be our smoothing parameter.

Training proceeds as follows. We begin by training the neural network f_θ to minimize $D_{\text{KL}}\big(h(y_i)\|g_\psi(f_\theta(x_i))\big)$ the Kullback–Leibler (KL) divergence between the true distributions $h(y_i)$ and our inferred distributions $g_\psi(f_\theta(x_i))$, on $\mathcal{D}_0$ in batches. For parameter updates, we use the Adam optimizer [23] that maintains different learning rates for each parameter like AdaGrad [14] and allows these rates to sometimes increase like Adadelta [47] but adapts them based on the first two moments from recent gradient updates. This optimization on labeled data produces parameters θ_0 for f and ψ_0 for g.

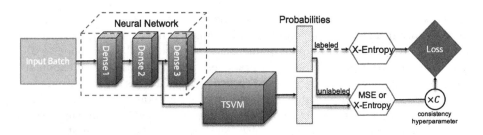

Fig. 3. A schematic for Deep Low-Density Separation. The first two layers of the neural network correspond to f, the last to g. The semi-supervised model corresponds to the TSVM segmentation. We optimize on the unlabeled dataset using mean square error (MSE) and on the labeled dataset using cross-entropy (X-Entropy).

2.2   Details: Low-Density Separation

Upon initializing f and g, f_{θ_0} is a mapping that produces features well-suited to differentiating between classes. We form $\tilde{\mathcal{D}}_0 = \{(f_{\theta_0}(x), y) : (x, y) \in \mathcal{D}_0\}$ and $\tilde{\mathcal{D}}_1 = \{f_{\theta_0}(x) : x \in \mathcal{D}_1\}$ by passing the feature-data through this mapping. We then train c TSVM's, one for each class, on the labeled data $\tilde{\mathcal{D}}_0$ and unlabeled data $\tilde{\mathcal{D}}_1$.

Our implementation follows Collobert et al.'s TSVM-CCCP method [11] and is based on the R implementation in RSSL [25]. The algorithm decomposes the TSVM loss function $J(w, b)$ from (1) into the sum of a concave function and a convex function by creating two copies of the unlabeled data, one with positive labels and one with negative labels. Using the concave-convex procedure (CCCP: [45, 46]), it then reduces the original optimization problem to an iterative procedure where each step requires solving a convex optimization problem similar to that of the supervised SVM. These convex problems are then solved using quadratic programming on the dual formulations (for details, see [5]). Collobert et al. argue that TSVM-CCCP outperforms previous TSVM algorithms with respect to both speed and accuracy [11].

2.3 Details: Iterative Refinement

Upon training the TSVM's, we obtain a probability vector $\hat{p}_i \in \mathbb{P}^c$ for each $i = \ell+1, \ldots, \ell+u$ with elements corresponding to the likelihood that x_i lies in a given class. We then form $\check{\mathcal{D}}_1 = \{(x_i, \hat{p}_i)\}$ and obtain a supervised training set for further refining $g \circ f$. We set the learning rate for our Adam optimizer to 1/10th of its initial rate and minimize the mean square error between $g(f(x_i))$ and $\hat{p}_i$ for $(x_i, \hat{p}_i) \in \check{\mathcal{D}}_1$ for 10 epochs (cf. "consistency loss" from [42]) and then minimize the KL-divergence between $h(y_i)$ and $g(f(x_i))$ for 10 epochs. This training starts with neural network parameters θ_0 and ψ_0 and produces parameters θ_1 and ψ_1. Then, f_{θ_1} is a mapping that produces features better suited to segmenting classes than those from f_{θ_0}. We pass our feature-data through this mapping and continue the iterative process for $T = 6$ iterations. Our settings for learning rate, number of epochs, and T were hand-chosen for our data and would likely vary for different applications.

As the algorithm progresses, we store the predictions $g_{\psi_t}(f_{\theta_t}(x_i))$ at each step t and form an exponential moving average (discount rate $\rho = 0.8$) over them to produce our final estimate for the probabilities of interest.

2.4 Remarks on Methodology

We view our algorithm as most closely related to the work of Iscen et al. [19] and Zhuang et al. [52]. Both their work and ours iterate between refining a neural network-based latent representation and applying a classical SSL method to that representation to produce labels for further network training. While their work concentrates on graph-based label propagation, ours uses low-density separation, an approach that we believe may be more suitable for the task. The representational embedding we learn is optimized to discriminate between class labels, and for this reason we argue it makes more sense to refine decision boundaries than it does to pass labels. Additionally, previous work on neural network-based classification suggests that an SVM loss function can improve classification accuracy [41], and our data augmentation step effectively imposes such a loss function for further network training.

By re-learning decision boundaries at each iterative step, we allow our algorithm to recover from mistakes it makes in early iterations. One failure mode of semi-supervised methods entails making a few false label assignments early in the iterative process and then having these mislabeled points pass these incorrect labels to their neighbors. For example, in pseudo-labelling [28], the algorithm augments the underlying training set $\mathcal{D}_0$ with pairs $(x_i, \hat{y}_i)$ for $x_i \in \mathcal{D}_1$ and predicted labels $\hat{y}_i$ for which the model was most confident in the previous iteration.

Similar error-reinforcement problems can occur with boosting [31]. It is easy to see how a few confident, but inaccurate, labels that occur in the first few steps of the algorithm can set the labeling process completely askew.

By creating an embedding $f : \mathbb{R}^d \to \mathbb{R}^m$ and applying linear separation to embedded points, we have effectively learned a distance metric $\kappa : \mathbb{R}^d \times \mathbb{R}^d \to \mathbb{R}_{\geq 0}$ especially suited to our learning problem. The linear decision boundaries we produce in $\mathbb{R}^m$ correspond to nonlinear boundaries for our original features in $\mathbb{R}^d$. Previously, Jean et al. [20] described using a deep neural network to embed features for Gaussian process regression, though they use a probabilistic framework for SSL and consider a completely different objective function.

3   Application to User Classification from Survey Data

In this section, we discuss the practical problem of segmenting users from survey data and compare the performance of our algorithm to other recently-developed methods for SSL on real data. We also perform an ablation study to ensure each component of our process contributes to the overall effectiveness of the algorithm.

3.1   Description of the Dataset

At Adobe, we are interested in segmenting users based on their work habits, artistic motivations, and relationship with creative software. To gather data, we administered detailed surveys to a select group of users in the US, UK, Germany, & Japan (just over 22 thousand of our millions of users). We applied Latent Dirichlet Allocation (LDA: [3,33]), an unsupervised model to discover latent topics, to one-hot encoded features generated from this survey data to classify each surveyed user as belonging to one of $c = 5$ different segments. We generated profile and usage features using an in-house feature generation pipeline (that could in the future readily be used to generate features for the whole population of users). In order to be able to evaluate model performance, we masked the LDA labels from our surveyed users at random to form the labelled and unlabelled training sets $\mathcal{D}_0$ and $\mathcal{D}_1$.

3.2   State-of-the-Art Alternatives

We compare our algorithm against two popular classification algorithms. We focus our efforts on other algorithms we might have actually used in practice instead of more similar methods that just recently appeared in the literature.

The first, LightGBM [22] is a supervised method that attempts to improve upon other boosted random forest algorithms (e.g. the popular xgBoost [10]) using novel approaches to sampling and feature bundling. It is our team's preferred nonlinear classifier, due to its low requirements for hyperparameter tuning and effectiveness on a wide variety of data types. As part of the experiment, we wanted to evaluate the conditions for semi-supervised learning to outperform supervised learning.

The second, Mean Teacher [42] is a semi-supervised method that creates two supervised neural networks, a teacher network and a student network, and trains both networks using randomly perturbed data. Training enforces a consistency loss between the outputs (predicted probabilities in $\mathbb{P}^c$) of the two networks: optimization updates parameters for the student network and an exponential moving averages of these parameters become the parameters for the teacher network. The method builds upon Temporal Ensembling [26] and uses consistency loss [34, 36].

3.3 Experimental Setup

We test our method with labelled training sets of successively increasing size $\ell \in \{35, 50, 125, 250, 500, 1250, 2500\}$. Each training set is a strict superset of the smaller training sets, so with each larger set, we strictly increase the amount of information available to the classifiers. To tune hyperparameters, we use a validation set of size 100, and for testing we use a test set of size 4780. The training, validation, and test sets are selected to all have equal class sizes.

For our algorithm, we perform $T = 6$ iterations of refinement, and in the TSVM we set the cost parameters $C = 0.1$ and $C^* = \frac{\ell}{u}C$. To reduce training time, we subsample the unlabeled data in the test set by choosing 250 unlabeled points uniformly at random to include in the TSVM training. We test using our own implementations of TSVM and MeanTeacher.

3.4 Numerical Results and Ablation

Table 1 reports our classification accuracy on five randomized shuffles of the training, validation, and test sets. These results are summarized in Fig. 4. The accuracy of our baseline methods are shown first, followed by three components of our model:

1. Initial NN: The output of the neural network after initial supervised training.
2. DeepSep-NN: The output of the neural network after iterative refinement with Algorithm 1.
3. DeepSep-Ensemble: Exponential moving average as described in Algorithm 1.

We find that Deep Low-Density Separation outperforms or matches Light-GBM in the range $\ell \leq 1250$. The relative increase in accuracy of Deep Separation is as much as 27%, which is most pronounced with a very small amount of training data ($\ell \leq 50$). Some of this increase can be attributed to the initial accuracy of the neural network; however, the iterative refinement of Deep Separation improves the accuracy of the initial network by up to 8.3% (relative). The addition of a temporal ensemble decreases variance in the final model, further increasing accuracy by an average of 0.54% across the range. Compared to Mean Teacher, the iterative refinement of Deep Separation achieves a larger increase in accuracy for $l \leq 500$.

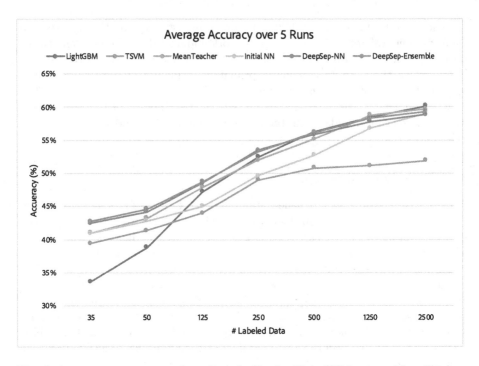

Fig. 4. Average accuracy over 5 random shuffles for LightGBM, TSVM, MeanTeacher and our proposed method. Random chance accuracy is 20%. We are primarily interested in the regime where few training examples exist – particularly when the number of labeled datapoints is 35–50.

To visualize how the iterative refinement process and exponential weighted average improve the model, Fig. 5 shows the accuracy of our model at each iteration. We see that for each random shuffle, the refinement process leads to increased accuracy compared to the initial model. However, the accuracy of the neural network fluctuates by a few percent at a time. Applying the exponential moving average greatly reduces the impact of these fluctuations and yields more consistent improvement, with a net increase in accuracy on average.

Regarding training time, all numerical experiments were performed on a mid-2018 MacBook Pro (2.6 GHz Intel Core i7 Processor; 16 GB 2400 MHz DDR4 Memory). Deep Separation takes up to half an hour on the largest training set ($\ell = 2500$). However, we note that for $\ell \leq 500$, the model takes at most three minutes, and this is the regime where our method performs best in comparison to other methods. In contrast, LightGBM takes under a minute to run with all training set sizes.

Table 1. Classification accuracy (in percent) for each of the methods tested. Shuffle # refers to the randomized splitting of the data into training, validation, and test sets. The final block contains the average accuracy over 5 random shuffles.

Shuffle #	Model	ℓ						
		35	50	125	250	500	1250	2500
1	LightGBM	30.98	34.73	47.45	51.55	55.99	59.39	60.65
	TSVM	38.65	38.26	40.26	46.84	48.54	51.02	52.94
	MeanTeacher	39.91	41.70	47.54	51.33	54.81	59.83	60.48
	Initial NN	38.65	40.09	41.92	47.89	51.15	58.13	61.09
	DeepSep-NN	39.04	41.79	44.97	53.55	54.51	57.60	60.13
	DeepSep-Ensemble	40.13	42.00	46.32	52.68	54.95	58.04	59.87
2	LightGBM	32.03	38.30	46.93	52.33	55.86	58.39	59.61
	TSVM	43.31	43.88	47.45	49.19	50.76	49.76	50.37
	MeanTeacher	43.14	42.75	48.58	53.03	54.12	58.08	59.35
	Initial NN	43.31	43.79	45.97	50.72	53.03	57.04	59.26
	DeepSep-NN	47.32	47.10	48.85	51.90	54.25	56.69	57.95
	DeepSep-Ensemble	46.45	46.58	49.06	51.94	54.47	57.60	58.56
3	LightGBM	32.33	40.31	47.63	50.94	56.34	57.82	60.13
	TSVM	30.37	34.55	37.30	49.93	51.59	52.42	51.42
	MeanTeacher	35.77	40.26	45.05	50.33	55.12	56.43	57.25
	Initial NN	37.12	40.87	43.05	48.15	52.72	55.82	57.86
	DeepSep-NN	36.69	40.48	46.88	52.33	55.60	56.95	57.82
	DeepSep-Ensemble	37.17	40.52	46.49	52.33	56.12	57.04	57.86
4	LightGBM	35.12	36.17	47.36	52.42	56.30	59.00	61.05
	TSVM	40.61	45.10	48.28	52.85	52.64	50.11	51.29
	MeanTeacher	41.96	44.31	49.54	51.76	55.56	59.56	60.96
	Initial NN	41.26	43.66	48.10	48.63	52.55	55.64	58.26
	DeepSep-NN	44.84	44.58	50.41	54.34	56.86	58.61	59.08
	DeepSep-Ensemble	44.49	44.88	50.33	53.46	56.86	59.39	60.44
5	LightGBM	37.60	44.44	46.67	55.16	56.60	57.95	59.30
	TSVM	44.14	45.14	46.71	46.06	50.41	52.24	53.51
	MeanTeacher	44.14	46.93	48.63	53.25	56.08	60.17	60.22
	Initial NN	44.44	45.62	45.88	52.85	54.29	57.39	58.69
	DeepSep-NN	44.44	46.93	51.46	55.38	58.00	59.39	59.17
	DeepSep-Ensemble	45.53	48.85	51.37	55.90	58.43	59.48	59.96
Average	LightGBM	33.61	38.79	47.21	52.48	56.22	58.51	60.15
	TSVM	39.42	41.39	44.00	48.98	50.79	51.11	51.90
	MeanTeacher	40.98	43.19	47.87	51.94	55.14	58.81	59.65
	Initial NN	40.96	42.81	44.98	49.65	52.75	56.80	59.03
	DeepSep-NN	42.47	44.17	48.51	53.50	55.84	57.85	58.83
	DeepSep-Ensemble	42.75	44.57	48.71	53.26	56.17	58.31	59.34

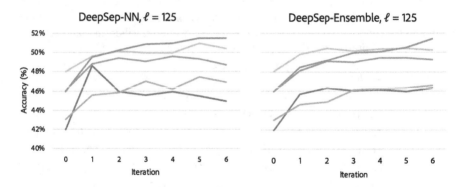

Fig. 5. Classification accuracy on the test set for all five random shuffles over the course of iterative refinement, using 125 labeled data, of (left) the refined neural network and (right) the exponential moving average of predictions. Here, different colors correspond to different choices for training set (different random seeds).

4   Conclusions

In this paper, we introduce a novel hybrid semi-supervised learning method, Deep Low-Density Separation, that iteratively refines a latent feature representation and then applies low-density separation to this representation to augment the training set. We validate our method on a multi-segment classification dataset generated from surveying Adobe's user base. In the future, we hope to further investigate the interplay between learned feature embeddings and low-density separation methods, and experiment with different approaches for both representational learning and low-density separation. While much of the recent work in deep SSL concerns computer vision problems and image classification in particular, we believe these methods will find wider applicability within academia and industry, and anticipate future advances in the subject.

References

1. Bennett, K.P., Demiri, A.: Semi-supervised support vector machines. In: Advances in Neural Information Processing System (1998)
2. Bishop, C.M.: Pattern Recognition and Machine Learning. Springer, New York (2006)
3. Blei, D.M., Ng, A.Y., Jordan, M.I.: Latent Dirichlet allocation. J. Mach. Learn. Res. **3**, 993–1022 (2003)
4. Blum, A., Chawla, S.: Learning from labeled and unlabeled data using graph min-cuts. In: International Conference on Machine Learning, pp. 19–26 (2001)
5. Boyd, S., Vandenberghe, L.: Convex Optimization. Cambridge University Press, Cambridge (2004)
6. Chapelle, O., Zien, A.: Semi-supervised classification by low density separation. In: Conference on Artificial Intelligence Statistics (2005)
7. Chapelle, O., Schölkopf, B., Zien, A. (eds.): Semi-Supervised Learning. MIT Press, Cambridge (2006)

8. Chapelle, O., Sindhwani, V., Keerthi, S.S.: Optimization techniques for semi-supervised support vector machines. J. Mach. Learn. Res. **9**, 203–233 (2008)
9. Chapelle, O., Weston, J., Schölkopf, B.: Cluster kernels for semi-supervised learning. In: Advances in Neural Information Processing System, pp. 601–608 (2003)
10. Chen, T., Guestrin, C.: XGBoost: a scalable tree boosting system. In: International Conference on Knowledge Discovery Data Mining, pp. 785–794 (2016)
11. Collobert, R., Sinz, F., Weston, J., Bottou, L.: Large scale transductive SVMs. J. Mach. Learn. Res. **7**, 1687–1712 (2006)
12. Dai, Z., Yang, Z., Yang, F., Cohen, W.W., Salakhutdinov, R.: Good semi-supervised learning that requires a bad GAN. In: Advances in Neural Information Processing System, pp. 6513–6523 (2017)
13. Ding, S., Zhu, Z., Zhang, X.: An overview on semi-supervised support vector machine. Neural Comput. Appl. **28**(5), 969–978 (2017)
14. Duchi, J., Hazan, E., Singer, Y.: Adaptive subgradient methods for online learning and stochastic optimization. J. Mach. Learn. Res. **12**, 2121–2159 (2011)
15. Gammerman, A., Vovk, V., Vapnik, V.: Learning by transduction. In: Uncertainity Artificial Intelligence, pp. 148–155 (1998)
16. Glorot, X., Bengio, Y.: Understanding the difficulty of training deep feedforward neural networks. In: Conference on Artificial Intelligence Statistics, vol. 9, pp. 249–256 (2010)
17. Grandvalet, Y., Bengio, Y.: Semi-supervised learning by entropy minimization. In: Advances in Neural Information Processing System, pp. 529–536 (2004)
18. Ioffe, S., Szegedy, C.: Batch normalization: accelerating deep network training by reducing internal covariate shift. In: International Conference on Machine Learning, pp. 448–456 (2015)
19. Iscen, A., Tolias, G., Avrithis, Y., Chum, O.: Label propagation for deep semi-supervised learning. In: Conference on Computer Vision Pattern Recognition (2019)
20. Jean, N., Xie, S.M., Ermon, S.: Semi-supervised deep kernel learning: regression with unlabeled data by minimizing predictive variance. In: Advances in Neural Information Processing System, pp. 5322–5333 (2018)
21. Joachims, T.: Transductive inference for text classification using support vector machines. In: International Conference on Machine Learning, pp. 200–209 (1999)
22. Ke, G., Meng, Q., et al.: LightGBM: a highly efficient gradient boosting decision tree. In: Advance in Neural Information Processing System, pp. 3146–3154 (2017)
23. Kingma, D.P., Ba, J.: Adam: a method for stochastic optimization. In: International Conference on Learning Represent (2015)
24. Kingma, D.P., Mohamed, S., Rezende, D.J., Welling, M.: Semi-supervised learning with deep generative models. In: Advances in Neural Information Processing System, pp. 3581–3589 (2014)
25. Krijthe, J.H.: RSSL: semi-supervised learning in R. In: Kerautret, B., Colom, M., Monasse, P. (eds.) RRPR 2016. LNCS, vol. 10214, pp. 104–115. Springer, Cham (2017). https://doi.org/10.1007/978-3-319-56414-2_8
26. Laine, S., Aila, T.: Temporal ensembling for semi-supervised learning. In: International Conference on Learning Represent (2017)
27. Lawrence, N.D., Jordan, M.I.: Semi-supervised learning via gaussian processes. In: Advances in Neural Information Processing System, pp. 753–760 (2005)
28. Lee, D.H.: Pseudo-label: the simple and efficient semi-supervised learning method for deep neural networks. In: ICML Workshop on Challenges in Representation Learning (2013)

29. Li, C., Xu, T., Zhu, J., Zhang, B.: Triple generative adversarial nets. In: Advances in Neural Information Processing System, pp. 4088–4098 (2017)
30. Li, Y., Zhou, Z.: Towards making unlabeled data never hurt. IEEE Trans. Pattern Anal. Mach. Intell. **37**(1), 175–188 (2015)
31. Mallapragada, P.K., Jin, R., Jain, A.K., Liu, Y.: Semiboost: boosting for semi-supervised learning. IEEE Trans. Pattern Anal. Mach. Intell. **31**(11), 2000–2014 (2009)
32. Murphy, K.P.: Machine Learning: A Probabilistic Perspective. MIT Press, Cambridge (2012)
33. Pritchard, J.K., Stephens, M., Donnelly, P.: Inference of population structure using multilocus genotype data. Genetics **155**(2), 945–959 (2000)
34. Rasmus, A., Berglund, M., Honkala, M., Valpola, H., Raiko, T.: Semi-supervised learning with ladder networks. In: Advances in Neural Information Processing System, pp. 3546–3554 (2015)
35. Real, E., Moore, S., Selle, A., Saxena, S., Suematsu, Y.L., Le, Q., Kurakin, A.: Large-scale evolution of image classifiers. In: International Conference on Machine Learning (2017)
36. Sajjadi, M., Javanmardi, M., Tasdizen, T.: Regularization with stochastic transformations and perturbations for deep semi-supervised learning. In: Advances in Neural Information Processing System, pp. 1163–1171 (2016)
37. Seeger, M.: Learning with labeled and unlabeled data. Technical Report, U. Edinburgh (2001)
38. Srivastava, N., Hinton, G., Krizhevsky, A., Sutskever, I., Salakhutdinov, R.: Dropout: a simple way to prevent neural networks from overfitting. J. Mach. Learn. Res. **15**, 1929–1958 (2014)
39. Szummer, M., Jaakkola, T.: Information regularization with partially labeled data. In: Advances in Neural Information Processing System, pp. 1049–1056 (2002)
40. Szummer, M., Jaakkola, T.: Partially labeled classification with markov random walks. In: Advances in Neural Information Processing System, pp. 945–952 (2002)
41. Tang, Y.: Deep learning using linear support vector machines. In: International Conference on Machine Learning: Challenges in Representation Learning Workshop (2013)
42. Tarvainen, A., Valpola, H.: Mean teachers are better role models: weight-averaged consistency targets improve semi-supervised deep learning results. In: Advances in Neural Information Processing System, pp. 1195–1204 (2017)
43. Tikhonov, A.N.: On the stability of inverse problems. Proc. USSR Acad. Sci. **39**(5), 195–198 (1943)
44. Tikhonov, A.N.: Solution of incorrectly formulated problems and the regularization method. Proc. USSR Acad. Sci. **151**(3), 501–504 (1963)
45. Yuille, A.L., Rangarajan, A.: The concave-convex procedure. Neural Comput. **15**(4), 915–936 (2003)
46. Yuille, A.L., Rangarajan, A.: The concave-convex procedure (CCCP). In: Advances in Neural Information Processing System, pp. 1033–1040 (2002)
47. Zeiler, M.D.: Adadelta: an adaptive learning rate method (2012). arXiv:1212.5701
48. Zhou, D., Bousquet, O., Lal, T.N., Weston, J., Schölkopf, B.: Learning with local and global consistency. In: Advances in Neural Information Processing System, pp. 321–328 (2004)
49. Zhu, J., Kandola, J., Ghahramani, Z., Lafferty, J.D.: Nonparametric transforms of graph kernels for semi-supervised learning. In: Advances in Neural Information Processing System, pp. 1641–1648 (2005)

50. Zhu, X.: Semi-supervised learning literature survey. Technical Report, U. Wisconsin-Madison (2005)
51. Zhu, X., Ghahramani, Z.: Learning from labeled and unlabeled data with label propagation. Technical Report, CMU-CALD-02-107, Carnegie Mellon U (2002)
52. Zhuang, C., Ding, X., Murli, D., Yamins, D.: Local label propagation for large-scale semi-supervised learning (2019). arXiv:1905.11581
53. Zoph, B., Le, Q.V.: Neural architecture search with reinforcement learning. In: International Conference Machine Learning (2017)

Learning Functions Using Data-Dependent Regularization: Representer Theorem Revisited

Qing Zou$^{(\boxtimes)}$ (iD)

Applied Mathematics and Computational Sciences,
University of Iowa, Iowa City, IA 52242, USA
zou-qing@uiowa.edu

Abstract. We introduce a data-dependent regularization problem which uses the geometry structure of the data to learn functions from incomplete data. We show another proof of the standard representer theorem when introducing the problem. At the end of the paper, two applications in image processing are used to illustrate the function learning framework.

Keywords: Function learning · Manifold structure · Representer theorem

1 Introduction

1.1 Background

Many machine learning problems involve the learning of multidimensional functions from incomplete training data. For example, the classification problem can be viewed as learning a function whose function values give the classes that the inputs belong to. The direct representation of the function in high-dimensional spaces often suffers from the issue of dimensionality. The large number of parameters in the function representation would translate to the need of extensive training data, which is expensive to obtain. However, researchers found that many natural datasets have extensive structure presented in them, which is usually known as manifold structure. The intrinsic structure of the data can then be used to improve the learning results. Nowadays, assuming data lying on or close to a manifold becomes more and more common in machine learning. It is called manifold assumption in machine learning. Though researchers are not clear about the theoretical reason why the datasets have manifold structure, it is useful for supervised learning and it gives excellent performance. In this work, we will exploit the manifold structure to learn functions from incomplete training data.

© Springer Nature Switzerland AG 2020
V. V. Krzhizhanovskaya et al. (Eds.): ICCS 2020, LNCS 12139, pp. 312–326, 2020.
https://doi.org/10.1007/978-3-030-50420-5_23

1.2   A Motivated Example

One of the main problems in numerical analysis is function approximation. During the last several decades, researchers usually considered the following problem to apply the theory of function approximation to real-world problems:

$$\min_{f} \|Lf\|^2, \quad s.t. \quad f(x_i) = y_i, \tag{1}$$

where L is some linear operator, $\{(x_i, y_i)\}_{i=1}^{n} \subset X \times \mathbb{R}$ are n accessible observations and $X \subset \mathbb{R}^d, d \geq 1$ is the input space. We can use the method of Lagrange multiplier to solve Problem (1). Assume that the searching space for the function f is large enough (for example $\mathcal{L}_2$ space). Then the Lagrangian function $C(f)$ is given by

$$C(f) = \langle Lf, Lf \rangle + \sum_{i=1}^{n} \lambda_i(f(x_i) - y_i).$$

Taking the gradient of the Lagrangian function w.r.t. the function f gives us

$$C'(f) = \lim_{\varepsilon \to 0} \frac{C(f + \varepsilon f) - C(f)}{\varepsilon} = \lim_{\varepsilon \to 0} \frac{2\varepsilon \langle Lf, Lf \rangle + \varepsilon \sum \lambda_i \varepsilon f(x_i)}{\varepsilon}$$

$$= 2\langle Lf, Lf \rangle + \sum_{i=1}^{n} \lambda_i f(x_i).$$

Setting $C'(f) = 0$, we have

$$2\langle Lf, Lf \rangle = -\sum_{i=1}^{n} \lambda_i f(x_i) = -\sum_{i=1}^{n} \lambda_i \langle f(x), \delta(x - x_i) \rangle,$$

where $\delta(\cdot - x)$ is the delta function. Suppose L^* is the adjoint operator of L. Then we have

$$2\langle f(x), (L + L^*)f \rangle = \langle f(x), -\sum_{i=1}^{n} \lambda_i \delta(x - x_i) \rangle,$$

which gives us $2(L + L^*)f = -\sum_{i=1}^{n} \lambda_i \delta(x - x_i)$. This implies $f = \sum_{i=1}^{n} a_i \ell(x, x_i)$, for some a_i and $\ell(\cdot, \cdot)$.

1.3   Kernels and Representer Theorem

As machine learning develops fast these years, kernel methods [1] have received much attentions. Researchers found that working in the original data space is somehow not well-performed. So, we would like to map the data to a high dimensional space (feature space) using some non-linear mapping (feature map). Then we can do a better job (e.g. classification) in the feature space. When we talk about feature map, one concept that is unavoidable to mention is the kernel, which easily speaking is the inner product of the features. With a kernel (positive definite), we can then have a corresponding reproducing kernel Hilbert space (RKHS) [2] $\mathcal{H}_K$. We can now solve the problem that is similar to (1) in the RKHS:

$$\min_{f \in \mathcal{H}_K} \|f\|_{\mathcal{H}_K}^2 \quad s.t. \quad f(x_i) = y_i.$$

A more feasible way is to consider a regularization problem in the RKHS:

$$\min_{f \in \mathcal{H}_K} \|f(x_i) - y_i\|^2 + \lambda \|f\|_{\mathcal{H}_K}^2. \tag{2}$$

Then the searching space of f becomes $\mathcal{H}_K$, which is a Hilbert space. Before solving Problem (2), we would like to recall some basic concepts about the RKHS. Suppose we have a positive definite kernel $K : X \times X \to \mathbb{R}$, i.e.,

$$\sum_{i=1}^{n} \sum_{j=1}^{n} a_i a_j K(x_i, x_j) \geq 0, \quad n \in \mathbb{N}, \ x_1, \cdots, x_n \in X, \ a_1, \cdots, a_n \in \mathbb{R},$$

then $\mathcal{H}_K$ is the Hilbert space corresponding to the kernel $K(\cdot, \cdot)$. It is defined by all the possible linear combination of the kernel $K(\cdot, \cdot)$, i.e., $\mathcal{H}_K = span\{K(\cdot, \cdot)\}$. Thus, for any $f(\cdot) \in \mathcal{H}_K$, there exists x_i and α_i such that

$$f(\cdot) = \sum_i \alpha_i K(\cdot, x_i).$$

Since $\mathcal{H}_K$ is a Hilbert space, it is equipped with an inner product. The principle to define the inner product is to let $\mathcal{H}_K$ have representer $K(\cdot, x)$ and the representer performs like the delta function for functions in $\mathcal{L}_2$ (note that delta function is not in $\mathcal{L}_2$). In other word, we want to have a similar result to the following formula:

$$f(x) = \langle f(\cdot), \delta(\cdot - x) \rangle_{\mathcal{L}_2}.$$

This is called reproducing relation or reproducing property. In $\mathcal{H}_K$, we want to define the inner product so that we have the reproducing relation in $\mathcal{H}_K$:

$$f(x) = \langle f(\cdot), K(\cdot, x) \rangle_{\mathcal{H}_K}.$$

To achieve this goal, we can define

$$\langle f, g \rangle_{\mathcal{H}_K} = \langle \sum_i \alpha_i K(\cdot, x_i), \sum_j \beta_j K(\cdot, x_j) \rangle_{\mathcal{H}_K}$$

$$=: \sum_i \sum_j \alpha_i \beta_j K(x_i, x_j).$$

Then we have

$$\langle K(\cdot, x), K(\cdot, y) \rangle_{\mathcal{H}_K} = K(x, y)$$

With the kernel, the feature map $\Phi.(x)$ can be defined as

$$\Phi.(x) = K(\cdot, x).$$

Having these knowledge about the RKHS, we can now look at the solution of Problem (2). It can be characterized by the famous conclusion named representer theorem, which states that the solution of Problem (2) is

$$f(x) = \sum_{i=1}^{n} \alpha_i K(x, x_i).$$

The standard proof of the representer theorem is well-known and can be found in many literatures, see for example [3,4]. While the drawback of the standard proof is that the proof did not provide the expression of the coefficients α_i. In the first part of this work, we will provide another proof of the representer theorem. As a by-product, we can also build the relation between Problem (1) and Problem (2).

2 Another Proof of Representer Theorem

To give another proof of the representer theorem, we first build some relations between $\langle \cdot, \cdot \rangle_{\mathcal{H}_K}$ and $\langle \cdot, \cdot \rangle_{\mathcal{L}_2}$. We endow the dataset X with a measure μ. Then the corresponding $\mathcal{L}_2(X)$ inner product is given by

$$\langle f, g \rangle_{\mathcal{L}_2} = \int_X f \cdot g d\mu.$$

Consider an operator L on f with respect to the kernel K:

$$Lf(x) = \int_X f(y) K(x, y) d\mu, \tag{3}$$

which is the Hilbert-Schmidt integral operator [5]. This operator is self-adjoint, bounded and compact. By the spectral theorem [6], we can obtain that the eigenfunctions $e_1(x), e_2(x), \cdots$ of the operator will form an orthonormal basis of $\mathcal{L}_2(X)$, i.e.,

$$\langle e_i, e_j \rangle_{\mathcal{L}_2} = \begin{cases} 1, & i = j \\ 0, & i \neq j \end{cases}.$$

With the operator L defined as (3), we can look at the relations between $\langle \cdot, \cdot \rangle_{\mathcal{L}_2(X)}$ and $\langle \cdot, \cdot \rangle_{\mathcal{H}_K}$. Suppose $e_i(x)$ are the eigenfunctions of the operator L and λ_i are the corresponding eigenvalues, then

$$\langle K(x, y), e_i(y) \rangle_{\mathcal{L}_2(X)} = \int_X e_i(y) K(x, y) d\mu = \lambda_i e_i(x). \tag{4}$$

But by the reproducing relation, we have

$$\langle K(x, y), e_i(y) \rangle_{\mathcal{H}_K} = e_i(x).$$

Now, let us look at how to represent $K(x, y)$ by the eigenfunctions. We have

$$K(x, y) = \sum_i \lambda_i e_i(x) e_i(y),$$

and λ_i can be computed by

$$\lambda_i = \int_X \int_X K(x,y)e_i(x)e_i(y)dxdy.$$

To see $K(x,y) = \sum_i \lambda_i e_i(x)e_i(y)$, we can just plug it into (4) to verify it:

$$\langle K(x,y), e_i(y)\rangle_{\mathcal{L}_2(X)} = \int_X e_i(y) \sum_j \lambda_j e_j(x)e_j(y)d\mu(y)$$

$$= \sum_j \lambda_j \int_X e_j(x)e_i(y)e_j(y)d\mu(y) = \sum_j \lambda_j e_j(x) \int_X e_i(y)e_j(y)dy = \lambda_i e_i(x).$$

Since the eigenfunctions of L form an orthogonal basis of $\mathcal{L}_2(X)$, then for any $f \in \mathcal{L}_2$, it can be written as $f = \sum_i a_i e_i(x)$. So we have

$$\langle K(x,\cdot), f(\cdot)\rangle_{\mathcal{H}_K} = f(x) = \sum_i a_i e_i(x).$$

While for the $\mathcal{L}_2$ norm, we have

$$\langle K(x,\cdot), f(\cdot)\rangle_{\mathcal{L}_2(X)} = \langle K(x,\cdot), \sum_i a_i e_i(\cdot)\rangle_{\mathcal{L}_2(X)}$$

$$= \sum_i a_i \langle K(x,\cdot), e_i(\cdot)\rangle_{\mathcal{L}_2(X)} = \sum_i a_i \lambda_i e_i(x).$$

Next we show that the orthonormal basis $e_i(x)$ are within $\mathcal{H}_K$. Note that

$$e_i(x) = \langle K(x,\cdot), e_i(\cdot)\rangle_{\mathcal{H}_K} = \langle \sum_j \lambda_j e_j(x)e_j(\cdot), e_i(\cdot)\rangle_{\mathcal{H}_K},$$

which implies

$$e_i(x) = \sum_j \lambda_j e_j(x)\langle e_j(\cdot), e_i(\cdot)\rangle_{\mathcal{H}_K}.$$

So we can get

$$\langle e_j, e_i\rangle_{\mathcal{H}_K} = \begin{cases} 0, & i \neq j \\ \frac{1}{\lambda_i}, & i = j \end{cases} < \infty.$$

Therefore, we get $e_i(x) \in \mathcal{H}_K$.

We now need to investigate that for any $f = \sum_i a_i e_i(x) \in \mathcal{L}_2(X)$, when will we have that $f \in \mathcal{H}_K$. To let $f \in \mathcal{H}_K$, we need to have $\|f\|_{\mathcal{H}_K}^2 \leq \infty$. So

$$\|f\|_{\mathcal{H}_K}^2 = \langle f, f\rangle_{\mathcal{H}_K} = \langle \sum_i a_i e_i(x), \sum_i a_i e_i(x)\rangle_{\mathcal{H}_K}$$

$$= \sum_i a_i^2 \langle e_j, e_i\rangle_{\mathcal{H}_K} = \sum_i a_i^2 \cdot \frac{1}{\lambda_i} < \infty.$$

This means that to let $f = \sum_i a_i e_i(x) \in \mathcal{H}_K$, we need to have $\sum_i \frac{a_i^2}{\lambda_i} < \infty$ [7].

Combining all these analysis, we can then get the following relation between $\langle \cdot, \cdot \rangle_{\mathcal{L}_2(X)}$ and $\langle \cdot, \cdot \rangle_{\mathcal{H}_K}$:

$$\langle f, g \rangle_{\mathcal{L}_2(X)} = \langle L^{1/2} f, L^{1/2} g \rangle_{\mathcal{H}_K}, \quad \forall f, g \in \mathcal{H}_K, \quad L = L^{1/2} \circ L^{1/2}.$$

According to which, we can have another proof of the representer theorem.

Proof. Suppose $e_1, e_2, \cdots$ are eigenfunctions of the operator L. Then we can write the solution as $f^* = \sum_i a_i e_i(x)$. To let $f^* \in \mathcal{H}_K$, we require $\sum_i \frac{a_i^2}{\lambda_i} < \infty$. We consider here a more general form of Problem (2):

$$\min_{f \in \mathcal{H}_K} \sum_{i=1}^n E\left((x_i, y_i), f(x_i) \right) + \lambda \|f\|_{\mathcal{H}_K}^2,$$

where $E(\cdot, \cdot)$ is the error function which is differentiable with respect to each a_i. We would use the tools in $\mathcal{L}_2(X)$ space to get the solution.

The cost function of the regularization problem is

$$C(f) = \sum_{i=1}^n E\left((x_i, y_i), f(x_i) \right) + \lambda \|f\|_{\mathcal{H}_K}^2.$$

By substituting f^* into the cost function, we have

$$C(f^*) = \sum_{i=1}^n E\left((x_i, y_i), \sum_j a_j e_j(x_i) \right) + \lambda \|f^*\|_{\mathcal{H}_K}^2.$$

Since

$$\|f^*\|_{\mathcal{H}_K}^2 = \| \sum_i a_i e_i(x) \|_{\mathcal{H}_K}^2 = \langle \sum_i a_i e_i(x), \sum_i a_i e_i(x) \rangle_{\mathcal{H}_K} = \sum_i \frac{a_i^2}{\lambda_i} (< \infty),$$

differentiating $C(f^*)$ w.r.t. each a_i and setting it equal to zero gives

$$\frac{\partial C(f^*)}{\partial a_k} = \sum_{i=1}^n e_k(x_i) \partial_2 E\left((x_i, y_i), \sum_j a_j e_j(x_i) \right) + 2\lambda \frac{a_k}{\lambda_k} = 0.$$

Solving a_k, we get

$$a_k = -\frac{\lambda_k}{2\lambda} \sum_{i=1}^n e_k(x_i) \partial_2 E\left((x_i, y_i), f^* \right).$$

Since $f^* = \sum_k a_k e_k(x)$, we have

$$
\begin{aligned}
f^* &= \sum_k \left(-\frac{\lambda_k}{2\lambda} \sum_{i=1}^n e_k(x_i) \partial_2 E\left((x_i, y_i), f^*\right) \right) e_k(x) \\
&= -\frac{1}{2\lambda} \sum_{i=1}^n \left(\sum_k \lambda_k e_k(x_i) e_k(x) \partial_2 E\left((x_i, y_i), f^*\right) \right) \\
&= -\frac{1}{2\lambda} \sum_{i=1}^n K(x, x_i) \cdot \partial_2 E\left((x_i, y_i), f^*\right) \\
&= \sum_{i=1}^n \underbrace{\left(-\frac{1}{2\lambda} \partial_2 E\left((x_i, y_i), f^*\right) \right)}_{:=\alpha_i} \cdot K(x, x_i) =: \sum_{i=1}^n \alpha_i K(x, x_i).
\end{aligned}
$$

This proves the representer theorem.

Note that this result not only proves the representer theorem, but also gives the expression of the coefficients α_i.

With the operator L, we can also build a relation between Problem (1) and Problem (2). Define the operator in Problem (1) to be the inverse of the Hilbert-Schmidt Integral operator. The discussion on the inverse of the Hilbert-Schmidt Integral operator can be found in [8]. Note that for the delta function, we have

$$
L\delta(x, x_i) = \int_X \delta(y, x_i) K(x, y) dy = K(x, x_i).
$$

Then the solution of Problem (1) becomes $2(2L^{-1})f = -\sum_{i=1}^n \lambda_i \delta(x, x_i)$. So we have $L^{-1}f = -\sum_{i=1}^n \frac{\lambda_i}{4} \delta(x, x_i)$. Applying L on both sides gives

$$
L(L^{-1}f) = \sum_{i=1}^n (-\frac{\lambda_i}{4}) L\delta(x, x_i).
$$

By which we obtain

$$
f = \sum_{i=1}^n (-\frac{\lambda_i}{4}) K(x, x_i) =: \sum_{i=1}^n \beta_i K(x, x_i).
$$

3   Data-Dependent Regularization

So far, we have introduced the standard representer theorem. While as we discussed at the very beginning, many natural datasets have the manifold structure presented in them. So based on the classical Problem (2), we would like to introduce a new learning problem which exploits the manifold structure of the data. We call it the data-dependent regularization problem. Regularization problem has a long history going back to Tikhonov [9]. He proposed the Tikhonov regularization to solve the ill-posed inverse problem.

To exploit the manifold structure of the data, we can then divide a function into two parts: the function restricted on the manifold and the function restricted outside the manifold. So the problem can be formulated as

$$\min_{f \in \mathcal{H}_K} ||f(x_i) - y_i||^2 + \alpha ||f_1||_{\mathcal{M}}^2 + \beta ||f_2||_{\mathcal{M}^c}^2, \tag{5}$$

where $f_1 = f|_{\mathcal{M}}$ and $f_2 = f|_{\mathcal{M}^c}$. The norms $|| \cdot ||_{\mathcal{M}}$ and $|| \cdot ||_{\mathcal{M}^c}$ will be explained later in details. α and β are two parameters which control the degree for penalizing the energy of the function on the manifold and outside the manifold. We will show later that by controlling the two balancing parameters (set $\alpha = \beta$), the standard representer theorem is a special case of Problem (5).

We now discuss something about the functions f_1 and f_2. Consider the ambient space $X \subset \mathbb{R}^n$ (or $\mathbb{R}^n$) and a positive definite kernel K. Let us first look at the restriction of K to the manifold $\mathcal{M} \subset X$. The restriction is again a positive definite kernel [2] and it will then have a corresponding Hilbert space. We consider the relation between the RKHS $\mathcal{H}_K$ and the restricted RKHS to explain the norms $|| \cdot ||_{\mathcal{M}}$ and $|| \cdot ||_{\mathcal{M}^c}$.

Lemma 1 ([10]). *Suppose $K : X \times X \to \mathbb{R}$ (or $\mathbb{R}^n \times \mathbb{R}^n \to \mathbb{R}$) is a positive definite kernel. Let $\mathcal{M}$ be a subset of X (or $\mathbb{R}^n$). $\mathcal{F}(\mathcal{M})$ denote all the functions defined on $\mathcal{M}$. Then the RKHS given by the restricted kernel $K_1 : \mathcal{M} \times \mathcal{M} \to \mathbb{R}$ is*

$$\mathcal{H}_1(\mathcal{M}) = \{f_1 \in \mathcal{F}(\mathcal{M}) : f_1 = f|_{\mathcal{M}} \text{ for some } f \in \mathcal{H}_K\} \tag{6}$$

with the norm defined as

$$||f_1||_{\mathcal{M}} =: \min\{||f||_{\mathcal{H}_K} : f \in \mathcal{H}_K, f|_{\mathcal{M}} = f_1\}.$$

Proof. Define the set

$$\mathcal{S}(\mathcal{M}) =: \{f_r \in \mathcal{F}(\mathcal{M}) : \exists f \in \mathcal{H}_K \quad s.t. \quad f_r = f|_{\mathcal{M}}\}.$$

We first show that the set $A = \{||f||_{\mathcal{H}_K} : f \in \mathcal{H}_K, f|_{\mathcal{M}} = f_r\}$ has a minuma for any $f_r \in \mathcal{S}(\mathcal{M})$. Choose a sequence $\{f_i\}_{i=1}^{\infty} \subset \mathcal{H}_K$. Then the sequence is bounded because the space $\mathcal{H}_K$ is a Hilbert space. It is reasonable to assume that $\{f_i\}_{i=1}^{\infty}$ is weakly convergent because of the Banach-Alaoglu theorem [11]. By the weakly convergence, we can obtain pointwise convergence according to the reproducing property. So the limit of the sequence $\{f_i\}_{i=1}^{\infty}$ attains the minima.

We further define $||f_r||_{\mathcal{S}(\mathcal{M})} = \min A$. We show that $(\mathcal{S}(\mathcal{M}), || \cdot ||_{\mathcal{S}(\mathcal{M})})$ is a Hilbert space by the parallelogram law. In other word, we are going to show that

$$2(||f_1||_{\mathcal{S}(\mathcal{M})}^2 + ||g_1||_{\mathcal{S}(\mathcal{M})}^2) = ||f_1 + g_1||_{\mathcal{S}(\mathcal{M})}^2 + ||f_1 - g_1||_{\mathcal{S}(\mathcal{M})}^2, \quad \forall f_1, g_1 \in \mathcal{S}(\mathcal{M}).$$

Since we defined $||f_r||_{\mathcal{S}(\mathcal{M})} = \min A$. Then for all $f_1, g_1 \in \mathcal{S}(\mathcal{M})$, there exists $f, g \in \mathcal{H}_K$ such that

$$2(||f_1||_{\mathcal{S}(\mathcal{M})}^2 + ||g_1||_{\mathcal{S}(\mathcal{M})}^2) \leq 2(||f||_{\mathcal{H}_K}^2 + ||g||_{\mathcal{H}_K}^2) = ||f + g||_{\mathcal{H}_K} + ||f - g||_{\mathcal{H}_K}.$$

By the definition of $\mathcal{S}(\mathcal{M})$, we can choose f_1, g_1 such that

$$||f_1 + g_1||^2_{\mathcal{S}(\mathcal{M})} = ||f + g||^2_{\mathcal{H}_K}$$

and

$$||f_1 - g_1||^2_{\mathcal{S}(\mathcal{M})} = ||f - g||^2_{\mathcal{H}_K}.$$

Thus, we have

$$2(||f_1||^2_{\mathcal{S}(\mathcal{M})} + ||g_1||^2_{\mathcal{S}(\mathcal{M})}) \leq ||f_1 + g_1||^2_{\mathcal{S}(\mathcal{M})} + ||f_1 - g_1||^2_{\mathcal{S}(\mathcal{M})}.$$

For the reverse inequality, we first choose f_1, g_1 such that $||f_1||^2_{\mathcal{S}(\mathcal{M})} = ||f||^2_{\mathcal{H}_K}$ and $||g_1||^2_{\mathcal{S}(\mathcal{M})} = ||g||^2_{\mathcal{H}_K}$. Then

$$2(||f_1||^2_{\mathcal{S}(\mathcal{M})} + ||g_1||^2_{\mathcal{S}(\mathcal{M})}) = 2(||f||^2_{\mathcal{H}_K} + ||g||^2_{\mathcal{H}_K})$$
$$= ||f + g||^2_{\mathcal{H}_K} + ||f - g||^2_{\mathcal{H}_K} \geq ||f_1 + g_1||^2_{\mathcal{S}(\mathcal{M})} + ||f_1 - g_1||^2_{\mathcal{S}(\mathcal{M})}.$$

Therefore, we get

$$2(||f_1||^2_{\mathcal{S}(\mathcal{M})} + ||g_1||^2_{\mathcal{S}(\mathcal{M})}) = ||f_1 + g_1||^2_{\mathcal{S}(\mathcal{M})} + ||f_1 - g_1||^2_{\mathcal{S}(\mathcal{M})}.$$

Next, we show (6) by showing that for all $f_r \in \mathcal{S}(\mathcal{M})$ and $x \in \mathcal{M}$,

$$f_r(x) = \langle f_r(\cdot), K_1(\cdot, x) \rangle_{\mathcal{S}(\mathcal{M})},$$

where $K_1 = K|_{\mathcal{M} \times \mathcal{M}}$.

Choose $f \in \mathcal{H}_K$ such that $f_r = f|_{\mathcal{M}}$ and $||f_r||_{\mathcal{S}(\mathcal{M})} = ||f||_{\mathcal{H}_K}$. This is possible because of the analysis above. Specially, we have

$$||K_1(\cdot, x)||_{\mathcal{S}(\mathcal{M})} = ||K(\cdot, x)||_{\mathcal{H}_K}, \quad \forall x \in \mathcal{M}.$$

Now, for any function $f \in \mathcal{H}_K$ such that $f|_{\mathcal{M}} = 0$, we have

$$||K(\cdot, x) + f||^2_{\mathcal{H}_K} = ||K(\cdot, x)||^2_{\mathcal{H}_K} + ||f||^2_{\mathcal{H}_K} + 2\langle f, K(\cdot, x) \rangle_{\mathcal{H}_K}$$
$$= ||K(\cdot, x)||^2_{\mathcal{H}_K} + ||f||^2_{\mathcal{H}_K} + 2f(x) = ||K(\cdot, x)||^2_{\mathcal{H}_K} + ||f||^2_{\mathcal{H}_K}.$$

Thus,

$$\langle f_r(\cdot), K_1(\cdot, x) \rangle_{\mathcal{S}(\mathcal{M})} = \langle f, K(\cdot, x) \rangle_{\mathcal{H}_K} = f(x) = f_r(x), \quad \forall x \in \mathcal{M}.$$

This completes the proof of the lemma.

With this lemma, the solution of Problem (5) then becomes easy to obtain. By the representer theorem we mentioned before, we know that the function satisfies

$$\min_{f \in \mathcal{H}_K} ||f(x_i) - y_i||^2 + \lambda ||f||^2_{\mathcal{H}_K}$$

is $f = \sum_{i=1}^{n} a_i K(x, x_i)$. Since we have

$$||f_1||^2_{\mathcal{M}} = \min\{||f||_{\mathcal{H}_K} : f|_{\mathcal{M}} = f_1\},$$

$$\|f_2\|_{\mathcal{M}^c}^2 = \min\{\|f\|_{\mathcal{H}_K} : f|_{\mathcal{M}^c} = f_2\}.$$

Thus, we can conclude that is solution of (5) is exactly

$$f = \sum_{i=1}^{n} a_i K(x, x_i),$$

where the coefficients a_i are controlled by the parameters α and β.

With the norms $\|\cdot\|_{\mathcal{M}}$ and $\|\cdot\|_{\mathcal{M}^c}$ being well-defined, we would like to seek the relation between $\|\cdot\|_{\mathcal{M}}, \|\cdot\|_{\mathcal{M}^c}$ and $\|\cdot\|_{\mathcal{H}_K}$. Before stating the relation, we would like to restate some of the notations to make the statement more clear. Let

$$K_1 = K|_{\mathcal{M} \times \mathcal{M}}, \qquad K_2 = K|_{(\mathcal{M} \times \mathcal{M})^c}$$

and

$$\mathcal{H}_1(\mathcal{M}) = \{f_1 \in \mathcal{F}(\mathcal{M}) : f_1 = f|_{\mathcal{M}} \text{ for some } f \in \mathcal{H}_K\},$$

$$\|f_1\|_{\mathcal{M}} =: \min\{\|f\|_{\mathcal{H}_K} : f \in \mathcal{H}_K, f|_{\mathcal{M}} = f_1\}.$$

$$\mathcal{H}_2(\mathcal{M}^c) = \{f_2 \in \mathcal{F}(\mathcal{M}^c) : f_2 = f|_{\mathcal{M}^c} \text{ for some } f \in \mathcal{H}_K\},$$

$$\|f_2\|_{\mathcal{M}^c} =: \min\{\|f\|_{\mathcal{H}_K} : f \in \mathcal{H}_K, f|_{\mathcal{M}^c} = f_2\}.$$

To find the relation between $\|\cdot\|_{\mathcal{M}}, \|\cdot\|_{\mathcal{M}^c}$ and $\|\cdot\|_{\mathcal{H}_K}$, we need to pullback the restricted kernel K_1 and K_2 to the original space. To do so, define

$$K_1^p = \begin{cases} K_1, & (x, y) \in \mathcal{M} \times \mathcal{M} \\ 0, & (x, y) \in (\mathcal{M} \times \mathcal{M})^c \end{cases}.$$

$$K_2^p = \begin{cases} K_2, & (x, y) \in (\mathcal{M} \times \mathcal{M})^c \\ 0, & (x, y) \in \mathcal{M} \times \mathcal{M} \end{cases}.$$

Then we have $K = K_1^p + K_2^p$. The corresponding Hilbert spaces for K_1^p and K_2^p are

$$\mathcal{H}_1^p(\mathcal{M}) = \{f_1^p \in \mathcal{F}(X) : f_1^p|_{\mathcal{M}} = f_1, f_1^p|_{\mathcal{M}^c} = 0\},$$

$$\mathcal{H}_2^p(\mathcal{M}^c) = \{f_2^p \in \mathcal{F}(X) : f_2^p|_{\mathcal{M}^c} = f_2, f_2^p|_{\mathcal{M}} = 0\}.$$

It is straightforward to define that

$$\|f_1^p\|_{\mathcal{H}_{K_1^p}} = \|f_1\|_{\mathcal{M}},$$

$$\|f_2^p\|_{\mathcal{H}_{K_2^p}} = \|f_2\|_{\mathcal{M}^c}.$$

The following lemma shows the relation between $\|\cdot\|_{\mathcal{H}_{K_1^p}}, \|\cdot\|_{\mathcal{H}_{K_2^p}}$ and $\|\cdot\|_{\mathcal{H}_K}$, which also reveals the relation between $\|\cdot\|_{\mathcal{M}}, \|\cdot\|_{\mathcal{M}^c}$ and $\|\cdot\|_{\mathcal{H}_K}$ by Moore-Aronszajn theorem [12].

Lemma 2. *Suppose $K_1, K_2 : Y \times Y \to \mathbb{R}$ (or $\mathbb{R}^n \times \mathbb{R}^n \to \mathbb{R}$) are two positive definie kernels. If $K = K_1 + K_2$, then*

$$\mathcal{H}_K = \{f_1 + f_2 : f_1 \in \mathcal{H}_{K_1}, f_2 \in \mathcal{H}_{K_2}\}$$

is a Hilbert space with the norm defined by

$$\|f\|^2_{\mathcal{H}_K} = \min_{f_1 \in \mathcal{H}_{K_1}, f_2 \in \mathcal{H}_{K_2}, f = f_1 + f_2} \|f_1\|^2_{\mathcal{H}_{K_1}} + \|f_2\|^2_{\mathcal{H}_{K_2}}.$$

The idea of the proof of this lemma is exactly the same as the one for Lemma 1. Thus we omit it here.

A direct corollary of this lemma is:

Corollary 1. *Under the assumption of Lemma 2, if the functions in $\mathcal{H}_{K_1}$ and $\mathcal{H}_{K_2}$ have no functions except for zero function in common. Then the norm of $\mathcal{H}_K$ is given simply by*

$$\|f\|^2_{\mathcal{H}_K} = \|f_1\|^2_{\mathcal{H}_{K_1}} + \|f_2\|^2_{\mathcal{H}_{K_2}}.$$

If we go back to our scenario, we can get the following result by Corollary 1:

$$\|f\|^2_{\mathcal{H}_K} = \|f_1\|^2_{\mathcal{M}} + \|f_2\|^2_{\mathcal{M}^c}.$$

This means that if we set $\alpha = \beta$ in Problem (5), it will reduce to Problem (2). Therefore, the standard representer theorem is a special case of our data-dependent regularization problem (5).

4 Applications

As we said in the introduction part, many engineering problems can be viewed as learning multidimensional functions from incomplete data. In this section, we would like to show two applications of functions learning: image interpolation and patch-based iamge denoising.

4.1 Image Interpolation

Image interpolation tries to best approximate the color and intensity of a pixel based on the values at surrounding pixels. See Fig. 1 for illustration. From function learning perspective, image interpolation is to learn a function from the known pixels and their corresponding positions.

We would like to use the Lena image as shown in Fig. 2(a) to give an example of image interpolation utilizing the proposed framework. The zoomed image is shown in Fig. 2(d). In the image interpolation example, the two balancing parameters are set to be the same and the Laplacian kernel [13] is used:

$$K(x, y) = \exp\left(-\frac{\|x - y\|}{\sigma}\right).$$

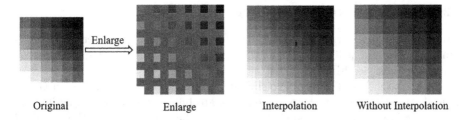

Original   Enlarge   Interpolation   Without Interpolation

Fig. 1. Illustration of image interpolation. The size of the original image is 6×6. We want to enlarge it as an 11×11 image. Then the blue shaded positions are unknown. Using image interpolation, we can find the values of these positions. (Color figure online)

Note that we can also use other kernels, for example, polynomial kernel and Gaussian kernel to proceed image interpolation. Choosing the right kernel is an interesting problem and we do not have enough space to compare different kernels in this paper.

In Fig. 2(b), we downsampled the original image by a factor of 3 in each direction. The zoomed image is shown in Fig. 2(e). The interpolation result with the zoomed image are shown in Fig. 2(c) and Fig. 2(f).

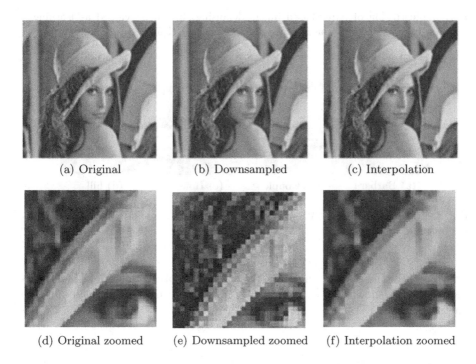

(a) Original   (b) Downsampled   (c) Interpolation

(d) Original zoomed   (e) Downsampled zoomed   (f) Interpolation zoomed

Fig. 2. Illustration of image interpolation. The original image is downsampled by a factor of 3 in each direction. We use the proposed function learning framework to obtain the interpolation function from downsampled image. From the results, we can see that the proposed framework works for image interpolation.

4.2   Patch-Based Image Denoising

From the function learning point of view, the patch-based image denoising problem can be viewed as learning a function from noisy patches to their "noise-free" centered pixels. See Fig. 3 for illustration.

In the patch-based image denoising application, we use the Laplacian kernel as well. We assume that the noisy patches are lying close to some manifold so we set the balancing parameter which controls the energy outside the manifold to be large enough. We use the images in Fig. 4 as known data to learn the function. Then for a given noisy image, we can use the learned function to do image denoising. To speed up the learning process, we randomly choose only 10% of the known data to learn the function.

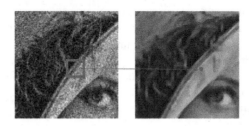

Fig. 3. Illustration of patch-based image denoising. It can be viewed as learning a function from the $m \times m$ noisy patches to the centered clean pixels.

(a) Barbara   (b) Couple   (c) House   (d) hill

Fig. 4. Four training images. We use noisy images and clean pixels to learn the denoising function.

We use the image Baboon to test the learned denoising function. The denoising results are shown in Fig. 5. Each column shows the result corresponding to one noise level.

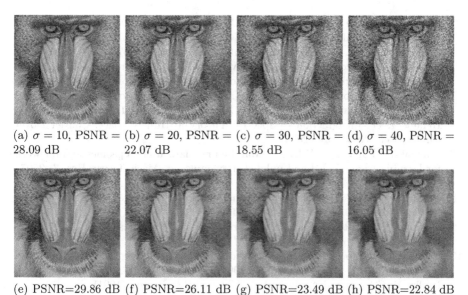

(a) $\sigma = 10$, PSNR = 28.09 dB

(b) $\sigma = 20$, PSNR = 22.07 dB

(c) $\sigma = 30$, PSNR = 18.55 dB

(d) $\sigma = 40$, PSNR = 16.05 dB

(e) PSNR=29.86 dB (f) PSNR=26.11 dB (g) PSNR=23.49 dB (h) PSNR=22.84 dB

Fig. 5. Illustration of the denoising results.

5   Conclusion and Future Work

In this paper, we introduced a framework of learning functions from part of the data. We gave a data-dependent regularization problem which helps us learn a function using the manifold structure of the data. We used two applications to illustrate the learning framework. While these two applications are just part of the learning framework. They are special cases of the data-dependent regularization problem. However, for the general application, we need to calculate $||f_1||^2_{\mathcal{M}}$ and $||f_2||^2_{\mathcal{M}^c}$, which is hard to do so since we only have partial data. So we need to approximate $||f_1||^2_{\mathcal{M}}$ and $||f_2||^2_{\mathcal{M}^c}$ from incomplete data and to propose a new learning algorithm so that our framework can be used in a general application. This is part of our future work. Another line for the future work is from the theoretical aspect. We showed that the solution of the data-dependent regularization problem is the linear combination of the kernel. It then can be viewed as a function approximation result. If it is an approximated function, then we can consider the error analysis of the approximated function.

References

1. Schölkopf, B., Smola, A.J., Bach, F.: Learning with Kernels: Support Vector Machines, Regularization, Optimization, and Beyond. MIT press, Cambridge (2002)
2. Aronszajn, N.: Theory of reproducing kernels. Trans. Am. Math. Soc. **68**(3), 337–404 (1950)

3. Schölkopf, B., Herbrich, R., Smola, A.J.: A generalized representer theorem. In: Helmbold, D., Williamson, B. (eds.) COLT 2001. LNCS (LNAI), vol. 2111, pp. 416–426. Springer, Heidelberg (2001). https://doi.org/10.1007/3-540-44581-1_27
4. Argyriou, A., Micchelli, C.A., Pontil, M.: When is there a representer theorem? vector versus matrix regularizers. J. Mach. Learn. Res. **10**(Nov), 2507–2529 (2009)
5. Gohberg, I., Goldberg, S., Kaashoek, M.A.: Hilbert-Schmidt operators. In: Classes of Linear Operators, vol. I, pp. 138–147. Birkhäuser, Basel (1990)
6. Helmberg, G.: Introduction to Spectral Theory in Hilbert Space. Courier Dover Publications, New York (2008)
7. Mikhail, B., Partha, N., Vikas, S.: Manifold regularization: a geometric framework for learning from labeled and unlabeled examples. J. Mach. Learn. Res. **7**, 2507–2529 (2006)
8. Pipkin, A.C.: A Course on Integral Equations, vol. 9. Springer, New York (1991). https://doi.org/10.1007/978-1-4612-4446-2
9. Tikhonov, A.N.: Regularization of incorrectly posed problems. Soviet Math. Doklady **4**(6), 1624–1627 (1963)
10. Saitoh, S., Sawano, Y.: Theory of Reproducing Kernels and Applications. Springer, Singapore (2016). https://doi.org/10.1007/978-981-10-0530-5
11. Rudin, W.: Functional Analysis. MA. McGraw-Hill, Boston (1991)
12. Amir, A.D., Luis, G.C.R., Yukawa, M., Stanczak, S.: Adaptive learning for symbol detection: a reproducing kernel hilbert space approach. Mach. Learn. Fut. Wirel. Commun., 197–211 (2020)
13. Kernel Functions for Machine Learning Applications. http://crsouza.com/2010/03/17/kernel-functions-for-machine-learning-applications/

Reduction of Numerical Errors in Zernike Invariants Computed via Complex-Valued Integral Images

Przemysław Klęsk$^{(\boxtimes)}$ ⓘ, Aneta Bera ⓘ, and Dariusz Sychel ⓘ

Faculty of Computer Science and Information Technology, West Pomeranian University of Technology, ul. Żołnierska 49, 71-210 Szczecin, Poland
{pklesk,abera,dsychel}@wi.zut.edu.pl

Abstract. Floating-point arithmetics may lead to numerical errors when numbers involved in an algorithm vary strongly in their orders of magnitude. In the paper we study numerical stability of Zernike invariants computed via complex-valued integral images according to a constant-time technique from [2], suitable for object detection procedures. We indicate numerically fragile places in these computations and identify their cause, namely—binomial expansions. To reduce numerical errors we propose *piecewise integral images* and derive a numerically safer formula for Zernike moments. Apart from algorithmic details, we provide two object detection experiments. They confirm that the proposed approach improves accuracy of detectors based on Zernike invariants.

Keywords: Zernike moments · Complex-valued integral images · Numerical errors reduction · Object detection

1 Introduction

The classical approach to object detection is based on sliding window scans. It is computationally expensive, involves a large number of image fragments (windows) to be analyzed, and in practice precludes the applicability of advanced methods for feature extraction. In particular, many *moment* functions [9], commonly applied in image recognition tasks, are often precluded from detection, as they involve inner products i.e. linear-time computations with respect to the number of pixels. Also, the deep learning approaches cannot be applied directly in detection, and require preliminary stages of prescreening or region-proposal.

There exist a few feature spaces (or descriptors) that have managed to bypass the mentioned difficulties owing to constant-time techniques discovered for them within the last two decades. Haar-like features (HFs), local binary patterns (LBPs) and HOG descriptor are state-of-the-art examples from this category [1,4,14] The crucial algorithmic trick that underlies these methods and allows

This work was financed by the National Science Centre, Poland. Research project no.: 2016/21/B/ST6/01495.

V. V. Krzhizhanovskaya et al. (Eds.): ICCS 2020, LNCS 12139, pp. 327–341, 2020.
https://doi.org/10.1007/978-3-030-50420-5_24

for constant-time—$O(1)$—feature extraction are *integral images*. They are auxiliary arrays storing cumulative pixel intensities or other pixel-related expressions. Having prepared them before the actual scan, one is able to compute fast the wanted sums via so-called 'growth' operations. Each growth involves two additions and one subtraction using four entries of an integral image.

In our research we try to broaden the existing repertoire of constant-time techniques for feature extraction. In particular, we have managed to construct such new techniques for Fourier moments (FMs) [7] and Zernike moments (ZMs) [2]. In both cases a *set* of integral images is needed. For FMs, the integral images cumulate products of image function and suitable trigonometric terms and have the following general forms: $\sum\sum_{j,k} f(j,k)\cos(-2\pi(jt/\text{const}_1 + ku/\text{const}_2))$, and $\sum\sum_{j,k} f(j,k)\sin(-2\pi(jt/\text{const}_1 + ku/\text{const}_2))$, where f denotes the image function and t, u are order-related parameters. With such integral images prepared, each FM requires only 21 elementary operations (including 2 growths) to be extracted during a detection procedure. In the case of ZMs, complex-valued integral images need to be prepared, having the form: $\sum\sum_{j,k} f(j,k)(k-ij)^t(k+ij)^u$, where i stands for the imaginary unit ($i^2 = -1$). The formula to extract a single ZM of order (p,q) is more intricate and requires roughly $\frac{1}{24}p^3 - \frac{1}{8}pq^2 + \frac{1}{12}q^3$ growths, but still the calculation time is not proportional to the number of pixels.

It should be remarked that in [2] we have flagged up, but not tackled, the problem of *numerical errors* that may occur when computations of ZMs are backed with integral images. ZMs are complex numbers, hence the natural data type for them is the `complex` type with real and imaginary parts stored in the double precision of the IEEE-754 standard for floating-point numbers (a precision of approximately 16 decimal digits). The main culprit behind possible numerical errors are *binomial expansions*. As we shall show the algorithm must explicitly expand two binomial expressions to benefit from integral images, which leads to numbers of different magnitudes being involved in the computations. When multiple additions on such numbers are carried out, digits of smaller-magnitude numbers can be lost.

In this paper we address the topic of numerical stability. The key new contribution are **piecewise integral images**. Based on them we derive a **numerically safer formula for** the computation of a single **Zernike moment**. The resulting technique introduces some computational overhead, but remains to be a constant-time technique.

2   Preliminaries

Recent literature confirms that ZMs are still being applied in many image recognition tasks e.g: human age estimation [8], electrical symbols recognition [16], traffic signs recognition [15], tumor diagnostics from magnetic resonance [13]. Yet, it is quite difficult to find examples of detection tasks applying ZMs directly. Below we describe the constant-time approach to extract ZMs within detection, together with the proposition of numerically safe computations.

2.1 Zernike Moments, Polynomials and Notation

ZMs can be defined in both polar and Cartesian coordinates as:

$$M_{p,q} = \frac{p+1}{\pi} \int_0^{2\pi} \int_0^1 f(r,\theta) \sum_{s=0}^{(p-|q|)/2} \beta_{p,q,s} r^{p-2s} e^{-iq\theta} \, r \, dr \, d\theta, \tag{1}$$

$$= \frac{p+1}{\pi} \iint_{x^2+y^2 \leqslant 1} f(x,y) \sum_{s=0}^{(p-|q|)/2} \beta_{p,q,s} (x+iy)^{\frac{1}{2}(p-q)-s} (x-iy)^{\frac{1}{2}(p+q)-s} \, dx \, dy, \tag{2}$$

where:

$$\beta_{p,q,s} = \frac{(-1)^s (p-s)!}{s!((p+q)/2 - s)!((p-q)/2 - s)!}, \tag{3}$$

i is the imaginary unit ($i^2 = -1$), and f is a mathematical or an image function defined over unit disk [2,17]. p and q indexes, representing moment order, must be simultaneously even or odd, and $p \geqslant |q|$.

ZMs are in fact the *coefficients* of an *expansion* of function f, given in terms of Zernike polynomials $V_{p,q}$ as the orthogonal base:[1]

$$f(r,\theta) = \sum_{\substack{0 \leqslant p \leqslant \infty \\ p-|q| \text{ even}}} \sum_{-p \leqslant q \leqslant p} M_{p,q} V_{p,q}(r,\theta), \tag{4}$$

where $V_{p,q}(r,\theta) = \sum_{s=0}^{(p-|q|)/2} \beta_{p,q,s} r^{p-2s} e^{iq\theta}$. As one can note $V_{p,q}$ combines a standard polynomial defined over radius r and a harmonic part defined over angle θ. In applications, finite partial sums of expansion (4) are used. Suppose ρ and ϱ denote the imposed maximum orders, polynomial and harmonic, respectively, and $\rho \geqslant \varrho$. Then, the partial sum that approximates f can be written down as:

$$f(r,\theta) \approx \sum_{\substack{0 \leqslant p \leqslant \rho \\ p-|q| \text{ even}}} \sum_{-\min\{p,\varrho\} \leqslant q \leqslant \min\{p,\varrho\}} M_{p,q} V_{p,q}(r,\theta). \tag{5}$$

2.2 Invariants Under Rotation

ZMs are invariant to scale transformations, but, as such, are not invariant to rotation. Yet, they do allow to build suitable expressions with that property. Suppose f' denotes a version of function f rotated by an angle α, i.e. $f'(r,\theta) = f(r, \theta + \alpha)$. It is straightforward to check, deriving from (1), that the following identity holds

$$M'_{p,q} = e^{iq\alpha} M_{p,q}, \tag{6}$$

---

[1] ZMs expressed by (1) arise as inner products of the approximated function and Zernike polynomials: $M_{p,q} = \langle f, V_{p,q} \rangle / \|V_{p,q}\|^2$.

where $M'_{p,q}$ represents a moment for the rotated function f'. Hence in particular, the *moduli* of ZMs are one type of rotational invariants, since

$$|M'_{p,q}| = |e^{iq\alpha}M_{p,q}| = |e^{iq\alpha}||M_{p,q}| = |M_{p,q}|. \tag{7}$$

Apart from the moduli, one can also look at the following *products* of moments

$$M_{p,q}{}^n M_{v,s} \tag{8}$$

with the first factor raised to a natural power. After rotation one obtains

$$M'_{p,q}{}^n M'_{v,s} = e^{inq\alpha}M_{p,q}{}^n e^{is\alpha}M_{v,s} = e^{i(nq+s)\alpha}M_{p,q}{}^n M_{v,s}. \tag{9}$$

Hence, by forcing $nq + s = 0$ one can obtain many rotational invariants because the angle-dependent term $e^{i(nq+s)\alpha}$ disappears.

2.3 ZMs for an Image Fragment

In practical tasks it is more convenient to work with rectangular, rather than circular, image fragments. Singh and Upneja [12] proposed a workaround to this problem: a square of size $w \times w$ in pixels (w is even) becomes *inscribed* in the unit disc, and zeros are "laid" over the square-disc complement. This reduces integration over the disc to integration over the square. The inscription implies that widths of pixels become $\sqrt{2}/w$ and their areas $2/w^2$ (a detail important for integration). By iterating over pixel indexes: $0 \leqslant j, k \leqslant w - 1$, one generates the following Cartesian coordinates within the unit disk:

$$x_k = \frac{2k - (w - 1)}{w\sqrt{2}}, \qquad y_j = \frac{w - 1 - 2j}{w\sqrt{2}}. \tag{10}$$

In detection, it is usually sufficient to replace integration involved in $M_{p,q}$ by a suitable summation, thereby obtaining a zeroth order approximation. In subsequent sections we use the formula below, which represents such an approximation (hat symbol) and is adjusted to have a convenient indexing for our purposes:

$$\widehat{M}_{2p+o,2q+o} = \frac{4p+2o+2}{\pi w^2} \sum_{0 \leqslant j,k \leqslant w-1} f(j,k) \sum_{q \leqslant s \leqslant p} \beta_{2p+o,2q+o,p-s}(x_k+iy_j)^{s-q}(x_k-iy_j)^{s+q+o} \tag{11}$$

— namely, we have introduced the substitutions $p := 2p + o$, $q := 2q + o$. They play two roles: they make it clear whether a moment is even or odd via the flag $o \in \{0, 1\}$; they allow to construct integral images, since exponents $\frac{1}{2}(p \mp s) - s$ present in (2) are now suitably reduced.

2.4 Proposition from [2]

Suppose a digital image of size $n_x \times n_y$ is traversed by a $w \times w$ sliding window. For clarity we discuss only a single-scale scan. The situation is sketched in Fig. 1. Let

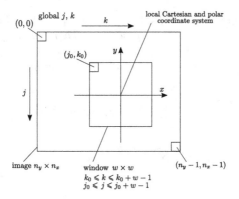

Fig. 1. Illustration of detection procedure using sliding window.

(j, k) denote global coordinates of a pixel in the image. For each window under analysis, its offset (top-left corner) will be denoted by (j_0, k_0). Thus, indexes of pixels that belong to the window are: $j_0 \leqslant j \leqslant j_0+w-1$, $k_0 \leqslant k \leqslant k_0+w-1$. Alse let (j_c, k_c) represent the *central index* of the window:

$$j_c = \frac{1}{2}(2j_0 + w - 1), \quad k_c = \frac{1}{2}(2k_0 + w - 1). \tag{12}$$

Given a global index (j, k) of a pixel, the local Cartesian coordinates corresponding to it (mapped to the unit disk) can be expressed as:

$$x_k = \frac{2(k-k_0) - (w-1)}{w\sqrt{2}} = \frac{\sqrt{2}}{w}(k-k_c), \quad y_j = \frac{(w-1) - 2(j-j_0)}{w\sqrt{2}} = \frac{\sqrt{2}}{w}(j_c-j). \tag{13}$$

Let $\{ii_{t,u}\}$ denote a set of **complex-valued integral images**[2]:

$$ii_{t,u}(l, m) = \sum_{\substack{0 \leqslant j \leqslant l \\ 0 \leqslant k \leqslant m}} f(j, k)(k-ij)^t(k+ij)^u, \quad \substack{0 \leqslant l \leqslant n_y-1 \\ 0 \leqslant m \leqslant n_x-1}; \tag{14}$$

where pairs of indexes (t, u), generating the set, belong to: $\{(t, u): 0 \leqslant t \leqslant \lfloor \rho/2 \rfloor, 0 \leqslant u \leqslant \min(\rho-t, \lfloor (\rho+\varrho)/2 \rfloor)\}$.

For any integral image we also define the **growth operator** over a rectangle spanning from (j_1, k_1) to (j_2, k_2):

$$\underset{\substack{j_1,j_2 \\ k_1,k_2}}{\Delta} (ii_{t,u}) = ii_{t,u}(j_2, k_2) - ii_{t,u}(j_1 - 1, k_2) - ii_{t,u}(j_2, k_1 - 1) + ii_{t,u}(j_1 - 1, k_1 - 1) \tag{15}$$

with two complex-valued substractions and one addition. The main result from [2] (see there for proof) is as follows.

---

[2] In [2] we have proved that integral images $ii_{t,u}$ and $ii_{u,t}$ are complex conjugates at all points, which allows for computational savings.

Proposition 1. *Suppose a set of integral images* $\{ii_{t,u}\}$, *defined as in* (14), *has been prepared prior to the detection procedure. Then, for any square window in the image, each of its Zernike moments* (11) *can be calculated in constant time* — $O(1)$, *regardless of the number of pixels in the window, as follows:*

$$\widehat{M}_{2p+o,2q+o} = \frac{4p+2o+2}{\pi w^2} \sum_{2q+o\leqslant 2s+o\leqslant 2p+o} \beta_{2p+o,2q+o,p-s}\left(\frac{\sqrt{2}}{w}\right)^{2s+o}$$

$$\cdot \sum_{t=0}^{s-q}\binom{s-q}{t}(-k_c+ij_c)^{s-q-t}\sum_{u=0}^{s+q+o}\binom{s+q+o}{u}(-k_c-ij_c)^{s+q+o-u}\underset{\substack{j_0,j_0+w-1\\k_0,k_0+w-1}}{\Delta}(ii_{t,u}). \quad (16)$$

3   Numerical Errors and Their Reduction

Floating-point additions or subtractions are dangerous operations [6,10] because when numbers of different magnitudes are involved, the right-most digits in the mantissa of the smaller-magnitude number can be lost when widely spaced exponents are aligned to perform an operation. When ZMs are computed according to Proposition 1, such situations can arise in *two* places.

The connection between the definition-style ZM formula (11) and the integral images-based formula (16) are expressions (13): $\frac{\sqrt{2}}{w}(k-k_c)$, $\frac{\sqrt{2}}{w}(j_c-j)$. They map global coordinates to local unit discs. When the mapping formulas are plugged into x_k and y_j in (11), the following subexpression arises under summations:

$$\cdots\left(\sqrt{2}/w\right)^{2s+o}(k-ij-k_c+ij_c)^{s-q}(k+ij-k_c-ij_c)^{s+q+o}. \quad (17)$$

Now, to benefit from integral images one has to explicitly expand the two binomial expressions, distinguishing two groups of terms: $k \mp ij$—dependent on the global pixel index, and $-k_c \pm ij_c$—independent of it . By doing so, coordinates of the particular window can be isolated out and formula (16) is reached. Unfortunately, this also creates two numerically fragile places. The first one are integral images themselves, defined by (14). Global pixel indexes j, k present in power terms $(k-ij)^t(k+ij)^u$ vary within: $0 \leqslant j \leqslant n_y - 1$ and $0 \leqslant k \leqslant n_x - 1$. Hence, for an image of size e.g. 640×480, the summands vary in magnitude roughly from $10^{0(t+u)}$ to $10^{3(t+u)}$. To fix an example, suppose $t + u = 10$ (achievable e.g. when $\rho = \varrho = 10$) and assume a roughly constant values image function. Then, the integral image $ii_{t,u}$ has to cumulate values ranging from 10^0 up to 10^{30}. Obviously, the rounding-off errors amplify as the $ii_{t,u}$ sum progresses towards the bottom-right image corner. The second fragile place are expressions: $(-k_c + ij_c)^{s-q-t}$ and $(-k_c - ij_c)^{s+q+o-u}$, involving the central index, see (16). Their products can too become very large in magnitude as computations move towards the bottom-right image corner.

In error reports we shall observe *relative errors*, namely:

$$\mathrm{err}(\phi, \phi^*) = |\phi - \phi^*|/|\phi^*|, \quad (18)$$

where ϕ denotes a feature (we skip indexes for readability) computed via integral images while ϕ^* its value computed by the definition. To give the reader an initial outlook, we remark that in our C++ implementation noticeable errors start to be observed already for $\rho=\varrho=8$ settings, but they do not yet affect significantly the detection accuracy. For $\rho=\varrho=10$, the frequency of relevant errors is already clear they do deteriorate accuracy. For example, for the sliding window of size 48×48, about 29.7% of all features have relative errors equal at least 25%. The numerically safe approach we are about to present reduces this fraction to 0.7%.

3.1 Piecewise Integral Images

The technique we propose for reduction of numerical errors is based on integral images that are defined *piecewise*. We partition every integral image into a number of adjacent pieces, say of size $W \times W$ (border pieces may be smaller due to remainders), where W is chosen to exceed the maximum allowed width for the sliding window. Each piece obtains its own "private" coordinate system. Informally speaking, the (j, k) indexes that are present in formula (14) become reset to $(0, 0)$ at top-left corners of successive pieces. Similarly, the values cumulated so far in each integral image $ii_{t,u}$ become zeroed at those points. Algorithm 1 demonstrates this construction. During detection procedure, global coordinates (j, k) can still be used in main loops to traverse the image, but once the window position gets fixed, say at (j_0, k_0), then we shall recalculate that position to new coordinates (j_0', k_0') valid for the current piece in the following manner:

$$N = \lfloor j_0/W \rfloor, \qquad\qquad M = \lfloor k_0/W \rfloor. \qquad (19)$$
$$j_0' = j_0 - N \cdot W, \qquad\qquad k_0' = k_0 - M \cdot W. \qquad (20)$$

Note that due to the introduced partitioning, the sliding window may cross partitioning boundaries, and reside in either: one, two or four pieces of integral images. That last situation is illustrated in Fig. 2. Therefore, in the general case, the outcomes of growth operations Δ for the whole window will have to be combined from four parts denoted in the figure by P_1, P_2, P_3, P_4. An important role in that context will be played by the central index (j_c, k_c). In the original approach its position was calculated only once, using global coordinates and formula (12). The new technique requires that we "see" the central index differently from the point of view of each part P_i. We give the suitable formulas below, treating the first part P_1 as reference.

$$j_{c,P_1} = (2j_0' + w - 1)/2, \qquad\qquad k_{c,P_1} = (2k_0' + w - 1)/2. \qquad (21)$$
$$j_{c,P_2} = j_{c,P_1}, \qquad\qquad k_{c,P_2} = k_{c,P_1} - W. \qquad (22)$$
$$j_{c,P_3} = j_{c,P_1} - W, \qquad\qquad k_{c,P_3} = k_{c,P_1}. \qquad (23)$$
$$j_{c,P_4} = j_{c,P_1} - W, \qquad\qquad k_{c,P_4} = k_{c,P_1} - W. \qquad (24)$$

Note that the above formulation can, in particular, yield negative coordinates. More precisely, depending on the window position and the P_i part, the coordinates of the central index can range within: $-W+w/2 < j_{c,P_i}, k_{c,P_i} < 2W-w/2$.

Algorithm 1. Piecewise integral image

procedure PIECEWISEINTEGRALIMAGE(f, t, u, W)
    create array $ii_{t,u}$ of size $n_x \times n_y$ and auxiliary array ii of size n_y
    $k := 0$, $j := 0$
    for $x := 0, \ldots, n_x - 1$ **do**
        if $x \bmod W = 0$ **then**
            $k := 0$
        for $y := 0, \ldots, n_y - 1$ **do**
            if $y \bmod W = 0$ **then**
                $j := 0$
            $s := f(x,y)(k - ij)^t (k + ij)^u$
            if $j > 0$ **then**
                $ii[y] := ii[y - 1] + s$
            else
                $ii[y] := s$
            if $k > 0$ **then**
                $ii_{t,u}[x, y] := ii_{t,u}[x - 1, y] + ii[y]$
            else
                $ii_{t,u}[x, y] := ii[y]$
            $j := j + 1$
        $k := k + 1$
    return $ii_{t,u}$

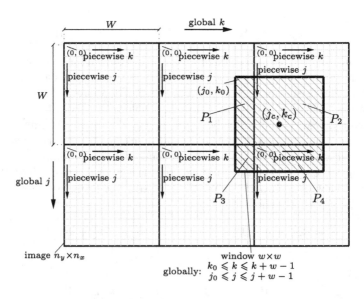

Fig. 2. Illustration of piecewise integral images. Growth operations must be in general combined from four summands corresponding to parts P_1, P_2, P_3, P_4 using redefined coordinates of the central index (j_c, k_c).

3.2   Numerically Safe Formula for ZMs

As the starting point for the derivation we use a variant of formula (11), taking into account the general window position with global pixel indexes ranging as follows: $j_0 \leqslant j \leqslant j_0+w-1$, $k_0 \leqslant k \leqslant k_0+w-1$. We substitute $\frac{4p+2o+2}{\pi w^2}$ into γ.

$$\widehat{M}_{2p+o,2q+o} = \gamma \sum_{\substack{j_0\leqslant j\leqslant j_0+w-1 \\ k_0\leqslant k\leqslant k_0+w-1}} f(j,k) \sum_{q\leqslant s\leqslant p} \beta_{2p+o,2q+o,p-s} \Big(\underbrace{\frac{\sqrt{2}}{w}(k-k_c)}_{x_k} + i\underbrace{\frac{\sqrt{2}}{w}(j_c-j)}_{y_j} \Big)^{s-q}$$

$$\cdot \Big(\underbrace{\frac{\sqrt{2}}{w}(k-k_c)}_{x_k} - i\underbrace{\frac{\sqrt{2}}{w}(j_c-j)}_{y_j} \Big)^{s+q+o} = \gamma \sum_{\substack{j_0\leqslant j\leqslant j_0+w-1 \\ k_0\leqslant k\leqslant k_0+w-1}} f(j,k) \sum_{q\leqslant s\leqslant p} \beta_{2p+o,2q+o,p-s} \Big(\frac{\sqrt{2}}{w}\Big)^{2s+o}$$

$$\cdot \big(k-k_c + i(j_c-j) \big)^{s-q} \big(k-k_c - i(j_c-j) \big)^{s+q+o} = \gamma \sum_{P\in\{P_1,...,P_4\}} \sum_{(j_P,k_P)\in P} f_P(j_P,k_P)$$

$$\cdot \sum_{q\leqslant s\leqslant p} \beta_{2p+o,2q+o,p-s} \Big(\frac{\sqrt{2}}{w}\Big)^{2s+o} \big(k_P-k_{c,P}+i(j_{c,P}-j_P) \big)^{s-q} \big(k_P-k_{c,P}-i(j_{c,P}-j_P) \big)^{s+q+o}$$

$$= \gamma \sum_{q\leqslant s\leqslant p} \beta_{2p+o,2q+o,p-s} \Big(\frac{\sqrt{2}}{w}\Big)^{2s+o} \sum_{t=0}^{s-q} \binom{s-q}{t} \sum_{u=0}^{s+q+o} \binom{s+q+o}{u} \sum_{P\in\{P_1,...,P_4\}} (-k_{c,P}+ij_{c,P})^{s-q-t}$$

$$\cdot (-k_{c,P}-ij_{c,P})^{s+q+o-u} \underbrace{\sum_{(j_P,k_P)\in P} f_P(j_P,k_P)\,(k_P-ij_P)^t\,(k_P+ij_P)^u}_{\underset{P}{\Delta}(ii_{t,u})} . \tag{25}$$

The second pass in the derivation of (25) splits the original summation into four smaller summations over window parts lying within different pieces of integral images, see Fig. 2. We write this down for the most general case when the window crosses partitioning boundaries along both axes. We remind that three simpler cases are also possible: $\{P_1\}$ (no crossings), $\{P_1,P_2\}$ (only vertical boundary crossed), $\{P_1,P_3\}$ (only horizontal boundary crossed). The parts are directly implied by: the offset (j_0,k_0) of detection window, its width w and pieces size W. For strictness, we could have written a functional dependence of form $P(j_0,k_0,w,W)$, which we skip for readability. Also in this pass we switch from global coordinates (j,k) to piecewise coordinates (j_P,k_P) valid for the given part P. The connection between those coordinates can be expressed as follows

$$j = \begin{cases} N{\cdot}W+j_P, & P \in \{P_1,P_2\}; \\ (N{+}1){\cdot}W+j_P, & P \in \{P_3,P_4\}; \end{cases} \quad k = \begin{cases} M{\cdot}W+k_P, & P \in \{P_1,P_3\}; \\ (M{+}1){\cdot}W+k_P, & P \in \{P_2,P_4\}. \end{cases} \tag{26}$$

In the third pass we apply two binomial expansions. In both of them we distinguish two groups of terms: $-k_{c,P}\pm ij_{c,P}$ (independent of the current pixel index) and $k_P \mp ij_P$ (dependent on it). Lastly, we change the order of summations and apply the constant-time Δ operation.

The key idea behind formula (25), regarding its numerical stability, lies in the fact that it uses decreased values of indexes j_P, k_P, $j_{c,P}$, $k_{c,P}$, comparing to the original approach, while maintaining the same differences $k_P - k_{c,P}$ and $j_{c,P} - j_P$. Recall the initial expressions $(k - k_c)\sqrt{2}/w$ and $(j_c - j)\sqrt{2}/w$ and note that one can introduce arbitrary shifts, e.g. $k := k \pm \alpha$ and $k_c := k_c \pm \alpha$ (analogically for the pair: j, j_c) as long as differences remain intact. Later, when indexes in such a pair become separated from each other due to binomial expansions, their numerical behaviour is safer because the magnitude orders are decreased.

Features with high errors can be regarded as damaged, or even useless for machine learning. In Table 1 and Fig. 3 we report percentages of such features for a given image, comparing the original and the numerically safer approach. The figure shows how the percentage of features with relative errors greater than 0.25, increases as the sliding window moves towards the bottom-right image corner.

Table 1. Percentages of damaged features (with relative errors at least 0.1, 0.25, 0.5) for a 640×480 image and different feature spaces and window sizes. Percentage values averaged over all possible window positions. Smaller percentages marked by gray color.

Type of integral images and window size	$\rho = \varrho = 6$ Relative errors:			$\rho = \varrho = 8$ Relative errors:			$\rho = \varrho = 10$ Relative errors:			$\rho = \varrho = 12$ Relative errors:		
	0.1	0.25	0.5	0.1	0.25	0.5	0.1	0.25	0.5	0.1	0.25	0.5
Original, 48×48	0.020	0.004	0.001	10.95	8.280	6.521	32.75	29.71	27.49	49.14	46.62	44.74
Piecewise, 48×48	0	0	0	0.0009	0	0	1.358	0.673	0.344	12.34	9.821	8.090
Original, 56×56	0.002	0	0	6.915	4.882	3.590	27.68	24.67	22.45	44.79	42.18	40.22
Piecewise, 56×56	0	0	0	0	0	0	0.370	0.128	0.049	8.039	5.929	4.539
Original, 68×68	0	0	0	3.187	1.887	1.120	21.08	18.07	15.86	38.79	35.95	33.80
Piecewise, 68×68	0	0	0	0	0	0	0.027	0.007	0.002	3.704	2.307	1.513
Original, 82×82	0	0	0	0.958	0.389	0.162	14.52	11.71	9.792	32.21	29.19	26.97
Piecewise, 82×82	0	0	0	0	0	0	0.001	0	0	1.179	0.560	0.258

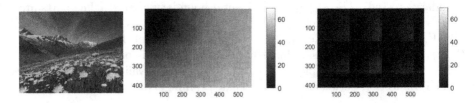

Fig. 3. Percentage of features with relative error at least 0.25 for each possible position of 68×68 sliding window: (left) input image of size 640×480, (middle) original approach, (right) piecewise approach with $W = 172$. Feature space settings: $\rho = \varrho = 12$.

4 Experiments

In subsequent experiments we apply *RealBoost* (RB) learning algorithm producing ensembles of weak classifiers: shallow *decision trees* (RB+DT) or *bins*

(RB+B), for details see [2,5,11]. Ensemble sizes are denoted by T. Two types of feature spaces are used: based solely on *moduli* of Zernike moments (ZMs-M), and based on *extended* Zernike product invariants (ZMs-E); see formulas (7), (8), respectively. The '-NER' suffix stands for: *numerical errors reduction*. Feature counts are specified in parenthesis. Hence, e.g. ZMs-E-NER (14250) represents extended Zernike invariants involving over 14 thousand features, computed according to formula (25) based on piecewise integral images.

Experiment: "Synthetic A letters" For this experiment we have prepared a synthetic data set based on font material gathered by T.E. de Campos et al. [2,3]. A subset representing 'A' letters was treated as positives (targets). In train images, only objects with limited rotations were allowed ($\pm 45°$ with respect to their upright positions). In contrast, in test images, rotations within the full range of $360°$ were allowed. Table 2 lists details of the experimental setup.

Table 2. "Synthetic A letters": experimental setup.

Train data		Test data		Detection procedure	
Quantity/Parameter	Value	Quantity/Parameter	Value	Quantity/Parameter	Value
No. of positives	20 384	No. of positives	417	Image resolution	600 × 480
No. of negatives	323 564	No. of negatives	3 745 966	No. of detection scales	5
Total set size	343 948	Total set size	3 746 383	Window growing coef.	1.2
		No. of images	200	Smallest window	100 × 100
				Largest window size	208 × 208
				Window jumping coef.	0.05

We report results in the following forms: examples of detection outcomes (Fig. 4), ROC curves over logarithmic FAR axis (Fig. 5a), test accuracy measures for selected best detectors (Table 4). Accuracy results were obtained by a batch detection procedure on 200 test images including 417 targets (within the total of 3 746 383 windows).

The results show that detectors based on features obtained by the numerically safe formula achieved higher classification quality. This can be well observed in ROCs, where solid curves related to the NER variant dominate dashed curves, and is particularly evident for higher orders ρ and ϱ (blue and red lines in Fig. 5a). These observations are also confirmed by Table 4

Fig. 4. "Synthetic A letters": detection examples. Two last images show error examples—a misdetection and a false alarm (yellow). (Color figure online)

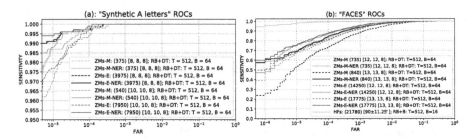

Fig. 5. (a): "Synthetic A letters": ROC curves for test data, (b) "Faces": ROC curves for test data for prescreeners and one example of angle-dependent classifier.

Experiment: "Faces" The learning material for this experiment consisted of 3 000 images with faces (in upright position) looked up using the *Google Images* and produced a train set with 7 258 face examples (positives). The testing phase was a two-stage process, consisting of preliminary and final tests. Preliminary tests were meant to generate ROC curves and make an initial comparison of detectors. For this purpose we used another set of faces (also in upright position) containing 1 000 positives and 2 000 000 negatives. The final batch tests were meant to verify the rotational invariance. For this purpose we have prepared the third set, looking up faces in unnatural rotated positions (example search queries: "people lying down in circle", "people on floor", "face leaning to shoulder", etc.). We stopped after finding 100 such images containing 263 faces in total (Table 3).

Table 3. "Faces": experimental setup.

Train data (upright faces)		Final test data (rotated faces)		Detection procedure	
Quantity/Parameter	Value	Quantity/Parameter	Value	Quantity/Parameter	Value
No. of positives	7 258	No. of positives	263	Image height	480
No. of negatives	100 000	No. of negatives	14 600 464	Window growing coef	1.2
Total set size	107 258	Total set size	14 600 564	Smallest window	48 × 48
		No. of images	100	Largest window	172 × 172
				Window jumping coefficient	0.05

Zernike Invariants as Prescreeners. Preliminary tests revealed that classifiers trained on Zernike features were *not* accurate enough to work as standalone face detectors invariant to rotation. ROC curves (Fig. 5b) indicate that sensitivities at satisfactory levels of ≈90% are coupled with false alarm rates ≈ $2 \cdot 10^{-4}$, which, though improved with respect to [2][3], is still too high to accept. Therefore, we decided to apply obtained detectors as *prescreeners* invariant to rotation. Candidate windows selected via prescreening[4] were then analyzed by angle-dependent classifiers. We prepared 16 such classifiers, each responsible

---

[3] In [2] the analogical false alarm rates were at the level ≈ $5 \cdot 10^{-3}$.

[4] We chose decision thresholds for prescreeners to correspond to 99.5% sensitivity.

for an angular section of $22.5°$, trained using $10\,125$ HFs and RB+B algorithm. Figure 5b shows an example of ROC for an angle-dependent classifier ($90° \pm 11.25°$). The prescreening eliminated about 99.5% of all windows for ZMs-M and 97.5% for ZMs-E.

Table 4. Batch detection results for: "Synthetic A letters" and "Faces".

"Synthetic A letters"					"Faces"				
Feature space (no. of feats.)	Sensitivity	FAR per image	FAR per window $[\cdot 10^{-6}]$	Accuracy per window	Feature space (no. of feats.)	Sensitivity	FAR per image	FAR per window $[\cdot 10^{-7}]$	Accuracy per window
ZMs-M (375)	.979	8/200	2.13	.999995734	ZMs-M (735)	.890	14/100	9.6	.999997054
ZMs-M-NER (375)	.979	8/200	2.13	.999995734	ZMs-M-NER (735)	.905	16/100	11.0	.999997192
ZMs-M (540)	.966	17/200	4.53	.999992001	ZMs-M (840)	.897	15/100	10.3	.999997123
ZMs-M-NER (540)	.987	8/200	2.13	.999996534	ZMs-M-NER (840)	.905	8/100	5.5	.999997740
ZMs-E (3975)	.982	3/200	0.80	.999997334	ZMs-E (14250)	.867	6/100	4.1	.999997192
ZMs-E-NER (3975)	.979	2/200	0.53	.999997334	ZMs-E-NER (14250)	.878	8/100	5.5	.999997260
ZMs-E (7950)	.971	9/200	2.40	.999994667	ZMs-E (17775)	.897	6/100	4.1	.999997740
ZMs-E-NER (7950)	.992	0/200	0	.999999200	ZMs-E-NER (17775)	.875	10/100	6.8	.999997055

Table 4 reports detailed accuracy results, while Fig. 6 shows detection examples. It is fair to remark that this experiment turned out to be much harder than the former one. Our best face detectors based on ZMs and invariant to rotation have the sensitivity of about 90% together with about 7% of false alarms per image, which indicates that further fine-tuning or larger training sets should be considered. As regards the comparison between standard and NER variants, as previously the ROC curves show a clear advantage of NER prescreeners (solid curved dominate their dashed counterparts). Accuracy results

Fig. 6. "Faces": detection examples. Images prescreened with ZMs-E features then processed with angle-dependent classifiers (HFs).

presented in Table 4 also show such a general tendency but are less evident. One should remember that these results pertain to combinations: 'prescreener (ZMs) + postscreener (HFs)'. An improvement in the prescreener alone may, but does not have to, influence directly the quality of the whole system.

5 Conclusion

We have improved the numerical stability of floating-point computations for Zernike moments (ZMs) backed with piecewise integral images. We hope this proposition can pave way to more numerous detection applications involving ZMs. A possible future direction for this research could be to analyze the computational overhead introduced by the proposed approach. Such analysis can be carried out in terms of *expected value* of the number of needed growth operations.

References

1. Acasandrei, L., Barriga, A.: Embedded face detection application based on local binary patterns. In: 2014 IEEE International Conference on High Performance Computing and Communications (HPCC, CSS, ICESS), pp. 641–644 (2014)
2. Bera, A., Klęsk, P., Sychel, D.: Constant-time calculation of Zernike moments for detection with rotational invariance. IEEE Trans. Pattern Anal. Mach. Intell. **41**(3), 537–551 (2019)
3. de Campos, T.E., et al.: Character recognition in natural images. In: Proceedings of the International Conference on Computer Vision Theory and Applications, Lisbon, Portugal, pp. 273–280 (2009)
4. Dalal, N., Triggs, B.: Histograms of oriented gradients for human detection. In: Conference on Computer Vision and Pattern Recognition (CVPR 2005), vol. 1, pp. 886–893. IEEE Computer Society (2005)
5. Friedman, J., Hastie, T., Tibshirani, R.: Additive logistic regression: a statistical view of boosting. Ann. Stat. **28**(2), 337–407 (2000)
6. Goldberg, D.: What every computer scientist should know about floating-point arithmetic. ACM Comput. Surv. **23**(1), 5–48 (1991)
7. Klęsk, P.: Constant-time fourier moments for face detection—can accuracy of haar-like features be beaten? In: Rutkowski, L., Korytkowski, M., Scherer, R., Tadeusiewicz, R., Zadeh, L.A., Zurada, J.M. (eds.) ICAISC 2017. LNCS (LNAI), vol. 10245, pp. 530–543. Springer, Cham (2017). https://doi.org/10.1007/978-3-319-59063-9_47
8. Malek, M.E., Azimifar, Z., Boostani, R.: Facial age estimation using Zernike moments and multi-layer perceptron. In: 22nd International Conference on Digital Signal Processing (DSP), pp. 1–5 (2017)
9. Mukundan, R., Ramakrishnan, K.: Moment Functions in Image Analysis - Theory and Applications. World Scientific, Singapore (1998)
10. Rajaraman, V.: IEEE standard for floating point numbers. Resonance **21**(1), 11–30 (2016). https://doi.org/10.1007/s12045-016-0292-x
11. Rasolzadeh, B., et al.: Response binning: improved weak classifiers for boosting. In: IEEE Intelligent Vehicles Symposium, pp. 344–349 (2006)
12. Singh, C., Upneja, R.: Accurate computation of orthogonal Fourier-Mellin moments. J. Math. Imaging Vision **44**(3), 411–431 (2012)

13. Sornam, M., Kavitha, M.S., Shalini, R.: Segmentation and classification of brain tumor using wavelet and Zernike based features on MRI. In: 2016 IEEE International Conference on Advances in Computer Applications (ICACA), pp. 166–169 (2016)
14. Viola, P., Jones, M.: Rapid object detection using a boosted cascade of simple features. In: Conference on Computer Vision and Pattern Recognition (CVPR 2001), pp. 511–518. IEEE (2001)
15. Xing, M., et al.: Traffic sign detection and recognition using color standardization and Zernike moments. In: 2016 Chinese Control and Decision Conference (CCDC), pp. 5195–5198 (2016)
16. Yin, Y., Meng, Z., Li, S.: Feature extraction and image recognition for the electrical symbols based on Zernike moment. In: 2017 IEEE 2nd Advanced Information Technology, Electronic and Automation Control Conference (IAEAC), pp. 1031–1035 (2017)
17. Zernike, F.: Beugungstheorie des Schneidenverfahrens und seiner verbesserten Form, der Phasenkontrastmethode. Physica **1**(8), 668–704 (1934)

Effect of Dataset Size on Efficiency of Collaborative Filtering Recommender Systems with Multi-clustering as a Neighbourhood Identification Strategy

Urszula Kużelewska$^{(\boxtimes)}$ (iD)

Faculty of Computer Science, Bialystok University of Technology,
Wiejska 45a, 15-351 Bialystok, Poland
u.kuzelewska@pb.edu.pl

Abstract. Determination of accurate neighbourhood of an active user (a user to whom recommendations are generated) is one of the essential problems that collaborative filtering based recommender systems encounter. Properly adjusted neighbourhood leads to more accurate recommendation generated by a recommender system. In classical collaborative filtering technique, the neighbourhood is modelled by kNN algorithm, but this approach has poor scalability. Clustering techniques, although improved time efficiency of recommender systems, can negatively affect the quality (precision or accuracy) of recommendations.

This article presents a new approach to collaborative filtering recommender systems that focuses on the problem of an active user's neighbourhood modelling. Instead of one clustering scheme, it works on a set of partitions, therefore it selects the most appropriate one that models the neighbourhood precisely. This article presents the results of the experiments validating the advantage of multi-clustering approach, $M - CCF$, over the traditional methods based on single-scheme clustering. The experiments particularly focus on the effect of great size of datasets concerning overall recommendation performance including accuracy and coverage.

Keywords: Multi-clustering · Collaborative filtering · Recommender systems

1 Introduction

Recommender Systems (RSs) are solutions to cope with information overload that is observed nowadays on the Internet. Their goal is to provide filtered data to the particular user [12]. As stated in [25], RSs are a special type of information retrieval to estimate the level of relevance of unknown items to a particular user and to order them according to the relevance.

There are non-personalized recommenders based on e.g. average customers' ratings as well as personalized systems predicting preferences based on analysing

© Springer Nature Switzerland AG 2020
V. V. Krzhizhanovskaya et al. (Eds.): ICCS 2020, LNCS 12139, pp. 342–354, 2020.
https://doi.org/10.1007/978-3-030-50420-5_25

users' behaviour. The most popular RSs are collaborative filtering methods (CF) that build a model on users and the items which the users were interested in [1]. The model's data are preferences e.g. visited or purchased items, ratings [20]. Then CF search for the similarities in the model to generate a list of suggestions that fit users' preferences [20].

They are based on either user-based or item-based similarity to make recommendations. The item-based approach usually generates more relevant recommendations since it uses user's ratings [23] - there are identified similar items to a target item, and the user's ratings on those items are used to extrapolate the ratings of the target. This approach is more resistant to changes in the ratings, as well, because usually the number of users is considerably greater than the number of items and new items are less frequently added to the dataset [2].

During recommendations generation, a huge amount of data is processed. To improve time efficiency and make it possible to generate proposition lists in real time, RSs reduce the search space around an active user to its closest neighbourhood. A traditional method for this purpose is k Nearest Neighbours (kNN) [4]. It calculates all user-user or item-item similarities and identifies the most k similar objects (users or items) to the target object as its neighbourhood. Then, further calculations are performed only on objects from the neighbourhood improving the time of processing. The kNN algorithm is a reference method used in order to determine the neighbourhood of an active user for the collaborative filtering recommendation process [8]. Simplicity and reasonably accurate results are its advantages; its disadvantages are low scalability and vulnerability to sparsity in data [24].

Clustering algorithms can be an efficient solution to the disadvantages of kNN approach due to the neighbourhood is shared by all cluster members. The problems are: the results can be different as the most of clustering methods are non-deterministic and usually significant loss of prediction accuracy. Multi-clustering approach, instead of one clustering scheme, works on a set of partitions, therefore it selects the most appropriate one that models the neighbourhood precisely, thus reducing the negative impact of non-determinism.

The article is organised as follows: the first section presents problems with scalability occurring in collaborative filtering Recommender Systems with a solution based on clustering algorithms, including their advantages and disadvantages. Next section describes the proposed multi-clustering algorithm on the background of alternative clustering techniques, whereas the following section contains results of performed experiments to compare multi-clustering and single-clustering approaches. The last section concludes the paper.

2 Background and Related Work

Clustering is a part of Machine Learning domain. The aim of clustering methods is to organize data into separate groups without any external information about their membership, such as class labels. They analyse only the relationship among the data, therefore clustering belongs to Unsupervised Learning techniques [13].

Due to independent *á priori* clusters identification, clustering algorithms are an efficient solution to the problem of RSs scalability, providing for recommendation process a pre-defined neighbourhood [21]. Recently, clustering algorithms have drawn much attention of researchers and there were proposed new algorithms, particularly developed for recommender systems application [6,16,22]. The efficiency of clustering techniques is related to the fact, that a cluster is a neighbourhood that is shared by all the cluster members, in contrast to kNN approach determining neighbours for every object separately [2]. The disadvantage of this approach is usually loss of prediction accuracy.

The explanation for decreasing recommendations accuracy is in the way how clustering algorithms work. A typical approach is based on a single partitioning scheme, which is generated once and then not updated significantly. There are two major problems related to the quality of clustering. The first is the clustering results depend on the input algorithm parameters, and additionally, there is no reliable technique to evaluate clusters before on-line recommendations process. Moreover, some clustering schemes may better suit to some particular applications, whereas other clustering schemes perform better in other solutions [28]. The other issue addressed to decreasing prediction accuracy is imprecise neighbourhood modelling of the data located on borders of clusters [14,18].

Popular clustering technique is $k - means$ due to its simplicity and high scalability [13]. It is often used in CF approach [21]. A variant of $k - means$ clustering, bisecting $k - means$, was proposed for privacy-preserving applications [7] and web-based movie RS [21]. Another solution, ClustKNN [19] was used to cope with large-scale RS applications. However, the $k - means$ approach, as well as many other clustering methods, do not always result in clustering convergence. Moreover, they require input parameters e.g. a number of clusters, as well.

The disadvantages described above can be solved by techniques called alternate clustering, multi-view clustering, multi-clustering or co-clustering. They include a wide range of methods which are based on widely understood multiple runs of clustering algorithms or multiple application of clustering process on different input data [5].

Multi-clustering or co-clustering have been applied to improve scalability in the domain of RSs. Co-clustering discovers samples that are similar to one another with respect to a subset of features. As a result, interesting patterns (co-clusters) are identified unable to be found by traditional one-way clusterings [28]. Multiple clustering approaches discover various partitioning schemes, each capturing different aspects of the data [3]. They can apply one clustering algorithm changing values of input parameters or distance metrics, as well as they can use different clustering techniques to generate a complementary result [28].

The role of multi-clustering in the recommendations generation process that is applied in the approach described in this article, is to determine the most appropriate neighbourhood for an active user. It means that the algorithm selects the best cluster from a set of clusters prepared previously (see the following Section).

A method described in [18] combines both content-based and collaborative filtering approaches. The system uses multi-clustering, however, it is interpreted as clustering of a single scheme on both techniques. It groups the ratings, to create an item group-rating matrix and a user group-rating matrix. As a clustering algorithm, it uses $k - means$ combined with a fuzzy set theory to represent the level of membership of an object to the cluster. Then a final prediction rating matrix is calculated to represent the whole dataset. In the last step of pre-recommendation process $k - means$ is used again on the new rating matrix to find a group of similar users. The groups represent the neighbourhood of users to limit the search space for a collaborative filtering method. It is difficult to compare this approach to the other techniques including single-clustering ones because the article [18] describes the experiments on the unknown dataset containing only 1675 ratings.

The other solution is presented in [26]. The authors observed, that users might have different interests over topics, thus they might share similar preferences with different groups of users over different sets of items. The method $CCCF$ (Co-Clustering For Collaborative Filtering) first clusters users and items into several subgroups, where the each subgroup includes a set of like-minded users and the set of items in which these users share their interests. The groups are analysed by collaborative filtering methods and the result recommendations are aggregated over all the subgroups. This approach has advantages like scalability, flexibility, interpretability and extensibility.

Other applications are: accurate recommendations of tourist attractions based on a co-clustering and bipartite graphs theory [27] and $OCuLaR$ (Overlapping co-CLuster Recommendation) [11] - an algorithm for processing very large databases, detecting co-clusters among users and items as well as providing interpretable recommendations.

There are some other methods, which can be generally called as multi-view clustering, that find partitioning schemes on different data (e.g. ratings and text description) combining results after all ([5,17]). The main objective of a multi-view partitioning is to provide more information about the data in order to understand them better by generating distinct aspects of the data and searching for the mutual link information among the various views [10]. It is stated, that single-view data may contain incomplete knowledge while multi-view data fill this gap by complementary and redundant information [9]. It is rather useful in interpretability aspect developing in Recommender Systems [11].

3 Description of M-CCF Algorithm

The approach presented in this article has a name Multi-Clustering Collaborative Filtering ($M - CCF$) and defines a multi-clustering process as generation of a set of clustering results obtained from an arbitrary clustering algorithm with the same data on its input. The advantage of this approach is a better quality of the neighbourhood modelling, leading to the high quality of predictions, keeping real-time effectiveness provided by clustering methods. The explanation is in imprecise neighbourhood modelling of the data located on borders of the clusters. The border objects have fewer neighbours in their closest area than the objects located in the middle of a cluster. The multi-clustering technique selects the most appropriate cluster to the particular data object. The most appropriate means the one that includes the active object in the closest distance to the cluster's center, thus delivering more neighbours around it. A more detailed description of this phenomenon is in [14,15].

The general algorithm $M - CCF$ is presented in Algorithm 1. The input set contains data of n users, who rated a subset of items - $A = \{a_1, \ldots, a_k\}$. The set of possible ratings - V - contains values $v_1, \ldots, v_c$. The input data are clustered ncs times into nc clusters every time, giving, as a result, a set of clustering schemes CS. Finally, the algorithm generates a list of recommendations R_{x_a} for the active user.

Algorithm 1: A general algorithm $M - CCF$ of a recommender system based on multi-clustering used in the experiments

---

Data:

- $U = (X, A, V)$ - matrix of clustered data, where $X = \{x_1, \ldots, x_n\}$ is a set of users, $A = \{a_1, \ldots, a_k\}$ is a set of items and $V = \{v_1, \ldots, v_c\}$ is a set of ratings values,
- $\delta : v \in V$ - a similarity function,
- $nc \in [2, n]$ - a number of clusters,
- $ncs \in [2, \infty]$ - a number of clustering schemes,
- $CS = \{CS_1, \ldots, CS_{ncs}\}$ - a set of clustering schemes,
- $CS_i = \{C_1, \ldots, C_{nc}\}$ - a set of clusters for a particular clustering scheme,
- $CS_r = \{c_{r,1}, \ldots, c_{r,nc \cdot ncs}\}$ - the set of cluster centres,

Result:

- A_{Rx_a} - a list of recommended items for an active user x_a,

begin

   $\delta_1 .. \delta_{ncs} \longleftarrow$ calculateSimilarity(CS_r, CS_i, δ);
   $C_{best_{x_a}} \longleftarrow$ findTheBestCluster$(x_a, CS_r, \delta_1 .. \delta_{ncs \cdot ncs}, CS_r, CS_i)$;
   $R_{x_a} \longleftarrow$ recommend$(x_a, C_{best_{x_a}}, \delta_1 .. \delta_{nc \cdot ncs})$;

The set of groups is identified by the clustering algorithm which is run several times with the same or different values of its input parameters. In the experiments described in this article, $k - means$ was used as a clustering method. The set of clusters provided for the collaborative filtering process was generated with the same parameter k (a number of clusters). This step, although time-consuming, has a minor impact on overall system scalability, because it is performed rarely and in an off-line mode.

After the neighbourhood identification, the following step, appropriate recommendation generations, is executed. This process requires, despite great precision, high time effectiveness. The multi-clustering approach satisfies these two conditions because it can select the most suitable neighbourhood area of an active user for candidates searching and the neighbourhood of all objects is already determined, as well.

One of the most important issues of this approach is to generate a wide set of input clusters that is not very numerous in the size, thus providing a high similarity for every user or item. The other matter concerns matching users with the best clusters as their neighbourhood. It can be obtained in the following ways. The first of them compares the active user's ratings with the cluster centers' ratings and searches for the most similar one using a certain similarity measure. The other way, instead of the cluster centers, compares the active user with all cluster members and selects the one with the highest overall similarity. Both solutions have their advantages and disadvantages, e.g. the first one works well for clusters of spherical shapes, whereas the other one requires higher time consumption. In the experiments presented in this paper, the clusters for active users are selected based on their similarity to the centers of groups (see Algorithm 2).

Algorithm 2: Algorithm of cluster selection of $M - CCF$ recommender system used in the experiments

---

Data:

- $U = (X, A, V)$ - matrix of clustered data, where x_a is an active user, $A = \{a_1, \ldots, a_k\}$ is a set of items and $V = \{v_1, \ldots, v_c\}$ is a set of ratings values,
- $\delta : v \in V$ - a similarity function,
- $CS = \{CS_1, \ldots, CS_{ncs}\}$ - a set of clustering schemes,
- $CS_i = \{C_1, \ldots, C_{nc}\}$ - a set of clusters for a particular clustering scheme,
- $CS_r = \{c_{r,1}, \ldots, c_{r,nc \cdot ncs}\}$ - the set of cluster centres,

Result:

- $C_{best_{x_a}}$ - the best cluster for an active user x_a,
- δ_{best} - a matrix of similarity within the best cluster

begin

$\quad \delta_1 .. \delta_{ncs \cdot ncs} \longleftarrow$ calculateSimilarity(x_a, CS_r, δ);
$\quad \delta_{best} \longleftarrow$ selectTheHighestSimilarity$(\delta_1 .. \delta_{ncs})$;
$\quad C_{best_{x_a}} \longleftarrow$ findTheBestCluster$(\delta_{best}, CS, CS_i)$;

Afterwards, a recommendations generation process works as a typical collaborative filtering approach, although the candidates are searched only within the selected cluster of the neighbourhood.

4 Experiments

Evaluation of the performance of the proposed algorithm M-CCF was conducted on the MovieLens dataset [30]. The original set is composed of 25 million ratings;

however two subsets were used in the experiments: a small dataset - $100k$ and a big dataset - $10M$. The parameters of the subsets are presented in Table 1.

Table 1. Description of the datasets used in the experiments.

Dataset	Number of ratings	Number of users	Number of items
Small dataset - $100k$	100 415	534	11109
Big dataset - $10M$	1 000 794	4537	16767

The results obtained with the algorithm $M - CCF$ were compared with the recommender system whose neighbourhood identification is based on a single-clustering. The attention was paid to the precision and completeness of recommendation lists generated by the systems. The evaluation criteria were related to the following baselines: Root Mean Squared Error ($RMSE$) described by (1) and $Coverage$ described by (2) (in %). The symbols in the equations, as well as the method of calculation are characterised in details below.

$$RMSE = \frac{\sum_{i=1}^{N} |r_{real}(x_i) - r_{est}(x_i)|}{N} \tag{1}$$

$$Coverage = \frac{\sum_{i=1}^{N} r_{est}(x_i) \in \mathbb{R}_+}{N} \cdot 100\% \tag{2}$$

where $\mathbb{R}_+$ stands for a set of positive real numbers. The performance of both approaches was evaluated in the following way. Before the clustering step, the whole input dataset was split into two parts: training and testing. In the case of $100k$ subset, the parameters of a testing part were as follows: 393 ratings, 48 users, 354 items, whereas the case of $10M$ subset: 432 ratings, 44 users and 383 items. This step provides the same testing data during experiments and makes the comparison more objective.

In the evaluation process, the values of ratings from the testing part were removed and estimated by the recommender systems. The difference between the original and the calculated value (represented respectively as $r_{real}(x_i)$ and $r_{est}(x_i)$ for user x_i and a particular item i) was taken for $RMSE$ calculation. The number of ratings is denoted as N in the equations. The lower value of $RMSE$ stands for a better prediction ability.

During the evaluation process, there were the cases in which estimation of ratings was not possible. It occurs when the item for which the calculations are performed, is not present in the clusters which the items with existing ratings belong to. It is considered in $Coverage$ index (2). In every experiment, it was assumed that the $RMSE$ is significant if the value of $Coverage$ is greater than 80%. It means that if the number of users to whom the recommendations were calculated was 48 and for each of them it was expected to estimate on average 5 ratings, therefore at least 192 values should be present in the recommendation lists.

The clustering method, similarity and distance measures were taken from Apache Mahout environment [29]. To achieve the comparable time evaluation, in implementation of the multi-clustering algorithm, data models (FileData-Model) and structures (FastIDMap, FastIDSet) derived from Apache Mahout were taken, as well. The following data models were implemented: ClusteringDataModel and MultiClusteringDataModel that implement the interface of DataModel. The appropriate recommender and evaluator classes were implemented, as well.

The first experiment was performed on $100k$ dataset, that was clustered independently five times into 10 groups. The clustering algorithm was $k - means$ and a distance measure - $cosine$ value between the vectors formed from data points. The number of groups (10) was determined experimentally as an optimal value that led to the highest values of $Coverage$ in the recommendations. In every case, a new recommender system was built and evaluated. Table 2 contains evaluation of the systems' precision that was run with the following similarity indices: $Cosine - based$, $LogLikelihood$, $Pearson\ correlation$, $Euclidean$ distance-based, $CityBlock$ distance-based and $Tanimoto$ coefficient. In the tables below they are represented by the following shortcuts respectively: $Cosine$, $LogLike$, $Pearson$, $Euclidean$, $CityBlock$, $Tanimoto$. The $RMSE$ values are presented with a reference value in brackets that stands for $Coverage$ in this case.

Table 2. RMSE of item based collaborative filtering recommendations with the neighbourhood determined by a single (5 different runs of $k - means$ algorithm) as well as multi-clustering ($k - means$ with $cosine - based$ distance measure) for a small dataset. The best values are in bold.

Similarity measure	Single clustering					Multi-clustering
Cosine	0.88(83%)	0.9(87%)	0.88(81%)	0.89(85%)	**0.87(85%)**	0.87(83%)
LogLike	0.89(87%)	0.88(81%)	0.89(85%)	0.88(81%)	**0.86(85%)**	0.9(83%)
Pearson	–	–	–	–	–	–
Euclidean	0.89(87%)	0.87(81%)	0.88(85%)	0.87(81%)	**0.87(85%)**	**0.85(83%)**
CityBlock	0.87(85%)	0.89(87%)	0.88(81%)	0.89(85%)	**0.86(87%)**	0.88(81%)
Tanimoto	**0.87(87%)**	0.86(81%)	**0.87(85%)**	0.87(81%)	**0.85(85%)**	–

It is visible that the values are different for different input data, although the number of clusters is the same in every result. As an example, the recommender system with $Cosine - based$ similarity has $RMSE$ in the range from 0.87 to 0.9. The difference in values may seem to be small, but the table contains only values whose $Coverage$ was high enough. Different values of $RMSE$ mean that the precision of a recommender system depends on the quality of a clustering scheme. There is no guarantee that the scheme selected for recommendation process is optimal. Table 2 contains performance results of the recommender system that has the neighbourhood determined by the multi-clustering approach. There is a case where the precision is better (for the *Euclidean* distance based similarity), but in the majority of cases it is slightly worse. Despite this, the multi-clustering approach has eliminated the ambiguity of clustering scheme selection.

The goal of the other experiment was to examine the influence of a distance measure used in the clustering process on a final recommender system performance. The dataset, as well as the similarity measures or a number of clusters, remained the same; however, the distance between the data points was measured by the *Euclidean* distance. The results are presented in Table 3. In this case, one can observe the same values of *RMSE* regardless of the similarity measure. Note, that the $M - CCF$ algorithm generated results identical to the values from the single-clustering approach.

Table 3. RMSE of item based collaborative filtering recommendations with the neighbourhood determined by a single (5 different runs of $k - means$ algorithm) as well as multi-clustering ($k - means$ with the *Euclidean* distance measure) for a small dataset. The best values are in bold.

Similarity measure	Single clustering					Multi-clustering
Cosine	0.85(83%)	0.85(83%)	0.85(83%)	0.85(83%)	0.85(83%)	0.85(83%)
LogLike	**0.84(83%)**	**0.84(83%)**	**0.84(83%)**	**0.84(83%)**	**0.84(83%)**	**0.84(83%)**
Pearson	–	–	–	–	–	–
Euclidean	**0.84(83%)**	**0.84(83%)**	**0.84(83%)**	**0.84(83%)**	**0.84(83%)**	**0.84(83%)**
CityBlock	0.85(83%)	0.85(83%)	0.85(83%)	0.85(83%)	0.85(83%)	0.85(83%)
Tanimoto	**0.84(83%)**	**0.84(83%)**	**0.84(83%)**	**0.84(83%)**	**0.84(83%)**	**0.84(83%)**

The following experiments were performed on the big dataset $(10M)$. By an analogy to the previous ones, the influence of a distance measure was examined, as well. In the first of them, the cosine value between data vectors was taken as a distance measure. The results, for all the similarity indices, are presented in Table 4. The overall performance is worse, although the size of the dataset is considerably greater. There are more cases with insufficient *Coverage* related to

Table 4. RMSE of item based collaborative filtering recommendations with the neighbourhood determined by a single (5 different runs of $k - means$ algorithm) as well as multi-clustering ($k - means$ with $cosine - based$ distance measure) for a big dataset. The best values are in bold.

Similarity measure	Single clustering					Multi-clustering
Cosine	0.99(98%)	–	0.97(93%)	0.98(91%)	–	–
LogLike	0.99(98%)	–	0.97(93%)	0.98(91%)	–	**0.95(95%)**
Pearson	–	–	–	0.98(91%)	–	–
Euclidean	0.98(98%)	–	0.96(93%)	0.97(91%)	–	**0.93(91%)**
CityBlock	0.98(98%)	0.96(95%)	0.98(91%)	0.96(93%)	–	0.97(91%)
Tanimoto	0.97(98%)	0.95(95%)	–	0.96(93%)	0.96(91%)	0.97(91%)

the great size of the data, as well. However, the phenomenon of different precision for various clustering schemes in the case of the single-clustering approach remained and the performance of the $M - CCF$ method improved. The table has bold values only for the multi-clustering column.

Finally, the last experiment was performed on the big dataset $(10M)$ which was clustered based on the *Euclidean* distance. Table 5 contains the results of *RMSE* and *Coverage*. *Coverage* values are visibly higher in this case, even for the *Pearson correlation* similarity index. The performance of the multi-clustering approach is still better than the method based on single-clustering - the *RMSE* values are lower in the majority of cases, in the case of the single-clustering, there is only one scheme that slightly outperforms the $M - CCF$ method.

Table 5. RMSE of item based collaborative filtering recommendations with the neighbourhood determined by a single (5 different runs of $k - means$ algorithm) as well as multi-clustering ($k - means$ with the *Euclidean* distance measure) for a big dataset. The best values are in bold.

Similarity measure	Single clustering					Multi-clustering
Cosine	0.96(93%)	0.97(91%)	0.96(93%)	0.96(93%)	0.95(93%)	**0.94(91%)**
LogLike	0.96(93%)	0.97(91%)	0.96(93%)	0.96(93%)	0.95(93%)	0.96(91%)
Pearson	1.42(93%)	1.09(91%)	0.96(93%)	2.7(91%)	0.99(93%)	**0.93(91%)**
Euclidean	0.95(93%)	0.96(91%)	0.95(93%)	0.95(93%)	**0.94(93%)**	**0.92(93%)**
CityBlock	0.96(95%)	0.96(93%)	0.96(95%)	0.95(95%)	**0.94(95%)**	0.96(93%)
Tanimoto	0.95(93%)	0.95(91%)	0.95(93%)	**0.94(93%)**	**0.93(93%)**	0.95(91%)

Taking into consideration all the experiments presented in this article, it can be observed, that the performance of a recommender system depends on the quality of a clustering scheme provided to the system by a clustering algorithm. In the case of the single-clustering and several schemes generated by this approach, the final precision of recommendations can differ. It means that in order to build a good neighbourhood model for a recommender system, a single run of a clustering algorithm is insufficient. A multi-clustering recommender system and the technique of dynamic selection the most suitable clusters, offers valuable results, particularly in the case of a great size of datasets.

5 Conclusions

In this paper, a developed version of a collaborative filtering recommender system based on multi-clustering neighbourhood modelling is presented. The algorithm $M - CCF$ dynamically selects the most appropriate cluster for every user whom recommendations are generated to. Properly adjusted neighbourhood leads to more accurate recommendations generated by a recommender system. The algorithm eliminates a disadvantage appeared in the case of the neighbourhood determination by a single-clustering method - dependence of the final performance of a recommender system on a clustering scheme selected for the recommendation process.

The experiments described in this paper validated the better performance of the recommender system when the neighbourhood is modelled by the $M - CCF$ algorithm. It was particularly evident in the case of the great dataset containing 10 million ratings. The experiments showed good scalability of the method and increased the competitiveness of the $M - CCF$ algorithm relative to a single-clustering approach in the case of the bigger dataset. Additionally, the technique is free from the negative impact on precision provided by selection of an inappropriate clustering scheme.

The future experiments will be performed to validate the proposed algorithm on different datasets, particularly focused on its great size. It is planned to check the impact of a type of a clustering method on the recommender system's final performance and a mixture of clustering schemes instead of one-algorithm output on an input of the recommender system.

Acknowledgment. The work was supported by the grant from Bialystok University of Technology and funded with resources for research by the Ministry of Science and Higher Education in Poland.

References

1. Tuzhilin, A., Adomavicius, G.: Toward the next generation of recommender systems: a survey of the state-of-the-art and possible extensions. IEEE Trans. Knowl. Data Eng. **17**(6), 734–749 (2005)
2. Aggarwal, C.C.: Time- and location-sensitive recommender systems. Recommender Systems, pp. 283–308. Springer, Cham (2016). https://doi.org/10.1007/978-3-319-29659-3_9
3. Dan, A., Guo, L.: Evolutionary parameter setting of multi-clustering. In: Proceedings of the 2007 IEEE Symposium on Computational Intelligence in Bioinformatics and Computational Biology, pp. 25–31 (2007)
4. Gorgoglione, M., Pannielloa, U., Tuzhilin, A.: Recommendation strategies in personalization applications. Inf. Manag. **56**(6), 103143 (2019)
5. Bailey, J.: Alternative Clustering Analysis: A Review. Intelligent Decision Technologies: Data Clustering: Algorithms and Applications, pp. 533–548. Chapman and Hall/CRC, Boca Raton (2014)
6. Berbague, C.E., Karabadji, N.E.I., Seridi, H.: An evolutionary scheme for improving recommender system using clustering. In: Amine, A., Mouhoub, M., Ait Mohamed, O., Djebbar, B. (eds.) CIIA 2018. IAICT, vol. 522, pp. 290–301. Springer, Cham (2018). https://doi.org/10.1007/978-3-319-89743-1_26
7. Bilge, A., Polat, H.: A scalable privacy-preserving recommendation scheme via bisecting k-means clustering. Inf. Process Manag. **49**(4), 912–927 (2013)
8. Bobadilla, J., Ortega, F., Hernando, A., Gutiérrez, A.: Recommender systems survey. Knowl.-Based Syst. **46**, 109–132 (2013)
9. Ye, Z., Hui, Ch., Qian, H., Li, R., Chen, C., Zheng, Z.: New approaches in multi-view clustering. Recent Appl. Data Clustering **197** (2018). InTechOpen

10. Zhang, G.Y., Wang, C.D., Huang, D., Zheng, W.S.: Multi-view collaborative locally adaptive clustering with Minkowski metric. Exp. Syst. Appl. **86**, 307–320 (2017)
11. Heckel, R., Vlachos, M., Parnell, T., Duenner, C.: Scalable and interpretable product recommendations via overlapping co-clustering. In: IEEE 33rd International Conference on Data Engineering, pp. 1033–1044 (2017)
12. Jannach, D.: Recommender Systems: An Introduction. Cambridge University Press, Cambridge (2010)
13. Kaufman, L.: Finding Groups in Data: An Introduction to Cluster Analysis. John Wiley & Sons, Hoboken (2009)
14. Kużelewska, U.: Collaborative filtering recommender systems based on k-means multi-clustering. In: Zamojski, W., Mazurkiewicz, J., Sugier, J., Walkowiak, T., Kacprzyk, J. (eds.) DepCoS-RELCOMEX 2018. AISC, vol. 761, pp. 316–325. Springer, Cham (2019). https://doi.org/10.1007/978-3-319-91446-6_30
15. Kużelewska, U.: Multi-clustering used as neighbourhood identification strategy in recommender systems. In: Zamojski, W., Mazurkiewicz, J., Sugier, J., Walkowiak, T., Kacprzyk, J. (eds.) DepCoS-RELCOMEX 2019. AISC, vol. 987, pp. 293–302. Springer, Cham (2020). https://doi.org/10.1007/978-3-030-19501-4_29
16. Pireva, K., Kefalas, P.: A recommender system based on hierarchical clustering for cloud e-learning. In: Ivanović, M., Bădică, C., Dix, J., Jovanović, Z., Malgeri, M., Savić, M. (eds.) IDC 2017. SCI, vol. 737, pp. 235–245. Springer, Cham (2018). https://doi.org/10.1007/978-3-319-66379-1_21
17. Mitra, S., Banka, H., Pedrycz, W.: Rough-fuzzy collaborative clustering. IEEE Trans. Syst. Man Cybern. Part B (Cybern.) **36**(4), 795–805 (2006)
18. Puntheeranurak, S., Tsuji, H.: A Multi-clustering hybrid recommender system. In: Proceedings of the 7th IEEE International Conference on Computer and Information Technology, pp. 223–238 (2007)
19. Rashid, M., Shyong, K.L., Karypis, G., Riedl, J.: ClustKNN: a highly scalable hybrid model - & memory-based CF Algorithm. In: Proceeding of WebKDD (2006)
20. Ricci, F., Rokach, L., Shapira, B.: Recommender systems: introduction and challenges. In: Ricci, F., Rokach, L., Shapira, B. (eds.) Recommender Systems Handbook, pp. 1–34. Springer, Boston, MA (2015). https://doi.org/10.1007/978-1-4899-7637-6_1
21. Sarwar, B.: Recommender systems for large-scale e-commerce: scalable neighborhood formation using clustering. In: Proceedings of the 5th International Conference on Computer and Information Technology (2002)
22. Selvi, C., Sivasankar, E.: A novel Adaptive Genetic Neural Network (AGNN) model for recommender systems using modified k-means clustering approach. Multimedia Tools Appl. **78**(11), 14303–14330 (2018). https://doi.org/10.1007/s11042-018-6790-y
23. Schafer, J.B., Frankowski, D., Herlocker, J., Sen, S.: Collaborative filtering recommender systems. In: Brusilovsky, P., Kobsa, A., Nejdl, W. (eds.) The Adaptive Web. LNCS, vol. 4321, pp. 291–324. Springer, Heidelberg (2007). https://doi.org/10.1007/978-3-540-72079-9_9
24. Singh, M.: Scalability and sparsity issues in recommender datasets: a survey. Knowl. Inf. Syst. **62**(1), 1–43 (2018). https://doi.org/10.1007/s10115-018-1254-2
25. Vargas, S.: Novelty and diversity enhancement and evaluation in recommender systems and information retrieval. In: Proceedings of the 37th International ACM SIGIR Conference on Research & Development in Information Retrieval - SIGIR '14, pp. 1281–1281 (2014)

26. Wu, Y., Liu, X., Xie, M., Ester, M., Yang, Q.: CCCF: improving collaborative filtering via scalable user-item co-clustering. In: Proceedings of the Ninth ACM International Conference on Web Search and Data Mining, pp. 73–82 (2016)
27. Xiong, H., Zhou, Y., Hu, C., Wei, X., Li, L.: A novel recommendation algorithm frame for tourist spots based on multi - clustering bipartite graphs. In: Proceedings of the 2nd IEEE International Conference on Cloud Computing and Big Data Analysis, pp. 276–282 (2017)
28. Yao, S., Yu, G., Wang, X., Wang, J., Domeniconi, C., Guo, M.: Discovering multiple co-clusterings in subspaces. In: Proceedings of the 2019 SIAM International Conference on Data Mining, pp. 423–431 (2019)
29. Apache Mahout. http://mahout.apache.org/. Accessed 24 Aug 2019
30. MovieLens 25M Dataset. https://grouplens.org/datasets/movielens/25m/. Accessed 18 Jul 2019

GCN-IA: User Profile Based on Graph Convolutional Network with Implicit Association Labels

Jie Wen[1,2], Lingwei Wei[1,2], Wei Zhou[1,2(✉)], Jizhong Han[1,2], and Tao Guo[1,2]

[1] Institute of Information Engineering, Chinese Academy of Sciences, Beijing, China
zhouwei@iie.ac.cn
[2] School of Cyber Security, University of Chinese Academy of Sciences, Beijing, China

Abstract. Inferring multi-label user profile plays a significant role in providing individual recommendations and exact-marketing, etc. Current researches on multi-label user profile either ignore the implicit associations among labels or do not consider the user and label semantic information in the social networks. Therefore, the user profile inferred always does not take full advantage of the global information sufficiently. To solve above problem, a new insight is presented to introduce implicit association labels as the prior knowledge enhancement and jointly embed the user and label semantic information. In this paper, a graph convolutional network with implicit associations (GCN-IA) method is proposed to obtain user profile. Specifically, a probability matrix is first designed to capture the implicit associations among labels for user representation. Then, we learn user embedding and label embedding jointly based on user-generated texts, relationships and label information. On four real-world datasets in Weibo, experimental results demonstrate that GCN-IA produces a significant improvement compared with some state-of-the-art methods.

Keywords: Implicit association labels · User profile · Graph convolutional networks

1 Introduction

With the growing popularity of online social networks including Weibo and Twitter, the "information overload" [1] come up and the social media platforms took more effort to satisfy users' more individualized demands by providing personalized services such as recommendation systems. User profile, the actual representation to capture certain characteristics about an individual user [25], is the basis of recommendation system [7] and exact-marketing [2, 3]. As a result, user profiling methods, which help obtaining accurate and effective user profiles, have drawn more and more attention from industrial and academic community.

A straightforward way of inferring user profiles is leveraging information from the user's activities, which requires the users to be *active*. However, in many real-world applications a significant portion of users are *passive* ones who keep following and reading

© Springer Nature Switzerland AG 2020
V. V. Krzhizhanovskaya et al. (Eds.): ICCS 2020, LNCS 12139, pp. 355–364, 2020.
https://doi.org/10.1007/978-3-030-50420-5_26

but do not generate any content. As a result, label propagation user profile methods [4–6] are widely studied, which mainly use the social network information rather than user's activities. In order to obtain user profile more accurately and abundantly, multi-label is applied in many researches to describe users' attributes or interests. Different labels were assumed independently [5] in some research, while the associations among labels were ignored and some implicit label features remained hidden. Meanwhile, several researches [1, 8, 9] considered the explicit associations among labels to get user profile and achieved better performance. Besides the explicit associations, there exists implicit association among labels that is beneficial to make user profile more accurate and comprehensive. The previous work [10] leveraged internal connection of labels, which is called implicit association. However, this work only considered the relation of labels, but ignored the user and label semantic information jointly based on user-generated texts, relationships and label information, which is also important for user profile.

To take advantage of this insight, a graph convolutional networks with implicit label associations (GCN-IA) is proposed to get user profile. A probability matrix is first designed to capture the implicit associations among labels for user representation. Then, we learn user embedding and label embedding jointly based on user-generated texts, relationships and label information. Finally, we make multi-label classification based on given user representations to predict unlabeled user profiles. The main contributions of this paper are summarized as follows:

- **Insight.** We present a novel insight about combination among implicit association labels, user semantic information and label semantic information. In online social networks, due to users' personalized social and living habits, there are still certain implicit associations among labels. At the same time, user and label information from user-generated texts, relationships and label information is significant for the construction of user profile.
- **Method.** A graph convolutional networks with implicit label associations (GCN-IA) method is proposed to get user profile. We first construct the social network graph with the relationship between users and design a probability matrix to record the implicit label associations, and then combine this probability matrix with the classical GCN method to embed user and label semantic information.
- **Evaluation.** Experiments evaluating GCN-IA method on 4 real Weibo data sets of different sizes are conducted. The comparative experiments evaluate the accuracy and effectiveness of GCN-IA. The results demonstrate that the performance is significantly improved compared with some previous methods.

The following chapters are organized as follows: In Sect. 2, related works are briefly elaborated. The Sect. 3 describes the details of GCN-IA, and experiments and results are described in Sect. 4. Finally, we summarize the conclusion and future work in Sect. 5.

2   Related Works

Label propagation method shows advantages of linear complexity and less required given user's labels, and disadvantages such as low accuracy and propagation instability. The existing label propagation methods in user profile can be divided into three parts. One is to optimize the label propagating process to obtain more stable and accurate profiles, the second part is to propagate multi-label through social network structure to get more comprehensive user profile, and the last part is to apply deep-learning methods such as GCN to infer multi-label user profile.

2.1 Propagation Optimization

Label propagation method was optimized by leveraging more user attributes information, specifying propagation direction and improving propagation algorithm. Subelj et al. proposed balanced propagation algorithm in which an increasing propagation preferences could decide the update order certain nodes, so that the randomness was counteracted by utilizing node balancers [14]. Ren et al. introduced node importance measurement based on the degree and clustering coefficient information to guide the propagation direction [15]. Li et al. leveraged user attributes information and user attributes' similarity to increase recall ratio of user profile [5]. Huang et al. redefined the label propagating process with a multi-source integration framework that considered content and network information jointly [16]. Explicit associations among labels also have been taken into consideration in some research, Glenn et al. [1] introduced the explicit association labels and the results proved the efficiency of the method.

We innovatively introduced the implicit association labels into multi-label propagation [10], the method was proved to be convergent and faster than traditional label propagation algorithm and its performance was significantly better than the state-of-the-art method on Weibo datasets. However the research [10] ignored user embedding and label embedding jointly based on user-generated texts, relationships and label information, which seemed very important for user profile.

2.2 Multi-label Propagation

The multi-label algorithms were widely applied to get abundant profile. Gregory et al. proposed COPRA algorithm and extended the label and propagation step to more than one community, which means each node could get up to v labels [17]. Zhang et al. used the social relationship to mine user interests, and discovered potential interests from his approach [6]. Xie et al. recorded all the historical labels from the multi-label propagation process, which make the profile result more stable [18]. Wu et al. proposed balanced multi-label propagation by introducing a balanced belonging coefficients p, this method improved the quality and stability of user profile results on the top of COPRA [19].

Label propagation algorithm has been improved in different aspects in the above work, however it's still difficult to get a high accuracy and comprehensive profile due to the lack of input information and the complex community structures.

2.3 GCN Methods

GCN [20] is one of the most popular deep learning methods, which can be simply understood as a feature extractor for graphs. By learning graph structure features through convolutional neural network, GCN is widely used in node classification, graph classification, edge prediction and other research fields. GCN is a semi-supervised learning method, which can infer the classification of unknown nodes by extracting the characteristics of a small number of known nodes and the graph structure. Due to the high similarity with the idea of label propagation, we naturally consider constructing multi-label user profile with GCN. Wu et al. proposed a social recommendation model based on GCN [21], in which both user embedding and item embedding were learned to study how users' interests are affected by the diffusion process of social networks. William et al. [22] and Yao et al. [23] applied GCN for text classification and recommendation systems respectively, with node label and graph structure considered to GCN modeling. However, the existing methods rarely consider the implicit relationships between labels in the GCN based methods.

3   Methodology

3.1   Overview

This section mainly focuses on the improvement of graph convolutional networks (GCN) based on implicit association labels. The goal of this paper is to learn user representation for multi-label user profile task by modeling user-generated text and user relationships.

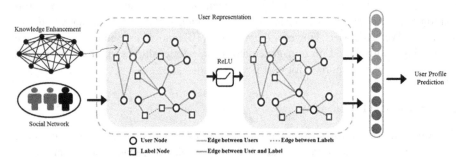

Fig. 1. Overall architecture of GCN-IA.

The overall architecture of GCN-IA is shown in Fig. 1. The model consists of three components: **Prior Knowledge Enhancement (PKE) module, User Representation module**, and **Classification module**. Similar with other graph-based method, we formulated the social network into a heterogeneous graph. In this graph, nodes represent the users in social network and edges represent user's multiple relationships such as following, supporting and forwarding. First, PKE captures the implicit associations among labels for user representation. Then, user representation module learns user embedding and label embedding jointly based on user-generated texts, relationships and label information. Classification module makes multi-label classification based on user representations to predict unlabeled user profiles.

3.2 Prior Knowledge Enhancement Module

Social networks are full of rich knowledge. According to [10], associations among implicit labels are very significant in user profile. In this part, we introduce the knowledge of implicit association among labels to capture the connections among users and their profile labels.

A priori knowledge probability matrix P is defined as Eq. (1). Probability of propagation among labels gets when higher P_{ij} gets a higher value.

$$P_{ij} = \frac{\left|\left\{t | t \in I \wedge \left(l_i, l_j\right) \subseteq t\right\}\right|}{\sum_{i=0}^{m} \sum_{j=0}^{m} \left|\left\{t | t \in I \wedge \left(l_i, l_j\right) \subseteq t\right\}\right|} \tag{1}$$

Associations in social network are complex due to uncertainty [12] or special events [13]. Therefore, we define the set of labels, where elements are sampled by co-occurrence, cultural associations, event associations or custom associations, as shown in Eq. (2).

$$I = I_1 \cup I_2 \cup I_3 \cup \ldots \tag{2}$$

Where $I_i (i = 1, 2, 3, \ldots)$ represents respectively a set of each user's interest label set.

3.3 User Representation Module

Generally, the key idea of GCNs is to learn the iterative convolutional operation in graphs, where each convolutional operation means generating the current node representations from the aggregation of local neighbors in the previous layer. A GCN is a multilayer neural network that operates directly on a graph and induces embedding vectors of nodes based on properties of their neighborhoods.

In the user representation module, we apply GCNs to embed users and profile labels into a vector space and learn user representation and label representation jointly from user-generated content information and social relationships. Specifically, the implicit associations as prior knowledge are introduced to improve the GCNs to model the associations among labels.

Formally, the model considers a social network $G = (V, E)$, where V and E are sets of nodes and edges, respectively. In our model, there are two types of nodes, *user node* and *label node*. The initialized embedding of user nodes and label nodes, denoted as X, is initialized with user name and their content via pre-trained word2vec model.

We build edges among nodes based on user relationships (*user-user edges*), users' profiles (*user-label edges*) and implicit associations among labels (*label-label edges*). We introduce an adjacency matrix A of G. and its degree matrix D, where $D_{ii} = \sum_{j=1,\ldots,n} A_{ij}$. The diagonal elements of A are set to 1 because of self-loops. The weight of the edges between a user node and a label node is based on user profile information, formulated as Eq. (3).

$$A_{ij} = \begin{cases} 1 & \text{if the user } i \text{ is with the label } j \\ 0 & \text{otherwise} \end{cases}, \text{ where } i \in \mathcal{U}_{gold}, j \in \mathcal{C} \tag{3}$$

Where $\mathcal{U}$ is the set of all users in the social network, $\mathcal{U}_{gold}$ denotes labeled users. And $\mathcal{C}$ is the set of labels of user profile.

To utilize label co-occurrence information for knowledge enhancement, we calculate weights between two label nodes as described in Sect. 3.2. The weights between two user nodes are defined as Eq. (4) according to user relationships.

$$A_{ij} = \begin{cases} 1 \times \text{Sim(i, j)} & \textit{if } (u_i, u_j) \in \mathcal{R} \\ 0 & \textit{otherwise} \end{cases}, \textit{where } i, j \in \mathcal{U} \tag{4}$$

Where $\mathcal{R} = \{(u_0, u_1), (u_1, u_3), ..\}$ is the set of relations between users and $Sim(i, j)$ indicates the similarity between user i and user j followed by [10]. The less the ratio of the value is, the closer the distance is.

GCN stacks multiple convolutional operations to simulate the message passing of graphs. Therefore, both the information propagation process with graph structure and node attributes are well leveraged in GCNs. For a one-layer GCN, the new k-dimensional node feature matrix is computed as:

$$L^{(1)} = \sigma\left(\tilde{A}XW_0\right) \tag{5}$$

Where $\tilde{A}$ ($\tilde{A} = D^{-1/2}WD^{-1/2}$) is a normalized symmetric adjacency matric, and W_0 is a weight matrix. $\sigma(\cdot)$ is an activation function, e.g. a ReLU function $\sigma(x) = \max(0, x)$. And the information propagation process is computed as Eq. (6) by stacking multiple GCN layers.

$$L^{(j+1)} = \sigma\left(\tilde{A}L^{(j)}W_j\right) \tag{6}$$

Where j denotes the layer number and $L^{(0)} = X$.

3.4 User Profile Prediction

The prediction of user profile is regarded as a multi-classification problem. After the above procedures, we obtain user representation according to user-generated content and relationships. The node embedding for user representation is fed into a *softmax* classifier to project the final representation into the target space of class probability:

$$Z = p_i(c|\mathcal{R}, \mathcal{U}; \Theta) = \textbf{softmax}\left(\tilde{A}\sigma\left(\tilde{A}XW_0\right)W_1\right) \tag{7}$$

Finally, the loss function is defined as the cross-entropy error over all labeled users as shown in Eq. (8).

$$\mathcal{L} = -\sum_{u \in y_u} \sum_{f=1}^{F} Y_{df} \ln Z_{df} \tag{8}$$

Where y_u is the set of user's indices with labels, and F is the dimension of the output features which is equal to the number of classes. Y is the label indicator matrix. The weight parameters W_0 and W_1 can be trained via gradient descent.

4   Experiments

4.1   Dataset

Weibo is the largest social network platform in China[1]. Followed by [10], we evaluate our method in different scale data sets in Weibo.

The datasets are sampled with different users in different time. And we select five classes as interest profiles of users, *Health, Women, Entertainment, Tourism, Society.*

The details of the datasets are illustrated in Table 1.

Table 1. The details of the datasets.

Dataset	Number of users	Health	Women	Entertainment	Tourism	Society
1#	4568	990	803	3397	1733	1031
2#	4853	1054	860	3592	1828	1088
3#	5140	1122	909	3811	1930	1163
4#	5711	1271	1014	4218	2146	1336

4.2   Comparisons and Evaluation Setting

To evaluate the performance of our method (GCN-IA), we compare it with some existing methods including textual feature-based method and relation feature-based method. In addition, to evaluate the implicit association labels for GCN, we compare GCN-IA with classical GCN. The details of these baselines are listed as follows:

SVM [26] uses the method of support vector machine to construct user profile based on user-generated context. In our experiment, we select username and blogs of users to construct user representation based on textual features. The textual features are obtained via pre-trained word2vec model.

MLP-IA [10] uses multi-label propagation method to predict user profiles. They capture relationship information by constructing probability transfer matrix. The labeled users are collected if the user is marked with a "V" which means his identity had been verified by Weibo. Analyzed by Jing et al. [24], these users were very critical in the propagation.

In the experiments, we will analyze the precision ratio (*P*) and recall ratio (*R*) of method which respectively represent the accuracy and comprehensiveness of user profile. And F1-Measure (*F1*) is a harmonic average of precision ratio and recall ratio, and it reviews the performance of the method.

---

[1] http://weibo.com/.

4.3   Results and Analysis

The experiment results are shown in Table 2. The results show that our method can make a significant increase in macro-F1 in all datasets.

Compared with feature-based method, our model makes a significant improvement. SVM fails since the method does not consider user relationships in the social networks. It only models the user-generated context, such as username and user's blogs.

Table 2. Experimental results of user profile task.

Method	1#	2#	3#	4#
SVM [26]	0.4334	0.2018	0.4418	0.4240
MLP-IA [10]	0.5248	0.4657	0.5030	0.5541
GCN-IA (Ours)	**0.5838**	**0.6040**	**0.5782**	**0.5708**

Compared with relation-based method, our model achieves improvements in all datasets, especially in dataset of 2#, we have improved 13.83% in macro-F1. MLP-IA [10] established user profiles based on user's relationships via label propagation. It suffers from leveraging the user-generated context, which contains semantic contextual features. Our model can represent users based on both relationships and context information via GCN module, which is more beneficial for identifying multi-label user profile task.

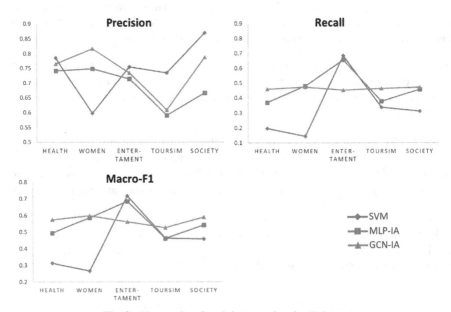

Fig. 2. The results of each interest class in 4# dataset.

The results of each interest class in 4# dataset are shown in Fig. 2. The results show that GCN-IA performs stably in all interest profiles, which demonstrate the good robustness of our model.

As shown in the results, the performance is little weak for the Entertainment interest class compared with baselines. In Weibo, there are much blogs with aspect to entertainment. Fake information exists in social network including fake reviews and fake accounts for specific purposes, which brings huge challenge for user profiles.

Our model constructs user profile via both textual features and relational features. The results can demonstrate that the user relationships can provide a beneficial signal for semantic feature extraction and the two features can reinforce each other.

5   Conclusion and Future Work

In this paper, we have studied the user profile by graph convolutional networks with implicit association labels, user information and label information embedding. We proposed a method to utilize implicit association among labels and then we take graph convolutional networks to embed the label and user information. On four real-world datasets in Weibo, experimental results demonstrate that GCN-IA produces a significant improvement compared with some state-of-the-art methods.

Future work will pay more attention to consider more prior knowledge to get higher performance.

References

1. Boudaer, G., Loeckx, J.: Enriching topic modelling with users' histories for improving tag prediction in Q and A systems. In: 25th International Conference Companion on World Wide Web, pp. 669–672. ACM, New York (2016)
2. Paulo, R.S., Frederico, A.D.: RecTwitter: a semantic-based recommender system for twitter users. In: Proceedings of the 24th Brazilian Symposium on Multimedia and the Web. ACM, New York (2018)
3. Nurbakova, D.: Recommendation of activity sequences during distributed events. In: Conference on User Modeling, Adaptation and Personalization, pp. 261–264 (2018)
4. Chang, P.S., Ting, I.H., Wang, S.L.: Towards social recommendation system based on the data from microblogs. In: International Conference on Advances in Social Networks Analysis and Mining, pp. 672–677. IEEE, Washington, D.C. (2011)
5. Li, R., Wang, C., Chang, C.C.: User profiling in an ego network: co-profiling attributes and relationships. In: International Conference on World Wide Web, pp. 819–830. ACM, New York (2014)
6. Zhang, J.L.: Application of tag propagation algorithm in the interest map of Weibo users. Programmer **5**, 102–105 (2014)
7. Fernando, A., Ashok, C., Tony, J., et al.: Artwork personalization at Netflix. In: Proceedings of the 12th ACM Conference on Recommender Systems, pp. 487–488. ACM, New York (2018)
8. Liang, S.S., Zhang, X.L., et al.: Dynamic embeddings for user profiling in twitter. In: Proceedings of the 24th SIGKDD Conference on Knowledge Discovery and Data Mining (KDD 2018), UK, pp. 1764–1773 (2018)

9. Roberto, P.S., Frederico, A.D.: RecTwitter: a semantic-based recommender system for Twitter users. In: Proceedings of the 24th Brazilian Symposium on Multimedia and the Web (WebMedia 2018), pp. 371–378. ACM, New York (2018)
10. Wei, L., Zhou, W., Wen, J., Lin, M., Han, J., Hu, S.: MLP-IA: multi-label user profile based on implicit association labels. In: Rodrigues, J.M.F., et al. (eds.) ICCS 2019. LNCS, vol. 11536, pp. 548–561. Springer, Cham (2019). https://doi.org/10.1007/978-3-030-22734-0_40
11. Huai, M., Miao, C., Li, Y., et al.: Metric learning from probabilistic labels. In: Proceedings of the 24th ACM SIGKDD International Conference on Knowledge Discovery and Data Mining (KDD 2018), pp. 1541–1550. ACM, New York (2018)
12. Peng, P., Wong, R.C.-W., Yu, P.S.: Learning on probabilistic labels. In: Proceedings of the 2014 SIAM International Conference on Data Mining, SIAM, pp. 307–315 (2014)
13. Iyer, A.S., Nath, J.S., Sarawagi, S.: Privacy-preserving class ratio estimation. In: Proceedings of the 22nd ACM SIGKDD International Conference on Knowledge Discovery and Data Mining, pp. 925–934. ACM (2016)
14. Šubelj, L., Bajec, M.: Robust network community detection using balanced propagation. Eur. Phys. J. B **81**(3), 353–362 (2011). https://doi.org/10.1140/epjb/e2011-10979-2
15. Ren, Z.-M., Shao, F., Liu, J.-G., et al.: Node importance measurement based on the degree and clustering coefficient information. Acta Phys. Sin **62**(12), 128901 (2013)
16. Huang, Y., Yu, L., Wang, X., Cui, B.: A multi-source integration framework for user occupation inference in social media systems. World Wide Web **18**(5), 1247–1267 (2014). https://doi.org/10.1007/s11280-014-0300-6
17. Gregory, S., et al.: Finding overlapping communities in networks by label propagation. New J. Phys. **12**(10), 103018 (2010)
18. Xie, J., Kelley, S., Szymanski, B.K.: Overlapping community detection in networks: the state-of-the-art and comparative study. ACM Comput. Surv. **45**(4), 1–35 (2011)
19. Wu, Z., Lin, Y., Gregory, S., et al.: Balanced multi-label propagation for overlapping community detection in social networks. J. Comput. Sci. Technol. **27**, 468–479 (2012)
20. Niepert, M., Ahmed, M., Kutzkov, K.: Learning convolutional neural networks for graphs. In: Proceedings of the 33rd International Conference on Machine Learning. ICML (2016)
21. Wu, L., Sun, P., Hong, R., Fu, Y., Wang, X., Wang, M.: SocialGCN: an efficient graph convolutional network based model for social recommendation. arXiv preprint arXiv:1811.02815 (2018)
22. Rex, Y., He, R., Chen, K., et al.: Graph convolutional neural networks for web-scale recommender systems. In: Proceedings of the 24th ACM SIGKDD Conference on Knowledge Discovery and Data Mining. KDD (2018)
23. Yao, L., Mao, C., Luo, Y.: Graph convolutional networks for text classification. In: Proceedings of the 33rd AAAI Conference on Artificial Intelligence. AAAI (2019)
24. Jing, M., Yang, X.X.: The characterization and composition analysis of Weibo"V". News Writ. **2**, 36–39 (2014)
25. Piao, G., Breslin, J.G.: Inferring user interests in microblogging social networks: a survey. User Model. User-Adap. Inter. **28**, 277–329 (2018). https://doi.org/10.1007/s11257-018-9207-8
26. Song, R., Chen, E., Zhao, M.: SVM based automatic user profile construction for personalized search. In: Huang, D.-S., Zhang, X.-P., Huang, G.-B. (eds.) ICIC 2005. LNCS, vol. 3644, pp. 475–484. Springer, Heidelberg (2005). https://doi.org/10.1007/11538059_50
27. Kipf, T.N., Welling, M.: Semi-supervised classification with graph convolutional networks. ICLR (Poster). arXiv preprint arXiv:1609.02907 (2017)

Interval Adjoint Significance Analysis for Neural Networks

Sher Afghan$^{(\boxtimes)}$ and Uwe Naumann

Software and Tools for Computational Engineering, RWTH Aachen University,
52074 Aachen, Germany
{afghan,naumann}@stce.rwth-aachen.de

Abstract. Optimal neural network architecture is a very important factor for computational complexity and memory footprints of neural networks. In this regard, a robust pruning method based on interval adjoints significance analysis is presented in this paper to prune irrelevant and redundant nodes from a neural network. The significance of a node is defined as a product of a node's interval width and an absolute maximum of first-order derivative of that node's interval. Based on the significance of nodes, one can decide how much to prune from each layer. We show that the proposed method works effectively on hidden and input layers by experimenting on famous and complex datasets of machine learning. In the proposed method, a node is removed based on its significance and bias is updated for remaining nodes.

Keywords: Significance analysis · Sensitivity analysis · Neural network pruning · Interval adjoints

1 Introduction

Neural networks and deep belief networks are powerful tools of machine learning for classification tasks. There are many things to consider for the construction of effective neural network architecture i.e., learning rate, optimization method, regularization, etc. But one of the most important hyper-parameter is network size. It is hard to guess the optimal size of a network. Large networks are good at memorization and get trained quickly but there is a lack of generalization in the large networks. We can end up in over-fitting our networks. We can solve this problem of generalization by constructing smaller networks and save the computational cost of classification but this approach can end up in under-fitting. Success is to come up with neural network architecture which can solve both problems [1].

Researchers have proposed different techniques such as; brute-force [2], growing [3] and pruning methods [4]. Out of these techniques, pruning results in effective compressed neural network architecture while not significantly hurting network accuracy. This technique starts with a well-trained network. Assuming

© Springer Nature Switzerland AG 2020
V. V. Krzhizhanovskaya et al. (Eds.): ICCS 2020, LNCS 12139, pp. 365–378, 2020.
https://doi.org/10.1007/978-3-030-50420-5_27

the network is oversized, it tries to remove irrelevant or insignificant parameters from the network. These parameters can be network's weights, inputs, or hidden units.

Over time multiple pruning methods have been proposed (see detailed surveys [1,2,5]). Among many methods, the sensitivity-based analysis technique is the most famous one [6–10]. It measures the impact of neural network parameters on the output. Our proposed method also utilizes the concept of sensitivity to define the significance of network parameters. Sensitivities of an objective function concerning weights of networks are used to optimize network's weights while sensitivities of output unit concerning input and hidden units are used to find the significance of the network's input and hidden units.

This paper presents a method for finding out the sensitivities of the network's output concerning the network's input and hidden units in a more robust and efficient way by using interval adjoints. Input and hidden unit values and their impact on output are used to define the significance of the input and hidden units. The significance analysis method defined in this paper takes care of all the information of the network units and the information stored during significance analysis is used to update the remaining parameters biased in the network.

The rest of the paper is organized as follows. Section 2 briefly describes algorithmic differentiation (AD) for interval data with examples. Section 3 presents our significance analysis method for pruning. Experimental results are given in Sect. 4. The conclusion is given in Sect. 5.

2   Interval Adjoint Algorithmic Differentiation

The brief introduction to AD [11,12] is given in this section along with the modes of AD and differentiation with one of the mode of AD commonly known as adoint mode of AD. Later, intervals for interval adjoint algorithmic differentiation are used.

2.1   Basics of AD

Let F be a differentiable implementation of a mathematical function F : $\mathbb{R}^{n+l} \longrightarrow \mathbb{R}^m : \mathbf{y} = (\mathbf{x}, \mathbf{p})$, computing an output vector $\mathbf{y} \in \mathbb{R}^m$ from inputs $\mathbf{x} \in \mathbb{R}^n$ and constant input parameter $\mathbf{p} \in \mathbb{R}^l$. Differentiating F with respect to $\mathbf{x}$ yields the Jacobian matrix $\nabla_{\mathbf{x}} F \in \mathbb{R}^{m*n}$ of F.

This mathematical function F can be transformed to coded form in some higher level programming language to apply AD on that code. AD works on the principle of the chain rule. It can be implemented using source transformation [13] or operator overloading [14] to change the domain of variables involved in the computation. It calculates the derivatives and different partials along with the each output (primal values) in a time similar to one evaluation of the function. There are multiple tools[1] available which can implement AD e.g. dco/c++ [15,16]. dco/c++ implements AD with the help of operator overloading and it has been successfully used for many applications [17,18].

---

[1] http://www.autodiff.org/?module=Tools.

2.2   Modes of AD

There are many modes of AD [11,12] but two of them are most widely used; one is forward mode AD (also; tangent linear mode of AD) and second is reverse mode AD (also; adjoint mode of AD). Because, we are interested in adjoints for our research so we are going to describe reverse mode AD briefly.

There are two phases in reverse mode of AD; forward and backward pass. In forward pass, function code is run forward yielding primal values for all intermediate and output variables and storing all the relevant information which are needed during backward pass. During backward pass, adjoints of outputs (for outputs, adjoint is evidently 1) will be propagated backwards through the computation of the original model to adjoints of inputs.

Example (AD Reverse Mode). Below is an example of adjoint mode AD evaluation on function $f(\mathbf{x}) = sin(x_o \cdot x_1)$.

Forward Pass: With intermediate variables $v_i \in \mathbb{R}, i = 1, 2$, a possible code list of f is

$$v_1 = x_1 \cdot x_2$$
$$v_2 = sin(v_1)$$
$$y = v_2$$

Backward Pass: With intermediate variables $v_i \in \mathbb{R}, i = 1, 2$ and associated adjoint variables $v_{i(1)}$, a possible adjoint code of f is

$$v_{2(1)} = y_{(1)}$$
$$v_{1(1)} = cos(v_1) \cdot v_{2(1)}$$
$$x_{2(1)} = x_1 * v_{1(1)}$$
$$x_{1(1)} = x_2 * v_{1(1)}$$

2.3   Interval Adjoints

Consider, lower case letters (e.g., $a, b, c, ...$) represents real numbers, uppercase letters (e.g., $A, B, C, ...$) represents interval data and an interval is represented by $X = [x^l, x^u]$, where l and u represents the lower and upper limit of the interval, respectively.

Interval Arithmetic (IA), evaluates a function $f[X]$ for the given range over a domain in a way that it gives guaranteed enclosure $f[X] \supseteq \{f[x] | x \ni [X]\}$ that contains all possible values of $f(x)$ for $x \ni [X]$. Similarly, interval evaluation yield enclosures $[V_i]$ for all intermediate variables V_i.

The adoint mode of AD can also be applied to interval functions for differentiation purpose [19]. The impact of individual input and intermediate variables on the output of an interval-valued function can easily be evaluated by using adjoint mode of AD over interval functions. AD not only computes the primal value of intermediate and output variables, it also computes their derivative with the help of chain rule. The first order derivatives $(\frac{\delta Y}{\delta X_i}, \frac{\delta Y}{\delta V_i})$ of output V with respect to all inputs X_i and intermediate variables V_i can be computed with the adjoint mode of AD in a single evaluation of function f. In the same way, IA and AD can be used to find out the interval-valued partial derivatives $(\nabla_{[V_i]}[Y], \nabla_{[X_i]}[Y])$ that contains all the possible derivatives of output Y with respect to intermediate variables V_i and input variables X_i over the given input interval X.

Example (AD Reverse Mode with Interval Data). Below is an example of adjoint mode AD evaluation on interval function $f(\mathbf{X}) = sin(X_1 . X_2)$ for calculation of interval adjoints. Let $X = \{X_1, X_2\} \in \mathbb{R}^2$, where $X_1 = [0.5, 1.5]$, $X_2 = [-0.5, 0.5]$.

Interval output of $f(X)$:

$$f(X) = [-0.6816, 0.6816]$$

Differentiation:

$$\nabla_X f(X) = \begin{pmatrix} \frac{\delta f(X)}{\delta X_1} \\ \frac{\delta f(X)}{\delta X_2} \end{pmatrix} = \begin{pmatrix} cos(X_1 * X_2) * X_2 \\ cos(X_1 * X_2) * X_1 \end{pmatrix}$$

Interval evaluation of $\nabla_X f(X)$:

$$\nabla_X f(X) = \begin{pmatrix} cos(X_1 * X_2) * X_2 \\ cos(X_1 * X_2) * X_1 \end{pmatrix} = \begin{pmatrix} [-0.5, 0.5] \\ [0.3658, 1.5] \end{pmatrix}$$

3   Significance Analysis

Sensitivity based method is most useful in defining the significance of the network parameters. In [7,10], researchers used the network's output to find the sensitivity of network parameters. Sensitivity is defined as the degree to which an output responds to the deviation in its inputs [7]. Deviation in output Δy due to deviation in inputs is the difference of deviated and non-deviated outputs $f((X + \Delta X) * w) - f(X * w)$. Meanwhile, inputs X_i is treated as an interval $[0, 1]$ for finding sensitivity not just on fixed points rather finding it for overall inputs range.

With the above-defined sensitivity, the significance is measured as a product of sensitivity of a node by the summation of the absolute values of its outgoing weights. This approach of significance defined by [7] has few limitations such as it can only be applied to hidden layer nodes, not to the network input layer. Secondly, one has to prune one layer first before moving to the second layer as it works by layer-wise.

To find out the significance for both input and hidden nodes of a network, another method proposed in [10], computes sensitivities by computing partial derivatives of network outputs to input and hidden nodes. Although this method is good in computing sensitivities of network's parameters, there is a high computational cost for this as it computes partial derivative at a given parameter and then finds out the average sensitivity for all training patterns. We proposed a new sensitivity method in Sect. 3.1 based on interval adjoints to tackle the shortcomings of earlier defined sensitivity based pruning methods.

3.1 Interval Adjoint Significance Analysis

Let us first define significance, before applying the proposed method of significance analysis on neural networks to obtain the ranking of nodes in the network. According to [20], significance can be defined by the product of width of an interval and absolute maximum of first order derivative of that interval.

$$S_Y(V_i) = w[V_i] * max|\nabla_{[v_i]}[y]| \tag{1}$$

The influence of all input variables X_i, $i = 1, ..., n$ is given in V_j. And this influence can be quantified by width $w[V_j] = v^u - v^l$. Larger the width of an interval larger the influence and vice versa. This means that variable V_j is highly sensitive to input variable X_i and vice versa. But this information alone is not sufficient to define the significance of a variable. Further operations, during the evaluation of Y and different intermediate variables V, may increase or decrease the influence of variable V_j. Therefore, it is necessary to find the influence of that variable V_j on output Y. The absolute maximum of first order partial derivative $max|\nabla_{[v_i]}[y]|$ of variable V_j gives us this influence of variable V_j over output Y.

3.2 Selection of Significant Nodes

Suppose an already trained neural network illustrated in Fig. 1, with four nodes on each hidden layer, four nodes on the input layer and one output node. Thus the problem in this subsection is to find out the significance of each node in the network to correctly find out the output. This concept of significance defined in (1) can be applied to already trained neural nets and deep belief nets to find out the influence of individual input and intermediate nodes in determining the output of the network. There is no change in the standard back-propagation neural network architecture [21,22] except all the variables in the network are of interval type instead of scalar type. Interval arithmetic [23] is used to calculate the input-output relations. We can calculate this significance layer-wise and rank them based on their magnitude. A node with a high significance value is very sensitive to the overall output of the network.

A single interval input vector is used for finding the significance. Let us assume that m training patters $x_k = \{x_{k1}, ..., x_{kn}\}$, $k = 1, 2, 3, ..., m$ are used to train the neural network. These m training patterns are used to generate the interval input vector for significance analysis by finding out the maximum and minimum value for each input (e.g. x_1) from all training patterns $(x_{11}, ..., x_{k1})$. These maximum and minimum values are used to construct the interval input vector $X = \{[min(x_{k1}), max(x_{k1})], ..., [min(x_{kn}), max(x_{kn})]\}$. As scalar is degenerated form of an interval whose upper and lower bounds are the

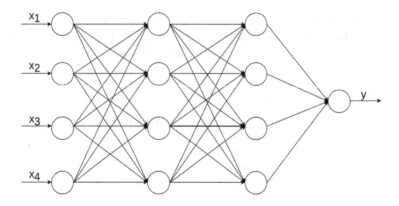

Fig. 1. Simple neural network.

same, one can use the trained weight and bias vector of the network and change them to interval vectors whose upper and lower bounds are the same.

A single forward and backward run of the network is required with new input, weight and bias vector for yielding the significance of intermediate and input nodes. Nodes on each layer can be ranked in decreasing order of significance and it's up to the user to select the number of nodes which will stay in the new architecture. Let's go back to our example of already trained network in Fig. 1. After the significance analysis, we find out middle two nodes on the first hidden layer, first and the last node on second hidden layer is not significant as shown in Fig. 2.

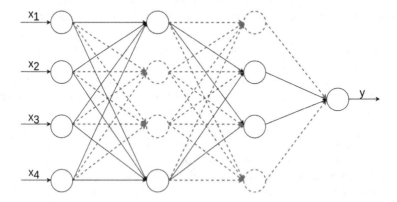

Fig. 2. Insignificant nodes and their connection identified by significance analysis.

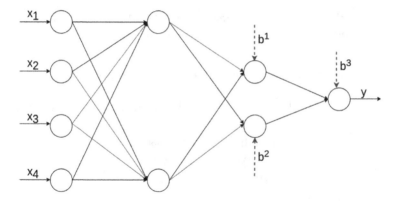

Fig. 3. Removal of insignificant nodes, and their incoming and outgoing connections.

3.3 Removal of Insignificant Nodes

After the selection of significant nodes in the network, it is necessary to preserve the information of insignificant nodes before throwing them out from the network to prune it, otherwise, we will be changing the inputs for next layer activations. Insignificant nodes have less impact on the network and the weight associated with its incoming and ongoing connections are mostly redundant and have very low values. Significance analysis together with interval evaluation not only gives us the influence of nodes but it also gives us a guaranteed enclosure that contains all the possible values for insignificant nodes and their outgoing connections. We can store all the information of the previous layer insignificant nodes into significant nodes of the next layer by calculating the midpoints ($v_j = \frac{(v_j^l + v_j^u)}{2}$) of all incoming connections from previous layer insignificant nodes to significant nodes of next layer. We can sum up all these midpoints and add them as the bias of significant nodes. This process is illustrated in Fig. 3.

4 Experimental Results

There are so many parameters to optimize in all the layers of fully connected neural networks and on the fully connected layers of most of the large-scale convolutional neural networks [24,25]. Currently, we analyzed the performance of our method on fully connected networks to reduce the number of parameters and obtain a compressed network yet achieving the same accuracy. For this purpose, we choose four datasets; MNIST [26], MNIST_ROT [27], Fashion-MNIST [29] and CIFAR-10 [28].

In all the networks, there were two hidden layers with a size of five hundred nodes on each layer, refer to Table 1 for more details. For the activation functions, the sigmoid activation function was used on the hidden layer and softmax on the output layer. Adam with weight decay commonly known as AdamW [30] and Adamax [31] optimization methods were used for parameter optimization. In the initial epochs, the network was trained with AdamW but in the end, Adamax was used for better accuracy.

Table 1. Datasets and networks used for experiments

Dataset	Architecture
MNIST	784-500-500-10
MNIST_ROT	784-500-500-10
Fashion-MNIST	784-500-500-10
CIFAR-10	3072-500-500-10

4.1   Experiments on MNIST

Significance analysis not only can be applied to hidden layers but our method is also effective for the input layer. It works as a feature selection method for inputs. Figure 4 shows test and train error plots for removed input features after significance analysis. As we can see from flatter curves, the error did not increase after removing the significant number of input features from the original network. But after a certain point error is almost increasing exponentially.

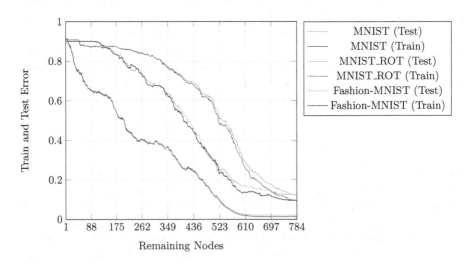

Fig. 4. Removal of insignificant nodes from input layer of MNIST, Fashion-MNIST, and MNIST_ROT data sets

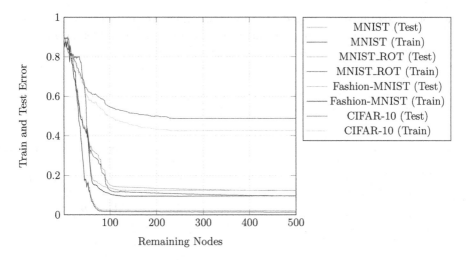

Fig. 5. Removal of insignificant nodes from first hidden layer of MNIST, Fashion-MNIST, MNIST_ROT, and CIFAR-10 data sets

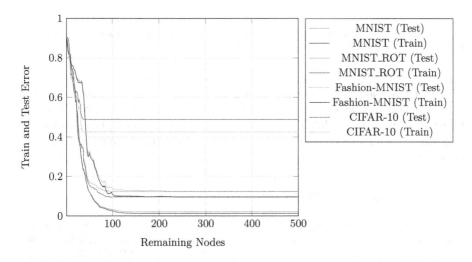

Fig. 6. Removal of insignificant nodes from second hidden layer of MNIST, Fashion-MNIST, MNIST_ROT, and CIFAR-10 data sets

Figure 5 and 6 are test and train error plots for the first and second hidden layer respectively. Significance analysis was performed on each layer separately. It is clearly shown in Fig. 5 that we can remove more than eighty percent of the nodes from the first layer without compromising accuracy. Figure 7 shows the result of the significance analysis applied on all hidden layers. If we observe the plots of significance applied to individual hidden layers in Fig. 5 and 6, we can clearly see the pattern when significance is applied on all hidden layers. The error increasing pattern is the same in Fig. 6 as it is in Fig. 7.

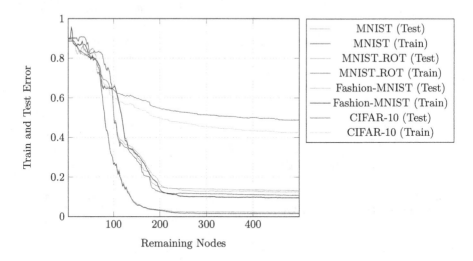

Fig. 7. Removal of insignificant nodes from first and second hidden layer of MNIST, Fashion-MNIST, MNIST_ROT, and CIFAR-10 data sets

Significance analysis works pretty well on CIFAR-10 dataset too and we can remove a few hundreds of input features from the network as shown in Fig. 8. Figure 5, 6 and 7 show plots of nodes removal from each hidden layer and nodes removal from all hidden layers in the network after the significance analysis respectively.

4.2   Experiments on CIFAR-10

From the last 11 years, CIFAR-10 has been the focus of intense research which makes it an excellent test case for our method. It has been most widely used as a test case for many computer vision methods. After MNIST, it has been ranked the second most referenced dataset [32]. CIFAR-10 is still the subject of current research [33–35] because of its difficult problems. All of its 50k train samples were used to train the network and 10k test samples were used to test the network.

Like MNIST, we can also see the pattern of error increasing when we apply significance analysis on all hidden layers. Error plot of first hidden layer nodes removal in Fig. 5 and error plot of all hidden layer nodes removal in Fig. 7 looks quite the same. This is because the second hidden layer is not contributing much to the overall performance of the network.

4.3   Experiments on MNIST_ROT and Fashion-MNIST

We performed the same experiments on rotated version of MNIST. Error plots for this dataset are given in the appendix. Furthermore, we also quantify the performance of our algorithm on the newly generated Fashion-MNIST dataset.

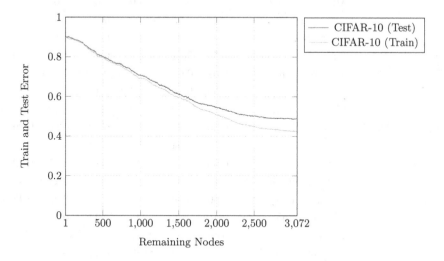

Fig. 8. Removal of insignificant nodes from input layer of CIFAR-10 data set

4.4   Network Performance Before and After Retraining

After the significance analysis and removing 50% of nodes from the network still, we can achieve the same accuracy on all the datasets used in experiments without additional retraining or fine-tuning. Table 2 lists the train and test error of different datasets on the different percentages of removed insignificant nodes in the network before and after retraining. In the case of MNIST and CIFAR-10, there is no increase in error if we remove 25% or 50% of all the hidden nodes from the network using significance analysis. If we further train them after removing the nodes there is no increase in the accuracy. But on the other two datasets (Fashion-MNIST and MNIST_ROT), we can increase the accuracy if we retrain them after removing the insignificant hidden nodes from the network.

Train and test error slightly increase if we remove 90% of the hidden nodes from the network and this is expected as we are taking away too much information from the original network. But we can improve the performance of network with retraining and using the remaining old weight vector and updated bias vector for significant nodes. It is better to use the remaining network connection for retraining than initializing the values again. After retraining, the error rate was decreased significantly for 90% nodes removal from the original network and in some cases decreasing the original error rate that was there before significance analysis.

Table 2. Training and test error before and after retraining for different percentage of removed insignificant nodes. Initially, all the neural nets have 1000 hidden nodes.

Removal of neurons from hidden layers		90%	80%	75%	50%	25%	0%
MNIST	Training error	0.26	0.03	0.01	0.01	0.01	0.01
	After retraining	0.002	0.001	0.0009	0.01	0.01	
	Test error	0.26	0.03	0.02	0.02	0.02	0.02
	After retraining	0.02	0.02	0.02	0.01	0.01	
MNIST_ROT	Training error	0.53	0.12	0.11	0.10	0.09	0.09
	After retraining	0.02	0.0001	0.0005	0.0004	0.003	
	Test error	0.53	0.15	0.14	0.13	0.12	0.12
	After retraining	0.13	0.10	0.10	0.10	0.10	
Fashion-MNIST	Training error	0.65	0.13	0.10	0.09	0.09	0.09
	After retraining	0.05	0.04	0.03	0.03	0.03	
	Test error	0.65	0.16	0.13	0.12	0.12	0.12
	After retraining	0.11	0.12	0.11	0.12	0.11	
CIFAR-10	Training error	0.60	0.49	0.47	0.42	0.42	0.42
	After retraining	0.46	0.42	0.43	0.42	0.42	
	Test error	0.64	0.54	0.52	0.48	0.48	0.48
	After retraining	0.50	0.49	0.49	0.48	0.48	

5   Future Work and Conclusion

A new method of finding and removing redundant and irrelevant nodes from the neural network using interval adjoints is proposed in this paper. Our method finds out the significance of hidden as well as input nodes. The significance depends upon two factors, the impact of a node on output and width of a node interval. The use of interval data and finding sensitivities with interval adjoints make our method more robust than multiple existing methods. The results presented in this paper indicate that the significance analysis correctly finds out irrelevant input and hidden nodes in a network and it also gives us much information to update the bias of relevant nodes so that performance of the network does not comprise by removing irrelevant nodes.

Our future work will be aimed at applying interval adjoint significance analysis on convolutional and fully connected layers of convolutional neural networks. Furthermore, investigation will be carried out on applying significance analysis during the training of a network and speed up the training process by eliminating the less significant nodes from the network.

References

1. Augasta, M.G., Kathirvalavakumar, T.: Pruning algorithms of neural networks — a comparative study. Cent. Eur. J. Comp. Sci. **3**(3), 105–115 (2013). https://doi.org/10.2478/s13537-013-0109-x
2. Reed, R.: Pruning algorithms-a survey. IEEE Trans. Neural Networks **4**(5), 740–747 (1993)
3. Fahlman, S.E., Lebiere, C.: The cascade-correlation learning architecture. In: Advances in Neural Information Processing Systems, pp. 524–532 (1990)
4. Castellano, G., Fanelli, A.M., Pelillo, M.: An iterative pruning algorithm for feedforward neural networks. IEEE Trans. Neural Networks **8**(3), 519–531 (1997)
5. Cheng, Y., Wang, D., Zhou, P., Zhang, T.: A survey of model compression and acceleration for deep neural networks. arXiv preprint arXiv:1710.09282 (2017)
6. Xu, J., Ho, D.W.: A new training and pruning algorithm based on node dependence and Jacobian rank deficiency. Neurocomputing **70**(1–3), 544–558 (2006)
7. Zeng, X., Yeung, D.S.: Hidden neuron pruning of multilayer perceptrons using a quantified sensitivity measure. Neurocomputing **69**(7–9), 825–837 (2006)
8. Lauret, P., Fock, E., Mara, T.A.: A node pruning algorithm based on a Fourier amplitude sensitivity test method. IEEE Trans. Neural Networks **17**(2), 273–293 (2006)
9. Hassibi, B., Stork, D.G., Wolff, G.J.: Optimal brain surgeon and general network pruning. In: IEEE International Conference on Neural Networks, pp. 293–299 (1993)
10. Engelbrecht, A.P.: A new pruning heuristic based on variance analysis of sensitivity information. IEEE Trans. Neural Networks **12**(6), 1386–1399 (2001)
11. Griewank, A., Walther, A.: Evaluating Derivatives: Principles and Techniques of Algorithmic Differentiation. SIAM, Philadelphia (2008)
12. Naumann, U.: The Art of Differentiating Computer Programs: An Introduction to Algorithmic Differentiation. SIAM, Philadelphia (2012)
13. Hascoet, L., Pascual, V.: The Tapenade automatic differentiation tool: principles, model, and specification. ACM Trans. Math. Softw. (TOMS) **39**(3), 1–43 (2013)
14. Corliss, G., Faure, C., Griewank, A., Hascoet, L., Naumann, U.: Automatic Differentiation of Algorithms. Springer, New York (2013)
15. Lotz, J., Leppkes, K., Naumann, U.: dco/c++ - derivative code by overloading in C++. https://www.stce.rwth-aachen.de/research/software/dco/cpp
16. Lotz, J., Naumann, U., Ungermann, J.: Hierarchical algorithmic differentiation a case study. In: Forth, S., Hovland, P., Phipps, E., Utke, J., Walther, A. (eds.) Recent Advances in Algorithmic Differentiation. LNCSE, vol. 87, pp. 187–196. Springer, Heidelberg (2012). https://doi.org/10.1007/978-3-642-30023-3_17
17. Towara, M., Naumann, U.: A discrete adjoint model for OpenFOAM. Procedia Comput. Sci. **18**, 429–438 (2013)
18. Lotz, J., Schwalbach, M., Naumann, U.: A case study in adjoint sensitivity analysis of parameter calibration. Procedia Comput. Sci. **80**, 201–211 (2016)
19. Schichl, H., Neumaier, A.: Interval analysis on directed acyclic graphs for global optimization. J. Global Optim. **33**(4), 541–562 (2005). https://doi.org/10.1007/s10898-005-0937-x
20. Deussen, J., Riehme, J., Naumann, U.: Interval-adjoint significance analysis: a case study (2016). https://wapco.e-ce.uth.gr/2016/papers/SESSION2/wapco2016_2_4.pdf

21. Kelley, H.J.: Gradient theory of optimal flight paths. ARS J. **30**(10), 947–954 (1960)
22. Rojas, R.: The backpropagation algorithm. In: Neural Networks, pp. 149–182. Springer, Heidelberg (1996). https://doi.org/10.1007/978-3-642-61068-4_7
23. Moore, R.E.: Methods and Applications of Interval Analysis. Society for Industrial and Applied Mathematics, Philadelphia (1979)
24. Krizhevsky, A., Sutskever, I., Hinton, G.E.: ImageNet classification with deep convolutional neural networks. In: Advances in Neural Information Processing Systems, pp. 1097–1105 (2012)
25. Simonyan, K., Zisserman, A.: Very deep convolutional networks for large-scale image recognition. arXiv:1409.1556 (2014)
26. LeCun, Y., Cortes, C.: MNIST handwritten digit database (2010). http://yann. lecun.com/exdb/mnist/
27. Larochelle, H., Erhan, D., Courville, A., Bergstra, J., Bengio, Y.: An empirical evaluation of deep architectures on problems with many factors of variation. In: ACM Proceedings of the 24th International Conference on Machine Learning, pp. 473–480 (2007)
28. Krizhevsky, A., Hinton, G.: Learning multiple layers of features from tiny images, vol. 1, no. 4, p. 7. Technical report, University of Toronto (2009)
29. Xiao, H., Rasul, K., Vollgraf, R.: Fashion-MNIST: a novel image dataset for benchmarking machine learning algorithms. arXiv:1708.07747 (2017)
30. Loshchilov, I., Hutter, F.: Fixing weight decay regularization in Adam. arXiv:1711.05101 (2017)
31. Kingma, D.P., Ba, J.: Adam: a method for stochastic optimization. arXiv:1412.6980 (2014)
32. Hamner, B.: Popular datasets over time. https://www.kaggle.com/benhamner/ populardatasets-over-time/code
33. Real, E., Aggarwal, A., Huang, Y., Le, Q.V.: Regularized evolution for image classifier architecture search. In: Proceedings of the AAAI Conference on Artificial Intelligence, vol. 33, pp. 4780–4789 (2019)
34. Miikkulainen, R., et al.: Evolving deep neural networks. In: Artificial Intelligence in the Age of Neural Networks and Brain Computing, pp. 293–312 (2019)
35. Su, J., Vargas, D.V., Sakurai, K.: One pixel attack for fooling deep neural networks. IEEE Trans. Evol. Comput. **23**(5), 828–841 (2019)

Ringer: Systematic Mining of Malicious Domains by Dynamic Graph Convolutional Network

Zhicheng Liu[1,2], Shuhao Li[1,2,4(✉)], Yongzheng Zhang[1,2,4], Xiaochun Yun[3], and Chengwei Peng[3]

[1] Institute of Information Engineering, Chinese Academy of Sciences, Beijing, China
{liuzhicheng,lishuhao,zhangyongzheng}@iie.ac.cn
[2] School of Cyber Security, University of Chinese Academy of Sciences, Beijing, China
[3] National Computer Network Emergency Response Technical Team/Coordination Center of China, Beijing, China
{yunxiaochun,pengchengwei}@cert.org.cn
[4] Key Laboratory of Network Assessment Technology, Chinese Academy of Sciences, Beijing, China

Abstract. Malicious domains are critical resources in network security, behind which attackers hide malware to launch the malicious attacks. Therefore, blocking malicious domains is the most effective and practical way to combat and reduce hostile activities. There are three limitations in previous methods over domain classification: (1) solely based on local domain features which tend to be not robust enough; (2) lack of a large number of ground truth for model-training to get high accuracy; (3) statically learning on graph which is not scalable. In this paper, we present Ringer, a scalable method to detect malicious domains by dynamic Graph Convolutional Network (GCN). Ringer first uses querying behaviors or domain-IP resolutions to construct domain graphs, on which the dynamic GCN is leveraged to learn the node representations that integrate both information about node features and graph structure. And then, these high-quality representations are further fed to the full-connected neural network for domain classification. Notably, instead of global statically learning, we adopt time-based hash to cut graphs to small ones and inductively learn the embedding of nodes according to selectively sampling neighbors. We construct a series of experiments on a large ISP over two days and compare it with state of the arts. The results demonstrate that Ringer achieves excellent performance with a high accuracy of 96.8% on average. Additionally, we find thousands of potential malicious domains by semi-supervised learning.

Keywords: Graph convolutional network · Malicious domain mining · Malware activities · Deep learning · Time-based hash

© Springer Nature Switzerland AG 2020
V. V. Krzhizhanovskaya et al. (Eds.): ICCS 2020, LNCS 12139, pp. 379–398, 2020.
https://doi.org/10.1007/978-3-030-50420-5_28

1   Introduction

Malicious domains are important platforms of launching malicious attacks in network security, such as spamming, phishing, botnet command and control (C2) infrastructure and so on. The cost of dollars is rising as a result of the growing prevalence of financial fraud based on domains. For instance, some ransomwares encrypt personal files through domain dissemination to extort money from individuals, which even cause more emotional and professional damage. Moreover, personal privacy and intellectual property rights caused by malicious domains are also arguably serious issues. Therefore, effective detection of malicious domains bears the utmost importance in fighting malwares.

Since blocking malicious domains can immediately lead to reduction in malwares, a wealth of techniques have been proposed to discover malicious domains. Those efforts can be divided into two categories, including feature-based and behavior-based methods. Traditional feature-based methods [9–11,26] basically use the network or domain name features to distinguish malicious domains from benign ones due to its efficiency and effectiveness. However, attackers can circumvent the detections through some simple manipulation on these features such as making the pronounceable feature conform to the normal domains' form. Advancements in machine learning make it possible to achieve superior performance on domain classification over the passive DNS statistical features. However, it hinders adversarial crafting data generated by the Generative Adversarial Network (GAN) [8] and needs a lot of labeled data with time-consuming manual inspections.

Behavior-based methods [17,19,21] make use of the querying relationship of the hosts or the parsing relationship of the domains to build the association graphs in which domains or hosts represent nodes. Some inference algorithms (e.g., Belief Propagation [31]) are leveraged to infer the reputation of unlabeled domains on the given labeled domains. However, they achieve poor precision when the number of ground truth is small and do nothing to the isolated nodes that have no relationship with the ground truth. Moreover, some methods [20,25] based on analyzing host behaviors are prone to evasion techniques such as fake-querying and sub-grouping.

In order to solve the limitations of the existing methods, we propose Ringer, a novel and effective system to classify malicious domains. Our insights are built on three observations: (1) Malicious domains often serve malicious users whose behaviors deviate from that of benign domains usually registered for benign services. Hence, it is very likely that multiple malicious domains are accessed by the overlapping client set; (2) Multiple malicious domains are commonly hosted on the joint server hosts due to the fact that malicious IPs are important and scarce resources which are generally reused. For example, attackers would typically place a number of rogue softwares on the same server hosts that they control to reduce the cost. Once identified, the malwares will be migrated to another host in chunks. The reuse mechanism and common querying behaviors of these malicious resources reflect the strong correlation among malicious domains; (3) Graph convolutional network (GCN) [18] can aggregate information of both graph structure and static features to generate the underlying representation of nodes for classification tasks. In turn, we use these two strong associations, combined with advanced graph deep learning technique, to discover malicious domains.

Ringer is designed to model the association among domains into graphs, on which dynamic GCN algorithm is applied to learn the high-quality node representations that are amenable to efficient domain classification. Instead of analyzing single domain in isolation, we study the relevance of multiple domains involved in malicious activities. The robustness of the system is enhanced owing to the unchangeable behavior relationship and global correlation. Both of them will not be eliminated or altered by the will of the human being. More specifically, the detection of malicious domains by Ringer consists of three main steps. Firstly, Ringer constructs the domain association graph according to the querying behavior and resolution relation. And then, Ringer takes advantage of dynamic GCN to quickly aggregate attribute information associated with each vertex through the neighborhood defined by the graph structure in sampling fashion, thus transferring the graph-represented domains to vectors. Finally, the node representations that embed high-dimensional structural information, as well as attributes, are fed to neural network for malicious domain mining. There are two challenges that need to be addressed: (1) the existing GCN operates on the full static graphs, which cannot be directly applied to large-scale graphs (especially our internet-scale domain data); (2) it is time-consuming to learn the entire graph every time as the graph is constantly evolving. Therefore, we introduce two innovations: (1) hashing the large graphs to form small ones according to time; (2) sampling the neighborhoods of each node to construct the graph dynamically and making the training of the model dependent only on the node and its sampled neighbors, but the entire input graphs. To improve performance, we simultaneously aggregate all intermediate representations using layer-wise attention encoder.

Our contributions are summarized as follows:

- We propose a robust malicious domain classification method named Ringer, which transforms the research problem from malicious domain mining to graph learning. We use time-based hash technology to split the graphs into small ones, which makes Ringer more scalable.
- Dynamic GCN is developed to automatically capture the structure and characteristics information of the sampled neighbors, thus generating the underlying representations for further domain classification.
- Our method has strong generalizability, which is independent of the input graph. Once the model is trained, it can be directly applied to the new domain association graphs or the newly-added nodes.
- We implement the prototype of Ringer and perform experiments on large-scale DNS traces from ISP to evaluate the effectiveness. The results show that our system has superior scalability and accuracy than state-of-the-art system, and a large number of potential malicious domains are found.

The remaining sections of this paper are organized as following. In Sect. 2, we review the related work. In Sect. 3, we describe the association among malicious domains and the original GCN. We provide a systematic overview of Ringer and detail each module in Sect. 4. The collection of the datasets and ground truth used in this paper is elaborated in Sect. 5. We highlight the experimental results and discuss the limitations of our systems in Sect. 6, while we conclude the whole study in Sect. 7.

2 Related Work

Malicious domain detection has been a hot issue and a great deal of strides have been made to discover malicious domains over the past years. The detection on DNS can dig out the malicious activities earlier due to the nature of DNS flow prior to attack traffic. DNS-based detection is also lighter than that based on all traffic and has no need to take into account the traffic encryption. Here, we present some representative works which can be divided into two categories: feature-based and behavior-based methods.

Feature-based methods adopt advanced machine learning combined with domain statistical features to detect malicious domains. Antonakakis et al. [9] put forward Notos, a system uses different characteristics of a domain to compute a reputation that indicates the domain malicious or legitimate. Bilge et al. [11] propose EXPOSURE, which uses fewer training datasets and less time to detect malicious domains. Moreover, the system is able to detect the malicious domains of the new categories without updating the data feeds. Babak et al. [25] present the Segugio, a system which tracks the co-occurrence of malicious domains by constructing a machine-domain bipartite graph to detect new malware-related domains. Antonakakis et al. [10] use the similarity of Non-Existent Domain responses generated by Domain Generation Algorithm (DGA) and querying behavior to identify DGA-related domains. Jehyun et al. [19] build domain travel graphs that represent DNS query sequences to detect infected clients and malicious domains which is robust with respect to space and time. Thomas et al. [28] analyze the NXDomains querying pattern on several TLD authoritative name servers to identify the strong connected groups of malware related domains.

Behavior-based approaches are to detect malicious domains by analyzing host behavior or network behavior. Pratyusa et al. [21] firstly use proxy logs to construct host-domain bipartite graphs on which belief propagation algorithm is used to evaluate the malicious probability of domains. Issa et al. [17] construct domain-resolution graph to find new malicious domains by using strong correlation between domain resolution values. Acar et al. [27] introduce a system AESOP, which uses a locality-sensitive hashing to measure the strength of the relationship between these files to construct a graph that is used to propagate information from a tagged file to an untagged file. Liu et al. [20] propose CCGA which first clusters NXdomains according to the hosts infected family DGA generate the same domain name set, and then extracts the cooperative behavior relationship of domains (including time, space and character similarity) to realize the classification of NXdomain cluster. Peng et al. [23] design a system MalShoot, which construct domain-resolution graph to learn the relationship among domains by using graph embedding technology. They use the node embeddings as features to identify malicious domains and defend illegal activities.

3 Background

In this section, we first describe malicious domain correlations used in malicious domain mining. Then we discuss the technique of interest GCN.

3.1 Malicious Domain Correlation

Our method is to use the relationship among domains combined with statistical characteristics for domain classification. It is suitable for detecting multi-domain malicious activities, which complements the detection of single-domain malicious activities. The associations among malicious domains mainly fall into two categories: client similarity and resolution similarity.

Client Similarity. Multiple malicious domains are accessed by a collection of overlapping clients, which is the client similarity. This similarity is determined by infected malware for the fact that the hosts infected with the same malware may have the same list of querying domains or seeds. In addition, they have the same attack strategy: get instructions from one domain, then download or update malware from another domain. Normal and malicious domains generally have no intersection on the client side. However, there are some exceptions. For example, some malwares now use fake-querying technique to query some benign domains deliberately which interfere with the detections. Fortunately, these benign domains have obvious recognizable features, such as being accessed by a large number of hosts. In that case, they can be easily removed.

Resolution Similarity. A plurality of malicious domains are hosted on the same malicious IPs which constitutes the resolution similarity. Resolution sharing reveals the essential association among domains which is not changed by the will of people. This is due to the limitation of malicious IP resources and the flexibility of multi-domain malware. Malicious IPs, as pivotal resources available for attackers, are very small in quantity. While, many malwares need to use multiple domains to evade detection, improve the usability, attacking ability or survivability. Given these points, multiple malicious domains are hosted on the same set of IPs controlled or maintained by attackers. For example, once a domain is blacklisted, malwares based on DGA will generate other domains which are also resolved to the same IP.

3.2 Graph Convolutional Network

The convolution in Convolutional Neural Network (CNN) essentially uses a shared parameter filter (kernel) to extract spatial features by calculating the weighted sum of central points and adjacent points. GCN is a kind of deep learning technology which applies CNN technology to graph. Given an attributed

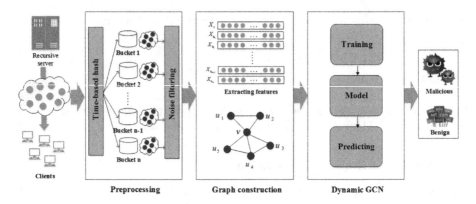

Fig. 1. The process flow diagram of malicious domain mining by using Ringer.

Table 1. The symbols used in the paper

Symbols	Meanings
TP	True positive; a malicious domain correctly identified as illicit domain
FP	False positive; a benign domain incorrectly identified as illicit domain
TN	True negative; a benign domain correctly identified as legitimate domain
FN	False negative; a malicious domain incorrectly identified as legitimate domain

un-directed graph $G = (V, E)$, where V is a set of nodes with the size of M and E is a set of edges. It is also assumed that $X \in R^{M \times N}$ is the feature matrix with each node $v \in V$ endued N-dimensional attributes, and $A \in R^{M \times M}$ is the adjacency matrix in which $A_{i,j} = a_{i,j}$ if there is an edge $e = <i, j>$ with the corresponding weight $a_{i,j}$, otherwise $A_{i,j} = 0$. Kipf et $al.$ [18] give the layer-wise propagation rule of multi-layer GCN:

$$H^{(l+1)} = \sigma \left(D^{-\frac{1}{2}} \tilde{A} D^{-\frac{1}{2}} H^{(l)} W^{(l)} \right) \tag{1}$$

Where $\tilde{L} = D^{-\frac{1}{2}} \tilde{A} D^{-\frac{1}{2}}$ is the symmetric normalized laplacian with $\tilde{A} = A + I$, I is M-demensional identity matrix, D is the diagonal degree matrix. $W^{(l)}$ is the weight matrix of layer l and $H^{(l)}$ is input activation of layer l with $H^{(0)} = X$. σ denotes activation function (eg., $ReLU(.) = max(0, .)$).

When we only take into account 2-layer GCN, the forward model becomes the following form:

$$Z = softmax \left(\tilde{L} ReLU(\tilde{L} X W^{(0)}) W^{(1)} \right) \tag{2}$$

Here, $W^{(0)} \in R^{M \times H}$ is an input-to-hidden weight matrix with H feature maps in hidden layer. $W^{(1)} \in R^{H \times F}$ is a hidden-to-output weight matrix assuming that the output has the F categories. Softmax activation function ranges each element of a vector x to $[0, 1]$ and the sum of all elements to 1 with definition

$softmax(x_i) = \frac{exp(z_i)}{\sum_i exp(x_i)}$. We evaluate the cross-entropy loss function over the labeled samples as:

$$loss = -\sum_{l \in y_L} \sum_{f=1}^{F} Y_{lf} ln Z_{lf} \qquad (3)$$

Where y_L is the indices of nodes that have labels in semi-supervised multi-class classification. The weight matrix $W^{(0)}$ and $W^{(1)}$ are trained by using gradient descent method to minimize the loss function (3).

GCN simultaneously learns the structure and attribute information, which have achieved great success in clustering and classification tasks [30,32]. However, from formula (1), it can be seen that the output of each node in the current layer needs the attribute support of its neighbors from the previous layer, and with the increase of the number of layers, the number of support required by each node increases in exponential explosion. Obviously, it is only suitable for working on relatively small graphs, not for large-scale domain data such as ours. Some works [12,15] provide a perspective on the effectiveness of sampling neighbors.

4   System Design

The goal of Ringer is to discover domains involved in malicious activities by analyzing passive DNS traffic (traces). As shown in the Fig. 1, the system architecture of Ringer consists of three modules: preprocessing, graph construction and dynamic GCN. In order to better describe our research, we introduce some notations listed in Table 1.

4.1   Preprocessing

The preprocessing module consists of two parts: time-based hash and noise filtering.

Time-Based Hash. The goal of time-based hash is to divide the big graphs into small ones according to the granularity of the time. Typically, domain-based malicious behaviors are generally relatively continuous and have short duration, such as domains generated by DGA. Intuitively, if the shared similarity between two domains happens within a short period of time, then their relevance will be strong. Relatively, if a period of time is very long, it has high probability to arise shared similarity for various reasons (such as server migration), which would introduce weak correlations. For example, the relevance intensity of two domains that have shared hosts within an hour and distinct weeks is different. Therefore, the length of the graph composition time is going to affect TPs and FPs. If the time is too short, the association between domains will not show up. If the time is too long, a large number of weak correlations will be introduced. Due to the page constraint, we will discuss the impact of time on performance in future studies. In this paper, we refer to [10] and empirically select the timespans as one hour. The time-based hash step is shown in the Fig. 1. Firstly, all records are hash to different time buckets according to the timestamp. Then, the noise filtering is used to remove some noises for each bucket. Finally, we construct the graphs in each hash bucket for the succedent operation.

Noise Filtering. Noise filtering is designed to remove the noises introduced by some normal users as well as reduce the FPs. We adopt two strategies to filter DNS traffic. Firstly, we remove domains based on the client degree. We define the client degree of a domain as the number of clients that query the domain at a given epoch. The more clients the domain is queried by, the more popular the domain is and the less likely it is to be malicious. We remove the domains queried by more than N clients. We will discuss the selection of the threshold N later. Similarly, we can discuss the resolution degree, remove the domains with more than M resolution values. In this way, we can remove some public domains, such as content delivery networks (CDN), cloud and other domains for public services. And then, we further get rid of domains generated by some normal applications such as BitTorrent due to the fact that they have obvious characteristics, for example, ending with 'tracker' or 'DHCP'.

4.2 Graph Construction

Ringer uses graph representation to express domains as the nodes and the relationship between two domains as an edge. If two domains have shared clients (resolutions) within a given period of time (in one hash bucket), then an edge exists between the corresponding nodes in the graph. The weight of the edge is the number of shared clients (resolutions) of the two domains. The graph construction algorithm is summarized as Algorithm 1. In order to save space, we store the attributes of nodes and edges as files. We can get them from the file system whenever we need. As a result, the graph construction process outputs of the correlation graph of domains and the graph-related property files. The construction of the graph aggregates the DNS query information on the graph in a cumulative manner, which makes the whole system more robust and scalable.

4.3 Dynamic GCN

Dynamic GCN is the key part of our method, which takes graph structure and attributes associated nodes as input to generate high-quality embedding for each node. These embeddings are then fed to full-connected neural network for domain classification. In this section, we discuss the technical details of Dynamic GCN.

Dynamic GCN uses localized graph convolution to generate embeddings from selectively sampled neighbors. The specific forward propagation procedure is shown as Algorithm 2.

More specifically, in order to generate the embedding of node v, we only need to consider the input features and the graph neighbors $N_v = \{u|(u,v) \in E\}$. We first selectionly sample neighbors depending on the weight of the edges. Then, we introduce node-wise attention aggregator to aggregate neighbors in attention

---

Algorithm 1. Algorithm to construct domain graph through passive DNS.

Input: DNS data D with each record r in the form of $\{sip,\ domain,\ type,\ response\}$.
Output: Domain graph G.
1: Initializing domain graph $G \leftarrow (V, E)$, $V \leftarrow \varnothing$ and $E \leftarrow \varnothing$, domain-clients$\leftarrow dict()$, client-domains $\leftarrow dict()$.
2: **for** each record $r \in D$ **do**
3:     domain-clients[domain].update(sip)
4:     client-domains[sip].update(domain)
5: **end for**
6: **for** each client in client-domains **do**
7:     **if** there is only one domain d in client-domains[client] **then**
8:         $V = V \cup \{d\}$
9:         continue
10:     **end if**
11:     **for** each pair $(d_1, d_2) \in$ client-domains[client] **do**
12:         $V = V \cup \{d_1, d_2\}$
13:         **if** edge $e(d_1, d_2) \in E$ **then**
14:             weight(e) = getEdgeWeight(e) + 1
15:         **else**
16:             $E = E \cup \{e(d_1, d_2)\}$
17:             weight(e) = 1
18:         **end if**
19:     **end for**
20: **end for**
21: **return** G

Algorithm 2. Algorithm to construct domain node representation.

Input: Current embedding z_v^k of node v at layer k, set of neighbor embeddings $\{z_u | u \in N(v)\}$, preceding embeddings $(z_v^{(0)}, z_v^{(1)}, ..., z_v^{(k)})$ of node v, AGG_{node} and AGG_{layer}.
Output: New vector representations $z_v^{(k+1)}$ at layer $k + 1$.
1: $h_v^{(k+1)} = AGG_{node}(z_v^{(k)}, \left\{ z_u^{(k)} \right\}_{u \in N_v})$
2: $z_v^{(k+1)} = \sigma \left(W * CONCAT(z_v^{(k)}, h_v^{(k+1)})) + b \right)$
3: $z_v^{(k+1)} = z_v^{(k+1)} / \left\| z_v^{(k+1)} \right\|_2$
4: $z_v^{(k+1)} = AGG_{layer}(z_v^{(0)}, z_v^{(1)}, ..., z_v^{(k)}, z_v^{(k+1)})$

---

fashion. Next, we concate the aggregation from the neighbors with the current representation of node to form the intermediate representation that will be encoded into a new node embedding; The output of the algorithm is the node representation that incorporates both information itself and the neighbors, on which we use the $L2$ normalization to prevent the model from over-fitting and make the training more stable. Finally, layer-wise attention module is implemented to encode embeddings generated by different layers for enhancement of representation learning.

Selectively Sample Neighbors. Instead of random sampling, we sample neighbors according to the weight of the edges. We deem to that the larger the weight of the edge is, the stronger the correlation between the two points is. As a consequence, we sample the top m nodes for each node with the largest weights as neighbors. It is important to set neighbor parameters for later computational models when we take into account the fixed number of nodes. Moreover, we can control the use of memory footprint during training with the certain nodes selected to aggregate.

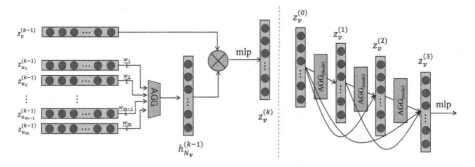

Fig. 2. Node-wise attention aggregator (left) and layer-wise attention encoder (right).

Node-Wise Attention Aggregator. With neighbor context, we can use GCN to aggregate neighbor features. However, neighbors have different influence on the target node due to the different relationship strength between nodes or some noises. Therefore, we propose node-wise attention aggregator which uses a weighted approach to aggregate neighbor nodes. As shown in Fig. 2 (left), we use the normalized weights as the attention weights which is in accordance with their relationship intensities. That is, for neighbors $N_v = \{u_1, u_2, ..., u_m\}$ with the corresponding weights $\{w_1, w_2, ..., w_m\}$, we can compute attention weights $\left\{\frac{w_1}{L}, \frac{w_2}{L}, ..., \frac{w_m}{L}\right\}$ and $L = |w_1| + |w_2| + ... + |w_m|$ is $L1$ normalization of weights. Compared with the original GCN, our specific node-wise attention aggregator formula is shown as follows:

$$h_v^{(k)} = AGG_{node}(z_v^{k-1}, \{z_u^{k-1}\}_{u \in N_v}) = \delta \left(W * (h_v^{k-1} + \sum_{i=1}^{m} \frac{w_i}{L} z_i^{k-1}) + b \right)$$
(4)

Here, k denotes the layer index k; w and b are trainable parameters of weights and bias respectively; z_v^0 is initialized by x_v.

Layer-Wise Attention Encoder. Encouraged by residual network, many works [32] adopt skip-connection to stack convolutions for high performance. Except for this, we also propose the layer-wise attention encoder to aggregate all intermediate representations, which is similar to DenseNet [16]. Figure 2 (right) schematically illustrates the layout of the layer-wise attention encoder which introduces direct connections from any layer to all subsequent layers. The node embedding z_v^k at kth layer receives all node feature maps $z_v^0, z_v^1, ..., z_v^{k-1}$ from previous layers as input. Formally, We use the following formula with $[z_v^{(0)} : z_v^{(1)} : ... : z_v^{(k-1)}]$ referring to the concatenation operation.

$$z_v^{(k)} = AGG_{layer}(z_v^{(0)}, z_v^{(1)}, ..., z_v^{(k-1)}) = \delta \left(W * [z_v^{(0)} : z_v^{(1)} : ... : z_v^{(k-1)}] + b \right) \tag{5}$$

Training Details. We train Ringer in a semi-supervised way. We define loss functions such as formula (3). Our goal is to optimize the parameters so that the output labels produced by model is close to the ground truth in labeled dataset. We use mini-batch gradient descent to train for each iteration, which make it fit in memory.

5 Data Collection

ISP Dataset. To verify the effectiveness of our scheme, we collect DNS traces in an ISP recursive server within two days from November 4th to November 5th, 2017. Many steps have been taken by our data provider to eliminate privacy risks for the network users. Due to the limitation of storage space and computing resources, we uniformly sample the data with the scale of 1/10. Previous work [22] has proved that uniform sampling makes it effective to evaluate and extrapolate the key attributes of the original dataset from the samples. This billion-level data contains 1.1 billion pieces of data per hour, which contains various types of DNS records. Our experiment is based only on the A records, thus ensuring the existence of domain-to-IP resolution except Non-Existent Domains (NXdomains). After time-based hash operation, our data is cut into 48 portions according to the time series. To better illustrate our data, we set forth the overview of dataset (shown as Fig. 3) in one bucket from 1:00 clock to 2:00 on November 4th, 2017, which only retains domains on A record. We can see that the distribution of querying host and resolution are heavy-tailed, with a few domains being accessed by a large number of hosts or resolved to a large number resolution IPs.

Ground Truth. We have collected 21558 distinct second-level domains from a number of authoritative blacklists, including malwaredomains.com [2], Zeustracker [7], malwaredomainlist.com [5], malc0de.com [4], bambenek consulting [1]. We also use all DGA data from DGArchive [24] until December 31, 2018, totaling 87 DGA families or variants, 88614672 distinct domain samples. To obtain benign domains, we use the domains that appear for one year (2017) in Alexa Top 1 Million Global Sites (https://www.alexa.com/topsites) as benign domains. The owners of these sites have always maintained these sites well, so they have good reputation. Motivated by this theory, we have collected benign domains with a number of 404,536. We further drop the malicious domains out

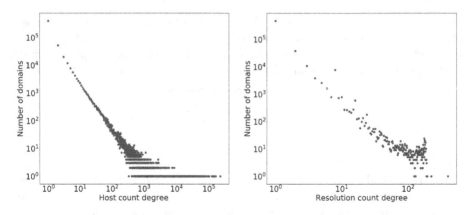

Fig. 3. Distribution of the number of domains with host count degree (left) and resolution count degree (right) in bucket 2 (within one hour from 1 o'clock to 2).

Table 2. Features extracted in this paper for learning

Type	Feature description
Structural	Domain name length, Number of subdomains, Subdomain length mean, Has www prefix, Has valid TLD, Contains single-character subdomain, Is exclusive prefix repetition, Contains TLD as subdomain, Ratio of digit-exclusive subdomains, Ratio of hexadecimal-exclusive subdomains, Underscore ratio, Contains IP address
Linguistic	Contains digits, Vowel ratio, Digit ratio, Alphabet cardinality, Ratio of repeated characters, Ratio of consecutive consonants, Ratio of consecutive digits, Ratio of meaningful words
Statistical	N-gram (N = 1, 2, 3) frequency distribution (mean, standard deviation, median, max, min, the lower quartile, the upper quartile), Entropy

of benign domain list. Although there are some FPs and FNs in the ground truth, it is relatively effective to evaluate our method.

Features for Learning. In the experiment, we extracted 42 statistical features to represent domain names with reference to FANCI [26], which are listed in Table 2. These features can be divided into three categories, including structural features, linguistic features and statistical features.

6   Experiments

In this section, we discuss the selection of parameters on real-world data and evaluate the effectiveness of our approach, as well as tracking ability.

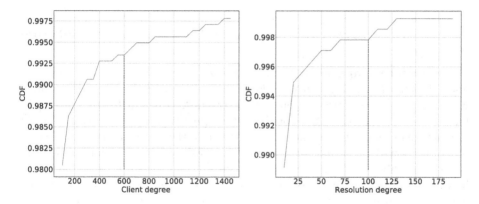

Fig. 4. The CDF distribution of the domain number over client degree (left) and resolution degree (right) in ground truth.

6.1 Selection of Threshold N and M

In order to select appropriate threshold N (client degree) and M (resolution degree), we count the client and resolution degree of each domain in ground truth. Figure 4 shows the Cumulative Distribution Function (CDF) of the domain number on client degree and resolution degree in one bucket. We find that more than 99% of malicious domains have client degree less than 600. We manually check these domains with relatively large client degree and find that they are some FPs, such as the CDN or Dynamic DNS. For example, the domain "ec2-52-207-234-89.compute-1.amazonaws.com" is used for Cerber ransomware reported by malwaredomainlist.com, however, it is unreasonable for us to take the second-level domain "amazonaws.com" that is in alexa top list as a malicious domain. In order to cover more malicious domains as well as reduce the FPs, we empirically choose a threshold $N = 600$. Through this threshold, we retain 99% of the malicious domains. Similarly, we have determined that the resolution degree is $M = 100$. We manually checked the domains we removed, such as "herokuapp.com", "igexin.com" and "cloudapp.net". All of them are domains used to provide public services, which are less likely to malicious domains. Therefore, it makes sense to remove those domains.

6.2 Detection Accuracy

We use three metrics to measure the performance of our approach, namely True Positive Rate (TPR), False Positive Rate (FPR) and accuracy (ACC) respectively. $TPR = \frac{|TPs|}{|TPs|+|FNs|}$, $FPR = \frac{|FPs|}{|TNs|+|FPs|}$ and $ACC = \frac{|TPs|+|TNs|}{|TPs|+|TNs|+|FPs|+|FNs|}$. For each of our buckets, there are around 3000 malicious domains and around 200,000 benign domains in ground truth. Total 136,827 malicious domains are labeled in two days (there are duplicate domains because they come from different buckets, but they have different neighbors). We randomly select the same number of benign domains (136,827) with malicious domains which are fed to Ringer to adopt K-fold cross validation (in this paper, $K = 5$). The results are shown in Table 3. Our system implements high TPR (0.957 on average) and low FPR (0.020 on average) in domain classification.

Baselines for Comparison. We make the comparison with the following state-of-art baselines including FANCI [26], node2vec [14] combined with features. FANCI is an open-source system that extracts 21 domain features and uses machine learning (such as Support Vector Machine (SVM)) to classify domains into benign or malicious instances. Node2vec maps the nodes into the feature vectors of the low-dimensional space by unsupervised learning, which preserves the network neighborhood relation of nodes. Node2vec is also open source and publicly available. Node2vec learns the node embeddings that only encodes the information of network structure. For this reason, we first concate the node embeddings with the extracted features, and then use SVM to classify the domains.

Table 3. Comparison of detection performance over Ringer, FANCI and Node2vec for domain classification using K-fold (for $K = 5$)

Systems	Metrics	1	2	3	4	5	Average
Ringer	TPR	0.956	0.955	0.958	0.958	0.956	**0.957**
	FPR	0.018	0.020	0.024	0.020	0.020	**0.020**
	ACC	0.972	0.965	0.970	0.968	0.966	**0.968**
FANCI	TPR	0.897	0.900	0.912	0.899	0.912	0.904
	FPR	0.027	0.027	0.019	0.023	0.019	0.023
	ACC	0.939	0.936	0.937	0.938	0.947	0.939
Node2vec	TPR	0.913	0.907	0.921	0.917	0.917	0.915
	FPR	0.027	0.029	0.025	0.038	0.026	0.030
	ACC	0.941	0.938	0.947	0.939	0.944	0.942

We applied these two methods on our dataset and the results are shown in the Table 3. Both of systems achieve promising results. However, FANCI ignores the structure information, and these relations embodied in time and space are of great significance to the classification of malicious domains. Node2vec only considers the structure relationship among the domains, and we concate the learned node embeddings with the static features, which has some improvement in classification performance. Yet, node2vec can only learn associated node embeddings by using unsupervised learning, and there is nothing to do with isolated nodes. To learn the node embedding of newly added nodes, all nodes need to learn in global fashion, which is time-consuming and labor-intensive. It is obvious that the performance of our system was significantly outperform the other two systems. In order to achieve optimal results, Ringer is capable of simultaneously learning statistical and structural features, and combining intermediate representations to make full use of all multi-order information of domains.

Domains detected by Ringer can be viewed as subgraphs. For example, Ringer detects 152 malicious domain subgraphs in bucket 2. In order to express the results of our method more intuitively, we select the top 8 subgraphs according to the size of connected components. The results drawed by software gephi [3] are showed as Fig. 5. The images vividly elaborate how attackers to organize and deploy malicious domains, which is more conducive to the future study of malicious resources.

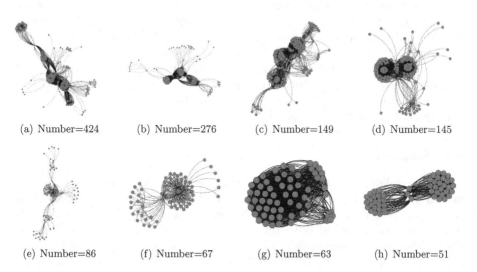

(a) Number=424   (b) Number=276   (c) Number=149   (d) Number=145

(e) Number=86   (f) Number=67   (g) Number=63   (h) Number=51

Fig. 5. The 8 samples of malicious domain subgraphs detected from bucket 2.

Table 4. The results of LSTM and FANCI in prolongation

Systems	Prolongation	TPs	FNs	Percent
LSTM	6109	5192	917	0.85
FANCI	6109	5387	722	0.88

6.3 Prolongation

One significant function of Ringer is to detect potential malicious domains. As we all know, the redundancy of multi-domain names provides better flexibility for malware. Therefore, detection of newly generated (zero-day) and rarely used malicious domains is an important metric of the system. We apply the model directly on domains that are not in ground truth, and we find 6109 potentially suspicious malicious domains.

For the thousands of potential malicious domains detected by our method, we use two heuristic methods to verify their wickedness conservatively. Firstly, we use two excellent systems, FANCI [26] and LSTM [29], both of which are open source and public available. FANCI adopts machine learning on the domain features and LSTM uses deep learning technology to distinguish between good and malicious domains. The results of the classification of new suspicious domains by the two systems are as shown as Table 4. Through two systems, there are still 530 unique domains that cannot be confirmed. Secondly, we try to find some historical snapshots of these domains or the IPs parsed from them by Virus-Total (www.virustotal.com). Although there are many public blacklists, they are not very comprehensive, some of which contain only malicious domains for the given day, while VirusTotal collects 66 authoritative blacklists. We deem to the domains that appear at least one blacklist as malicious domains by using the public API [6]. At a result, 108 new malicious domains we have found are exposed in the form of domains or IPs. We observed that there are still 422 domain names without qualitative judgment. Through analysis, a total of 5687 malicious domain names are found, and we have 93% confidence to believe that our system have ability to respond against new threats.

6.4   Scalability

Ringer is scalable to large-scale DNS data, such as DNS traces from ISP. To further illustrate the scalability of our system, we analyze the complexity of Ringer. Suppose that we have N records and the number of unique domain with return value on A record is V, the complexity of Ringer is shown as follows. During the graph construction, all DNS records need to be scanned which takes $O(N)$. The graph convolution is our main computational workhorse. Previously spectral convolutions defined on the graph is multiplication using a filter $g\theta(L) = diag(\theta)$ in Fourier domain with the given signal x that is a scalar for every node:

$$g\theta(L)x = g\theta(U\Lambda U^T)x = Ug\theta(\Lambda)U^T x \tag{6}$$

Where $\theta \in R^V$, U is the matrix of eigenvectors of the normalized graph Laplacian matrix. $L = I_V - D^{-\frac{1}{2}}AD^{-\frac{1}{2}} = U\Lambda U^T$ with a diagonal matrix of its eigenvalues Λ. The computational complexity of formula (6) is $O(V^2)$. And for large graphs, the eigendecomposition of L is prohibitively expensive. In order to alleviate the computational cost, we adopt K-order truncated Chebyshev polynomials as [13] approximate $g\theta(\Lambda)$: $g\theta(\Lambda) \approx \sum_{k=0}^{K-1} \theta_k T_k\left(\tilde{\Lambda}\right)$. Where $\tilde{\Lambda} = \frac{2}{\lambda_{max}}\Lambda - I_V$, and λ_{max} is the largest eigenvalue of L. The Chebyshev ploynomials is recursively defined as $T_k(x) = 2xT_{k-1}(x) - T_{k-2}(x)$, with $T_0 = 1$ and $T_1 = x$. Therefore, we have the approximations:

$$g\theta(L)x \approx \sum_{k=0}^{K-1} \theta_k T_k\left(\tilde{L}\right)x \tag{7}$$

Where θ_k is a vector of Chebyshev coefficients and $\tilde{L} = \frac{2}{\lambda_{max}}L - I_V$ is the scaled Laplacian matrix. The complexity of formula (7) is $O(E)$, linear in the number of edges. Through the above analysis, the complexity is mainly related to the number of unique edges and the $E \leq m * V$ where the m is the number of sampled neighbors (constant).

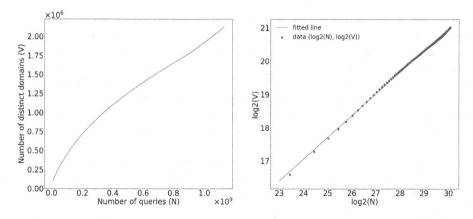

Fig. 6. The distribution of the number of distinct domains over the number of DNS queries (left) and the straight line fitted with the Least Squares approximation on points $(log2(N), log2(V))$ (right). (Color figure online)

Figure 6 (left) shows the distribution of the number of distinct domains to be analyzed changing with the number of DNS queries in real-world data. As an illustration, the size of unique domains does not increase linearly with the size of queries, but follows Heaps' law $V(N) = \alpha N^\beta$. In order to find the appropriate parameters α and β, we take the log2 on both sides of the equation. Then, we get $log_2(V) = log_2(\alpha) + \beta log_2(N)$. Figure 6 (right) shows the scattered distribution of points $(log2(N), log2(V))$ (blue), and the straight line (red) fitted with the Least Squares approximation. Finally, in our dataset, the parameters $\alpha = 2.577$, $\beta = 0.6536$. To sum up, the computation overhead of whole system is $O(N)$ linear in the number of input records, which proves scalable.

6.5 Limitation

We discuss about two drawbacks that need to be considered. One potential limitation is that Ringer cannot distinguish specific service categories that the malicious domains is used for, such as fishing, spamming, C2 and so on. We lack more detailed knowledge base or ground truth to cover and label them, thus obtaining the distribution of different malicious categories. Another limitation is that for unassociated or weakly associated domains, our approach is equivalent to using only its static statistical features without neighbors relevance.

7   Conclusion

The intelligence of attackers using malicious domains makes it more resilient for existing detection methods. In this paper, we propose a malicious domain detection mechanism Ringer, which uses graphs to represent the strong correlation among domains including client similarity and resolution similarity. Dynamic

GCN is used to learn the node representations that combines structural information and statistical feature information inductively. The dynamic learning depending on neighbors and itself enables great gains in both effectiveness and efficiency. Exposing the relevance of malicious domains by graph enhances the robustness of the system irrespective of some evading techniques. We use DNS data from ISP to evaluate our system, the results show that our system perform higher precision and scalability. Our approach, as a promising effort, helps to prevent illegal domain-based activities more effectively in practice.

Acknowledgments. This work is supported by the National Key Research and Development Program of China (Grant No. 2016YFB0801502), and the National Natural Science Foundation of China (Grant No. U1736218). We are grateful to the anonymous reviewers for their insightful comments and many thanks to Daniel Plohmann for providing us access to the DGArchive. And, I also owe my sincere gratitude to Ms. Zhou Yue, who helped me a lot during the writing of this thesis. The corresponding author of this paper is Shuhao Li.

References

1. Bambenek Consulting (2019). https://www.bambenekconsulting.com/
2. DNS-BH (2019). http://www.malwaredomains.com/
3. Gephi: The Open Graph Viz Platform (2019). https://gephi.org/
4. Malc0de.com (2019). https://malc0de.com/bl/ZONES
5. Malware Domain List (2019). http://www.malwaredomainlist.com/
6. VirusTotal public API (2019). https://github.com/clairmont32/VirusTotal-Tools
7. Zeustracker (2019). https://zeustracker.abuse.ch/blocklist.php
8. Anderson, H.S., Woodbridge, J., Filar, B.: DeepDGA: adversarially-tuned domain generation and detection. In: Proceedings of the 2016 ACM Workshop on Artificial Intelligence and Security, pp. 13–21. ACM (2016)
9. Antonakakis, M., Perdisci, R., Dagon, D., Lee, W., Feamster, N.: Building a dynamic reputation system for DNS. In: USENIX Security Symposium, pp. 273–290 (2010)
10. Antonakakis, M., et al.: From throw-away traffic to bots: detecting the rise of DGA-based malware. In: Presented as Part of the 21st USENIX Security Symposium (USENIX Security 12), pp. 491–506 (2012)
11. Bilge, L., Kirda, E., Kruegel, C., Balduzzi, M.: EXPOSURE: finding malicious domains using passive DNS analysis. In: NDSS, pp. 1–17 (2011)
12. Chen, J., Ma, T., Xiao, C.: FastGCN: fast learning with graph convolutional networks via importance sampling. arXiv preprint arXiv:1801.10247 (2018)
13. Duvenaud, D.K., et al.: Convolutional networks on graphs for learning molecular fingerprints. In: Advances in Neural Information Processing Systems, pp. 2224–2232 (2015)
14. Grover, A., Leskovec, J.: node2vec: scalable feature learning for networks. In: Proceedings of the 22nd ACM SIGKDD International Conference on Knowledge Discovery and Data Mining (2016)
15. Hamilton, W., Ying, Z., Leskovec, J.: Inductive representation learning on large graphs. In: Advances in Neural Information Processing Systems, pp. 1024–1034 (2017)

16. Huang, G., Liu, Z., Van Der Maaten, L., Weinberger, K.Q.: Densely connected convolutional networks. In: Proceedings of the IEEE Conference on Computer VIsion and Pattern Recognition, pp. 4700–4708 (2017)

17. Khalil, I., Yu, T., Guan, B.: Discovering malicious domains through passive DNS data graph analysis. In: Proceedings of the 11th ACM on Asia Conference on Computer and Communications Security, pp. 663–674. ACM (2016)

18. Kipf, T.N., Welling, M.: Semi-supervised classification with graph convolutional networks. arXiv preprint arXiv:1609.02907 (2016)

19. Lee, J., Lee, H.: GMAD: graph-based malware activity detection by DNS traffic analysis. Comput. Commun. **49**, 33–47 (2014)

20. Liu, Z., Yun, X., Zhang, Y., Wang, Y.: CCGA: clustering and capturing group activities for DGA-based botnets detection. In: 2019 18th IEEE International Conference on Trust, Security and Privacy in Computing and Communications/13th IEEE International Conference on Big Data Science and Engineering (TrustCom/BigDataSE), pp. 136–143. IEEE (2019)

21. Manadhata, P.K., Yadav, S., Rao, P., Horne, W.: Detecting malicious domains via graph inference. In: Kutyłowski, M., Vaidya, J. (eds.) ESORICS 2014. LNCS, vol. 8712, pp. 1–18. Springer, Cham (2014). https://doi.org/10.1007/978-3-319-11203-9_1

22. Papalexakis, E.E., Dumitras, T., Chau, D.H., Prakash, B.A., Faloutsos, C.: Spatio-temporal mining of software adoption & penetration. In: 2013 IEEE/ACM International Conference on Advances in Social Networks Analysis and Mining (ASONAM 2013), pp. 878–885. IEEE (2013)

23. Peng, C., Yun, X., Zhang, Y., Li, S.: MalShoot: shooting malicious domains through graph embedding on passive DNS data. In: Gao, H., Wang, X., Yin, Y., Iqbal, M. (eds.) CollaborateCom 2018. LNICST, vol. 268, pp. 488–503. Springer, Cham (2019). https://doi.org/10.1007/978-3-030-12981-1_34

24. Plohmann, D., Yakdan, K., Klatt, M., Bader, J., Gerhards-Padilla, E.: A comprehensive measurement study of domain generating malware. In: USENIX Security Symposium, pp. 263–278 (2016)

25. Rahbarinia, B., Perdisci, R., Antonakakis, M.: Segugio: efficient behavior-based tracking of malware-control domains in large ISP networks. In: 2015 45th Annual IEEE/IFIP International Conference on Dependable Systems and Networks, pp. 403–414. IEEE (2015)

26. Schüppen, S., Teubert, D., Herrmann, P., Meyer, U.: FANCI: feature-based automated NXDomain classification and intelligence. In: 27th USENIX Security Symposium (USENIX Security 18), pp. 1165–1181 (2018)

27. Tamersoy, A., Roundy, K., Chau, D.H.: Guilt by association: large scale malware detection by mining file-relation graphs. In: Proceedings of the 20th ACM SIGKDD International Conference on Knowledge Discovery and Data Mining, pp. 1524–1533. ACM (2014)

28. Thomas, M., Mohaisen, A.: Kindred domains: detecting and clustering botnet domains using DNS traffic. In: Proceedings of the 23rd International Conference on World Wide Web, pp. 707–712. ACM (2014)

29. Woodbridge, J., Anderson, H.S., Ahuja, A., Grant, D.: Predicting domain generation algorithms with long short-term memory networks. arXiv preprint arXiv:1611.00791 (2016)

30. Wu, Z., Pan, S., Chen, F., Long, G., Zhang, C., Yu, P.S.: A comprehensive survey on graph neural networks. arXiv preprint arXiv:1901.00596 (2019)

31. Yedidia, J.S., Freeman, W.T., Weiss, Y.: Understanding belief propagation and its generalizations. In: Exploring Artificial Intelligence in the New Millennium. vol. 8, pp. 236–239 (2003)

32. Ying, R., He, R., Chen, K., Eksombatchai, P., Hamilton, W.L., Leskovec, J.: Graph convolutional neural networks for web-scale recommender systems. In: Proceedings of the 24th ACM SIGKDD International Conference on Knowledge Discovery & Data Mining, pp. 974–983. ACM (2018)

An Empirical Evaluation of Attention and Pointer Networks for Paraphrase Generation

Varun Gupta[1] and Adam Krzyżak[1,2(✉)]

[1] Department of Computer Science and Software Engineering, Concordia University,
Montreal H3G 1M8, Canada
varun9208@yahoo.com, krzyzak@cs.concordia.ca
[2] Department of Electrical Engineering, Westpomeranian University of Technology,
70-313 Szczecin, Poland

Abstract. In computer vision, one of the common practices to augment the image dataset is by creating new images using geometric transformation preserving similarity. This data augmentation was one of the most significant factors for winning the Image Net competition in 2012 with vast neural networks. Unlike in computer vision and speech data, there have not been many techniques explored to augment data in natural language processing (NLP). The only technique explored in the text data is lexical substitution, which only focuses on replacing words by synonyms.

In this paper, we investigate the use of different pointer networks with the sequence-to-sequence models, which have shown excellent results in neural machine translation (NMT) and text simplification tasks, in generating similar sentences using a sequence-to-sequence model and the paraphrase dataset (PPDB). The evaluation of these paraphrases is carried out by augmenting the training dataset of IMDb movie review dataset and comparing its performance with the baseline model. To our best knowledge, this is the first study on generating paraphrases using these models with the help of PPDB dataset.

Keywords: Paraphrase generation · Data augmentation · Attention networks · Pointer networks

1 Paraphrase Theory

A standard ability of human language communication is the ability of humans to communicate the equivalent information in multiple ways. These dialects are known as paraphrases, which in the literature have been also referred to as reformulations, restating and other diversity of phenomena.

Supported by the Natural Sciences and Engineering Research Council of Canada. Part of this research was carried out by the second author during his visit of the Westpomeranian University of Technology while on sabbatical leave from Concordia University.

© Springer Nature Switzerland AG 2020
V. V. Krzhizhanovskaya et al. (Eds.): ICCS 2020, LNCS 12139, pp. 399–413, 2020.
https://doi.org/10.1007/978-3-030-50420-5_29

There can be many variations of paraphrases, but in this work, we try to limit generation of paraphrases to such, which can be carried out by linguistic and semantic knowledge to produce similar sentences. Here are some examples[1]

1. **S:** The couple wants to **purchase** a home.
   P: The couple wants to **buy** a home.
2. **S: It was** a Honda **that** John **sold to** Aman.
   P: John **sold** a Honda **to** Aman.
3. **S:** Aman **bought** a house **from** John.
   P: John **sold** a house **to** Aman.
4. **S:** The flood **last year** was a terrible catastrophe in which many people died.
   P: The flood **in 2006** was a terrible catastrophe in which many people died.

In all the examples mentioned above, we only require linguistic, lexical, referential and structural knowledge to generate paraphrases. Example (1), is generated using knowledge of synonym words which comes under the lexical category. Example (2) is generated using structural information, which comes under syntactic knowledge. This type of transformation is described in the theory of transformational grammar [6]. Example (3) is an illustration of alternation which can be carried out by syntactic transformation. Example (4) is an instance of referential paraphrase.

One common thing about all the above-generated paraphrases is that we do not need any domain knowledge or domain is common in both the original sentence and in its paraphrase sentence, i.e., 'English literature.' These things become more tricky when we try to generate paraphrases where the original sentence is in one domain, but we want to generate paraphrase in a domain other than the original domain. Here is an example of these kinds of paraphrases.

5. **S:** Nearest neighbor is good.
   P (Literature Domain): The closest neighbor is good.
   P (Machine learning Domain): The closest neighbor is good.

As we can see in the above example when generating paraphrase in one domain for example in 'English literature' (as described in sample 5) 'nearest neighbour' is a synonym of the 'closest neighbour.' However, when generating paraphrase in machine learning domain, it might not be a good idea to convert 'nearest neighbour' to 'closest neighbour' as 'nearest neighbour' has a technical or reserved meaning in machine learning context. This means context or domain knowledge is also required in generating paraphrases.

In this work, we focus on evaluating past methods on generating similar sentences using linguistic, lexical, referential and structural knowledge.

---

[1] Here 'S' represents the original sentence, and 'P' represents paraphrase of it. Furthermore, bold words are the primary information used in generating paraphrase.

2   Similarity Measures

Before evaluating the existing methods first, we should clarify the conditions which should be fulfilled by the constructed sentence to be considered as a paraphrase. Here are a few criteria for assessing paraphrases.

1. **Syntax Level:** The first minimal requirement for generating paraphrase from the source sentence is to have a valid syntax in the given language. In other words, it should follow all the syntax rules defined by the given language while generating a natural language paraphrase.
2. **Semantics Level:** The second minimal requirement which must be followed by the generated paraphrase in natural language generation is its meaning and interpretation of output or target sentence. The output sentence must be meaningful.
3. **Lexical Level:** Lexical level is a way to convert characters or words into tokens, which can be identified by machines (numbers). These tokens dimensions capture several likeness measures of a word, a context of a word and other things. There can be many ways to convert these characters or words into these tokens, for example, n-grams, similarity coefficients, alter remove, etc. This type of measure is useful to find the similarity or to generate more interpretations from a given source sentence. However, in our case, while generating similar sentences, they should also be checked for contextual meaning of the source sentence.
4. **Same Meaning:** The main property of paraphrase sentences is to have the same meaning in a given context. To better understand how two sentences can have the same meaning, let us describe two key terms: **Connotation** and **Denotation**. Connotation is the emotional and imaginative association surrounding a word. For example, connotations for the word snake could include evil or danger. Denotation is the strict dictionary meaning of word. For example, if we look up the word "snake" in a dictionary, we discover that one of its denotative meanings is "any of numerous scaly, legless, sometimes venomous reptiles having a long, tapering, cylindrical body and found in most tropical and temperate regions" [7].

In this work, we assess some of the existing methods which performed well in NMT and text simplification by generating paraphrases in the given context.

3   Paraphrase Generation

Paraphrase generation task is a particular case of neural machine translation (NMT) task in which, given the source sentence we need to generate an output sentence which has the same meaning. The only anomaly in a paraphrase generation and NMT is that in the former case, output sentence is also in the same language as the source sentence.

Paraphrase generator models are given an input interpretation as a source sentence, and they produce more than one (depending on the beam size) similar interpretations which are then given a score based on some criteria.

Next we describe several strategies for paraphrase generation.

Bootstrapping. This method does not need any machine translation. We generate paraphrases using templates. This technique can only be used when the input and output sentences are templates and it is applied on a large monolingual corpus. We start with retrieving the sentences in the corpus that contain seed pairs which match to the template we wish to generate. Filtering techniques are used to filter out candidate paraphrases, which are not useful enough. Next, after obtaining these templates, we look into the corpus again for the sentences which match these new templates. More seed values are extracted from new sentences, and more iterations are used to generate more templates and more seeds. This process is repeated until no new seed can be obtained or limitation on number of iterations is reached.

In this method, if the slot values can be identified reliably, then one can obtain initial seed slot values automatically by retrieving direct sentences that match the original templates. There are many well-known methods for bootstrapping; one of them is template extraction anchor set extraction (TEAS). It has been used in many information extraction patterns [1]. There are other methods which require corpora annotated with instances of particular types of events to be extracted [3].

Statistical Machine Translation (SMT). As mentioned earlier, the paraphrase generation can be seen as a particular case of the machine translation problem. In general, most of the generation tasks rely on statistical machine translation (SMT), which is based on a large corpus. Next we define SMT.

Let S be an input sentence, whose words are w_1, w_2, w_3 .... $w_{|S|}$ and let N be an instance of one candidate translation or in our case it is a candidate for good paraphrase which has words a_1, a_2, a_3 .... a_N. If we have more than one instance of such N, our aim is to find the best N^* from the list of N, which has maximum probability of being a translation or paraphrase of S (Source) sentence. This can be represented as follows:

$$N^* = \mathrm{argmax}_N P(N|S) = \mathrm{argmax}\frac{P(N)P(S|N)}{P(S)} = \mathrm{argmax}_N P(N)P(S|N) \quad (1)$$

In Eq. (1), using the conditional probability formula $\mathrm{argmax}P(N|S)$ can be further expanded as shown below. The source sentence, i.e., S is fixed, so, $P(S)$ is fixed across all translations N, hence can be removed from the denominator. $P(N|S)$ is the probability of translation given source sentence. $P(N)$ is the language model which is used to find out the probability of being a correct sentence of output sentences. Also, $P(S|N)$ is probability of translation or paraphrase model.

In the candidate sentence, each word probability is dependent on its precedence word. So, the total probability of P(N) becomes:

$$P(N) = P(a_1) * P(a_2|a_3) * P(a_3|a_1, a_2).....P(a_N|a_{N-2}, a_{N-1}) \qquad (2)$$

This language model has a smoothing mechanism. Smoothing mechanism is needed to handle cases, where n-grams are unique or do not exist in the corpus, which can lead to the language model where probability of the whole sentence is 0, i.e., $P(N) = 0$. There is some progress seen in utilizing long short-term memory (LSTM) models to produce paraphrases in this case [11]. The model consists of encoder and decoder, both utilizing varieties of the stacked remaining LSTM.

These models are ubiquitous and are very generic for a wide variety of different generation tasks in natural language processing, for example, in question answering, paraphrase generation and text summarizing. Also, they are the state-of-the-art in most of the generation task. In this work, we have used these models for generating paraphrases of the input sentence.

Parsing. Syntactic transfer in machine translation may also be used [9] to generate paraphrases. In this approach, we first need to parse the input expression. Then to generate output paraphrase sentence, these parse tree or expression are modified in a way which preserves the meaning of the syntax. There may be errors induced while parsing the input sentence.

4   Encoder-Decoder Networks for NLP

Encoder-decoders are the neural network approaches, which are genuinely ongoing models for deep learning in NLP. These models in some cases outperform classical statistical machine translation methods. The Encoder-Decoder architecture has become an effective and standard approach for both neural machine translation (NMT) and sequence-to-sequence (seq2seq) prediction tasks which involve paraphrase generation, question answering and language modeling. Encoder-decoder normally consists of two components: (1) an encoder to encode input sentence into a context vector, (2) the decoder which decodes the context vector to the output sequence. The key advantage of seq2seq model is the capacity to train a solitary end-to-end model right on the source and target sentences, and the capacity to deal with sentences of variable length to yield sequence of content. They were first introduced autonomously as Recurrent Neural Network (RNN) encoder-decoder [4] and sequence-to-sequence [12] nets for machine interpretation. This group of encoder-decoder models are regularly referred to as seq2seq, regardless of their particular execution. The seq2seq model tries to learn the conditional distribution

$$p(y_1, y_2, ..., y_{T'}|x_1, x_2, ..., x_T) \qquad (3)$$

where, y is the output sequence conditioned on the input sequence x or source sequence, $y_{T'}$ denotes the word generated by the model at time step T', where T' is the length of the output sentence and T is the length of the input sentence. T' and T are not necessarily the same. A seq2seq model first encodes the entire variable x input with its encoder RNN into a fixed size vector c known as context vector. Then, a decoder RNN generates output $y_1, ...; y'_T$ conditioned on previous predictions and the context vector c:

$$p(y_1, y_2, ..., y_{T'} | x_1, x_2, ..., x_T) = \prod_{t=1}^{T'} p(y_t | y_1, y_2, ..., y_{t-1}, c) \tag{4}$$

There are two different ways to define dependency of output sequence y on the context vector c. One way is to condition y on c at the first output from the decoder [12]. Another way is to condition every generation of $y_{T'}$ on the same context vector c, thus forming the basis to our model. For the sake of simplicity, we modify equation for vanilla RNN version to get the hidden state s at time step t, denoted by s_t. Modifying hidden state equation leads to

$$s_t = f_h(W_s x_t^d + U_s s_{t-1} + Cc + b_s). \tag{5}$$

Here and elsewhere C is the parameter matrix.

As demonstrated in [12] performance of seq2seq model while generating text can be improved by feeding the input sentence in reverse order. The framework accomplished a bilingual evaluation understudy (BLEU) score of 34.81, which is a decent score contrasted with the standard score of 33.30 achieved by STM. This is the first case of a neural machine interpretation framework that defeated a phrase-based statistical machine translation baseline on a large scale problem. However, this work has not accounted for reverse order of sentences.

The encoder-decoder framework has one disadvantage, namely that the length of a sentence increases as the performance of seq2seq model decreases. We also use other variations of RNN like Long Short Term Memory (LSTM) or Gated Recurrent unit (GRU) for better performance on long sentences.

5   Encoder-Decoder with Attention

In this section, we present the attention mechanism to improve the poor performance of seq2seq model on longer sentences [2,8].

The problem occurs while generating the word in decoder network. It looks at the entire input sentence every time while generating a new word. The basic concept of attention is to only focus on a particular part of the sentence. Each time the model predicts an output word, it only uses parts of an input where the most relevant information is concentrated instead of an entire sentence. In other words, it only pays attention to some input words.

The basic structure of the model is the same as the encoder-decoder discussed in the previous section. However, the main difference after adding attention mechanism in seq2seq model can be observed when generating the next word in the decoder network. In seq2seq model with attention mechanism, we determine the hidden state of the decoder at the current time by taking the previous output, previous hidden state and context vector. Further, note that here we are not using the single context vector c for generating all the words in the decoder network, but a separate context vector c_i for each target word $y_{T'}$.

The encoder first encodes input sentence represented by its word embedding sequence x, into a single context vector c (which is a combination of all the hidden units in encoder and represented by $c = q(h_1, ..., h_{T_x})$) and a hidden state $h_t = f(x_t, h_{t-1})$. Typically decoder network predicts the sequence by predicting one word at a time denoted by y_t, where each y_t output is conditioned on the previous outputs $y_1, ..., y_{t-1}$ and the context vector c, maximizing the following joint probability

$$p(y) = \prod_{t=1}^{T'} p(y_t | y_1, ..., y_{t-1}, c).$$ (6)

In the context of RNNs, the conditional probability of each y_t in the joint probability of (6) is modeled as a nonlinear function g with input y_{t-1} context vector c and hidden state s_{t-1}:

$$p(y_t | y_1, ..., y_{t-1}, c) = g(y_{t-1}, s_{t-1}, c).$$ (7)

Then [2] proposes to use unique vector c_t for each decoding time step, redefining the decoder conditional probability for each word y_t as:

$$p(y_t | y_1, ..., y_{t-1}, x) = g(y_{t-1}, s_{t-1}, c_t),$$ (8)

where the context vector c_t is a weighted sum over all input hidden states $(h_1, ..., h_T)$:

$$c_t = \sum_{j=1}^{T} a_{tj} h_j.$$ (9)

Here, attention weights a_{tj} are calculated as

$$a_{tj} = \frac{exp(e_{tj})}{\sum_{k=1}^{T_x} exp(e_{tk})},$$ (10)

$$e_{tj} = a(s_{t-1}, h_j).$$ (11)

The scoring function (11) is a pairwise scoring function which is used for scoring the relation between decoder hidden state s_{t-1} and encoder hidden state h_j. This scoring is learned jointly while training the whole seq2seq model with the help of a feedforward network.

There are many different kinds of attention mechanism, out of which in this work we have tried two different variations proposed in [8] and [2]. Moreover, we have used seq2seq model with an attention mechanism to compare with seq2seq model with pointer network in generating similar sentences.

6 Encoder-Decoder with Pointer Network

Encoder-Decoder network also suffers from two other problems, which are reproducing factual details or unknown words or rare word inaccurately, and also they tend to repeat themselves by generating the same words again and again. The second problem can be resolved by attention and coverage mechanisms [13].

Typically, we create a seq2seq model when we have to define the maximum value of vocabulary length, which is represented by their word embedding. Usually, this vocabulary length varies from 10,000 to 50,000, containing the maximum number of most frequent words. Note that an increase in the maximum length of vocabulary also increases computation complexity of seq2seq model and makes the training process slower. All the other words or tokens which are not incorporated under vocabulary are marked as '<UNK>' or unknown words and all these tokens have the same word embedding. Therefore whenever decoder is generating an output word of embedding <UNK> token, then decoder outputs <UNK> as a token. This is known as unknown words problem and can be very problematic in the case of paraphrase generation. To solve this unknown words problem, we use pointer network in seq2seq model.

6.1 COPYNET Network

CopyNet was proposed in [5] to incorporate copying mechanism in seq2seq model. This mechanism has shown good results on text summarization tasks on different datasets. The model uses bidirectional RNN as an encoder which transforms or encodes the variable length of the sentence into a fixed size of context vector. It has the same setup as proposed in [2]. The difference is in the way the model copies words from the input sentence in a decoder network.

The model has two sets of vocabularies: $V = V_1, ..., V_n$, which also includes <UNK> for out of vocabulary (OOV) words and the source vocabulary $A = a_1, ..a_N$, which includes unique vocabulary from input sentence. Source vocabulary makes COPYNET to copy OOV words in the output sentence.

At time t, the decoder RNN state is represented by s_t. The probability of a generated word y_t is given by

$$p(y_t|s_t, y_{t-1}, c_t, H) = p(y_t, g|s_t, y_{t-1}, c_t, H) + p(y_t, c|s_t, y_{t-1}, c_t, H), \qquad (12)$$

where H is a combination of hidden states h_r of the encoder network, c_t is a context vector at t. Here g stands for generative mode and c stands for copy mode, and these probabilities are calculated as follows

$$p(y_t, g|\cdot) = \begin{cases} \frac{1}{F}e^{\Psi_g(y_t)} & y_t \in V \\ 0 & y_t \in A \cap \bar{V} \\ \frac{1}{F}e^{\Psi_g(UNK)} & y_t \notin V \cap A \end{cases}$$

$$p(y_t, c|\cdot) = \begin{cases} \frac{1}{F} \sum\limits_{j:x_j=y_t} e^{\Psi_c(x_j)} & y_t \in A \\ 0 & \text{otherwise,} \end{cases}$$

where F is a normalization term, and $e^{\Psi_c(\cdot)}$ and $e^{\Psi_g(\cdot)}$ are scoring functions for copy mode and generate mode respectively. Because of the shared normalization term both generate mode and copy mode probabilities that are competing through the softmax function (12). The scoring functions are calculated as follows

$$\Psi_g(y_t = v_i) = v_i^T W_o s_t, v_i \in V \cap UNK \tag{13}$$

and

$$\Psi_c(y_t = x_j) = \sigma(h_j^T W_c), x_j \in X, \tag{14}$$

where X is an input sentence, x_j is a word at j position, v_i and W_0 are one-hot indicator vector for word v_i from the vocabulary. Here σ is a nonlinear activation function.

COPYNET updates decoder RNN state at every time step t, using previous hidden state s_{t-1}, predicted word y_{t-1} and the context vector c as follows

$$s_t = f(y_{t-1}, s_{t-1}, c). \tag{15}$$

However, if y_{t-1} is copied over to the output sentence then the decoder RNN states are updated by changing y_{t-1} to $[w(y_{t-1}); S(y_{t-1})]^T$, where $w(y_{t-1})$ is the word embeddings of y_{t-1} and $S(y_{t-1})$ is the weighted sum of hidden states in H or in encoder RNN network corresponding to y_t.

$$S(y_{t-1}) = \sum_{r=1}^{T} ptr \cdot h_r \tag{16}$$

$$ptr = \begin{cases} \frac{1}{K} p(x_i, c|s_{t-1}, H) & x_i = y_{t-1} \\ 0 & \text{otherwise.} \end{cases}$$

Here K is a normalizing term. Pointer network (ptr) is only concentrated on one location from source sentence. Although $S(y_{t-1})$ helps decoder to copy over subsequence from the source sentence and is named "selective read."

The COPYNET network is fully differentiable and can be trained end-to-end exactly like seq2seq model. It minimizes the negative log-likelihood as an objective loss function given by

$$J = -\frac{1}{N} \sum_{k=1}^{N} \sum_{t=1}^{T} \log[p(y_t^{(k)}|y_1^{(k)}, y_2^{(k)}, ... y_{t-1}^{(k)}, X^{(k)})]. \tag{17}$$

6.2 Pointer Softmax Network

The Pointer Softmax Network (PS) was proposed in [14]. The idea is to use attention mechanism and attention weights to select a word or token from the input sequence as the output instead of using it to blend hidden units of an encoder to a context vector at each decoder step. This setup was able to copy a word from the input sentence to the output sentence, which is not present in seq2seq model vocabulary or is not seen by the model in the training process. This approach shows the improvement in two tasks, i.e., neural machine translation on the Europarl English to French parallel corpora and text summarization on the Gigaword dataset.

The model learns two main things: 1) to predict whether the pointing mechanism is required at each time step 't'; 2) to point to the correct location of the word in source sentence which needs to be copied over to target sentence. The model uses two different softmax output layers, first, is *shortlist softmax layer*, and the second one is the *location softmax layer*. The first one is the softmax layer, which is used in the attention mechanism over all the vocabulary to generate a word from the model vocabulary. The second softmax is a location softmax, which is a pointer network in seq2seq model, where each of the output dimension corresponds to the location of a word in the context sequence. Consequently, the output dimension of the location softmax varies according to the length of the given source sequence. The decision whether the pointer word should be used or shortlisted is made by a switching network. The switching network is a multilayer perceptron network which takes the representation of source sequence and the previous hidden state from decoder RNN as an input and outputs the value of binary variable z_t which decides whether to shortlist softmax layer (when $z_t == 1$) or location softmax layer (when $z_t == 0$). When the word is not in the shortlist softmax and not even in the location softmax layer, this switching network chooses to shortlist softmax and gives <UNK> as the next token.

7 Experiments

7.1 Dataset

In the first part of the experiments, we compared seq2seq with attention model, pointer softmax model (PS) and the COPYNET network described in Sect. 5 and 6 for paraphrase generation. We train our model on the PPDB dataset [10]. Note that for higher precision, we have used medium size of PPDB dataset which has almost 9,103,492 pair sentences, which means their score for being a paraphrase is high. Before training the model, we preprocessed the PPDB dataset. We only took sentence pairs where the maximum number of words in a sentence is 50. We also removed all the punctuation signs from the source and target sentences. Consequently the word "I'm" becomes "I m," and "I haven't" becomes "I havent" and we consider them as two different words. After preprocessing we partition the dataset into three different categories with 80% of samples in the training set and 10-10% of samples in testing and validation sets, respectively. The same dataset was used for both models for training, testing and validation.

7.2 Models

Here we compare three different models for paraphrase generation described in Sect. 5 and 6. The first model described in Sect. 5 only uses attention to generate paraphrases. The model is trained on the preprocessed PPDB dataset [10]. We have used word-level embedding in the model to represent the whole sentence. We first created a vocabulary of 30,000 most frequent words in training samples and represented them with a unique index. This vocabulary is also augmented with four extra tokens that are <UNK> for representing unknown words or the words which are not covered in the vocabulary of the model, <PAD> used to add extra spacing to a sentence to make it equal to the length of source sentence if the target or source sentence is smaller than length 50, <SOS> and <EOS> representing the start of sentence and end of sentence, respectively. Both <SOS> and <EOS> were added before the sentence starts and at the end of the sentence respectively. Using these unique indices, the words in source text sentence are converted into the list of integers, which is then input to the seq2seq model described in 5.

Furthermore, the weights and bias parameters of the model were learned by backpropagating the error at training time. This model was then tested and validated using test samples and validation dataset with different hyper-parameters like the number of hidden units, type of RNN cell used in the encoder-decoder model, batch size and different attention types. The output from the decoder is an index value of a generated token which is then converted back to a word by matching it to the model vocabulary and then combined to form one complete sentence. We have used the beam of size 3, which means we picked the top 3 sentences with the highest probabilities.

We followed the same preprocessing for the COPYNET and the pointer softmax model with different variants of coping mechanism in seq2seq model. Therefore while training the PS model with the encoder-decoder network, a separate multi-layer perceptron model was also trained and used for binary classification. We used standard binary cross-entropy as a loss function for backpropagating error in the model. We have also tried this model with different hyper-parameters of the model described in the previous section.

All three models were fine-tuned on hyperparameters and then compared against each other for paraphrase generation by finding the loss in dev dataset, BLEU and METEOR scores. To save time in training we have fixed some parameters like we used teacher forcing while training encoder-decoder model, dropout ratio was fixed at 0.2, vocabulary size was set to 30,000 most frequent words in training time, and the batch size was set to 250. Note that to our best knowledge, these models were previously compared only on summarizing of paragraphs and not on paraphrase generation of sentences.

As codes for all these model were not made publicly available, we have implemented all these models in Pytorch. We trained our models on GPU provided by Helios Calcul Quebec, which has 15 compute nodes, each of which has eight K20 GPU's from NVIDIA, and 6 compute nodes and eight NVIDIA K80 boards each. Each K80 board contains two GPU, for a total of 216 GPUs for the cluster.

8 Results and Analysis

Due to time limitation, to compare models, we first trained the attention, PS and COPYNET models with only one epoch and with different hyper-parameters. It was done to find out the best configuration for the models to proceed using the first iteration. Then we trained every model over several epochs and after looking at the training and validation perplexity, we concluded that the models converged after 15 iterations. Tables below summarize the results.

Table 1. Results of seq2seq with Attention model with different hyper parameters on PPDB test dataset. Smaller perplexity indicates better performance.

Seq2Seq model with attention			
Hidden layer	Number of layers in RNN	Type of RNN cell	Valid perplexity
128	1	GRU	27.1790
128	1	LSTM	27.3762
256	1	GRU	28.1144
512	1	GRU	26.5589
128	**2**	**GRU**	**26.5401**
128	2	LSTM	26.7232

Table 2. Results for seq2seq with Pointer softmax model with different hyper parameters on PPDB test dataset. Smaller perplexity indicates better performance.

Seq2Seq model with pointer softmax network			
Hidden layer	Number of layers in RNN	Type of RNN cell	Valid perplexity
128	1	LSTM	29.9218
128	**1**	**GRU**	**26.5936**
256	1	GRU	27.4747
512	1	GRU	26.8019
128	2	GRU	28.2140

Tables 1, 2 and 3 demonstrate the validation perplexity on the PPDB dataset. Looking at the results, COPYNET outperforms the other two models by a small margin. Switching layer in the pointer softmax model did not help much in the

Table 3. Results for seq2seq with COPYNET Pointer network with different hyper parameters on PPDB test dataset. Smaller perplexity indicates better performance.

Seq2Seq model with COPYNET network			
Hidden layer	# Layers in RNN	Type of RNN cell	Valid perplexity
128	1	LSTM	26.6721
128	1	GRU	26.7842
256	1	GRU	26.9701
512	1	GRU	26.6891
128	**2**	**GRU**	**25.9625**
128	2	GRU	26.3713

Table 4. BLEU and METEOR score on test dataset of PPDB dataset with attention, COPYNET and Pointer softmax. Higher scores indicate better performance.

Model comparison		
Model	BLEU-Score	METEOR-Score
$Seq2Seq_{attn}$	0.4538	0.3035
$Seq2Seq_{attn+COPYNET}$	**0.4547**	**0.3464**
$Seq2Seq_{attn+PS}$	0.2922	0.3219

generation of paraphrases. Table 4 compares performance of all different models on the test dataset consisting of 25,000 examples. The BLEU and METEOR scores were slightly better for COPYNET network than for other models.

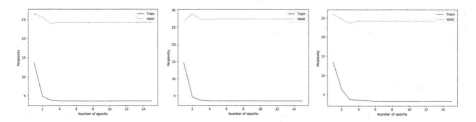

Fig. 1. Plots of perplexity of seq2seq model with attention model (left), pointer softmax model (middle) and COPYNET pointer network model on PPDB training and test data set.

Figure 1 presents training and validation perplexity curves. Seq2seq model with attention mechanism and with COPYNET network model both show the best performance at iteration 3, and they have minimum validation perplexity at this point, i.e., 23.9142 and 23.6172 respectively. On the other hand, the pointer softmax model gave the best result at one iteration, where we got minimum validation perplexity of 26.6837.

We next show examples of paraphrases generated by different models. Note, that source sentences were picked randomly and were not in the PPDB test dataset. Below, 'sentence' represents the original sentence which was given as an input to the model, while Paraphrase(X) represents the paraphrase generated by "X".

1. **Sentence:** Economy is a big problem for the Bush administration
   Paraphrase(Attn): <UNK> government problem is <UNK> <EOS>
   Paraphrase(COPYNET): Bush government problem is economy <EOS>
   Paraphrase(PS): George's government problem is big <EOS>
2. **Sentence:** Language is complex and the process of reading and understanding language is difficult for many groups of people
   Paraphrase(Attn): Language is difficult for <UNK> and <UNK> <eos>
   Paraphrase(COPYNET): Language is difficult for reading and understanding <eos>
   Paraphrase(PS): Speech is complex for understanding for many people <eos>

Glancing at the examples above, we conclude that seq2seq model with COPY-NET network model generated better paraphrases, which is consistent with our earlier results.

9 Conclusions

We performed experiments on seq2seq model with attention and two different variants of pointer networks under the supervision of PPDB dataset and compared the results using BLEU and METEOR score metrics. In this experiment, COPYNET outperforms pointer softmax pointer network by a small margin. Observing examples of paraphrases generated by these models it can be concluded that COPYNET pointer network generates the best paraphrases among compared models.

References

1. Androutsopoulos, I., Malakasiotis, P.: A survey of paraphrasing and textual entailment methods. J. Artif. Intell. Res. **38**, 135–187 (2010)
2. Bahdanau, D., Cho, K., Bengio, Y.: Neural machine translation by jointly learning to align and translate. In: Proceedings of The International Conference on Learning Representations (ICLR 2015), pp. 1–15 (2015)
3. Califf, M.E., Mooney, R.J.: Bottom-up relational learning of pattern matching rules for information extraction. J. Mach. Learn. Res. **4**, 177–210 (2003)
4. Cho, K., et al.: Learning phrase representations using RNN encoder-decoder for statistical machine translation. In: Proceedings of the 2014 Conference on Empirical Methods in Natural Language Processing, pp. 1724–1734 (2014)
5. Gu, J., Lu, Z., Li, H., Li, V.O.: Incorporating copying mechanism in sequence-to-sequence learning. In: Proceedings of the 54-th Annual Meeting of the Association for Computational Linguistics, pp. 1631–1640 (2016)

6. Harris, Z.S.: Transformational Theory. Springer, Dordrecht (1970). https://doi.org/10.1007/978-94-017-6059-1_27
7. Inkpen, D.: Building a lexical knowledge-base of near-synonym differences. Comput. Linguist. **32**, 223–262 (2004)
8. Luong, M., Pham, H., Manning, C.D.: Effective approaches to attention based neural machine translation. In: Proceedings of the 2015 Conference on Empirical Methods in Natural Language Processing, pp. 1412–1421 (2015)
9. McKeown, K.R.: Paraphrasing questions using given and new information. J. Comput. Linguist. **9**(1), 1–10 (1983)
10. Pavlick, E., Callison-Burch, C.: Simple PPDB: a paraphrase database for simplification. In: Proceedings of the 54-th Annual Meeting of the Association for Computational Linguistics, pp. 143–148 (2016)
11. Prakash, A., et al.: Neural paraphrase generation with stacked residual LSTM networks. In: Proceedings of the 26-th International Conference on Computational Linguistics COLING 2016, pp. 2923–2934 (2016)
12. Sutskever, I., Vinyals, O., Le, Q.V.: Sequence to sequence learning with neural networks. In: Advances in Neural Information Processing Systems (NIPS 2014), pp. 3104–3112 (2014)
13. Tu, Z., Lu, Z., Liu, Y., Liu, X., Li, H.: Coverage-based neural machine translation. In: Proceedings of the 54-th Annual Meeting of the Association for Computational Linguistics, pp. 76–85–430 (2016)
14. Vinyals, O., Fortunato, M., Jaitly, N.: Pointer networks. In: Advances in Neural Information Processing Systems (NIPS 2015), pp. 2692–2700 (2015)

Interval Methods for Seeking Fixed Points of Recurrent Neural Networks

Bartłomiej Jacek Kubica$^{(\boxtimes)}$, Paweł Hoser, and Artur Wiliński

Institute of Information Technology, Warsaw University of Life Sciences – SGGW,
ul. Nowoursynowska 159, 02-776 Warsaw, Poland
{bartlomiej_kubica,pawel_hoser,artur_wilinski}@sggw.pl

Abstract. The paper describes an application of interval methods to train recurrent neural networks and investigate their behavior. The HIBA_USNE multithreaded interval solver for nonlinear systems and algorithmic differentiation using ADHC are used. Using interval methods, we can not only train the network, but precisely localize all stationary points of the network. Preliminary numerical results for continuous Hopfield-like networks are presented.

Keywords: Interval computations · Nonlinear systems · HIBA_USNE · Recurrent neural network · Hopfield network

1 Introduction

Artificial neural networks (ANN) have been used in many branches of science and technology, for the purposes of classification, modeling, approximation, etc. Several training algorithms have been proposed for this tool. In particular, several authors have applied interval algorithms for this purpose (cf., e.g., [5,6,19]). Most of these efforts (all known to the authors) have been devoted to feedforward neural networks.

Nevertheless, in some applications (like prediction of a time series or other issues related to dynamical systems, but also, e.g., in some implementations of the associative memory), we need the neural network to remember its previous states – and this can be achieved by using the feedback connections. In this paper, we apply interval methods to train this sort of networks.

2 Hopfield-Like Network

Let us focus on a simple model, similar to popular Hopfield networks, described, i.a., in [18]. There is only a single layer and each neuron is connected to all other ones. Figure 1 illustrates this architecture.

Let us present the mathematical formulae. Following [10], we denote vectors by small letters and matrices – by capital ones.

© Springer Nature Switzerland AG 2020
V. V. Krzhizhanovskaya et al. (Eds.): ICCS 2020, LNCS 12139, pp. 414–423, 2020.
https://doi.org/10.1007/978-3-030-50420-5_30

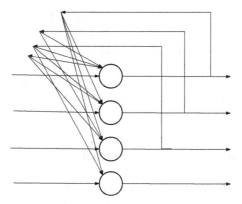

Fig. 1. A Hopfield-type neural network

The output of the network is the vector of responses of each neuron, which, for the i'th one, is:

$$y_i = \sigma\Big(\sum_{j=1}^{n} w_{ij}x_j\Big), \quad \text{for } i = 1,\dots,n\,, \tag{1}$$

where $\sigma(\cdot)$ is the activation function, described below.

The weights can have both positive and negative values, i.e., neurons can both attract or repel each other. Also, typically, it is assumed that $w_{ii} = 0$, i.e., neurons do not influence themselves directly, but only by means of influencing other neurons. Unlike most papers on Hopfield networks, we assume that the states of neurons are not discrete, but continuous: $x_i \in [-1,1]$. As the activation function, the step function has been used originally:

$$\sigma(t) = \mathcal{H}(t) = \begin{cases} 1 \text{ for } t \geq 0, \\ -1 \text{ for } t < 0. \end{cases} \tag{2}$$

but sigmoid functions can be used, as well; for instance:

$$\sigma(t) = \frac{2}{1 + \exp(-\beta \cdot t)} - 1, \tag{3}$$

the hyperbolic tangent or the arctan. Please mind that in both above functions, (2) and (3), the value of the activation function ranges from -1 to 1 and not from 0 to 1, like we would have for some other types of ANNs.

In our experiments, we stick to using activation functions of type (3) with $\beta = 1$, but other values $\beta > 0$ would make sense, as well.

What is the purpose of such a network? It is an associative memory that can store some patterns. These patterns are fixed points of this network: when we feed the network with a vector, being one of the patters, the network results in the same vector on the output.

What if we give another input, not being one of the remembered vectors? Then, the network will find the closest one of the patterns and return it. This may take a few iterations, before the output stabilizes.

Networks presented in Fig. 1, very popular in previous decades, has become less commonly used in practice, nowadays. Nevertheless, it is still an interesting object for investigations and results obtained for Hopfield-like networks should be easily extended to other ANNs.

3   Training Hopfield-Like Networks

How to train a Hopfield network? There are various approaches and heuristics.

Usually, we assume that the network is supposed to remember a given number of patterns (vectors) that should become its stationary points.

An example is the Hebb rule, used when patterns are vectors of values $+1$ and -1 only, and the discrete activation function (2) is used. Its essence is to use the following weights:

$$w_{ij} = \begin{cases} \sum_{k=1}^{N} x_i^k \cdot x_j^k & \text{for } i \neq j, \\ 0 & \text{for } i = j. \end{cases} \tag{4}$$

which results in the weights matrix of the form:

$$W = \sum_{k=1}^{N} x^k (x^k)^T - \text{diag}\big((x_i^k)^2\big).$$

Neither the above rules, nor most other training heuristics take into account problems that may arise while training the network:

- several "spurious patterns" will, in fact, be stationary points of the network, as well as actual patterns,
- capacity of the network is limited and there may exist no weight matrix, responding to all training vectors properly.

Let us try to develop a more general approach.

4   Problems Under Solution

In general, there are two problems we may want to solve with respect to a recurrent ANN, described in Sect. 2:

1. We know the weights of the network and we want to find all stationary points.
2. We know all stationary points the network should have and we want to determine the weights, so that this condition was satisfied.

In both cases, the system under consideration is similar:

$$x_i - \sigma\left(\sum_{j=1}^{n} w_{ij}x_j\right) = 0, \quad \text{for } i = 1, \ldots, n , \tag{5}$$

but different quantities are unknowns under the search or the given parameters. In the first case, we know the matrix of weights: $W = [w_{ij}]$ and we seek x_i's and in the second case – vice versa.

Also, the number of equations differs in both cases. The first problem is always well-determined: the number of unknowns and of equations is equal to the number of neurons n. The second problem is not necessarily well-determined: we have $n \cdot (n - 1)$ unknowns and the number of equations is equal to $n \cdot N$, where N is the number of vectors to remember.

To be more explicit: in the first case, we obtain the following problem:

Find x_i, $i = 1, \ldots, n$, such that:

$$x_i - \sigma\left(\sum_{j=1}^{n} w_{ij}x_j\right) = 0, \quad \text{for } i = 1, \ldots, n . \tag{6}$$

In the second case, it is:

Find w_{ij}, $i, j = 1, \ldots, n$, such that:

$$x_i^k - \sigma\left(\sum_{j=1}^{n} w_{ij}x_j^k\right) = 0, \quad \text{for } k = 1, \ldots, N . \tag{7}$$

But in both cases, it is a system of nonlinear equations. What tools shall we apply to solve it?

5   Interval Tools

Interval analysis is well-known to be a tractable approach to finding a solution – or all solutions of a nonlinear equations system, like the above ones.

There are several interval solvers of nonlinear systems (GlobSol, Ibex, Realpaver and SONIC are representative examples). In our research, we are using HIBA_USNE [4], developed by the first author. The name HIBA_USNE stands for Heuristical Interval Branch-and-prune Algorithm for Underdetermined and well-determined Systems of Nonlinear Equations and it has been described in a series of papers (including [11,12,14–16]; cf. Chap. 5 of [17] and the references therein).

As the name states, the solver is based on interval methods (see, e.g., [8, 9,20]), that operate on intervals instead of real numbers (so that result of an operation on numbers always belongs to the result of operation on intervals that contain the numerical inputs). Such methods are robust, guaranteed to enclose all solutions, even if they are computationally intensive and memory demanding. Their important advantage is allowing not only to locate solutions of well-determined and underdetermined systems, but also to *verify* them, i.e., prove that in a given box there is a solution point (or a segment of the solution manifold).

Details can be found in several textbooks, i.a., in these quoted above.

5.1   HIBA_USNE

Let us present the main algorithm (the standard interval notation, described in [10], will be used). The solver is based on the branch-and-prune (B&P) schema that can be expressed by pseudocode presented in Algorithm 1.

Algorithm 1. Interval branch-and-prune algorithm

Require: $L, \mathsf{f}, \varepsilon$
1: $\{L$ – the list of initial boxes, often containing a single box $\mathbf{x}^{(0)}\}$
2: $\{L_{ver}$ – verified solution boxes, L_{pos} – possible solution boxes$\}$
3: $L_{ver} = L_{pos} = \emptyset$
4: $\mathbf{x} = \mathrm{pop}\,(L)$
5: **loop**
6:     process the box $\mathbf{x}$, using the rejection/reduction tests
7:     **if** ($\mathbf{x}$ does not contain solutions) **then**
8:        discard $\mathbf{x}$
9:     **else if** ($\mathbf{x}$ is verified to contain a segment of the solution manifold) **then**
10:       push ($L_{ver}, \mathbf{x}$)
11:    **else if** (the tests resulted in two subboxes of $\mathbf{x}$: $\mathbf{x}^{(1)}$ and $\mathbf{x}^{(2)}$) **then**
12:       $\mathbf{x} = \mathbf{x}^{(1)}$
13:       push ($L, \mathbf{x}^{(2)}$)
14:       **cycle loop**
15:    **else if** ($\mathrm{wid}\,\mathbf{x} < \varepsilon$) **then**
16:       push ($L_{pos}, \mathbf{x}$) {The box $\mathbf{x}$ is too small for bisection}
17:    **if** ($\mathbf{x}$ was discarded **or** $\mathbf{x}$ was stored) **then**
18:       **if** ($L == \emptyset$) **then**
19:          **return** L_{ver}, L_{pos} {All boxes have been considered}
20:       $\mathbf{x} = \mathrm{pop}\,(L)$
21:    **else**
22:       bisect ($\mathbf{x}$), obtaining $\mathbf{x}^{(1)}$ and $\mathbf{x}^{(2)}$
23:       $\mathbf{x} = \mathbf{x}^{(1)}$
24:       push ($L, \mathbf{x}^{(2)}$)

The "rejection/reduction tests", mentioned in the algorithm are described in previous papers (cf., e.g., [14–16] and references therein):

- switching between the componentwise Newton operator (for larger boxes) and Gauss-Seidel with inverse-midpoint preconditioner, for smaller ones,
- a heuristic to choose whether to use or not the BC3 algorithm,
- a heuristic to choose when to use bound-consistency,

- a heuristic to choose when to use hull-consistency,
- sophisticated heuristics to choose the bisected component,
- an additional second-order approximation procedure,
- an initial exclusion phase of the algorithm (deleting some regions, not containing solutions) – based on Sobol sequences.

It is also worth mentioning that as Algorithm 1, as some of the tests performed on subsequent boxes are implemented in a multithreaded manner. Papers [11–16] discuss several details of this implementation and a summary can be found in Chap. 5 of [17].

5.2   ADHC

The HIBA_USNE solver collaborates with a library for algorithmic differentiation, also written by the first author. The library is called ADHC (Algorithmic Differentiation and Hull Consistency enforcing) [3]. Version 1.0 has been used in our experiments. This version has all necessary operations, including the exp function, used in (3), and division (that was not implemented in earlier versions of the package).

6   Numerical Experiments

Numerical experiments have been performed on a machine with two Intel Xeon E5-2695 v2 processors (2.4 GHz). Each of them has 12 cores and on each core two hyper-threads (HT) can run. So, $2 \times 12 \times 2 = 48$ HT can be executed in parallel. The machine runs under control of a 64-bit GNU/Linux operating system, with the kernel 3.10.0-123.e17.x86_64 and glibc 2.17. They have non-uniform turbo frequencies from range 2.9–3.2 GHz. As there have been other users performing their computations also, we limited ourselves to using 24 threads only.

The Intel C++ compiler ICC 15.0.2 has been used. The solver has been written in C++, using the C++11 standard. The C-XSC library (version 2.5.4) [1] was used for interval computations. The parallelization was done with the packaged version of TBB 4.3 [2].

The author's HIBA_USNE solver has been used in version Beta2.5 and ADHC library, version 1.0.

We consider the network with n neurons ($n = 4$ or $n = 8$) and storing 1 or 3 vectors. The first vector to remember is always $(1, 1, \ldots, 1)$. The second one consists of $\frac{n}{2}$ values $+1$ and $\frac{n}{2}$ values -1. The third one consists of $n - 2$ values $+1$ and 2 values -1 (Tables 1 and 2).

The following notation is used in the tables:

- fun.evals, grad.evals, Hesse evals – numbers of functions evaluations, functions' gradients and Hesse matrices evaluations (in the interval automatic differentiation arithmetic),
- bisecs – the number of boxes bisections,

Table 1. Computational results for Problem (6)

Problem	$n = 4, N = 1$	$n = 8, N = 1$	$n = 4, N = 3$	$n = 8, N = 3$	$n = 12, N = 3$
fun.evals	1133	365,826	2,010	432,551	4,048,010,515
grad.evals	1629	483,302	2,484	689,774	6,298,617,714
Hesse evals	4	3,568	124	6,974	398,970,087
bisections	37	29,210	117	41,899	245,816,596
preconds	68	32,970	149	45,813	252,657,916
Sobol excl.	14	62	15	63	143
Sobol resul.	321	1,541	346	1,548	3,869
pos.boxes	1	2	4	0	3
verif.boxes	2	1	0	5	6
Leb.poss.	3e−36	4e−70	3e−27	0.0	4e−77
Leb.verif.	6e−30	2e−92	0.0	5e−71	1e−129
time (sec.)	< 1	1	<1	2	12,107

Table 2. Computational results for Problem (7)

Problem	$n = 4, N = 1$	$n = 8, N = 1$	$n = 4, N = 3$	$n = 8, N = 3$	$n = 12, N = 3$
fun.evals	7376	369100	8870	420,484	3,303,590
grad.evals	16	64	48	192	432
Hesse evals	4	8	12	24	36
bisections	0	0	0	0	0
preconds	0	0	0	0	0
Sobol excl.	144	3136	144	3,136	17,424
Sobol resul.	0	0	0	0	0
pos.boxes	0	0	0	0	0
verif.boxes	0	0	0	0	0
Leb.poss.	0.0	0.0	0.0	0.0	0.0
Leb.verif.	0.0	0.0	0.0	0.0	0.0
time (sec.)	<1	< 1	<1	< 1	2

- preconds – the number of preconditioning matrix computations (i.e., performed Gauss-Seidel steps),
- Sobol excl. – the number of boxes to be excluded generated by the initial exclusion phase,
- Sobol resul. – the number of boxes resulting from the exclusion phase (cf. [13,14]),
- pos.boxes, verif.boxes – number of elements in the computed lists of boxes containing possible and verified solutions,
- Leb.pos., Leb.verif. – total Lebesgue measures of both sets,
- time – computation time in seconds.

7   Analysis of the Results

The HIBA_USNE solver can find solutions of Problem (6) pretty efficiently. The solutions get found correctly. For instance, in the case of four neurons and a single stored pattern, three solutions are quickly found; two of the solutions are guaranteed:

```
x = [-1.030683E-008,1.031144E-008]
[-2.702852E-010,2.703469E-010]
[-3.596875E-009,3.598297E-009]
[-1.189070E-009,1.188578E-009]

x = [  0.858559,  0.858560]
[  0.858559,  0.858560]
[  0.858559,  0.858560]
[  0.858559,  0.858560]
```

and one is a possible solution:

```
x = [ -0.858560, -0.858559]
[ -0.858560, -0.858559]
[ -0.858560, -0.858559]
[ -0.858560, -0.858559]
```

Because of the properties of the sigmoid function (3), that nowhere reaches the values ± 1, the actual pattern $(1, 1, 1, 1)$ cannot be the solution of the equations. Yet, the solution is a (relatively crude), approximation of the pattern. Another solutions are the point $(0, 0, 0, 0)$, and minus the first solution. The number of solutions that get verified or found as possible solutions only, varies (cf. Problem 6), but all of them get bounded correctly.

For problems of small dimensionality, all solutions get found immediately. Unfortunately, the time increases quickly with the number of neurons (but not with the number of stored patterns!) in the network. This is partially because Hopfield networks are 'dense': each neuron is connected to all other ones. Multi-layer networks have a more 'sparse' structure, that may improve the scalability of the branch-and-prune method.

For Problem (7) of computing the weights matrix, the HIBA_USNE solver was less successful. This is not surprising: Problem (7) is underdetermined, and can have uncountably many solutions.

Actually, the solver has been successful on (7) when there had been no solutions: this can be verified easily, in many cases. As the sigmoid function (3) does not reach values ± 1 for finite arguments, there are no weights for which sequences of ± 1's are stationary points of the network, and the solver verifies it easily.

Unfortunately, seeking weights for a feasible solution is not that efficient. For instance, seeking weights for a network with a single stationary point at $(0.858, 0.858, 0.858, 0.858)$, had to be interrupted after three hours, without obtaining the results!

Possibly, it would make sense to seek solutions of Problem (7) with some additional constraints, but this has not been determined yet. In such case, it might be beneficial to transform the equations to the form:

$$\sum_{j=1}^{n} w_{ij} x_j^k = -\frac{1}{\beta} \cdot \ln\left(\frac{2}{1+x_i^k} - 1\right), \quad i = 1, \ldots, n, \ k = 1, \ldots, N \ .$$

Now, the equations are linear with respect to w_{ij}.

Also, interval methods can naturally be applied to seek *approximate* fixed points, instead of precise ones, but such experiments have not been performed yet.

8   Conclusions

The paper presents a promising application of interval methods and the HIBA_USNE solver. It can be used both to train and investigate the behavior of a recurrent neural network. The interval solver of nonlinear systems can potentially be applied to determining the weights matrix of the network, but more importantly: to localizing all stationary points of the network.

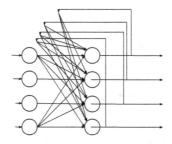

Fig. 2. A Hamming-type neural network

We have considered single-layer continuous Hopfield-like networks, but generalization to Hamming networks (Fig. 2) or convolutional multilayer ANNs (e.g., [7]) seems straightforward. This will be the subject of our further research, as well as further studies about Hopfield network: seeking for periodic states, seeking for approximate stationary points, and more sophisticated interval algorithms to train the network.

References

1. C++ eXtended Scientific Computing Library (2017). http://www.xsc.de
2. Intel TBB (2017). http://www.threadingbuildingblocks.org
3. ADHC, C++ Library (2018). https://www.researchgate.net/publication/316610415_ADHC_Algorithmic_Differentiation_and_Hull_Consistency_Alfa-05

4. HIBA_USNE, C++ library (2018). https://www.researchgate.net/publication/316687827_HIBA_USNE_Heuristical_Interval_Branch-and-prune_Algorithm_for_Underdetermined_and_well-determined_Systems_of_Nonlinear_Equations_-_Beta_25

5. Adam, S.P., Karras, D.A., Magoulas, G.D., Vrahatis, M.N.: Solving the linear interval tolerance problem for weight initialization of neural networks. Neural Netw. **54**, 17–37 (2014)

6. Beheshti, M., Berrached, A., de Korvin, A., Hu, C., Sirisaengtaksin, O.: On interval weighted three-layer neural networks. In: 1998 Proceedings of the 31st Annual Simulation Symposium, pp. 188–194. IEEE (1998)

7. Goodfellow, I., Bengio, Y., Courville, A.: Deep Learning. MIT Press, Cambridge (2016)

8. Jaulin, L., Kieffer, M., Didrit, O., Walter, E.: Applied Interval Analysis. Springer, Heidelberg (2001). https://doi.org/10.1007/978-1-4471-0249-6

9. Kearfott, R.B.: Rigorous Global Search: Continuous Problems. Kluwer, Dordrecht (1996)

10. Kearfott, R.B., Nakao, M.T., Neumaier, A., Rump, S.M., Shary, S.P., van Hentenryck, P.: Standardized notation in interval analysis. Vychislennyie Tiehnologii (Comput. Technol.) **15**(1), 7–13 (2010)

11. Kubica, B.J.: Interval methods for solving underdetermined nonlinear equations systems. Reliable Comput. **15**, 207–217 (2011). Proceedings of SCAN 2008

12. Kubica, B.J.: Tuning the multithreaded interval method for solving underdetermined systems of nonlinear equations. In: Wyrzykowski, R., Dongarra, J., Karczewski, K., Waśniewski, J. (eds.) PPAM 2011. LNCS, vol. 7204, pp. 467–476. Springer, Heidelberg (2012). https://doi.org/10.1007/978-3-642-31500-8_48

13. Kubica, B.J.: Excluding regions using Sobol sequences in an interval branch-and-prune method for nonlinear systems. Reliable Comput. **19**(4), 385–397 (2014). Proceedings of SCAN 2012 (15th GAMM-IMACS International Symposium on Scientific Computing, Computer Arithmetic and Validated Numerics)

14. Kubica, B.J.: Presentation of a highly tuned multithreaded interval solver for underdetermined and well-determined nonlinear systems. Numer. Algorithms **70**(4), 929–963 (2015). https://doi.org/10.1007/s11075-015-9980-y

15. Kubica, B.J.: Parallelization of a bound-consistency enforcing procedure and its application in solving nonlinear systems. J. Parallel Distrib. Comput. **107**, 57–66 (2017)

16. Kubica, B.J.: Role of hull-consistency in the HIBA_USNE multithreaded solver for nonlinear systems. In: Wyrzykowski, R., Dongarra, J., Deelman, E., Karczewski, K. (eds.) PPAM 2017. LNCS, vol. 10778, pp. 381–390. Springer, Cham (2018). https://doi.org/10.1007/978-3-319-78054-2_36

17. Kubica, B.J.: Interval Methods for Solving Nonlinear Constraint Satisfaction, Optimization and Similar Problems. SCI, vol. 805. Springer, Cham (2019). https://doi.org/10.1007/978-3-030-13795-3

18. Mańdziuk, J.: Hopfield-type neural networks. Theory and applications, Akademicka Oficyna Wydawnicza EXIT (2000). (in Polish)

19. Saraev, P.V.: Numerical methods of interval analysis in learning neural network. Autom. Remote Control **73**(11), 1865–1876 (2012)

20. Shary, S.P.: Finite-dimensional interval analysis. Institute of Computational Technologies, Sibirian Branch of Russian Academy of Science, Novosibirsk (2013)

Fusion Learning: A One Shot Federated Learning

Anirudh Kasturi$^{(\boxtimes)}$, Anish Reddy Ellore, and Chittaranjan Hota

BITS Pilani, Hyderabad Campus, Hyderabad, India
anirudh.kasturi@gmail.com, anishreddy.ellore@gmail.com,
hota@hyderabad.bits-pilani.ac.in

Abstract. Federated Learning is an emerging distributed machine learning technique which does not require the transmission of data to a central server to build a global model. Instead, individual devices build their own models, and the model parameters are transmitted. The server constructs a global model using these parameters, which is then re-transmitted back to the devices. The major bottleneck of this approach is the communication overhead as all the devices need to transmit their model parameters at regular intervals. Here we present an interesting and novel alternative to federated learning known as Fusion Learning, where the distribution parameters of the client's data along with its local model parameters are sent to the server. The server regenerates the data from these distribution parameters and fuses all the data from multiple devices. This combined dataset is now used to build a global model that is transmitted back to the individual devices. Our experiments show that the accuracy achieved through this approach is in par with both a federated setup and a centralized framework needing only one round of communication to the central server.

Keywords: Federated Learning · Feature distributions · Communication efficiency · Distributed machine learning

1 Introduction

Smartphones and smart devices have become the norm in society. They are now an integral part of many people [11]. With more advancements in technology, they are all the more powerful. These devices have enhanced the user experience by collecting massive amounts of data through various sensors and are providing meaningful feedback to the user. With this increase in the computational power of devices and concerns over privacy, while transmitting data to servers, researchers have focused on storing data locally and perform network computations on the edge. Several works such as [6, 11] were published where machine learning models are trained centrally and then pushed to the local devices. This approach led to personalizing models for users. With an increase in the computational capabilities of the devices, it is now possible to make use of this computational power within a distributed network. With this possibility, it had created

© Springer Nature Switzerland AG 2020
V. V. Krzhizhanovskaya et al. (Eds.): ICCS 2020, LNCS 12139, pp. 424–436, 2020.
https://doi.org/10.1007/978-3-030-50420-5_31

a new research direction coined as Federated Learning (FL) [16], where models are trained directly on mobile devices. The local models are then aggregated on a central location, which is passed back to the clients. One fundamental example is that of a next word predictor on mobile devices where each mobile device computes a model locally instead of transmitting the raw data to the server. These models are aggregated at a central server to generate a global model. The global model after each communication round is transmitted back to all the devices in the network. The communication between client and server continues until a required convergence is achieved. In this paper, we try to address one of the main challenges in Federated Learning i.e. communication overheads. Transmission overhead is a crucial blockade in a typical federated learning scenario, where model parameters need to be exchanged at regular intervals. Federated networks are potentially made up of large number of devices to the tune of millions and communications at that scale can potentially make the network slower [8]. Therefore, in order to make federated learning work in such a scenario, it is important to come up with innovative methods that are communication efficient. Two major contributions that have been made in this area were to: (i) reduce the total number of contact rounds, (ii) minimizing the size of exchanged messages in each round [16]. The composition of a federated network is highly varied because the computational power, network connectivity, memory capacities, and power usage varies with each device type. Due to these limitations, only a fraction of devices actively participate in the exchange of data. An active system may also drop out during an exchange due to either a network issue or a possible power outage. These system-level features significantly intensify problems such as prevention and acceptance of failure. Consequently, federated learning methods built and evaluated must: (i) expect a low level of involvement in the federated process; (ii) open to variability in hardware; and (iii) resilient to underlying network equipment.

Client nodes also produce and collect data non-identically across the network, e.g., in the context of the next word prediction, users make different use of the language on a mobile phone. Also, the amount of data collected across devices can vary considerably, and the possibility of finding a fundamental design capturing the relationship between devices and their related distributions is unlikely. In distributed optimization, the data generation approach challenges independent and IDD principles commonly used and can add to the uncertainty of modeling, analysis, and evaluation. Alternate learning techniques such as transfer learning and multi-task learning frameworks [21] have been proposed to counter these issues in federated learning.

Our contribution in this paper is a novel learning technique termed *Fusion Learning* in which each device computes its data distribution parameters along with its model parameters. These data distribution parameters are specific to each feature of the dataset. If a dataset has ten features, each feature might follow a different distribution. We find out the distributions of individual features, and these distribution parameters are transmitted to the server. These are sent only once, thereby requiring only one communication round. The server generates artificial data using the distribution parameters received from the client, creating a corpus of data for each client. The individual datasets are then combined to form a larger dataset. The server now computes a global model on this cumulative data, and the final global model is passed back to all the clients.

2   Related Work

Federated Learning difficulties at first glance mimic traditional problems in areas like confidentiality, robust machine learning, and distributed optimization. Throughout machine learning, optimization, and signal processing communities, for example, several approaches were proposed to tackle costly communication. Nevertheless, the size of federation networks, in terms of complexities of the system and statistical heterogeneity, is usually much larger and are not fully covered by these approaches.

The prevalent methodology for distributed machine learning in data center environments has been mini-batch optimization, which involves expanding conventional stochastic methods for processing multiple data points [3,20]. Nevertheless, in practice, there was little versatility to respond to the trade-offs between communication and computation that maximizes distributed data processing [22]. Also, a large number of probable approaches have been proposed to reduce the transmission costs in distributed settings through simultaneous application of a variety of local updates to each computer at each communication round, becoming considerably more versatile. Distributed primal-dual local updating methods have become a popular way to solve such convex optimization problems [12,24]. These approaches utilize a dual-format to efficiently divide the parent goal into smaller problems. These can now be solved in parallel during every round of communication. There have also been several distributed local primal updating methods that add the benefit of applying to non-convex purposes [19]. Such techniques improve performance significantly and have shown that they reach higher order-of-magnitude speeds in real-world data center environments over conventional mini-batch approaches like ADMM [1]. Optimization approaches for adaptive local notifications and weak customer engagement are de facto resolvers in federated settings [16,21]. Federated Averaging (FedAvg) [16], which averages stochastic gradient descent(SGD) components from local devices, is the most common method used for federated learning. FedAvg has proved to operate extremely well empirically and specifically for non-convex issues. However, it does not have any guarantees of convergence and can differ in realistic settings when heterogeneous data is used [13]. While local updating methods can decrease the total number of contact rounds, models such as sparse sampling, subsampling, and quantization can significantly minimize the size of message transmissions during each exchange. These approaches have been widely studied in [25].

Decentralized training was shown to be quicker than centralized training in data center settings while running on high or low bandwidth networks. [7] explains in great detail on both pros and cons of such an approach. These algorithms can also lessen the connectivity costs of the central server in the federated setup. They examined heterogeneous data with local update schemes as decentralized learning. Either these approaches are limited to linear models, or they require all devices to be part of the exercise. Eventually, hierarchical models developed in [14,15] were also suggested to decrease the load on the central server by using edge servers. Only a small subset of devices in federated networks usually take part in every training round. Nishio and Yonetani, for example in [17], are exploring new sampling rules for devices based on system resources with a view to aggregate a maximum number of device updates within a predefined time window. Likewise, in creating the motivation mechanisms to enable high-qualified devices to engage in the learning process, Kang et al. [9] take into account seven overhead systems for each computer. Such techniques, however, presume a static machine model of network characteristics; how to expand those strategies to manage device-specific fluctuations in real time remains open. However, while these approaches mainly concentrate on system variability for active sampling, we note that a collection of limited but reasonably representative devices based on the underlying statistical structure should also be taken actively for sampling. Information that is not distributed identically across devices emerge when training federated models, both in terms of data modeling and the study of the integration of related training procedures.

MOCHA [21], a federated setting optimization framework, can personalize each device by learning separate models but linked to each device while using multi functional learning to leverage shared representation. The size is limited to large networks and convex targets. [2] forms the network of stars as a Bayesian network and offers a variance while learning. Generalizing to large networks is expensive using this approach even though it can handle non-convex models. Khodak et al. [10] have tested the use of multi-task information to meta-learn a task-by-task learning rating (where each task corresponds to one device). Eichner et al. [5] are exploring a pluralistic approach to resolve cyclical trends in data samples during the federated training process (adaptively selecting a global model and device-specific models). Despite these recent developments, the major challenges remain in building robust, scalable, and automated methods of heterogeneous modeling in federated environments.

3   Fusion Learning Algorithm

We introduce our proposed one-shot Federated Learning Algorithm called *Fusion Learning* in this section, which has three modules: (1) finding the distribution of each feature in the dataset on the local device. Locally training the model with the available data. (2) aggregating both the distribution parameters and the model parameters at a central server. Generating artificial data points from the distribution. (3) building a global model from the generated points and transmitting the new global model parameters back to the clients. These steps have been depicted pictorially in Fig. 1.

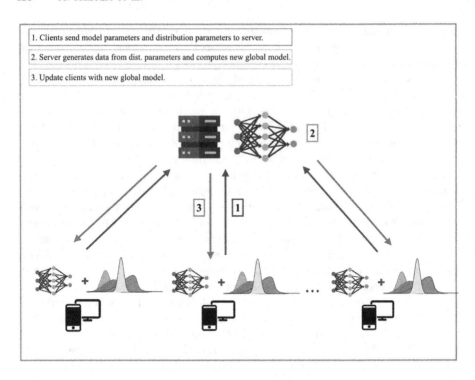

1. Clients send model parameters and distribution parameters to server.

2. Server generates data from dist. parameters and computes new global model.

3. Update clients with new global model.

Fig. 1. Architectural diagram of a Fusion Learning system.

3.1   Distribution of Individual Features at Client

It is important to have our data being reflected accurately in the distribution. Different distributions are usually evaluated against data to determine which one matches the data best. The parameters of different distributions are calculated using statistical techniques. Distribution is generally defined by four parameters: *location, scale, shape, and threshold.* Fitting for distribution involves estimating these parameters, which define different distributions. A distribution's *location* parameter specifies where the distribution lies along the x-axis (the horizontal axis). *Scale* parameter determines how much spread there is within the distribution. *Shape* parameter lets the distribution take different shapes. *Threshold* parameter defines the distribution's minimum value along the x-axis. The parameters for distribution can be calculated using a variety of statistical techniques. One such technique being the Maximum likelihood estimator where negative log-likelihood is minimized. Upon completion of this calculation, we use the goodness of fit techniques to help us determine which distribution best fits our data. The next step is to find out to which distribution our data fits into. We have used the *stats* library from SciPy [23] to fit the data into various distributions.

To determine which distribution fits the data best, we use the *p-values* generated using *Kolmogorov-Smirnov* test. The distribution with the greatest *p* value is considered to be the right fit for that data. Using these steps, we find out

Table 1. Different types of distributions used to verify the distribution of individual feature set

norm	pareto	genextreme	gamma	uniform
exponweib	lognorm	expon	logistic	vonmises
weibull max	beta	cauchy	lomax	wald
weibull min	chi	cosine	maxwell	wrapcauchy
hi2	pearson3	powerlaw	rdist	erlang

all the distributions for every feature. Each feature for every dataset is tested against the 25 most commonly used distributions. These distributions have been listed in Table 1.

Once we find the distribution parameters, we build a machine learning model from the available data. On the contrary to the Federated Averaging model, where each client updates the server with its gradient after every epoch, we transmit the parameters *only once* when the complete training of the local model is completed.

3.2   Generating Data at Server

The server, instead of aggregating gradients from all the clients, it first generates data from the distributions it receives from them. For each client, based on the distribution parameters of each feature, we randomly generate data points, thereby creating a repository of sample training points. The predicted values for these features are generated using the weights that are also transmitted by the client resulting in an artificially generated dataset that follows a similar distribution as that present on the client node. These steps are presented in Algorithm 1.

3.3   Model Building at Server

Once the data from all the clients is combined, we run a multi-layer perceptron model on this dataset. Multi-layer perceptrons (MLP) [18] is a widely used feedforward ANN with a minimum of three layers: input, hidden, and an output layer. All nodes in each layer are connected with those in the other layer without loops. Each node uses an activation function for non-linear projections and extraction features on previous layer outputs. The gradients or the model parameters that are derived from this model are passed back to the client.

This approach significantly reduces the communication cost as we need only one round of communication to transfer the model and distribution parameters of the client and then receive back the updated global parameters.

Algorithm 1. Fusion Learning

1: **Client Update:**
2: **for** $i \in \{1 \ \ to \ \ F \ \}\forall \ \ features$ **do**
3:     a. calculate the p value of each distribution using K-S test
4:     b. find the maximum value from the above list to indicate its feature
5:     c. store the distribution parameters for that feature
6: **end for**
7: **for** $e \in \{1 \ \ to \ \ E \ \}\forall \ \ epochs$ **do**
8:     **for** $x \in \{1 \ \ to \ \ X \ \}\forall \ \ inputs$ **do**
9:         Update weights given by:
10:

$$\theta^k = \theta^k - \eta \delta L_k(\theta^k, b)$$

$$where \ \ \theta = weightvector, \eta = learningrate, L_k = Loss$$

11:     **end for**
12: **end for**
13: store the final weights
14: send distribution parameters and model parameters to server

1: **Server Update:**
2: **for** $i \in \{1 \ \ to \ \ C \ \}\forall \ \ clients$ **do**
3:     a. generate points for each distribution feature
4:     b. find predicted value for these points using model parameters
5: **end for**
6: $D_s = \bigcup\limits_{i=1}^{C} D_i \ //merge \ data \ points \ from \ all \ clients$
7: build a neural network model on the above dataset
8: transmit back the new global model parameters to the clients

---

4 Experimental Results

In this section, we present our experimental results obtained using fusion learning. The accuracies obtained through our approach have been compared against both a federated setup and a centralized learning system. A centralized learning system is where all the clients transmit their data to a central server. The server then builds a global model from this data, and the global model parameters are sent back to the clients. A significant issue with such an approach is the amount of data that needs to be transmitted across the network. The data increases with

Table 2. Dataset description

Dataset	Instances	Features
Credit Card	30000	24
Breast Cancer	569	9
Gender Voice	3169	20
Audit Data	777	18

an increase in the number of clients, and the other being the privacy concerns associated with the transmission of sensitive data over the network. The experiments are performed on four different datasets, namely Credit Card, Breast Cancer, Gender Voice, and Audit Data sets. These datasets have been retrieved from the commonly used UCI Repository [4].

The parameters of these datasets, which include the number of features and the number of data points, are depicted in Table 2. The initial results show that the accuracies obtained through fusion learning are almost similar to those achieved with federated and centralized frameworks.

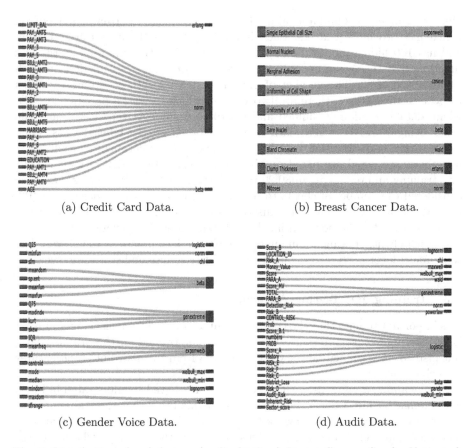

(a) Credit Card Data.

(b) Breast Cancer Data.

(c) Gender Voice Data.

(d) Audit Data.

Fig. 2. Distribution of each feature for Credit Card, Breast Cancer, Gender Voice and Audit datasets

4.1   Feature Distributions

Every dataset is made up of a number of features, and since each feature might follow a different distribution, the first step of the algorithm is to find out these distributions. The steps to obtain these distributions are explained in the previous section. The distributions for each of the four datasets can be seen in Fig. 2. We can observe that the features of the credit card dataset map to only three distributions, whereas breast cancer, gender voice, and audit data map to six, ten, and thirteen features. The precision of distribution detection can be further improved by considering more distributions during the initial phase of the algorithm.

4.2   Local and Global Models

Our experiments to build a local model include a simple multi-layer perceptron which has two hidden layers. Each hidden layer uses ReLu as the activation and has 100 hidden nodes. Sparse categorical cross-entropy is used as the loss function, and the Adam optimizer is used for stochastic gradient descent with a batch size of 32. The number of parameters varies with each dataset. With respect to the dataset, there are two ways in which it can be partitioned: IID, where data is randomly shuffled and distributed amongst ten clients, and the other being Non-IID, where data is divided based on the labeled class. Each labeled data is distributed to a different client. The experimental results that have been presented are based on IID data. Each dataset that is used is split into training and testing in an 80:20 ratio. For all the three frameworks, we have considered the number of clients to be ten and the number of epochs to be 100. Once the local model is built, and distributions are transmitted to the server, the server regenerates the points from these distributions. We generate 1000 points from each client, creating a cumulative of 10,000 data points at the server. The same multi-layer perceptron is used to build the global model at the server.

4.3   Training and Testing Accuracies

It is important to note that the training accuracy of the fusion learning approach is the testing accuracy because the model is not trained on the original data, but instead, it is trained on the data generated from the distribution of features of each client.

The training accuracies of all three frameworks have been illustrated in Fig. 3 and summarized in Table 3. We can see from this table that the training accuracies of fusion learning framework fall slightly below those obtained from both federated and a centralized setup. This is because the quality of the data generated is not on par with the original data. We can also notice that there is a subtle difference in accuracies of Credit Card, Breast Cancer, and Audit Data sets between Federated and Fusion Learning algorithms, whereas the accuracy of the Gender Voice dataset, is slightly lesser. The accuracies of such datasets can be increased by adding more distributions because determining the right distribution plays an important role in generating artificial data. Also, more data at the client node helps in determining the corresponding feature distribution parameters with more confidence, which results in an increase in the quality of

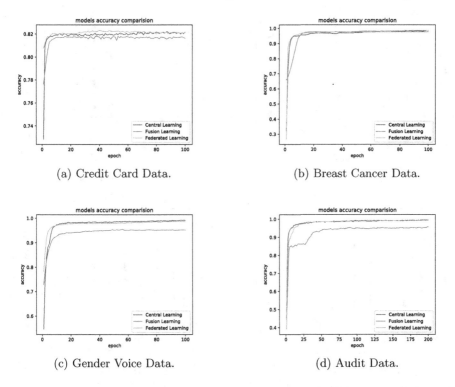

(a) Credit Card Data.

(b) Breast Cancer Data.

(c) Gender Voice Data.

(d) Audit Data.

Fig. 3. Comparison of training accuracies between Centralized Learning, Federated Learning and Fusion Learning algorithms

Table 3. Comparison of training accuracies (in %) between Central Learning, Federated Learning and Fusion Learning

Dataset	Central Learning	Federated Learning	Fusion Learning
Credit Card	81.11	81.60	81.09
Breast Cancer	97.08	96.35	95.62
Gender Voice	96.84	97.31	94.32
Audit Data	98.06	98.71	97.42

the generated data. As can be seen from Fig. 4, we have also compared the accuracies on each client node obtained using the local model and the global model built using the fusion learning framework. We see that in all the datasets, for all clients, the global model either outperforms the local model or achieves similar accuracies, which is also the case for a federated setup.

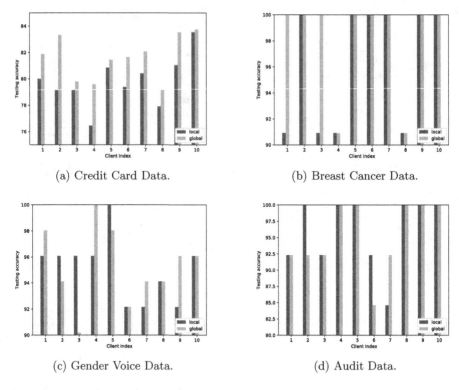

(a) Credit Card Data.   (b) Breast Cancer Data.

(c) Gender Voice Data.   (d) Audit Data.

Fig. 4. Comparison of testing accuracies of initial local model vs final global model at each client

4.4 Communication Efficiency

The main aim of this work is to reduce the number of communication rounds in a federated setup. A federated approach typically takes 'E' rounds to converge to a global model where 'E' is the number of epochs. In our case, the number of rounds is just one as we send both the model parameters and the distribution parameters at one shot. The server sends back the global parameters to the clients once it is built. This is summarized in Table 4.

Table 4. Network usage of Federated Learning and Fusion Learning for E epochs for a single client

Approach	Network calls	Data exchanged
Federated Learning	2 * E	Model parameters
Fusion Learning	2	Model params + feature distr parameters

5 Conclusions and Future Work

We have presented a new approach for distributed learning termed as *Fusion Learning*, which is able to achieve similar accuracies as compared to a Federated setup using only one communication round. This approach throws up a new direction for research in distributed learning and has its own set of challenges that needs to be addressed in greater detail. An important next step is to examine the proposed solution on broader datasets, which truly captures the massively distributed complexity of real-world issues. Another important direction would be to apply this technique to image datasets. Experimenting with this approach with different machine learning models on the server is an interesting direction for future work.

References

1. Boyd, S., Parikh, N., Chu, E., Peleato, B., Eckstein, J., et al.: Distributed optimization and statistical learning via the alternating direction method of multipliers. Found. Trends® Mach. Learn. **3**(1), 1–122 (2011)
2. Corinzia, L., Buhmann, J.M.: Variational federated multi-task learning. arXiv preprint arXiv:1906.06268 (2019)
3. Dekel, O., Gilad-Bachrach, R., Shamir, O., Xiao, L.: Optimal distributed online prediction using mini-batches. J. Mach. Learn. Res. **13**, 165–202 (2012)
4. Dua, D., Graff, C.: UCI machine learning repository (2017). http://archive.ics.uci.edu/ml
5. Eichner, H., Koren, T., McMahan, H.B., Srebro, N., Talwar, K.: Semi-cyclic stochastic gradient descent. arXiv preprint arXiv:1904.10120 (2019)
6. Garcia Lopez, P., et al.: Edge-centric computing: vision and challenges. SIGCOMM Comput. Commun. Rev. **45**(5), 37–42 (2015). https://doi.org/10.1145/2831347.2831354
7. He, L., Bian, A., Jaggi, M.: COLA: decentralized linear learning. In: Advances in Neural Information Processing Systems, pp. 4536–4546 (2018)
8. Huang, J., et al.: An in-depth study of lte: effect of network protocol and application behavior on performance. ACM SIGCOMM Comput. Commun. Rev. **43**(4), 363–374 (2013)
9. Kang, J., Xiong, Z., Niyato, D., Yu, H., Liang, Y.C., Kim, D.I.: Incentive design for efficient federated learning in mobile networks: a contract theory approach. In: 2019 IEEE VTS Asia Pacific Wireless Communications Symposium (APWCS), pp. 1–5. IEEE (2019)
10. Khodak, M., Balcan, M.F.F., Talwalkar, A.S.: Adaptive gradient-based meta-learning methods. In: Advances in Neural Information Processing Systems, pp. 5915–5926 (2019)
11. Kuflik, T., Kay, J., Kummerfeld, B.: Challenges and solutions of ubiquitous user modeling. In: Krüger, A., Kuflik, T. (eds.) Ubiquitous Display Environments. CT, pp. 7–30. Springer, Heidelberg (2012). https://doi.org/10.1007/978-3-642-27663-7_2
12. Lee, C.P., Roth, D.: Distributed box-constrained quadratic optimization for dual linear SVM. In: International Conference on Machine Learning, pp. 987–996 (2015)

13. Li, T., Sahu, A.K., Zaheer, M., Sanjabi, M., Talwalkar, A., Smith, V.: Federated optimization in heterogeneous networks. arXiv preprint arXiv:1812.06127 (2018)

14. Lin, T., Stich, S.U., Patel, K.K., Jaggi, M.: Don't use large mini-batches, use local SGD. arXiv preprint arXiv:1808.07217 (2018)

15. Liu, L., Zhang, J., Song, S., Letaief, K.B.: Edge-assisted hierarchical federated learning with non-IID data. arXiv preprint arXiv:1905.06641 (2019)

16. McMahan, H.B., Moore, E., Ramage, D., Hampson, S., et al.: Communication-efficient learning of deep networks from decentralized data. arXiv preprint arXiv:1602.05629 (2016)

17. Nishio, T., Yonetani, R.: Client selection for federated learning with heterogeneous resources in mobile edge. In: ICC 2019–2019 IEEE International Conference on Communications (ICC), pp. 1–7. IEEE (2019)

18. Pham, D.: Neural networks in engineering. In: Rzevski, G., et al. (eds.) Applications of Artificial Intelligence in Engineering IX, AIENG 1994, Proceedings of the 9th International Conference, Computational Mechanics Publications, Southampton, pp. 3–36 (1994)

19. Reddi, S.J., Konečný, J., Richtárik, P., Póczós, B., Smola, A.: AIDE: fast and communication efficient distributed optimization. arXiv preprint arXiv:1608.06879 (2016)

20. Shamir, O., Srebro, N.: Distributed stochastic optimization and learning. In: 2014 52nd Annual Allerton Conference on Communication, Control, and Computing (Allerton), pp. 850–857. IEEE (2014)

21. Smith, V., Chiang, C.K., Sanjabi, M., Talwalkar, A.S.: Federated multi-task learning. In: Advances in Neural Information Processing Systems, pp. 4424–4434 (2017)

22. Stich, S.U.: Local SGD converges fast and communicates little. arXiv preprint arXiv:1805.09767 (2018)

23. Virtanen, P., et al.: SciPy 1.0: fundamental algorithms for scientific computing in Python. Nat. Methods (2020). https://doi.org/10.1038/s41592-019-0686-2

24. Yang, T.: Trading computation for communication: distributed stochastic dual coordinate ascent. In: Advances in Neural Information Processing Systems, pp. 629–637 (2013)

25. Zhang, H., Li, J., Kara, K., Alistarh, D., Liu, J., Zhang, C.: ZipML: training linear models with end-to-end low precision, and a little bit of deep learning. In: Proceedings of the 34th International Conference on Machine Learning, vol. 70, pp. 4035–4043 (2017). JMLR.org

The Concept of System for Automated Scientific Literature Reviews Generation

Anton Teslyuk[1,2]([envelope])

[1] National Research Center "Kurchatov Institute", Moscow, Russia
`anthony.teslyuk@grid.kiae.ru`
[2] Moscow Institute of Physics and Technology, Dolgoprudny, Moscow Region, Russia

Abstract. We present a concept of system which is aimed to create a literature review of scientific articles having a small sketch of statements as the input. Key elements of the system include transformer-based BERT encoder, deep LSTM decoder and a loss function which combines auto-encoder loss and forces generated summaries to be in the input text domain. We propose to use PMC open access subset for model learning.

Keywords: Text summarization · Auto-encoder · NLP · BERT · LSTM

1 Introduction

Recent advances in machine learning methods demonstrate impressive results in a wide range of areas including generation of a new content. State of the art methods techniques based on generative adversarial networks (GANs), variational auto-encoders (VAE) and autoregressive models allow to generate images, videos, voice, texts which are very close or even indistinguishable from those create by humans. The success of such algorithms is determined by the availability of large structured datasets which allow to train complex models. Among promising sources of structured data which can be used for development of generative models one can distinguish scientific papers. A number of large paper collections are available including arXiv.org [1] and Pubmed Central [18]. Paper texts are well structured, labeled with keywords, summarized by abstract and title and are organized in a citation network. Such data seems to be very attractive source of information to train sophisticated algorithms for text processing and generation.

In this paper we present a design of the algorithm which will help to create a literature review based on a draft sketch. The idea is correlated with sketch to image and text to image synthesis algorithms, when the model learns to generate photo realistic images having simple sketch or text description as an input. Our model learns to generate blocks of text with a summary about some topic guided by a simple description of the topic as the input. The model includes encoder to map text sequences to latent space, decoder to do the reverse mapping and a loss function which shapes the latent space and forces representations of text blocks about similar topics to be grouped together.

© Springer Nature Switzerland AG 2020
V. V. Krzhizhanovskaya et al. (Eds.): ICCS 2020, LNCS 12139, pp. 437–443, 2020.
https://doi.org/10.1007/978-3-030-50420-5_32

2   Related Research

The task of creation new literature review based on a small sketch is a special case of text summarization task when large text or collection of texts are to be compressed in a more compact summary containing most important points [3]. There are two main approaches to do text summarization: extractive when the most important sentences are extracted from original text and abstractive when a new summary content is created.

Extractive summarization is a classification problem, the input text is split into sentences and each sentence is classified either to be selected for summary or not. It can be trained in either supervised mode when both source text and extracted summary is available in input data or in unsupervised when only source text is used. A common approach for this type of problems is to construct a mapping of sentences to some vector space and then use general purpose classifier to select important sentences. The mapping from sentences to vectors can be done using manually constructed features like TF-IDF [5], convolutional neural network [24], recurrent neural network [14], recurrent network with attention mechanism [23]. Current state of the art methods use transformer-based models [22] for sentence encoding with a stack of self-attention layers [7,10] which significantly outperforms previous methods. Extractive summarization can be done in unsupervised mode. In this case sentences are also mapped into some vector space and a clustering algorithm is applied to select sentences which are nearest to cluster centers [17]. Each cluster is supposed to contain sentences about some distinct topic. Finally a single sentence from each cluster is included in summary.

Abstractive summarization problem is a special case of sequence to sequence learning problem. A common approach for this sort of problems is to train encoder-decoder model which will encode input text into a vector or a sequence of vectors in latent space and then decode it into smaller sequence. Sutsveker [20] proposed to use deep recurrent neural networks (RNN) for a similar problem in machine translation where input sequence in one language needs to be translated into output sequence in target language. Later RNN approach was significantly improved by introducing attention mechanism [2,13] which enriches sequence representation by the use of information from intermediate hidden states of recurrent neural network. Vaswani [22] extended this idea in transformer model by using only a deep stack of self-attention layers without using recurrent networks in encoder and decoder. These methods were successfully applied for abstractive summarization problem including recurrent networks with attention [15] and transformer model [11].

The peculiarity of the summation task is that we need to transform multiple sequences from the input into a single summary sequence in the output. For this task a similar encoder-decoder model is used. The main difference is that the output of the encoder is not sent to the decoder directly. Rather one can use sampling from distribution in latent space inferred by encoding input sequences. One can encode a number of text sequences to be summarized into latent space vectors, then do some averaging to sample latent representation of summary and then decode summary latent space vector into a summary text. There are several approaches to do the sampling and training the encoder-decoder. Liu [9] proposed to use generative adversarial networks (GANs) having LSTM encoder, attention-based LSTM decoder which generate summaries for input text and a convolutional network as descriminator to classify synthetic summaries from those generated by human. Although GANs are often used to sample from latent space Liu only used it to train encoder-decoder pair, not for sampling. Analysis of features of latent space constructed by GAN is an interesting topic for further research.

Another approach to construct latent space is to use variational auto-encoders (VAEs) which impose additional regularization constraints for latent space. Latent space representation constructed by VAE have important feature for sampling: near points in latent space lead to similar or "meaningfull" sequences in data space. Shumann [19] demonstrated applicability of VAE for abstractive summarization task. A similar approach was presented by Chu [6] who trained encoder-decoder pair with a combination of auto-encoder loss forcing decoder output to be similar encoder input and similarity loss which forces representation of averaged summary in latent space to be close to latent vectors of original texts.

3   Input Data

To build the model we have used collection of scientific papers from Pubmed Central Open Access Subset collection [18]. This dataset includes more than 643k full-text papers from biomedical and life science journals with total volume of 75.5 Gb. Papers from this collection follow Pubmed Central Tagging guidelines which introduce a rich set of metadata and a standardized structure. The full texts are available in a structured form in XML format.

From the paper, we extract citations and a text content around every citation: N-sentences before and after it. In this way we obtain a number of text blocks containing content related to a paper being cited. Later when constructing the encoder from text sequences to latent space we will use this information to reduce distance between latent vectors for text content related to the same article. This idea can be further improved if we use additional information from paper citation network as a similarity measure in encoder latent space. One can use graph node similarity between paper nodes in citation network [12] as a distance factor between text blocks representations in latent space or use neural network mapping from citation graph space to latent space, e.g. graph2vec [16]. This is a topic for our further research.

We used Pubmed parser library [21] to extract paragraphs and reference entities from XML files. Then we applied NLTK toolkit [4] to split paragraphs into sentences and selected up to five sentences around every citation (two before and two after the citation). In this way we obtained dataset containing more than 12M text blocks labeled with citations of 643k papers.

4 Model Description

The general scheme of our model in presented in Fig. 1. It includes transformer encoder (BERT pre-trained model), recurrent network decoder, averaging of text block representations with the same label in latent space and a feedback from summary decoder to minimize distance between text blocks with the same label and their summary in latent space.

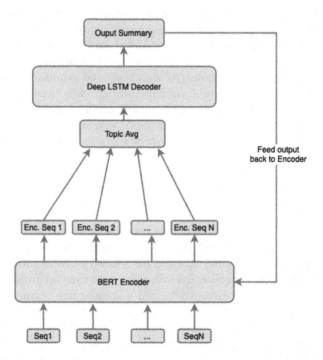

Fig. 1. Model scheme. A set of sentences on the same topic is passed through encoder and then is averaged to calculate topic representation in latent space. After that LSTM decoder in used to generate topic summary. The summary sequence is passed through encoder again to estimate the distance in latent space between summary representation and averaged topic representation. This distance is added to model loss function.

The input is a labeled sequence of text blocks, each text block includes up to five sentences around citation and a label with ID of a paper being cited. Then each text block is encoded using transformed-based BERT encoder [7]. We use BioBERT [8] version of BERT, pre-trained using biomedical text corpora which shows better results for life science specific texts. BERT output is a set of context-aware representations of every token in input sequences. Token representation is a 768-dimension vector. For further processing a set of token representations have to be squeezed into a single vector representation of the whole sequence. To do this a number of methods are proposed: using single token representation, average and max pooling of all tokens, hierarchical and attentive pooling. We use representation of the first <CLS> token, indicating start of the sequence as a representation of the whole sequence. This method is used in original BERT model for next sequence prediction task, when two sequences are fed into BERT encoder and the task is to classify if they are neighbors in the text. During training BERT encoder is fine-tuned to produce embedding of the first <CLS> token optimized for use as a hidden initial state in the first layer of LSTM decoder.

After encoding we average representations of sequences corresponding to the same label. In this way we obtain averaged representations of summaries to be created. This requires us to put enough examples of every label in the batch. In every batch we stack 16 text blocks of two labels, having total batch size of 32 and two summary vectors in the output.

Then averaged representations are fed into recurrent decoder. As a decoder we use stack of four LSTM layers, each having 768 hidden units and a fully connected layer of vocabulary size with softmax activation which outputs probabilities for output words. We initialize hidden state of the first LSTM layer with the output sequence from encoder. The rest LSTM layers hidden states are initialized using transformations of encoder output with fully connected layer with RELU activation. In this way each LSTM layer receives its own optimized initialization vector.

To compute loss we use the idea of Chu [6]. The loss is a sum of auto-encoder loss, which trains encoder and decoder to learn effective representation of sequences in latent space and the averaged similarity between encoder output for generated summary and encoder output for original texts. The idea can be further improved if use additional discriminator classifier which learns to distinguish generated summary output from auto-encoder output of initial text blocks.

When the model is trained it can be used to generate summaries. A brief sentence describing the topic is sent to the model input. Then we calculate encoder output for the input sentence and use K-nearest neighbors algorithm to get nearest K vectors in latent space from training data. Then set of K-vectors from training dataset is averaged and the mean vector is sent to decoder to generate output sequence for the summary.

5   Summary

In this paper we presented a general concept of a system to create a literature reviews. The system is based on abstractive text summarization methods: auto-encoder which combines transformer BERT encoder with LSTM decoder and additional loss factor which shapes latent space to be suitable for summary representation sampling. We believe that having focused to a particular area of life science literature make the task easier than trying to build a general purpose summarization system. Additional advantage is the availability of a big corpora of research papers in PMC collection which are well structured and enhanced with rich metadata. Currently the system is under our intensive development and testing.

Acknowledgements. This work has been supported by NRC Kurchatov institute project "Development of modular platform for scientific data processing and mining" (Project No. 1571).

References

1. arxiv.org e-print archive. https://arxiv.org/
2. Bahdanau, D., Cho, K., Bengio, Y.: Neural machine translation by jointly learning to align and translate. arXiv preprint arXiv:1409.0473 (2014)
3. Basiron, H., Jaya Kumar, Y., Ong, S.G., Ngo, H.C., Suppiah, P.C.: A review on automatic text summarization approaches. J. Comput. Sci. **12**, 178–190 (2016)
4. Bird, S., Klein, E., Loper, E.: Natural Language Processing with Python: Analyzing Text with the Natural Language Toolkit. O'Reilly Media, Inc., Sebastopol (2009)
5. Christian, H., Agus, M.P., Suhartono, D.: Single document automatic text summarization using term frequency-inverse document frequency (TF-IDF). ComTech: Comput. Math. Eng. Appl. **7**(4), 285–294 (2016)
6. Chu, E., Liu, P.J.: Meansum: a neural model for unsupervised multi-document abstractive summarization. arXiv preprint arXiv:1810.05739 (2018)
7. Devlin, J., Chang, M.W., Lee, K., Toutanova, K.: BERT: pre-training of deep bidirectional transformers for language understanding. arXiv preprint arXiv:1810.04805 (2018)
8. Lee, J., et al.: BioBERT: a pre-trained biomedical language representation model for biomedical text mining. Bioinformatics (2019). https://doi.org/10.1093/bioinformatics/btz682
9. Liu, L., Lu, Y., Yang, M., Qu, Q., Zhu, J., Li, H.: Generative adversarial network for abstractive text summarization. In: Thirty-Second AAAI Conference on Artificial Intelligence (2018)
10. Liu, Y.: Fine-tune BERT for extractive summarization. arXiv preprint arXiv:1903.10318 (2019)
11. Liu, Y., Lapata, M.: Text summarization with pretrained encoders. arXiv preprint arXiv:1908.08345 (2019)
12. Lu, W., Janssen, J., Milios, E., Japkowicz, N., Zhang, Y.: Node similarity in the citation graph. Knowl. Inf. Syst. **11**(1), 105–129 (2007)
13. Luong, M.T., Pham, H., Manning, C.D.: Effective approaches to attention-based neural machine translation. arXiv preprint arXiv:1508.04025 (2015)
14. Nallapati, R., Zhai, F., Zhou, B.: SummaRuNNer: a recurrent neural network based sequence model for extractive summarization of documents. In: Thirty-First AAAI Conference on Artificial Intelligence (2017)
15. Nallapati, R., Zhou, B., Gulcehre, C., Xiang, B., et al.: Abstractive text summarization using sequence-to-sequence RNNs and beyond. arXiv preprint arXiv:1602.06023 (2016)
16. Narayanan, A., Chandramohan, M., Venkatesan, R., Chen, L., Liu, Y., Jaiswal, S.: graph2vec: Learning distributed representations of graphs. arXiv preprint arXiv:1707.05005 (2017)
17. Nomoto, T., Matsumoto, Y.: A new approach to unsupervised text summarization. In: Proceedings of the 24th Annual International ACM SIGIR Conference on Research and Development in Information Retrieval, pp. 26–34 (2001)
18. PMC open access subset. https://www.ncbi.nlm.nih.gov/pmc/tools/openftlist/
19. Schumann, R.: Unsupervised abstractive sentence summarization using length controlled variational autoencoder. arXiv preprint arXiv:1809.05233 (2018)

20. Sutskever, I., Vinyals, O., Le, Q.V.: Sequence to sequence learning with neural networks. In: Advances in Neural Information Processing Systems, pp. 3104–3112 (2014)
21. Titipat, A., Acuna, D.: Pubmed parser: a python parser for pubmed open-access XML subset and MEDLINE XML dataset (2015). https://doi.org/10.5281/zenodo. 159504. http://github.com/titipata/pubmed_parser
22. Vaswani, A., et al.: Attention is all you need. In: Advances in Neural Information Processing Systems, pp. 5998–6008 (2017)
23. Yu, H., Yue, C., Wang, C.: News article summarization with attention-based deep recurrent neural networks. Technical report, Stanford University (2017)
24. Zhang, Y., Er, M.J., Zhao, R., Pratama, M.: Multiview convolutional neural networks for multidocument extractive summarization. IEEE Trans. Cybern. **47**(10), 3230–3242 (2016)

A Proposed Machine Learning Model for Forecasting Impact of Traffic-Induced Vibrations on Buildings

Anna Jakubczyk-Gałczyńska [ID] and Robert Jankowski[(✉)] [ID]

Faculty of Civil and Environmental Engineering, Gdansk University of Technology,
80-233 Gdansk, Poland
{annjakub,jankowr}@pg.edu.pl

Abstract. Traffic-induced vibrations may cause various damages to buildings located near the road, including cracking of plaster, cracks in load-bearing elements or even collapse of the whole structure. Measurements of vibrations of real buildings are costly and laborious. Therefore the aim of the research is to propose the original numerical algorithm which allows us to predict, with high probability, the negative dynamic impact of traffic-induced vibrations on the examined building. The model has been based on machine learning. Firstly, the experimental tests have been conducted on different buildings using specialized equipment taking into account six factors: distance from the building to the edge of the road, type of surface, condition of road surface, condition of the building, the absorption of soil and the type of vehicle. Then, the numerical algorithm based on machine learning (using support vector machine) has been created. The results of the conducted analysis clearly show that the method can be considered as a good tool for predicting the impact of traffic-induced vibrations on buildings, being characterized by high reliability.

Keywords: Buildings · Vibrations · Machine learning · Numerical algorithm

1 Introduction

There a number of different dynamic loads, including the effects of earthquakes, wind, vibrating machines, jumping of spectators, piling works or passing of vehicles, which can induce vibrations of buildings (see [1–3] for example). Traffic-induced vibrations, even that they are not as severe as vibrations caused by wind or earthquakes [4, 5], can also lead to major problems. They can cause plaster scratches and cracks, scratches on the structure, structural elements cracking or even collapsing of the building (see [6]). The harmfulness of vibrations on structural elements is influenced by many factors regarding the road on which vehicles move. The dynamic parameters of the building are also important. Vibration measurements on real structures are labour-intensive and costly projects, and what is important, they are not justified in every case. Nowadays, there are many decision support systems, research tools and computer programs necessary for their application. The problem is to find solutions that are both useful and economical.

© Springer Nature Switzerland AG 2020
V. V. Krzhizhanovskaya et al. (Eds.): ICCS 2020, LNCS 12139, pp. 444–451, 2020.
https://doi.org/10.1007/978-3-030-50420-5_33

Therefore, approximate methods are increasingly used. These methods allow the engineering problem to be solved accurately enough and the result is satisfactory, even if it is approximate (for example, see [7]). An example of this approach is the use of machine learning (ML). An example of this methodology is the support vector machine (SVM). The idea of using ML appeared as early as in 1952, when Arthur Samuel from IBM began building a computer program for training chess players [8]. The concept of ML appeared for the first time in 1983 in publication [9]. The purpose and sense of operation of ML was described by Tom Michael Mitchell in the basic publication regarding ML algorithms [10]. In turn, Chen et al. [11] used SVM to detect burglars. ML was also used in building operation problems. An example is the publication [12] which presents a system based on ML techniques supporting the detection of a threat, for example in the form of a fire, based on the analysis of the image from monitoring. Related works also include the paper [13] aiming to create a model of technical degradation of buildings located in mining areas and subjected to paraseismic tremors. An interesting application can also be found in the publication by Martínez-Rego et al. [14]. The authors presented the idea of detecting defects in a wind mill.

A review of the literature indicates that ML can be widely used. The implementation of this method in various fields can positively affect the development of technology and reduce costs while minimizing the risk of adverse effects. However, ML has not been used to forecast impact of traffic-induced vibrations on buildings. Therefore, the aim of this research is to implement this methodology for such a purpose.

2 Problem of Traffic-Induced Vibrations

Vibrations are described by movement of building structure particles, most often caused by waves propagated in the ground and reaching the foundations [15]. It should be noted that vibrations and noise are defined as environmental pollution. A very important factor in analysing the impact of traffic-induced vibrations on buildings is the soil in which the wave propagates. They are two wave parameters (damping and absorption) of significant importance described in [16]. Both of them depend mainly on the type of medium in which the wave propagates. Therefore, the soil impact parameter was also included in this ML model. It is also important to determine the source of vibration, propagation path and indicate the vibration receiver, which can be a structure itself, people or equipment located inside a building.

Special attention should be paid to the possibility of traffic-induced vibrations already during the design phase of building, and also in the exploitation phase. According to the standard [15], the load of the building caused by vibrations transmitted through the ground can generally be omitted if a building is located at a distance of more than 25 m from the axis of the railway line or at a distance of more than 15 m from the axis of the tram line, the axis of the first category road or a thoroughfare street. Therefore, if a building is located closer than the standard indicates, it is recommended to examine the impact of vibrations. However, after performing field tests using specialized equipment, it often appears that there is no immediate danger to the structure. Moreover, carrying out such tests for all buildings located along the road may not be economically justified. Therefore, there is a need to develop effective methods for forecasting impact of traffic-induced vibrations on buildings. The experience and acquired knowledge of experts are the basis of systems and calculation programs. Therefore, the aim of the operation is to create an expert model based on ML which allows us to predict with high probability the threat of negative dynamic impact on the tested building without performing laborious and expensive field measurements.

It should be added that vibrations may also have significant impact on people in buildings. This problem is particularly noticeable when people are exposed to long-term vibrations (see [6, 16]). Traffic-induced vibrations can cause discomfort to people in the affected area. Test results show that the threshold of human vibration perception is lower than the limit, after which damage may occur in the building itself [16]. In European standards (see [17–20]), three parameters are described for assessing the impact of vibrations: the corrected value of acceleration or velocity of vibrations in the frequency range 1–80 Hz, spectrum (frequency structure) for effective acceleration or velocity values in 1/3 octave bands and the dose of vibration.

3   Experimental Study

The extensive experimental study has been firstly carried out for 11 buildings (see Table 1). The measurements have been conducted for different types of passing vehicles in accordance with standard [15]. A separate detailed analysis has been performed for each building. The extreme amplitude values of measured vibrations have been obtained and compared with the values of the dynamic influence scale so as to determine to which zone the structure is assigned to (see [15] for details).

Table 1. Summary of data on experimental vibration testing on buildings.

Building no.	Condition of building B_C	Distance of the building B_{DR}	Soil absorption S_A	Type of road surface R_S	Condition of road surface R_C
1	Standard	7 m	Good	Bitumen	Good
2	Good	10 m	Good	Bitumen	Good
3	Good	9 m	Standard	Bitumen	Bad
4	Good	15 m	Standard	Bitumen	Standard
5	Good	20 m	Standard	Bitumen	Good
6	Standard	8 m	Standard	Bitumen	Good
7	Good	15 m	Good	Reinforced concrete slabs	Bad
8	Good	15 m	Good	Reinforced concrete slabs	Bad
9	Good	23 m	Bad	Bitumen	Standard
10	Good	20 m	Bad	Bitumen	Standard
11	Good	12 m	Standard	Ground	Standard

The measurement results (see [21] for details) have been used as data for creating the ML model. All tested buildings in experimental study have been divided into 2 sets: firstly, buildings with small external dimensions of the horizontal projection (maximum length of 15 m) with one or two stories and secondly, buildings up to five stories high. To create the ML model, all cases have been divided into 2 groups with descriptions of zones defined in [15]:

- zone I: vibrations unnoticed by a building; the lower limit of impact of vibrations on a building and the lower limit of taking into account the dynamic influences;
- zones II–V: vibrations noticed by a building; minimum, mean and significant impact of vibrations on a building.

4 Forecasting Algorithm Based on SVM

4.1 Input and Output Information

The purpose of the SVM algorithm in the present paper is to classify the variables and assign them to certain zones. The SVM task is to set a separator that will split the data into obvious subsets. The aim is to create a way to classify new data for which the assignment is not known. The SVM method assumes that the input variables are mutually independent and have the same probability distribution. The following input signals have been adopted on the basis of standard [15] and publication [21]: the distance of the building from the road edge (quantitative variable), condition and type of pavement, condition of the building, type of vehicle and absorption of vibrations in the ground (quality variables). Before the final set of parameters has been adopted, different combinations of these parameters have been examined. The output signal has been defined in this way that it contains the information whether there is a risk of negative impact of vibrations on buildings or not. The danger has been determined on the basis of the criteria described in standard [15]. No impact indicates zone I of dynamic influence scale, i.e. no influence on a building. If the algorithm predicts a possible threat, it is an indication for performing *in situ* measurements, as it may mean that the building falls in zone II–V of dynamic influence scale.

The construction of SVM has been based on the principles described in the literature [22–29]. Firstly, the database necessary to start the algorithm has been created. 63 samples have been collected, including 33 input data samples by independent own measurements (see chapter 3 and [21]) and 30 samples based on measurements of other researchers [30, 31]. The input signals have been defined as factors (independent variables) determined during field measurements, i.e. B_C - condition of building, B_{DR} - distance from the edge of the building, S_A - soil absorption, R_S - type of road surface, R_C - condition of road surface, V_T - type of vehicle. The input vector has therefore been adopted as follows: $x_{(6 \times 1)} = \{B_C, B_{DR}, S_A, R_S, R_C, V_T\}$, quantitative variable: $B_{DR} \in\ < 1.91$–22.5 m $>$, qualitative variables: $B_C \in \{bad, standard, good\}$; $S_A \in \{bad, standard, good\}$; $V_T \in \{bitumen, ground, reinforced concrete slabs, granite cube\}$; $R_C \in \{bad, standard, good\}$; $V_T \in \{type1, type2, type3\}$.

Two classes as the output signal have been adopted: 0 - no impact of vibrations on the building; 1 - probable impact of vibrations on the building. Thus, the output vector has been assumed to have the form: $\mathbf{y}_{(1 \times 1)} = \{y\}$; $y \in \{0, 1\}$.

The SVM algorithm has been tested for four different kernel functions. The optimization problem has been solved, in which the margin (weak margin) has been minimized because it had been not linearly separable. A binary classification algorithm has been used since two classes of sets have been assumed: a set of cases for which there is a threat of impact of vibrations on buildings and a set of safe cases. All cases have been randomly divided into 2 sets: learning and testing. 47 samples have been randomly assigned to the learning set (74.6% of all samples) and 16 samples for the test set (25.4% of the total).

Because the cases have been unevenly distributed into classes (for 47 samples empirically determined during measurements, no risk of impact of vibrations on building - class "0" and for 16 samples a result indicating the risk of vibration impact - class "1") a penalty has been used to avoid incorrectly classifying cases of class "1" into a larger class, that is class "0". The kernel functions have been adopted in turn as a linear, polynomial, radial base and sigmoidal function [24]. To determine the optimal values of the learning constants, a cross-validation has been used for each machine with each type of kernel according to formulas (1–4). The algorithm's reliability has been assessed on the basis of errors recognizing the assignment of structures from the set of all data and the set of learning and testing data to previously defined patterns.

In the first case, the linear function of the form described in [24] has been adopted as the kernel function $K(x, y)$ according to formula (1):

$$K(x, y) = x^T y + c \tag{1}$$

where:

c - optional constant;

In accordance with the principle described in [26], a 10-fold cross-validation has been carried out, the validity of which has been equal to 78.72%. 21 support vectors have been determined (9 for the "0" class, 12 for the "1" class), including 11 associated ones. Classification accuracy in the learning set has been found to be equal to 82.98%, and in the testing set 81.25%. Overall validity (taking into account all samples) has been equal to 82.54% (this is a weighted average taking into account the size of the sample in a given set). The algorithm's reliability has been assessed on the basis of errors recognizing the assignment of structures from the set of all data and the set of learning and testing data to previously defined patterns.

In the next stage, the machine with polynomial function $K(x, y)$ of the third degree has been analysed [25]:

$$K(x, y) = (\alpha x^T y + c)^d \tag{2}$$

where:

c - optional constant;
α - slope, $\alpha = 1/N$, where N is the data dimension;
d - degree of polynomial;

The cross-validation and reliability of the algorithm have been carried out in the same way as for the previous machine. The accuracy of the cross test has been found to be equal to 78.72%. 27 support vectors have been designated (14 for the "0" class, 13 for the "1" class), including 10 associated ones. Classification accuracy in the learning set has been found to be equal to 85.11%, and in the testing set 81.25%. Overall validity has been equal to 84.13%.

In the third step to build the algorithm, the radial base function has been adopted as the function $K(x, y)$ [24]:

$$K(x, y) = \exp\left(-\frac{\|x - y\|^2}{2\sigma^2}\right) \tag{3}$$

where:

σ - parameter regulating data noise and ensuring non-linearity of functions.

The accuracy of the cross-validation test has been found to be equal to 78.72%. 24 supporting vectors have been determined (11 for the "0" class, 13 for the "1" class), including 11 associated ones. Classification accuracy in the training set has been found to be equal to 85.11%, and in the test set 81.25%. Overall validity has been equal to 84.13%.

In the last case, the sigmoidal function has been adopted according to the formula (4) for the algorithm [24]:

$$K(x, y) = \tanh\left(\alpha x^T y + c\right) \tag{4}$$

where:

c - optional constant;

The accuracy of the cross-validation has been found to be equal to 76.60%. 21 supporting vectors have been determined (8 for the "0" class, 13 for the "1" class), including 13 associated ones. Classification accuracy in the learning set has been found to be equal to 74.47%, and in the testing set 81.25%. Overall relevance has been equal to 76.19%.

The application of the classification algorithm using SVM has allowed us to obtain results for all cases: learning and testing. The result for each sample has been compared with the result obtained from the experiment. The comparison is presented in Table 2 in the form of classification reliability. The overall reliability of the machines has been calculated as a weighted average taking into account the size of the learning and testing sets. It can be seen from the table that, considering the four kernel functions used for construction, they all have achieved the same correctness (81.25%) in the testing set. However, in the case of learning samples, the third degree polynomial kernel and radial base function have achieved better prediction, equal to 85.11%. Table 2 also shows that the best overall validity (84.13% of the prediction) has been obtained using the third degree polynomial and the radial base function.

Table 2. Summary of the prediction: classification reliability [%].

Kernel function	Correctly determined cases - reliability [%]		
	Learning set	Testing set	General (all samples)
Linear	82.98	81.25	82.54
Polynomial	85.11	81.25	84.13
Radial basis functions	85.11	81.25	84.13
Sigmoidal	74.47	81.25	76.19

5 Conclusions

In this paper, the model for forecasting the impact of traffic-induced vibrations on buildings using machine learning method has been considered. Firstly, a database has been created on the results of vibration measurements for buildings using both own research and other researchers. In the next stage, the rules for creating ML have been presented. Finally, the model proposed by authors has been presented.

The SVM algorithm has been tested for four different kernel functions: linear, polynomial, radial base and sigmoidal functions. The model has been created for the input variable vector $x_{(6 \times 1)} = \{B_C, B_{DR}, S_A, R_S, R_C, V_T\}$. Other combinations of input parameters have been also considered at the testing stage, but the presented model achieved the best results. This means that the best combination of input factors for the analysed cases of traffic-induced vibrations prediction are: technical condition of the building, distance between the building and the road edge, absorption of vibrations in the ground, type of surface, technical condition of the surface and type of vehicles moving on the way. The results clearly show that ML, in this case SVM can be considered as a very effective method for forecasting impact of traffic-induced vibrations on buildings. The results obtained for the model proposed by the authors are satisfactory in terms of credibility. The algorithm is capable to forecast both existing and designed cases.

References

1. Mahmoud, S., Jankowski, R.: Elastic and inelastic multi-storey buildings under earthquake excitation with the effect of pounding. J. Appl. Sci. **9**(18), 3250–3262 (2009)
2. Elwardany, H., Seleemah, A., Jankowski, R.: Seismic pounding behavior of multi-story buildings in series considering the effect of infill panels. Eng. Struct. **144**, 139–150 (2017)
3. Sołtysik, B., Jankowski, R.: Non-linear strain rate analysis of earthquake-induced pounding between steel buildings. Int. J. Earth Sci. Eng. **6**(3), 429–433 (2013)
4. Miari, M., Choong, K.K., Jankowski, R.: Seismic pounding between adjacent buildings: identification of parameters, soil interaction issues and mitigation measures. Soil Dyn. Earthq. Eng. **121**, 135–150 (2019)
5. Falborski, T., Jankowski, R.: Experimental study on effectiveness of a prototype seismic isolation system made of polymeric bearings. Appl. Sci. **7**(8), 808 (2017)
6. Hunaidi, O.: Traffic vibrations in buildings. Constr. Technol. Update **39**, 1–6 (2000)
7. Jankowski, R., Walukiewicz, H.: Modeling of two-dimensional random fields. Probab. Eng. Mech. **12**(2), 115–121 (1997)

8. Michalski, B.J., Proudfoot, D.: What turing did after he invented the universal turing machine. J. Logic Lang. Inform. **9**(4), 491–509 (2000)
9. Michalski, R.S., Carbonell, J.G., Mitchell, T.M.: Machine Learning: An Artificial Intelligence Approach. Springer, New York (1983). https://doi.org/10.1007/978-3-662-12405-5
10. Mitchell, T.M.: Machine Learning. McGraw-Hill Science, Columbus (1997)
11. Chen, W.H., Hsu, S.H., Shen, H.P.: Application of SVM and ANN for intrusion detection. Comput. Oper. Res. **32**(10), 2617–2634 (2005)
12. Wabik, W.: Monitoring system to detect potential dangerous situations. Stud. Informatica **33**(2B), 497–508 (2012)
13. Firek, K., Rusek, J., Wodyński, A.: Wybrane metody eksploracji danych i uczenia maszynowego w analizie stanu uszkodzeń oraz zużycia technicznego zabudowy terenów górniczych. Przegląd Górniczy **72**(1), 50–55 (2016). (in Polish)
14. Martínez-Rego, D., Fontenla-Romero, O., Alonso-Betanzos, A.: Power wind mill fault detection via one-class ν-SVM vibration signal analysis. In: The 2011 International Joint Conference, California, USA, pp. 511–518 (2011)
15. PN-B-02170:2016-12: Ocena szkodliwości drgań przekazywanych przez podłoże na budynki: Evaluation of the harmfulness of building vibrations due to ground motion, Polish Committee for Standardization of Measurement and Quality (2016). (in Polish)
16. Dulińska, J., Kawecki, J., Kozioł, K., Stypuła, K., Tatara, T.: Oddziaływania parasejsmiczne przekazywane na obiekty budowlane. Wydawnictwo Politechniki Krakowskiej, Kraków (2014). (in Polish)
17. ISO 10137:2007: Bases for design of structures - serviceability of buildings and walkways against vibrations (2007)
18. Directive 2002/49/EC of the European parliament and the Council of 25 June 2002 relating to the assessment and management of environmental noise. Official J. **189** (2002)
19. DIN 4150-2: Structural vibration, Part 2: Human exposure to vibration in buildings (1999)
20. BS 6472-1:2008: Guide to evaluation of human exposure to vibration in buildings, Part 1: Vibration sources other than blasting (2008)
21. Siemaszko, A., Jakubczyk-Gałczyńska, A., Jankowski, R.: The idea of using Bayesian networks in forecasting impact of traffic-induced vibrations transmitted through the ground on residential buildings. Geosciences **9**(8), 339 (2019)
22. Cortes, C., Vapnik, V.: Support-vector networks. Mach. Learn. **20**, 273–297 (1995). https://doi.org/10.1007/BF00994018
23. Haykin, S.: Neural Networks and Machine Learning. Pearson Prentice Hall, Upper Saddle River (2009)
24. Souza, C.R.: Kernel functions for machine learning applications. Creative Commons Attribution-Noncommercial-Share Alike **3**, 29 (2010)
25. Cristianini, N., Shawe-Taylor, J.: An Introduction to Support Vector Machines and Other Kernel-based Learning Methods. Cambridge University Press, Cambridge (2000)
26. Conway, D., White, J.: Machine Learning for Hackers. O'Reilly Media Inc., Sebastopol (2012)
27. Vapnik, V., Lerner, A.J.: Generalized portrait method for pattern recognition. Autom. Remote Control **24**, 774–780 (1963)
28. Boser, B.E., Guyon, I.M., Vapnik, V.N.: A training algorithm for optimal margin classifiers. In: Proceedings of the Fifth Annual Workshop on Computational Learning Theory, Pittsburgh, USA, pp. 144–152 (1992)
29. Bennett, K.P., Campbell, C.: Support vector machines: hype or hallelujah? ACM SIGKDD Explor. Newsl. **2**(2), 1–13 (2000)
30. Czech, R., Miedziałowski, C., Chyży, T.: Wpływ poprawy stanu drogi na redukcję drgań w zabytkowym kompleksie budynków. Materiały Budowlane **6**, 105–106 (2015). (in Polish)
31. MOIIB Homepage. http://www.map.piib.org.pl/materialy-szkoleniowe. Accessed 17 Dec 2017

Retrain or Not Retrain? - Efficient Pruning Methods of Deep CNN Networks

Marcin Pietron[(⊠)] and Maciej Wielgosz

AGH - University of Science and Technology, Krakow, Poland
{pietron,wielgosz}@agh.edu.pl

Abstract. Nowadays, convolutional neural networks (CNN) play a major role in image processing tasks like image classification, object detection, semantic segmentation. Very often CNN networks have from several to hundred stacked layers with several megabytes of weights. One of the possible techniques to reduce complexity and memory footprint is pruning. Pruning is a process of removing weights which connect neurons from two adjacent layers in the network. The process of finding near optimal solution with specified and acceptable drop in accuracy can be more sophisticated when DL model has higher number of convolutional layers. In the paper few approaches based on retraining and no retraining are described and compared together.

Keywords: Deep learning · CNN · Pruning · Image processing

1 Introduction

The convolutional neural networks are the most popular and efficient model used in many AI tasks. They achieve best results in image classification, semantic segmentation, object detection etc. The reduction of memory capacity and complexity can make use of them in real-time applications like self driving cars, humanoid robots, drones etc. Therefore compression CNN models is a important step in adapting them in embedding systems and hardware accelerators. One of the methods to decrease memory footprint is a pruning process. In case of small convolutional network, the complexity of this process is much lower than in larger ones. In very deep CNN models which have several to few hundreds of convolutional layers the process of finding near global optimum solutions which guarantee acceptable drop in accuracy is quite a complex task. Genetic/memetic algorithms, reinforcement learning, random hill climbing or simulated annealing are good candidates to solve this problem. In paper, algorithm based on RMHC and simulated annealing methods is presented.

The pruning process can be done by two major methodologies. First one is a pruning a pre-trained networks, the second one is pruning using retraining. The first one is much faster. It needs only an inference step run on a test dataset in each stage/iteration of the algorithm, [2]. In case of mode with retraining

© Springer Nature Switzerland AG 2020
V. V. Krzhizhanovskaya et al. (Eds.): ICCS 2020, LNCS 12139, pp. 452–463, 2020.
https://doi.org/10.1007/978-3-030-50420-5_34

pruning can be done after every weights updated in training process. This paper describes and compares the approaches using both the methodologies.

The squeezenet [9] model was one of the first approach in which compression by reducing the filters size was used. In this approach, architecture of alexnet was modified to create less complex model with same accuracy. Later approaches were concentrated more on quantization and pruning [2,6] as a steps that enables compression. In [6] authors present approaches for CNN compression including pruning with retraining. The results for older architectures VGG and AlexNet are presented. In paper [8] authors describe reinforcement learning as a method for choosing channels for structural pruning. In article [7] the SNIP algorithm is described. The algorithm computes gradients during retraining and assigns priorities to weights based on gradients values. The pruning is done using knowledge about importance of weights in a training process. In papers [4,5] compression for other machine learning models are described in NLP tasks. It is shown that by especially using sparse representations, it is possible to achieve better results than in baseline models. The paper is organized as follows. The Sect. 2 presents the methods for pruning pre-trained networks. There is a basic method and its further enhancements using more complex models analysis. The next Sect. 3 is about pruning with retraining on imagenet, CIFAR10 and CIFAR100 datasets and structural pruning. Finally, in Sect. 4 and 5 further work and conclusions are described.

2 Pruning with No Retraining

After the process of training neural model we acquire a set of weights for each trainable layer. These weights are not evenly distributed over the range of possible values for a selected data format. Majority of weights are concentrated around 0 or very close to it. Therefore, their impact on the resulting activation values is not significant. Depending on network implementation specifications, storing weights may require a significant amount of memory. Applying pruning process to remove some weights has a direct impact on lowering storage requirements. In this section the approaches based on pre-trained networks are presented. The first one is memetic approach which is based on random hill climbing with few extensions. The parameters to the heuristic were added to optimize and speed up the process of finding local optima solutions. In this algorithm pruning is a function that set weights values to zero whose magnitudes are below specific threshold (Eq. 1 and Eq. 2).

$$W_p = pruning(threshold, W) \tag{1}$$

$$W_p = \{\forall w_i \in W_p : w_i = 0, |w_i| < threshold \lor w_i = w_i \in W\} \tag{2}$$

Next, additionally more sophisticated analysis was incorporated to previous approach to improve obtained results. Presented method analyses energies/contributions of 2D filters in layers and heat maps to increase sparsity further.

2.1 Incremental Pruning Based on Random Hill Climbing

The presented approach for fast pruning is based on random hill climbing and simulated annealing local search. In each iteration, it chooses specified number of layers to be pruned. The layers are chosen using probability distribution based on layers' complexities and sensitivities (Eq. 3, Eq. 4, line 4). If a layer is more complex and less sensitive than others, it has more probability to be chosen. In each iteration, layers are pruned by the step which can be different and computed independently for each layer (line 7). If drop in accuracy is higher than given threshold reverse pruning is applied (the step can be cancelled or sparsities of different layers are decreased). Fitness function is a weighted sparsity which is overall memory capacity of current pruned model (line 11). Solution is a simple genotype where each layer is represented as a percentage of weights that were pruned in this layer. Algorithm can use as an option simulated annealing strategy which accepts worse solutions (exploration phase) to have the possibility to escape from local optima (line 18–22). In this case, in line 21 a next created solution can be worse than previous solution and will be accepted with specified probability which decreases in each iteration. Algorithm has a ranked list of all k-best solution already found (line 14). It helps to overcome algorithm stagnation by giving opportunity to return to good solutions (line 19). Each layer as it was mentioned earlier has sensitivity parameter which measures the latest impacts (number of impacts is defined by window size parameter) of this layer to the drop in accuracy of the model (Eq. 5, line 13). The layer sensitivity is updated after each iteration in which given layer is pruned (line 13). The step size which indicates percentage of weights to be pruned for a given layer is computed using current sensitivity value of a layer. If sensitivity is less than acceptable drop in accuracy (threshold) algorithm increases step size and vice versa using Eq. 6, line 24.

$$probability_i = size_i \times (threshold - sensitivity_i) \tag{3}$$

$$policy = categorical(probability) \tag{4}$$

$$sensitivity_i^t = \frac{\sum_{t-|window|}^{t}(baseline\_acc - pruned\_acc_t)}{window} \tag{5}$$

$$step_i^t = step_i^{t-1} + k \times step_i^{t-1} \times (threshold - sensitivity_i^{t-1}) \tag{6}$$

The presented algorithm can be run in multi-layer mode in which, in one iteration more than one layer can be pruned. In Table 1 there are results achieved using Algorithm 1 with constant policy by running 150 iterations. Table contains weighted sparsities of pruned models and their drops from baseline accuracies. The threshold drop was set to 1.0. The Table 2 presents results using prioritization mode in which largest layers in given models were chosen for pruning in the first stage of the algorithm till the drop in accuracy is higher than given threshold. After that rest of the layers are pruned. We can observe significant

Algorithm 1. Pruning algorithm

1: **Input:** number_of_iterations
2: **Input:** drop_in_accuracy_threshold
3: **for** number_of_iterations **do**
4:    update_policy()
5:    layer = choose_layer_for_pruning(policy)
6:    **if (top_1-baseline) < drop_in_accuracy_threshold then**
7:       prune_layer_by_step($step_l$)
8:    **else**
9:       reverse_prune_by_step($step_l$)
10:   **end if**
11:   fitness = compute_new_fitness()
12:   top1 = compute_accuracy()
13:   update_layer_sensitivity(layer)
14:   update_ranked_list()
15:   **if fitness < best_fitness then**
16:      next_solution = current_solution
17:   **else**
18:      **if SA_Probability < threshold then**
19:         next_solution = solution_from_ranked_list()
20:      **else**
21:         next_solution = current_solution
22:      **end if**
23:   **end if**
24:   $step_l$ = update_steps(layer)
25: **end for**

Table 1. Pruning results with constant policy

Name	Weighted sparsity	T1
vgg16	35.3%	−0.8
resnet50	32.1%	−1
vgg19	32.6%	−1
inception_v3	18.1%	−0.8

improvements in achieved results. Table 3 shows results of using dynamic policy updates during algorithm.

2.2   2D Filter and Its Activation Analysis for Further Pruning Improvements

Improvement presented in this subsection does additional analysis that can explain the internal representation of the model and removes more weights with high probability to not decrease its accuracy. First approach is to compute 2D average filter contributions in a final answer of the network (Eq. 7, Eq. 8, Eq. 9). The next one is to analyze filter contribution in a process of recognition specific class. Each class is analyzed separately and average neurons activations are measured. Then in each layer we can extract region of weights that are less important in the whole process of recognition using some threshold of importance. In Table 4 and Table 5 there are results presented for these two steps performed on the last layer before softmax in VGG16 after running Algorithm 1. It shows that is possible to do further pruning to increase sparsity without drop in the accuracy.

$$F = \{f_{mnhw} \in R : m \in M, n \in N, h \in H, w \in W\} \tag{7}$$

where: M, N, H, W are number of channels, kernels, height and width respectively of layer filter

$$F\_2D = \{f_{mn} \in R :\in R : m \in M, n \in N\} \tag{8}$$

$$f\_2D\_energy_{ij} = mean(sum((f\_2D_{ij}) \odot receptive\_field_{ij})) \tag{9}$$

Table 2. Pruning results with specified prioritization

Name	Weighted sparsity	T1 drop	Prioritization list
vgg16	67%	−0.9	Layers 14, 15, 16
resnet50	37%	−1.1	5 largest layers
vgg19	65	−1.0	Layers 17, 18, 19
inception_v3 25	35%	−0.9	8 largest layers

Table 3. Results of pruning with dynamic policy

Name	Weighted sparsity	T1 drop
vgg16	65%	−1.0
resnet50	35%	−1.0
vgg19	60%	−0.9
inception_v3	24%	−0.9

Table 4. VGG16 with 2D filter analysis

Name	T1	Pruning	With 2D	T1 drop
CIFAR10	90.76%	50%	52%	−1.0
CIFAR100	77.6%	45%	47%	−1.0

Table 5. VGG16 with 2D filter analysis and filter contributions in a classes recognition

Name	T1	Pruning	+Filter contributions	T1 drop
CIFAR10	90.55 %	50%	54%	−1.0
CIFAR100	77.5 %	45%	50%	−1.0

3 Pruning with Retraining

The methods described in the previous section have one main drawback, their weight can be fine tuned during the pruning process to boost model accuracy. The training step can improve accuracy of pruned network by learning weights that were not removed before. In this section, results of these methods are presented.

3.1 Methods

Retraining is recognized as an effective method for regaining performance of the pruned model. However, it is important to pick a right protocol and retraining parameters. We have examined three different schemes of pruning and retraining.

- simple retraining which without masking,
- simple retraining with masking,
- adaptive retraining with boosting.

The first two methods apply a simple retraining procedure after each step of pruning. The procedure can be interleaved with masking operation. It is implemented by zeroing gradient which otherwise would be applied to the pruned weights. It is worth noting that even without masking the pruned weights are mode prone to be pruned again in the next epoch because they are small. Consequently, the masking operation makes the pruning process more stable since a pool of pruned weights is progressively enlarged without change of coefficients. The simple method is limited in its effectiveness mostly because it lacks ability to adopt pruning both in terms of layers of the model and the retraining time. Some layers during selected training epochs are more prone to pruning, which is not taken into account in the simple method. Therefore, we have proposed the retraining with boosting procedure which is given by Algorithm 2.

The proposed approach Algorithm 2 relies on a choice of priority list of the layers which is supposed to be set at the very beginning of the process. The rest of the parameters *steps* decide how many steps are taken before *scale* is changes. This gradually reduces pruning factor. The scale (refer to Algorithm 2) decides how many time the step is reduced. Once the model is pruned it is validated with a small dataset to check if the performance drop is not to large. If this is the case the process of pruning is stopped for the given layer in this iteration (epoch) and the algorithm goes to the next layer on the priority list. The pruning process may terminate in a regular fashion when all the steps and scale rates are exhausted. In order to speedup the process a layer which was skipped several times due to the performance drop after pruning is marked as permanently skipped. It is worth noting that a number of epochs should be picked properly in order to satisfy the number of the protocol interactions (number of steps and scale changes).

Algorithm 2. Retraining with boosting

```
1: Input: scales
2: Input: steps
3: Input: step_size
4: for number_of_epochs do
5:     layer = choose_high_priority_layer_for_pruning()
6:     for layers do
7:         pick_the_next_layer_from_the_priority_list()
8:         for scales do
9:             for steps do
10:                prune()
11:                validate()
12:                if performance_drop < threshold0 then
13:                    if skipped_no < threshold1 then
14:                        mark_layer_done_for_this_iteration()
15:                    else
16:                        mark_layer_done_for_all_iteration()
17:                    end if
18:                    mark_layer_done_for_this_iteration()
19:                    exit()
20:                end if
21:            end for
22:            step_value ← step_value/reduction_factor
23:        end for
24:    end for
25: end for
```

3.2 Results of the Pruning and Retraining Experiments on Imagenet

Table 6. Results of Resnet-50 simple pruning and retraining

Layers pruned	Sparsity	Masking	Best T1	T1 drop	Best T5	Epoch	lr (reduction)
None	0	None	76.13%	0	92.862%	103	0.1 (30)
all-0.2	All layers 0.2	False	76.83%	0.7	93.15%	14	1.00E−03
all-0.3	All layers 0.3	False	76.95%	0.82	93.21%	68	1.00E−03
all-0.7	All layers 0.7	True	59.55%	−16.58	83.62%	26	0.1

Table 7. Results of Resnet-50 progressive pruning and retraining

Layers pruned	Sparsity	Masking	Best T1	T1 drop	Best T5	Epoch	lr (reduction)
None	0	None	76.13%	0	92.862%	103	0.1 (30)
all-0.2 + 0.1 * epoch	all layers 0.3	False	76.48%	0.35	93.07%	1	1.00E−03
all-0.1 + 0.1 * epoch	all layers 0.2	True	76.56%	0.43	93.1%	1	1.00E−03
all-0.1 + 0.01 * epoch	all_layers_0.53	True	76.09%	−0.04	93%	43	0.01(30)

There was series of experiments conducted as presented in Table 6, 7 and 8. Different parameters were chosen as well as different strategies were tested. In the first a naive approach was explored as a baseline. The results are presented in Table 6. We can see that equal pruning of all the layers for 0.2 and 0.3 sparsity led to the boost of the model performance. However, more aggressive pruning of 0.7 equal sparsity resulted in a significant decline of the sparsity. The proposed simple method may be useful when treated as a form of regularization and slight increase of the model sparsity.

Table 8. Results of Resnet-50 boosted pruning and retraining

Layers pruned	Sparsity	Masking	Best T1	T1 drop	Best T5	Epoch	lr (reduction)
None	0	None	76.13%	0	92.862%	103	0.1 (30)
Steps: 12, scales: 2, step: 0.05	Weighted: 0.37	True	75.12%	−1.01	92.57%	47	1.00E−03
Steps: 12, scales: 2, step: 0.05	Weighted: 0.42	True	75%	−1.13	93.09%	30	1.00E−03
Steps: 12, scales: 2, step: 0.05	Weighted: 0.427	True	79.88%	3.75	94.96%	100	1.00E−03
Steps: 12, scales: 2, step: 0.05	Weighted: 0.57	True	75.35%	−0.78	92.59%	299	1.00E−03
Steps: 4, scales: 2, step: 0.05	Weighted: 0.5137	True	75.61%	−0.52	92.67%	431	1.00E−03
Steps: 6, scales: 2, step: 0.05	Global: 0.57	True	75.52%	−0.61	92.68%	279	1.00E−03
Steps: 10, scales: 2, step: 0.05	Global: 0.44	True	76.23%	0.1	93%	141	1.00E−03
Steps: 2, scale: 4, step: 0.2	Global: 0.648	True	74.51%	−1.62	92.19%	88	1.00E−03

It is worth noting that progressive pruning which results are presented in Table 7 is much more effective. For instance, the experiment with starting point of 0.1 and progress of 0.01 every epoch (see the last row in Table 7) allowed to reach equal sparsity of 53% after 43 epochs with negligible loss of performance. This method despite its benefits is limited in its capacity to reduce sparsity. Method saturates at about 60% of sparsity. The most advanced approach of pruning and retraining in the boosting method given by Algorithm 2. Its results are presented in Table 8. We can in Table 8 that different values of steps and scales lead to huge discrepancies in the results in terms of sparsity. The highest sparsity of 64.8% was achieved for steps: 2, scale: 4 and step value: 0.2. This was achieved at the expanse of noticeable loss of the performance. On the other hand small step value, large number of steps and training epochs lead to much lower performance degradation as proved by the experiment with steps: 6, scales: 2, step value: 0.05 and 279 epochs of training. However, such large number of epochs required approx. 10 days of training time on 8 Nvidia GTX 1080 GPUs.

Choice of a proper number of steps, scales and step values should be done individually for each model and ideally facilitated with an optimization algorithm.

During a pruning and retraining operation of a pretrained model with high learning rate, there is a huge degradation of the performance (t1 and t5) in the very first epoch as presented in Fig. 1. In the next epochs the model regains it original performance quite fast. The presented in Fig. 1 resembles in terms of a training pattern most of the experiments showed in Table 8.

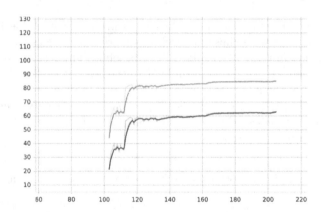

Fig. 1. Retraining of the pretrained Resnet50 with global sparsity of 20%. Retraining starts at 104 epoch. Top5 is marked in blue and Top1 in red. (Color figure online)

3.3 Pruning with Retraining on CIFAR Datasets

The similar approach as described in the previous section was performed on a CIFAR10 and CIFAR100 datasets. The main difference is that in each step, the weights for pruning were chosen using its gradient values. This information gives feedback how important the weight was in the former training step (Algorithm 3). If its significance is less then it is safer to remove it. Table 9 and Table 10 present results obtained using Algorithm 3. They show significant improvement in the sparsity obtained when compared to fast pruning approach.

Algorithm 3. Pruning algorithm with retraing

1: **Input:** number_of_epochs
2: **Input:** drop_in_accuracy_threshold
3: **Input:** init_sparsity
4: **Input:** init_step
5: **for** number_of_epochs **do**
6:     layer = choose_layer_for_pruning(policy)
7:     analyze_gradients_and_update_statistics()
8:     **if (top_1-baseline) < drop_in_accuracy_threshold then**
9:         prune_layer_by_step($step_l$)
10:    **else**
11:        reverse_prune_by_step($step_l$)
12:    **end if**
13:    top1 = compute_accuracy()
14:    update_layer_sensitivity(layer)
15:    $step_l$ = update_layer_step(layer)
16:    masking()
17:    retrain()
18: **end for**

Table 9. Results of fine-grain pruning with retraining (CIFAR10)

Name	Baseline T1	Pruned T1	Pruned size
vgg19	92.37%	91.8%1	2%
resnet50	95.26%	94.99%	8%

Table 10. Results of fine-grain pruning with retraining (CIFAR100)

Name	Baseline T1	Pruned T1	Pruned size
vgg19	70.62%	70.12%	5%
resnet50	78.21%	77.56%	22%

3.4   Structural Pruning

Structural pruning is a process where blocks of weights are removed. One of the most popular is reducing number of channels in a filter. Using this approach, straightforward implementation on many hardware accelerators can speed up original network without any software modification. Reducing the number of channels (chunk of weights) in a pretrained network usually affects, significantly, model accuracy. This approach should be mixed with training steps to minimize the accuracy drop. In the presented approach, the channels with lowest L1 norm and lowest variance among 2D filters inside given channel were chosen to be removed. The subset of such channels were extracted in each iteration. Then retraining process was run to increase accuracy. The process was performed till drop in accuracy was higher than given threshold (1%). The results are presented in Table 11, Table 12. It is worth noting that results achieved using this approach are significantly worse than in fine grain pruning and the process is significantly slower than presented fast pruning algorithm.

Table 11. Results of structural pruning with retraining (CIFAR10)

Name	Baseline T1	Pruned T1	Pruned size
vgg19	92.37%	92.42%	52%
resnet50	95.26%	94.98%	72%

Table 12. Results of structural pruning with retraining (CIFAR100)

Name	Baseline T1	Pruned T1	Pruned size
vgg19	70.62%	70.71%	54%
resnet50	78.21%	77.50%	78%

4   Conclusions

The results presented in this paper show quite high disparities in sparsities between pruning with retraining or without retraining. Retraining can significantly improve the drop in accuracy after pruning. During retraining process, other aspects like masking, step size of the pruning at a current stage of pruning process are very important to achieve better results. The same effect can be observed in fast pruning on pretrained networks. It is worth noting about the time difference between these two pruning approaches. In case of pruning without retraining, it is possible to prune the very deep networks from several minutes to 2–3 h. The time depends on the size of testing the dataset. In case of using retraining, many epochs should be run to achieve satisfactory level of sparsity with a very small drop in accuracy. In case of Imagenet, one epoch lasts for approximately one hour. The overall process takes a few days. Choosing the method depends on hardware accelerator which will be used after pruning. If given hardware can make use of lower sparsity then pruning without retraining can be fast and efficient. In case of accelerator, it needs very high sparsity, slow pruning with retraining should be performed. The last conclusion is that structure pruning without retraining doesn't guarantee low drop in accuracy. It should be run with retraining.

5   Further Work

Further work will concentrate on tuning hyper-parameters in pruning algorithms which were described in a paper. It is still an open question if it is possible or how to find common rules for pruning all CNN networks to achieve satisfactory result. The next issue to focus on will be speeding up the pruning with retraining process by using more knowledge and statistics about the network. The proposed pruning methods of Deep Learning architectures can also be optimized and tested on a system level by taking data into consideration. This can be pronounced especially in latency critical systems [10].

References

1. Pietron, M., Karwatowski, M., Wielgosz, M., Duda, J.: Fast compression and optimization of deep learning models for natural language processing. In: Proceedings of the CANDAR 2019, Nagasaki. IEEE Explore (2019)
2. Al-Hami, M., Pietron, M., Casas, R., Wielgosz, M.: Methodologies of compressing a stable performance convolutional neural networks in image classification. Neural Process. Lett. **51**(1), 105–127 (2019). https://doi.org/10.1007/s11063-019-10076-y
3. Al-Hami, M., Pietron, M., Casas, R., Hijazi, S., Kaul, P.: Towards a stable quantized convolutional neural networks: an embedded perspective. In: 10th International Conference on Agents and Artificial Intelligence (ICAART), vol. 2, pp. 573–580 (2018)

4. Wróbel, K., Wielgosz, M., Pietroń, M., Karwatowski, M., Duda, J., Smywiński-Pohl, A.: Improving text classification with vectors of reduced precision. In: Proceedings of the ICAART 2018: 10th International Conference on Agents and Artificial Intelligence, vol. 2, pp. 531–538 (2018)

5. Wróbel, K., Pietroń, M., Wielgosz, M., Karwatowski, M., Wiatr, K.: Convolutional neural network compression for natural language processing. arXiv preprint arXiv:1805.10796 (2018)

6. Han, S., Mao, H., Dally, W.J.: Deep compression: compressing deep neural networks with pruning, trained quantization and Huffman coding. In: ICLR 2016. arXiv preprint arXiv:1510.00149

7. Lee, N., Ajanthan, T., Torr, P.H.S.: SNIP: single-shot network pruning based on connection sensitivity. In: ICLR 2019. arXiv preprint arXiv:1810.02340 (2018)

8. Huang, Q., Zhou, K., You, S., Neumann, U.: Learning to prune filters in convolutional neural networks. arXiv preprint arXiv:1801.07365 (2018)

9. SqueezeNet. arXiv preprint arXiv:1804.09028 (2018)

10. Wielgosz, M., Marutiz, P., Jiang, W., Rønningen, L.A.: An FPGA-baed platform for a network architecture with delay guarantee. J. Circuits Syst. Comput. **22**(06) (2013). https://doi.org/10.1142/S021812661350045X

Hidden Markov Models and Their Application for Predicting Failure Events

Paul Hofmann[1](✉) and Zaid Tashman[2](✉)

[1] Los Gatos, CA 95033, USA
paul@paul.email
[2] San Francisco, CA 94118, USA
zaid.tashman@accenture.com

Abstract. We show how Markov mixed membership models (MMMM) can be used to predict the degradation of assets. We model the degradation path of individual assets, to predict overall failure rates. Instead of a separate distribution for each hidden state, we use hierarchical mixtures of distributions in the exponential family. In our approach the observation distribution of the states is a finite mixture distribution of a small set of (simpler) distributions shared across all states. Using tied-mixture observation distributions offers several advantages. The mixtures act as a regularization for typically very sparse problems, and they reduce the computational effort for the learning algorithm since there are fewer distributions to be found. Using shared mixtures enables sharing of statistical strength between the Markov states and thus transfer learning. We determine for individual assets the trade-off between the risk of failure and extended operating hours by combining a MMMM with a partially observable Markov decision process (POMDP) to dynamically optimize the policy for when and how to maintain the asset.

Keywords: Hidden Markov model · Markov mixed membership model · Tied-mixture hidden markov model · HMM · Reinforcement learning · POMDP · Partially observable markov decision process · Time-series prediction · Asset degradation · Predictive maintenance

1 Introduction

Predictive maintenance is an important topic in asset management. Up-time improvement, cost reduction, lifetime extension for aging assets and the reduction of safety, health, environment and quality risk are some reasons why asset intensive industries are experimenting with machine learning and AI based predictive maintenance.

Traditional approaches to predictive maintenance fall short in today's data-intensive and IoT-enabled world [16]. In this paper we introduce a novel machine learning based approach for predicting the time of occurrence of rare events using Markov mixed membership models (MMMM) [5,12,22]. We show how we use these models to learn complex stochastic degradation patterns from data by introducing a terminal state that represents the failure state of the asset,

© Springer Nature Switzerland AG 2020
V. V. Krzhizhanovskaya et al. (Eds.): ICCS 2020, LNCS 12139, pp. 464–477, 2020.
https://doi.org/10.1007/978-3-030-50420-5_35

whereas other states represent health-states of the asset as it progresses towards failure. The probability distribution of these non-terminal states and the transition probabilities between states are learned from non-stationary time-series data gathered as historic data, as well as real time streaming data (e.g. IoT sensors).

Our approach is novel in two ways. First, we use an end-to-end approach combining dynamic failure prediction of individual assets with optimization under uncertainty [14,15] to find optimal replacement and repair policies. Typically, reinforcement learning approaches to predictive maintenance are satisfied with simple nondynamic prediction models [11]. Dynamic and more accurate failure prediction models are motivated by extending asset operating hours and are enabled by low cost cloud compute power. In Sect. 5 we explain this in detail.

Secondly, we found several advantages using dynamic mixed membership modeling for remaining useful life estimates, over recurrent networks (specifically LSTM-based [21]) and classical statistical approaches, like Cox-proportional hazard regression (CPHR) [17,20].

Adopting a Bayesian approach allows for starting with an estimate of the probability that can be subsequently refined by observation, as more sensor data is revealed in real time. In particular, our approach allows task specific knowledge to be included into the model. For example, the number of health-states, the number of mixtures of (topics, or archetypes) and the structure of the transition matrix may be modeled explicitly using engineering knowledge of practitioners.

Typically, the data structure for LSTM is fixed, e.g. a matrix, or time-series, while MMMM is more flexible allowing different sampling frequencies for example. MMMM can also work with missing data out of the box, using expert knowledge as priors. A typical LSTM approach has to rely on transformation models using PCA for ad-hoc feature extraction for example, before being able to input time-series data into LSTM [21], thus separating the feature extraction part from the prediction part.

Further, LSTM-based approaches require complete episodic inputs to learn the prediction task, and thus can not be directly applied to right-censored data. Right-censored data, or absorbing Markov chains, are needed for modeling degradation time-series unlike predicting classical time-series like for stock trends.

Traditionally, Cox-proportional hazard regression (CPHR) models with time-varying covariates are extensively used to represent stochastic failure processes [17,20]. Though CPHR works well for right-censored data, it lacks the health-state representation of the asset. In other words, the proposed generative MMMM model can infer the probability of failure and the most likely health-state, whereas regression based models can only produce a probability estimate. This is particularly relevant in domains where the interpretability of the model results is important like in engineering. Further, the CPHR analysis allows only modeling relationships between covariates and the response variable, while MMMM enables modeling the relationships between any variable. That means, dynamic Bayseian networks and MMMM in particular, allow to model not only the relationships among covariates and the response variable, but allow to capture the relationships among the covariates too [13]. Understanding the full relationship between all covariates is important to understand asset degradation patterns.

The paper is structured as follows. Section 2 and 3 give a high level intro to HMM and HMM for failure prediction respectively. We are using the terminology of HMMs sharing hierarchical mixtures over their states; this being as a special case of MMMM. Section 4 explains how we use reinforcement learning. Section 5 is a tutorial explaining by example how this can be applied to a large scale system of hundreds of degrading assets.

2   A Brief Introduction of the Hidden Markov Model

A hidden Markov model is a generative graph model that represents probability distributions over sequences of observations [7]. It involves two interconnected models. The state model consists of a discrete-time, discrete-state first-order Markov chain $z_t \in \{1,...,N\}$ that transitions according to $p(z_t|z_{t-1})$, while the observation model is governed by $p(x_t|z_t)$, where x_t are the observations. The corresponding joint distribution of a sequence of states and observations can be factored as:

$$p(z_{1:T}, x_{1:T}) = p(z_1)p(x_1|z_1)\prod_{t=2}^{T} p(z_t|z_{t-1})p(x_t|z_t) \tag{1}$$

Therefore, to fully define this probability distribution, we need to specify a probability distribution over the initial state $p(z_1)$, a $N \times N$ state transition matrix defining all transition probabilities $p(z_t|z_{t-1})$, and the emission probability model $p(x_t|z_t)$. To summarize, the HMM generative model has the following assumptions:

1. Each observation x_t is generated by a hidden state z_t.
2. Transition probabilities between states $p(z_t|z_{t-1})$ represented by a transition matrix are constant.
3. At time t, an observation x_t has a certain probability distribution corresponding to possible hidden states.
4. States are finite and satisfy first-order Markov property.

The observation model specified by $p(x_t|z_t)$ can be represented by a discrete distribution (Bernoulli, Binomial, Poisson,...etc), a continuous distribution (Normal, Gamma, Weibull,...etc), or a joint distribution of many components assuming individual components are independent. In the work discussed in this paper we use a mixture distribution to represent the observation model. That is given a state z_t, the mixture component y_t is drawn from a discrete distribution whose parameters θ are determined by the state z_t, denoted by $y_t \sim Discrete(\theta_{z_t})$, where θ_{z_t} is the vector of mixing weights associated with state z_t. The observation x_t is then drawn from one of a common set of K distributions determined by component y_t, denoted as $x_t \sim p(.|\mu_{y_t})$, where μ_k is the parameters of the kth distribution. It is important to note that the mixture components μ are common and shared across states while the mixing weights θ vary across states. Coupling the mixture components across states provides a balanced trade-off between *Heterogeneity* and *Homogeneity* [1,2,19] allowing for information pooling across states, see Fig. 1. This compromise is also beneficial when fitting HMM models with large number of states especially when certain states don't have enough observations (imbalanced data), as it provides a way of regularizing the model avoiding over-fitting.

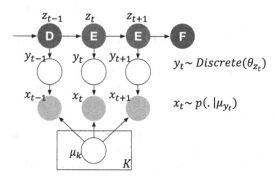

Fig. 1. States share common distributions providing a way of regularizing the model to avoid overfitting

2.1 Inference

Once the parameters of a hidden Markov model distribution are learned from data, there are several relevant quantities that can be inferred from existing and newly observed data. For instance, given a partially observed data sequence $X = \{x_t, t = 1, ..., \tau\}$, what is the posterior distribution over the hidden states $p(z_t|x_{1:\tau})$ up to time $\tau < T$, the end of the sequence. This is a filtering task and can be carried out using the *forward* algorithm. This posterior distribution will enable us to uncover the hidden health-state of an asset as we observe data streaming in. Additionally, one can be interested in computing the most probable state sequence path, z^*, given the entire data sequence. This is the maximum a posteriori (MAP) estimate and can be computed through the *viterbi* algorithm [4]. Readers can find more information about the *forward, forward-backward, Viterbi, and Baum-Welch* algorithms in [12].

3 Hidden Markov Model for Failure Time Prediction

The model will represent the data as a mixture of different stochastic degradation processes. Each degradation process, a hidden Markov model, is defined by an initial state probability distribution, a state transition matrix, and a data emission distribution. Each of the hidden Markov models will have a terminal state that represents the failure state of the factory equipment. Other states represent health-states of the equipment as it progresses towards failure and the probability distribution of these non-terminal states are learned from data as well as the transition probabilities between states. Forward probability calculations enable prediction of failure time distributions from historical and real time data. Note that the data rate and the Markov chain evolution are decoupled allowing for some observations to be missing. Domain knowledge about the degradation process is important. Therefore, expert knowledge of the failure mechanism is incorporated in the model by enforcing constraints on the structure of the transition matrix. For example, not allowing the state to transition from an "unhealthy" state to a "healthy" state can be incorporated by enforcing a zero probability in the entries of the transition matrix that represent the probability of transition from an "unhealthy" state to a "healthy" state (see Fig. 4 for an example of a transition matrix with zeros to the left of the diagonal representing an absorbing Markov chain). Enforcing constraints on the transition matrix also reduces the computational complexity during model training as well as when the model is running in production for online prediction and inference. An important property of data generated from a fleet of factory equipment is right censoring. In the context of failure data, right censoring means that the failure times are only known in a few cases because for the vast majority of the equipment the failure time is unknown. Only information up to the last time the equipment was operational is known. Right censored observations are handled in the model by conditioning on the possible states the equipment can be in at each point in time, i.e. all non-terminal states. Once the model parameters are estimated, the model is used for different inferential tasks. As new data streams in from the asset, the state belief is calculated online in real time or recursively, which gives an estimate of the most probable health-state the asset is in. This is a filtering operation, which applies Bayes rule in a sequential fashion.

Another inference task is failure time prediction. As new data is streaming in, an estimate of the asset health-state over a certain future horizon is calculated as well as the most probable time at which the asset will enter the "failure" state (terminal state). Both of those inferential tasks are important as they provide a picture of the current state of the factory, as well as a forecast of when each asset will most likely fail; see Fig. 2. This information will then be used to optimize the decision-making process, to maintain or replace assets.

4   Optimal Decision Making Using Partially Observable Markov Decision Process

At each time step we are confronted with a maintenance decision. Choosing the best action requires considering not only immediate effects but also long-term effects, which are not known in advance. Sometimes action with poor immediate effects can have better long-term ramifications. An "optimal" policy is a policy

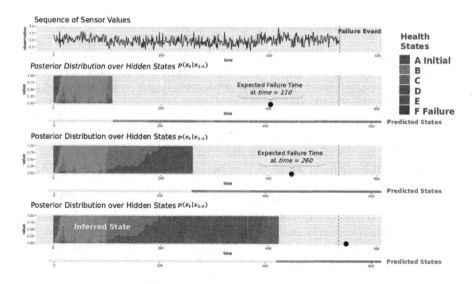

Fig. 2. Failure time prediction using HMM

that makes the right trade-off between immediate effects and future rewards [3] and [10]. This is a dynamic problem due to the uncertainty of variables that are only revealed in the future. For example, sensors are not always placed in the right location on the equipment making the inference of the health-state of the asset noisy. There is also uncertainty about how the equipment will evolve over time, or how operators will use it.

The goal of the optimal policy is to determine the best maintenance action to take for each asset at any given point in time, given the uncertainty of current and future health-states. We derive the policy from a value function, which gives a numerical value for each possible maintenance action that can be taken at each time step. In other words, a policy is a function that maps a vector of probability estimates of the current health-state to an action that should be taken at that time step. There is no restriction on this value function, which can be represented by neural networks, multi-dimensional hyper planes, or decision trees.

In this paper we focus on a local policy that is derived from a value function represented by multi-dimensional hyper planes. A hyper plane can be represented by a vector of its coefficients; therefore, the value function can be represented by a set of vectors; see Fig. 3.

In order to solve for our maintenance policy computing the value function, we assume that the model used for the degradation process of the asset is a hidden Markov model (HMM). Combining the dynamic optimization with the HMM enables us to use the parameters of our HMM to construct a partially observable Markov decision process (POMDP). A POMDP, in the context of asset modeling, is defined by:

$$POMDP = < S, A, T, R, X, O, \gamma, \pi > \tag{2}$$

A set of health-states S, a set of maintenance actions A, an initial health-state belief vector π, a set of conditional transition probabilities T between health-states, a cost or reward function R, a set of observations X, a set of conditional probabilities for the distribution of the observations O, and a discount factor $\gamma \in [0,1]$. Since our model of degradation is assumed to be a hidden Markov model, the states S, the transition probabilities T and the initial probabilities π of the POMDP are the same as the hidden Markov model parameters. The set of actions A can be defined as $a_0 =$ "Do Nothing", $a_1 =$ "Repair", and $a_2 =$ "Replace" for instance; see Fig. 8 for example. This set is configurable based on the maintenance policy for the asset and how it is operated. Similar to A, the cost function R is also configured based on maintenance policy and asset operation. The cost function R typically includes the cost of failure, the cost of replacement, the cost of repair, and the negative cost of non-failure to name a few. In addition to financial cost, one can include other forms of cost like social cost of failure if the equipment failure could cause disruption to the environment for example, or cause shortage of supply. R can be any type of function, production rules set up by the operator, look up tables, etc.

Once the POMDP is defined like in 2, we solve for the policy by computing the value function using a value iteration algorithm finding a sequence of intermediate value functions, each of which is derived from the previous one. The first iteration determines the value function for a time horizon of 1 time step, then the value function for a time horizon of 2 is computed from the horizon 1 value function, and so forth [18].

Once the value function is computed, the best action to take on each asset at time t is determined by finding the action with the highest value given the current state probabilities at time t.

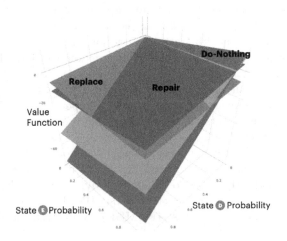

Fig. 3. Value function

5   Examples

There are many use cases for the methodology described. For example, the results of the failure prediction for a fleet of wind turbines may be input to reinforcement learning (POMDP) to optimize the schedule and route of the maintenance crews on a wind farm [9].

We illustrate our approach by going step by step through a real world example. We start with historic observations like multi-sensor time-series data for individual assets. We use expectation maximization (EM) to find the maximum likelihood or maximum a posterior (MAP) estimates of the parameters of our model from Fig. 1. Typically, we use EM to solve larger problems with a few hundred sensors and more than 20 states. In our experience, EM is computationally more tractable for larger real world problems than Markov chain Monte Carlo (MCMC), or its modern variant, Hamilton Monte Carlo (HMC) [8]), or the even more expensive variational inference approaches.

When using EM, the number of states for the HMM and the number of distinct distributions that make up our degradation processes are treated as hyper-parameters, which we input into our model. Posterior predictive checks (PPC) [6] is then used to find the right set of hyper-parameters.

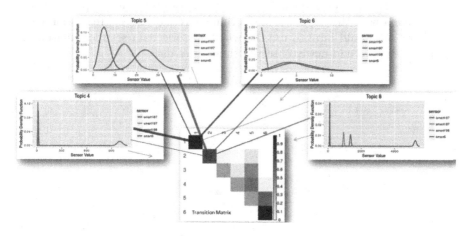

Fig. 4. Example of transition matrix for absorbing MMMM

In Fig. 4 we see an example of a transition matrix of a 6-state model. The off diagonal elements show the probabilities of transitioning between the HMM states, whereas the diagonals represent probabilities of remaining in each state. The lower triangular part of the matrix is set to zero to enforce an absorbing Markov chain where states move only from left to right towards the failure state. The Figure shows how the first two states are composed of mixtures, i.e. mixtures of common distributions. The thickness of the lines between the pdfs and the transition matrix indicates how much each common distribution contributes to the pertinent state distribution. For example, state 1 consists mainly of distribution 4 and 6, while distribution 5 contributes mainly to state 2. This shows an example of how the observation distributions of the states are mixtures of some set of common (simpler) distributions shared across all the states of the HMM. This approach can be seen as a generalization of tied-mixture HMM [12], where the shared distributions are limited to be Gaussian, while we allow for hierarchical mixtures of all distributions of the exponential family. In our approach any topic, or archetype can a priori transition to any other archetype. Using a sparse Dirichlet prior on transition distributions we learn a meaningful dependence between archetypes through posterior inference [22].

As mentioned before, expert knowledge of the failure mechanism maybe incorporated in the model by enforcing constraints on the structure of the transition matrix.

Figure 2 shows an example, of how we infer the hidden states given the sequence of observations, in our case, a time-series of sensor data. We see how the prediction of the expected failure time changes with additionally revealed sensor data over time. Using a Bayesian model like the MMMM enables us to calculate a new posterior for each newly observed data point thus gaining statistical strength and better prediction accuracy. Calculating the posterior is simple and short. It could be even done on edge devices. Contrast the simple solving of Bayes formula with the approach of discriminative models (i.e. regression models), where one would have to use the whole historic data set recalculating an improved model to add newly observed time series data.

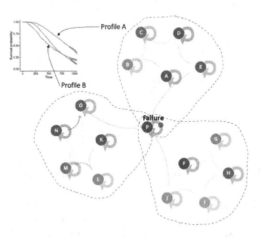

Fig. 5. Degradation profiles share states

Further, there is no obvious way using LSTM to get to a better prediction based on more real-time data points, since the posterior distribution is hidden in the weights. Similar to regression models, LSTM typically requires to run new back-propagation to learn from additional real time values. Another advantage of HMM of mixtures vs. LSTM is, it captures the natural (hidden) groupings within the data. Each group represents different asset profiles, i.e. distinct degradation processes, and thus failure curves (top left insert of Fig. 5).

Figure 5 shows an example of degradation states evolving over time. The thickness of the lines between the states indicates the probability of transitioning from one state to another. All assets start in state A, the initial state. After a certain amount of time, they end up in the terminal failure state P. See Fig. 2 for an example how the states evolve over time (in this Figure the final state is called F). The other health-states (B to E, L to O, and F to J, for example) represent states of an asset as it progresses towards failure.

The data frequency and Markov chain evolution are decoupled allowing for real time data arriving at different rates.

Once the model is fit, one can calculate the survival curves for the different degradation profiles which gives a summarized view of how assets fail as a baseline. See right plot in Fig. 6. The Figure also shows how the model can be used to "infer" which degradation profile the asset belongs to as new data arrives (colored graph on the left in the middle). The left bottom plot shows the entropy of the model's belief for a specific asset as more data is observed. Entropy, as a measure of uncertainty, decreases over time after more and more data points of the time-series have been reveled. The decreased entropy shows that after about 50 observations we can already be rather sure (entropy about 0.5) to which profile the asset at hand belongs. Thus the prediction for the profile and the life expectancy is rather reliable, after observing only a third of its life time (for this particular example of a degrading pump).

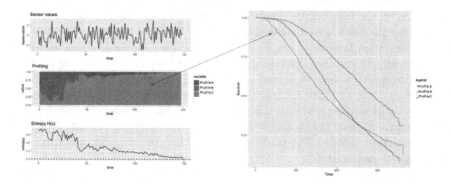

Fig. 6. Profiles evolve as data is revealed leading to different survival curves

Having a measure of accuracy for failure prediction is very important for practitioners. Obviously, the traditional ROC curves are not a good choice since they do not capture the dynamic nature of our approach, i.e. recalculating new posteriors when new data points arrive. Typically, practitioners are facing trade-off questions. For example, what is the right point in time to replace a part.

Replacing assets too early leads to unnecessary expenses. On average, parts are being replaced before they break. Running assets too long risks unforeseen down-time. To use such trade-offs as a measure for model quality is often more meaningful than ROC-type accuracy curves.

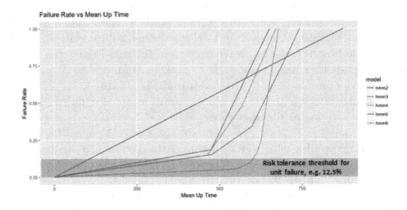

Fig. 7. Model performance measured by complexity

Risk tolerance is a typical constraint for operations managers. Using a trade-off diagram she can choose the model that predicts the longest operating hours given a certain risk level. Figure 7 shows the trade-off between risk of failure vs. operating hours (mean up-time). Typically, the model that produces the fewest false negatives, i.e. the steepest hockey stick failure curve, is the most efficient. To the operator, the onset of the hockey stick indicates the latest point in time for exchanging the asset, given a chosen risk level (12.5% in the pictured-example). Flat hockey stick failure curves, i.e. those with higher number of false positives, lead typically to reduced operating hours, since they indicate to the operator exchanging parts before their end of life-time. Sometimes, the steepest hokey stick curves come with less accuracy. The operator could mitigate reduced model accuracy by increasing spare parts inventory for example, thus still profiting from longer hours of asset operation. We see, ROC-type accuracy is not always the most important metric. The trade-off between failure rate and operating time can be more meaningful.

So far, we have shown in our example how we determine asset health evolving over time (Fig. 2). We are able to predict the degradation of individual assets by deriving profiles, which lead to different survival curves (Fig. 6).

Next, we have to derive actionable insights from the predictions by finding an optimal maintenance and resource allocation policy. We support the practitioner by determining the best action to take on an asset at any given moment, and assigning the right repair task to the right resource, i.e. who is to repair what and how.

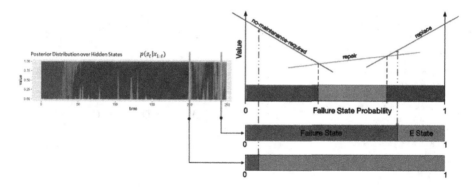

Fig. 8. Optimal action changing over time. (Color figure online)

Before we can make decisions about repairing individual assets we need to understand which part of the value function (Fig. 3) to use, depending on a given asset health-state. Figure 8 shows how the best action changes over time depending on the transition state probabilities and the pertinent value function (green, yellow or red). States of assets are not observed directly but our model can be used to infer the posterior distribution over the hidden states. For example, the asset of Fig. 8 has a low probability of failure around time point 200. According to the pertinent value function (policy) the best action is to "take no action". Around time point 230 the asset has high probability of being in a failure state (brown), thus, the value function recommends a "replace" action as the optimal take at this point in time.

More quantitatively, not knowing the current (hidden) state we generalize a Markov decision process (MDP) to a partially observable MDP. Using POMDP we observe the state only indirectly relating it to the underlying hidden state probabilistically, Fig. 8. Being uncertain about the state of the asset, we introduce rewards R (e.g. costs to repair, to replace, or the cost of down time), a set of actions A, and transition probabilities between health-states T; for details see Eq. 2. The transition probabilities T and the initial probabilities π of the POMDP are the same as our HMM parameters since we have used the hidden states S of our HMM to model the degradation. The set of actions A are $a_0 =$ "Do Nothing", $a_1 =$ "Repair", and $a_2 =$ "Replace", see Fig. 9.

We do not know the states, which are hidden. We only observe time dependent sensor data. From the observed sensor data we construct a belief, i.e. a posterior distribution over the states. From the belief we use the optimal policy, i.e. the solution of the POMDP, to find the optimal action to take, given the level of uncertainty. The POMDP solution is represented by a piecewise linear and convex value function calculated by the value iteration algorithm [18]. Once the value function is computed, the best action to take for each asset at time t is determined by finding the action with the highest value given the current state probabilities at time t, as shown in Fig. 9.

POMDP = < $S, A, T, R, \Omega, O, \gamma$ >

- S is a set of states,
- A is a set of actions,
- T is a set of conditional transition probabilities between states,
- $R : S \times A \to \mathbb{R}$ is the reward function,
- Ω is a set of observations,
- O is a set of conditional observation probabilities, and
- $\gamma \in [0, 1]$ is the discount factor.
- π is the initial state belief

$S = \{$ 🅐 🅑 🅒 🅓 🅔 🅕 $\}$

$A = \{$ do-nothing, repair, replace $\}$

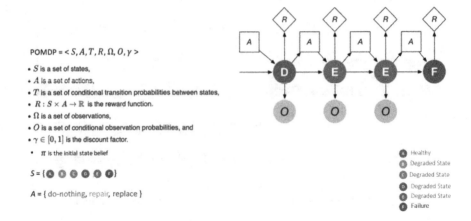

🅐 Healthy
🅑 Degraded State
🅒 Degraded State
🅓 Degraded State
🅔 Degraded State
🅕 Failure

Fig. 9. Selecting the best actions based on learned optimal policy

6 Conclusions

We showed how Markov mixed membership models (MMMM) can be used to predict the degradation of assets. From historic observations and real time data, we modeled the degradation path of individual assets, as well as predicting overall failure rates. Using MMMM models has several advantages. Mixing over common shared distributions acts as a regularization for a typically very sparse problem, thus avoiding overfitting and reducing the computational effort for learning. Further, hierarchical mixtures (topics, or archetypes) allow for transfer learning through sharing of statistical strength between Markov states. We used a dual approach combining the MMMM failure prediction with a partially observable Markov decision process (POMDP) to optimize the policy for when and how to repair assets by determining the dynamic optimum between the risk of failure and extended operating hours of individual assets. We showed how to apply this approach using tutorial type of examples.

References

1. Blei, D.M.: Probabilistic topic models. Commun. ACM **55**(4), 77–84 (2012)
2. Blei, D.M., Ng, A.Y., Jordan, M.I.: Latent dirichlet allocation. J. Mach. Learn. Res. 3(Jan), 993–1022 (2003)
3. Cassandra, A.R., Kaelbling, L.P., Littman, M.L.: Acting optimally in partially observable stochastic domains. AAAI **94**, 1023–1028 (1994)
4. Forney, G.D.: The viterbi algorithm. Proc. IEEE **61**(3), 268–278 (1973)
5. Fox, E.B., Jordan, M.I.: Mixed membership models for time series (2013). arXiv preprint arXiv:1309.3533
6. Gelman, A., Carlin, J.B., Stern, H.S., Dunson, D.B., Vehtari, A., Rubin, D.B.: Bayesian Data Analysis. CRC Press, Boca Raton (2013)

7. Ghahramani, Z.: An introduction to hidden Markov models and Bayesian networks. In: Hidden Markov Models: Applications in Computer Vision, pp. 9–41. World Scientific (2001)
8. Hoffman, M.D., Gelman, A.: The no-u-turn sampler: adaptively setting path lengths in hamiltonian monte carlo. J. Mach. Learn. Res. **15**(1), 1593–1623 (2014)
9. Hofmann, P.: Machine Learning on IoT Data (2017). Accessed 21 Mar 2020. https://www.slideshare.net/paulhofmann/machine-learning-on-iot-data
10. Kaelbling, L.P., Littman, M.L., Cassandra, A.R.: Planning and acting in partially observable stochastic domains. Artifi. Intell. **101**(1–2), 99–134 (1998)
11. Koprinkova-Hristova, P.: Reinforcement learning for predictive maintenance of industrial plants. Inf. Technol. Control **11**(1), 21–28 (2013)
12. Murphy, K.P.: Machine Learning: A Probabilistic Perspective. MIT press, Cambridge (2012). Exercise 17.3
13. Onisko, A., Druzdzel, M.J., Austin, R.M.: How to interpret the results ofmedical time series data analysis: classical statistical approaches versusdynamic bayesian network modeling. J. Pathol. Inf. **7** (2016)
14. Powell, W.B.: A unified framework for optimization under uncertainty. In: Optimization Challenges in Complex, Networked and Risky Systems, pp. 45–83. INFORMS (2016)
15. Powell, W.B., Meisel, S.: Tutorial on stochastic optimization in energy–part i: modeling and policies. IEEE Trans. Power Syst. **31**(2), 1459–1467 (2015)
16. Ran, Y., Zhou, X., Lin, P., Wen, Y., Deng, R.: A survey of predictive maintenance: systems, purposes and approaches (2019). arXiv preprint arXiv:1912.07383
17. Satten, G.A., Datta, S., Williamson, J.M.: Inference based on imputed failure times for the proportional hazards model with interval-censored data. J. Am. Stat. Assoc. **93**(441), 318–327 (1998)
18. Shani, G., Pineau, J., Kaplow, R.: A survey of point-based pomdp solvers. Auton. Agents Multi-Agent Syst. **27**(1), 1–51 (2013)
19. Teh, Y.W., Jordan, M.I., Beal, M.J., Blei, D.M.: Sharing clusters among related groups: hierarchical dirichlet processes. In: Advances in Neural Information Processing Systems, pp. 1385–1392 (2005)
20. Wei, L.J., Lin, D.Y., Weissfeld, L.: Regression analysis of multivariate incomplete failure time data by modeling marginal distributions. J. Am. Stat. Assoc. **84**(408), 1065–1073 (1989)
21. Wu, Y., Yuan, M., Dong, S., Lin, L., Liu, Y.: Remaining useful life estimation of engineered systems using vanilla lstm neural networks. Neurocomputing **275**, 167–179 (2018)
22. Zhang, A., Paisley, J.: Markov mixed membership models. In: International Conference on Machine Learning, pp. 475–483 (2015)

Biomedical and Bioinformatics Challenges for Computer Science

Reference-Based Haplotype Phasing with FPGAs

Lars Wienbrandt[✉], Jan Christian Kässens, and David Ellinghaus

Institute of Clinical Molecular Biology, University Medical Center Schleswig-Holstein,
Kiel University, Am Botanischen Garten 11, 24118 Kiel, Germany
{l.wienbrandt,j.kaessens,d.ellinghaus}@ikmb.uni-kiel.de

Abstract. Haplotype phasing of individual samples is commonly carried out as a precursor step before genotype imputation to reduce the runtime complexity of the imputation step and to improve imputation accuracy. The phasing process is time-consuming and generally exceeds hours even on server-grade computing systems. Loh et al. recently introduced a fast and effective reference-based haplotype phasing software named EAGLE2 which scales linearly with the number of reference samples and variants to phase. We discovered that from the several steps of the EAGLE2 phasing process, data preparation for the internally used *HapHedge* data structure already consumes about half of the total runtime in general use cases. We addressed this problem by introducing a new design for reconfigurable architectures that accelerates this part of the software by a factor of up to 29 on a Xilinx Kintex UltraScale FPGA, resulting in a total speedup of the complete phasing process of almost 2 (the theoretical limit according to Amdahl's Law) when compared to a server-grade computing system with two Intel Xeon CPUs. As a result, we reduced the EAGLE2 runtime of genome-wide phasing of 520,000 variants in 2500 samples using the *1000 Genomes Project* reference panel from 68 min to 39 min on our system while maintaining quality.

Keywords: Haplotype phasing · Genotype imputation · EAGLE2 · PBWT · Hardware accelerator · Reconfigurable architecture

1 Introduction

An individual's *diplotype* sample consists of two strands of unique nucleotide content for each chromosome, one inherited from the mother (*maternal*) and one inherited from the father (*paternal*). *Single Nucleotide Polymorphism (SNP)* microarrays, known as *genome-wide association study (GWAS)* SNP arrays, e.g. the *Global Screening Array (GSA)* from Illumina [8], allow for the genotyping of more than 650,000 biallelic SNPs (two observed alleles; counting the reference as one, and allowing for one alternative allele) across the human genome. Each pair of alleles of a SNP is experimentally measured and subsequently encoded as a genotype (thereby ignoring phase information), for example, 0 for *homozygous*

© Springer Nature Switzerland AG 2020
V. V. Krzhizhanovskaya et al. (Eds.): ICCS 2020, LNCS 12139, pp. 481–495, 2020.
https://doi.org/10.1007/978-3-030-50420-5_36

reference (reference allele on both strands), 1 for *heterozygous* (one strand with reference allele, the other strand with an alternative allele), and 2 for *homozygous alternative* (alternative allele on both strands). *Haplotype* information refers to the alleles found on a single strand, e.g. 0 for the reference allele and 1 for the alternative allele. Since the genotyping process does not separate the maternally and paternally derived alleles of each SNP, the haplotype phasing problem can simply be described as to guess for each heterozygous marker of a SNP genotype on which strand the alternative allele is located (e.g. on the maternal strand), implying the reference allele to be found on the other strand (e.g. the paternal strand in this event). Technically, this means to find the best suited pair of haplotype strings (encoded with 0 and 1) to a given genotype string (encoded with 0, 1 and 2).

Phase information is important for human genetics research for several reasons, among others, to screen for the occurence of compound heterozygosity, i.e. the presence of two deleterious variants located at different locations in the same gene but on different chromosome copies (paternal and maternal) of an individual. Further, many studies have linked specific haplotypes to drug response, clinical outcomes in transplantations and to susceptibility or resistance to disease [11]. Because of the lack of cost-effective experimental approaches for obtaining phase information, the haplotype phase is commonly estimated using computational approaches [2]. Haplotype phasing is also a very important preprocessing step for *genotype imputation* which relies on pre-phased genotypes. Genotype imputation has become a standard tool in genome-wide association studies and describes an approach of predicting (or imputing) the genotypes of variants that were observed in a reference panel but were not directly genotyped on a SNP microarray [4]. Imputation servers, such as the Sanger Imputation Server [16] or the Michigan Imputation server [14], enable users to upload GWAS SNP array data to these servers to deploy well-equipped high-performance computing clusters in order to carry out phasing and imputation.

Recently, methods have been exploited that reduce the computational requirements of phasing and imputation [5]. However, Das and colleagues demand in their article that continued computational improvements in phasing and imputation tools are necessary for imputation to remain practical as millions of samples are genotyped and reference panels increase in size to tens of thousands of sequenced genomes. Following this demand, Loh et al. presented EAGLE2 [9]. According to the article, EAGLE2 is 10% more accurate and up to 20 times faster than the preliminary state-of-the-art tool SHAPEIT2 [6], due to its linear complexity in the number of samples, reference samples and variants to phase. Thus, EAGLE2 became the default phasing method in the Sanger and the Michigan Imputation Services. (As an alternative SHAPEIT2 and the similar performing HAPI-UR [17] can be chosen for the phasing step.)

In this paper, we present a method to accelerate EAGLE2 [9] by using reconfigurable hardware. We discovered that for general use cases target preparation as the first step in the phasing process already consumes around half of the total runtime, due to the large amount of data that has to be processed here. This step includes the analysis of the reference panel by creating a *condensed reference* individually for each target. The condensed reference mainly consists of 1 bit haplotype information for every reference sequence at each heterozygous site in the target plus 1 bit of consistency information marking a segment between heterozygous sites inconsistent or not. Furthermore, an *identity-by-descent (IBD)* check is performed and the condensed reference information is required in transposed format for efficient usage. We show that these steps are perfectly suited for a hardware design by implementing this part of EAGLE2 on a *Field Programmable Gate Array (FPGA)*. We run a performance and quality benchmark of our implementation using the *1000 Genomes Project (1000G)* [12] haplotype data as reference panel.

To our knowledge, phasing and imputation processes in general have been rarely addressed with alternative architectures such as FPGAs or GPUs. An at that time successful approach was made by Chen et al. with Mendel-GPU [3] in 2012 using a GPU-accelerator for imputation, but we have not found adequate literature describing the utilization of FPGAs to target this problem.

2 Haplotype Phasing with EAGLE2

2.1 EAGLE2 Core Overview

Po-Ru Loh et al. introduced EAGLE2 as a haplotype phasing software that uses a phased reference panel to determine the phase of the genotypes of a diploid target sample [9]. The phasing process in EAGLE2 consists of several parts. We give an overview here with details following in the subsequent sections. The input data (reference panel and target data) is required in packed and indexed *Variant Call Format (VCF)* (`.vcf.gz` or `.bcf`), whereby `.bcf` is the packed binary format, which is fastest.

The first part after reading the input data is to prepare an adapted reference for each target. We call this the *condensed reference* of a target. In the following, we will focus on this part for FPGA acceleration. It is divided into several steps:

1. Find the k best fitting haplotype sequences from the reference.
2. Reduce the remaining sequences to *split sites*, i.e. all sites containing a heterozygous genotype in the target plus some sites required to split large segments without any heterozygous sites. This step also preserves the information if segments between split sites are consistent with the target, or not.
3. Check for *Identity-by-descent (IBD)*, i.e. parts of a reference sample that are too similar with the target sample are masked. This keeps the process from creating a bias on certain samples from the reference.

The second part is to generate a data structure based on a *Position-based Burrows Wheeler Transformation (PBWT)* [7] from the resulting condensed reference. Loh et al. call this a *HapHedge*, which is based on a graph that allows quick access to all information from the PBWT plus meta information. For more information on the HapHedge, we refer to the supplementary material of [9].

The third and final part of the phasing is to perform a *beam search* through possible haplotype sequences fitting the target, following the probability model introduced in Sect. 2.2. The beam search is also performed in a series of substeps:

1. Fast pre-phasing. The parameters of the beam search are adjusted to create a fast rough estimation of the phased haplotypes, whereby the most confident positions are used as constraints for the next step.
2. Fine phasing. The beam search is performed again, but with parameters forcing a better estimation for the cost of runtime. To keep the runtime growth to a minimum, this step uses the constrained sites from the first step, such that only uncertain positions are phased again.
3. Reverse fine phasing. The previous step is repeated in reverse order, i.e. started from the end and forwarded until the beginning. The phase confidences of each heterozygous site are compared and the phase with the higher probability is chosen.

Previously described steps are repeated for each target. The phased output is written in VCF format again, whereby the user can choose if the output should be packed or binary packed.

2.2   Probability Model

To understand why preliminary steps are necessary, we describe the underlying probability model for the phase estimation first. We focus on the main equations here, further details can be found in the original article [9].

The core probability of two haplotype paths h^{mat} and h^{pat} forming the current target genotypes g up to a site m is defined as:

$$P\left(h_{1\ldots m}^{\mathrm{mat}}, h_{1\ldots m}^{\mathrm{pat}} | g_{1\ldots m}\right) \approx \underbrace{P\left(g_{1\ldots m} | h_{1\ldots m}^{\mathrm{mat}}, h_{1\ldots m}^{\mathrm{pat}}\right)}_{\varepsilon^{n_{\mathrm{err}}}} P\left(h_{1\ldots m}^{\mathrm{mat}}\right) P\left(h_{1\ldots m}^{\mathrm{pat}}\right) \quad (1)$$

n_{err} is the number of consistency errors between the pair of haplotype paths to the genotypes, e.g. if both paths show a common allele at a site (homozygous) while the genotype should be heterozygous at that site, this counts as an error. ε is a fixed error probability set to 0.003 per default.

The haplotype path probability is assumed to be:

$$P\left(h_{1\ldots m}\right) \approx \sum_{x=m-H}^{m-1} P\left(h_{1\ldots x}\right) f\left(h_{x+1\ldots m}\right) P\left(\mathrm{rec}\ m | x\right) \quad (2)$$

The parameter H indicates the size of a "history", i.e. how many sites should be looked back at in order to calculate the actual probability of the current site m. This is set to 30 for the fast pre-phasing step and set to 100 in the fine phasing steps. $P\left(\mathrm{rec}\ m | x\right)$ is the recombination probability between the two sites m and x. It is computed as a function dependent on the genetic distance between the sites, the effective population size and the number of references (see [9] for details).

The function $f(h_{x\ldots m})$ returns the frequency of the given sequence $h_{x\ldots m}$ in all reference sequences from sites x to m. Since this function is called H times for each site and for each active haplotype path in the beam search at that site, it is crucial to get this information as fast as possible. The HapHedge data structure in combination with the PBWT allow to access this information in approximately constant time (see [9] and [7] for details) while a naive implementation would be linear in the number of references.

2.3   Beam Search

In order to estimate the phase on a heterozygous target site, EAGLE2 computes the probability to keep the phase (P_m^{keep}) at such a site m. For this purpose, it performs a beam search. It starts at the first heterozygous site with a probability of 0.5 and creates two consistent haplotype paths up to that site. This pair of haplotype paths is called a *diplotype* by Loh et al. The process continues over all heterozygous sites and extends each haplotype path in a diplotype by 0 and 1, i.e. for each diplotype, it creates four new diplotypes with the corresponding extension. A fixed parameter P for the beam width controls the maximum number of diplotypes in the beam search, i.e. only the P most probable diplotypes (according to Eq. 1) are kept in the beam. P is set to 30 in the fast pre-phasing step and set to 50 during fine phasing. Loh et al. use a forward extension before doing the final phase call for a site. The beam front is extended over a number of Δ heterozygous sites in advance before doing the phase call on the current site. Δ is set to 10 in the fast pre-phasing and set to 20 in the fine phasing steps. The keep phase probability is now easily computed as the sum of the probabilities of those diplotypes that keep their phase at the current heterozygous site in comparison to the last heterozygous site divided by the sum of all probabilities from diplotypes that are heterozygous at the current site.

$$P_m^{\text{keep}} = \frac{\sum \left\{ P\left(h_{1\ldots m}^{\text{mat}}, h_{1\ldots m}^{\text{pat}}|g_{1\ldots m}\right) \mid \text{diplotype keeps phase} \right\}}{\sum \left\{ P\left(h_{1\ldots m}^{\text{mat}}, h_{1\ldots m}^{\text{pat}}|g_{1\ldots m}\right) \mid h_m^{\text{mat}} \neq h_m^{\text{pat}} \right\}} \tag{3}$$

In the final phase call, the phase will be switched if that probability is less than 0.5. Note, that the extension of the diplotype may also be homozygous and thus, its probability has to be ignored for the current phase call.

EAGLE2 performs further optimizations on the beam search, such as merging similar diplotypes. We omit explanations here and refer to [9] again.

2.4   Condensed Reference

Phase calling is only necessary for heterozygous target sites since the maternal and the paternal haplotype paths carry the same information at homozygous sites. For this reason a *condensed reference* is created from the original reference panel for each target. The condensed reference carries only information of the reference samples at the target's heterozygous sites (and some other split sites in areas of large homozygosity) and combines the information on segments between

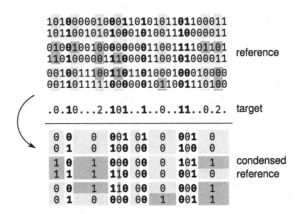

Fig. 1. Process of creating a condensed reference from a series of reference sequences. Heterozygous target sites are highlighted in bold. For homozygous sites inconsistent haplotypes in the reference are highlighted in dark grey, consistent ones in light grey. Consequently, consistent and inconsistent segments in the condensed reference are highlighted likewise.

two split sites to one bit indicating consistency with the target for each reference haplotype.

In order to reduce the computational burden from a large reference panel, a preliminary step of creating a condensed reference is to choose the k best fitting reference haplotypes from the panel, i.e. the k haplotype sequences with the lowest number of consistency errors when compared to the actual target. k is a user-defined parameter that defaults to 10,000. Thus, in the default configuration, this step is only necessary when choosing a large reference panel with more than 5,000 samples (i.e. 10,000 haplotypes), such as the HRC1.1 panel [13].

The second step is to copy the information from the reference only at split sites. Since the haplotypes are encoded in sequences of 0 and 1 the size is one bit per reference and split site. Another bit is used to encode the condensed information of the following segment. In particular, 0 encodes a consistent segment and 1 encodes an inconsistent segment, i.e. at least one consistency error is found between this haplotype sequence and the target genotypes in this segment. So, the condensed reference is represented as a bit array with a bit sequence for every reference where each odd bit represents a heterozygous site of the target and its reference information, and each even bit (the *inconsistency bit*) represents the segment between two heterozygous sites and the information if the segment is consistent with the target or not. Note, that the frequency lookup $f(h_{x...m})$ in Eq. 2 requires this information for counting sequences in the condensed reference. Only sequences over consistent segments will be counted. Figure 1 depicts the process of condensing the reference.

A third step added to the creation of the condensed reference is to check reference samples for *Identity-by-descent (IBD)*. Regions of a sample are considered to be in IBD with the target if the region is large enough and the reference sample and the target are exactly equal in that region. If an IBD situation is encountered the program would generate a bias towards this sample. So, the region in this sample is excluded in the analysis by setting the corresponding inconsistency bits to 1 in the condensed reference. EAGLE2 distinguishes between two types of IBD. If a region of 10 to 20 split sites is found to be equal to the target only the region in the reference haplotype with the latest error will be masked. If the equality spreads over more than 20 split sites, both haplotype sequences belonging to the reference sample will be masked in that region.

3   Implementation

Due to the large amount of data that has to be processed, our analyses revealed that in normal use cases the process to create a condensed reference for each target already consumes around 50% of the runtime. We targeted this part of the haplotype phasing for acceleration on reconfigurable architectures, in particular FPGAs. Our computing architecture consists of a server-grade mainboard hosting two Intel Xeon E5-2667v4 8-core CPUs @ 3.2 GHz and 256 GB of RAM, and an Alpha Data ADM-PCIE-8K5 FPGA accelerator card equipped with a recent Xilinx Kintex UltraScale KU115 FPGA with two attached 8 GB SODIMM memory modules, connected via PCI Express Gen3 x8 allowing high-speed communication with the host. The system runs a Ubuntu 19.04 Linux OS (Kernel version 5.0).

3.1   Target Preparation on the Host System

Firstly, the host system loads the input data, i.e. the target data and the reference data into memory. VCF files are stored in a variant major format, i.e. one line of a file contains all haplotype/genotype information of the included samples for one variant. Both files are read concurrently variant after variant, and only bi-allelic sites common in target and reference are kept for the analysis. The data is implicitly transposed to sample-major as required for the target pre-processing.

Since all targets can be handled independent of each other, EAGLE2 spawns multiple threads with each thread phasing a single target at a time. The number of threads running in parallel is defined by a user parameter. In the case of the FPGA acceleration, we prepared the FPGA design to handle several targets concurrently as well. Our design is capable of creating the condensed reference for 48 targets in parallel. For each bunch of 48 targets, the genotype data is encoded in two bits per genotype and aligned in 256 bit words (the PCIe transmission bus width of the FPGA). The host allocates page-locked transmission buffers to ensure fast data transfers between host and FPGA. The target data is copied into the buffers and transferred to the FPGA. For initialization of the 48 processing pipelines, the host determines target specific constants. These are the number of split sites (mainly heterozygous sites) in each target as well as the pre-computed size of the condensed reference. Since the size of the reference is too large to be kept in local FPGA RAM in general, this data is prepared as a compact bit array in page-locked memory on the host. The haplotype data is encoded in one bit per haplotype and each haplotype sequence is again aligned to the FPGA's PCIe transmission word width of 256 bit. The process of creating the condensed reference of the current targets starts with streaming the reference data to the FPGA.

3.2   Creating the Condensed Reference on the FPGA

The central part of the FPGA design contains 48 parallel processing pipelines, each producing a stream of condensed reference data for each target. One unit controls the pipeline input. It distributes the raw target data from the host to the corresponding pipeline which stores it in a local buffer implemented in the FPGA's *block RAM (BRAM)*. When the host starts sending the reference haplotype data, the first haplotype sequence is stored in a local BRAM buffer. The second sequence is directly streamed together with the first sequence from the buffer through all pipelines at the same time such that in each clock cycle a pair of haplotypes (one bit maternal and one bit paternal) is processed in the pipeline. This pair of haplotypes forms a phased genotype from the reference. It is directly compared with the corresponding unphased genotype in the target in the *checkIncon* unit. The unit produces two streams (one for the maternal reference and one for the paternal reference) of condensed data, i.e. as long as the target genotypes are homozygous it only records an inconsistency to the corresponding reference by a boolean flag. Whenever the target shows a heterozygous site, *checkIncon* generates a 2-bit output for each stream consisting of the current inconsistency flag and the current reference haplotype at that site.

The resulting pair of 2-bit condensed reference data continues to the *check-IBD* unit. The unit keeps a buffer for the data of the last 64 split sites in a simple shift register implemented in logic. This way, we can set up to 64 preliminary sites to inconsistent in one clock cycle if an IBD situation occurs (which would add a certain bias towards the current sample, see Sect. 2.4). The actually applied numbers depend on the type of IBD and can be set by the host via a runtime constant. For detection of an IBD situation the unit simply counts the number of past pairwise (maternal and paternal) consistent sites, and it keeps a flag that marks the sequence where the last inconsistency occurred. After at least 10 consistent sites (first IBD type) the unit marks the sequence with the last error as inconsistent along these sites. If the number of consistent sites extends to 20 (second IBD type) the unit marks both sequences along these sites as inconsistent. This process continues until the first inconsistent site is reported from the *checkIncon* unit.

All pipeline outputs are connected via a simple shared bus system with a 512-bit word width transferring the condensed data directly to the first DRAM module. In order to convert the data stored in sample-major format to variant-major, as required by the phasing process on the host system (see Sect. 3.3), we designed a data transmission unit that transposes blocks of 512 × 512 bit data. It is implemented as a shift register that is able to shift data horizontally and vertically. The data is organized in a mesh of 512 × 512 cells that are

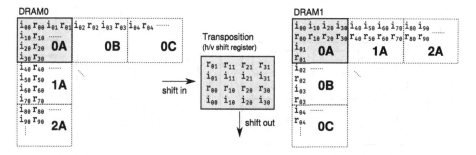

Fig. 2. Example of data transposition. The transposition unit represents a horizontal and vertical shift register. The condensed reference in the first DRAM module is transposed block-wise and stored in the second DRAM module. r_{nm} refers to the reference allele of haplotype n at call site m, i_{nm} refers to the inconsistency bit from haplotype n for the segment between call sites $m-1$ and m.

each connected with its horizontal and vertical neighbors. So, 512 sample-major RAM words from 512 subsequent samples shifted horizontally into the mesh can be read out vertically in variant-major format afterwards, such that each word contains the information of 512 samples for one site. In particular, since reference information and inconsistency information is stored in alternating bits for each sample, the transposed output contains alternating words for reference and inconsistency information for subsequent segments and sites. The output is stored in the second DRAM module from where it can directly be fetched from the host (via an output buffer).

The transposition unit also controls the way the condensed reference is distributed in blocks, i.e. the source addresses in the first DRAM module of the 512 words to be shifted into the unit. And it calculates the destination addresses in the second DRAM module of the output words, such that subsequent words belong to the same variant and the host can directly read variant-major sequences containing all information on the samples for one variant in one sequence. See Fig. 2 for an example. The complete FPGA design is illustrated in Fig. 3.

3.3 Final Target Processing on the Host System

For target phasing we adapted the procedure of EAGLE2 described in Sect. 2. However, it is worth to mention that in the original EAGLE2 software the condensed reference information is kept in sample-major format. Thus, for creating the *HapHedge* data structure the information is read "vertically", i.e. the current site information is picked from each corresponding sample words by bit-masking. We omitted this time-consuming implicit transposition by directly providing transposed data from the FPGA. The rest of the process is implemented as explained in the supplementary material of the original article [9].

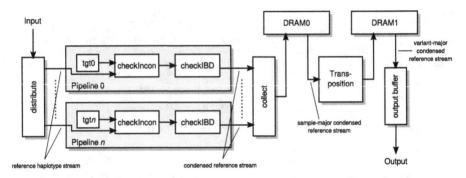

Fig. 3. FPGA design for creating the condensed reference. Several targets are handled concurrently in separate pipelines. The streamed reference data is continuously compared to the targets and the condensed information including IBD check is extracted. The raw sample-major condensed data is temporarily stored in the first local DRAM module before it is transposed to variant-major. The transposition output is stored in the second local DRAM module before it is transferred to the host, where it can directly be used for the phasing process.

4   Performance Evaluation

We performed a quality check and a runtime evaluation on our architecture from Sect. 3 as described in the following. In all tests we used the binary packed *variant call format* (.bcf) for input and output data, and ran the benchmark without an additional system load.

4.1   Phasing Quality

We measure phasing quality by calculating the mean *switch error rate*, i.e. we phase a number of targets with already known phase and compare the output to the original afterwards by counting *switch errors*, i.e. the number of phase switches different to the original. We chose the publicly available 1000 Genomes reference panel [12], containing more than 31 million phased markers from 2504 unrelated diploid samples of different anchestry. To create a real-world oriented test data set, we created chromosome-wise subsets with all samples from the panel but reduced the markers to those that can also be found on the *Global Screening Array (GSA)* [8], resulting in about 520,000 quality-controlled markers in total. We used the software *bcftools* [15] for creating the test data.

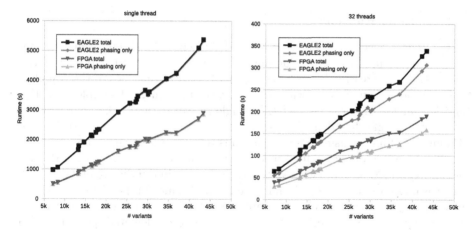

Fig. 4. Runtimes (in s) of single-threaded (left) and multi-threaded (32 threads, right) EAGLE2 and FPGA-accelerated runs. The graphs seperate total and phasing-only runtimes for each run.

Firstly, we phased the test data against the original 1000 Genomes Project reference panel using the EAGLE2 software in its currently latest build, version 2.4.1 [10]. Note, that although the reference and the target data contain the same samples, we did not create a bias towards copying the phase information from the original sample because the IBD check (see Sect. 2.1) ensures that for each targeted sample the same sample in the reference is masked out. So, each target sample was phased using only the phase information of all other samples. We ran the phasing fully parallelized with 32 threads (--numThreads 32) and with a single thread (--numThreads 1). The EAGLE2 software set the number of phasing iterations automatically to 1, other parameters were left as default. The EAGLE2 software reports its total runtime in its output log divided into reading, phasing and writing time. For evaluation, we used the total runtime and the phasing runtime.

Secondly, we phased the test data against the 1000 Genomes Project reference panel again using our FPGA-accelerated version of EAGLE2. As before, we performed the phasing twice, once fully parallelized with 32 threads and once with only a single phasing thread. We reported the runtimes distributed by phasing and IO in the tool using the C standard library. As for the EAGLE2 run, we ran a single phasing iteration.

Finally, we counted the switch errors in the output data sets by comparing them to the original reference. As a result, we counted an average of 4491.5 genome-wide switch errors per sample for the EAGLE2 output. Our FPGA acceleration created 4528.8 switch errors on average, which is slightly larger. Taking the 520,789 variants into account, we computed mean switch error rates of 0.00862 and 0.0087 respectively. Concluded, our FPGA acceleration of EAGLE2 phasing maintains the mean switch error rate with a negligible difference.

4.2   Phasing Runtime

In the previous section (Sect. 4.1, we recorded the phasing and total runtimes for each chromosome-wise subset for single- and multi-threaded (32 threads) original EAGLE2 and FPGA-accelerated runs. The total runtimes of the complete genome-wide input dataset with 520,789 variants from 2504 samples was 17 h and 53 min for the original EAGLE2 single-threaded run (17 h 44 min phasing only). The single-threaded FPGA-accelerated run was measured with 9 h and 37 min (9 h 31 min phasing only). This leads to a total speedup of 1.86.

The speedup for the multi-threaded run is similar. With 32 phasing threads on our 2x Intel Xeon computing system the total runtime of the original EAGLE2 run was measured with 1 h and 8 min (1 h 1 min phasing only) and 39 min (32 min phasing only) for the FPGA-accelerated run. This time, the difference between total and phasing only runtimes has an impact on the calculated speedup, which dropped to 1.73 due to IO. Without IO it results to 1.88.

As we described in Sect. 3, the FPGA acceleration targets only 50% of the total phasing process. Thus, according to Amdahl's Law [1], we have not expected a speedup exceeding 2 for the phasing. In order to quantify the speedup for the part which was accelerated by the FPGA alone, we introduced a time measure in our single-threaded run of that part which was not accelerated (t_{notacc}). We assumed this part to be nearly the same as in the original EAGLE2 run. So we calculated our speedup of the FPGA part by putting the differences to the complete phasing only times from the original run (t_{PhEAGLE}) and the FPGA accelerated run (t_{PhFPGA}) into relation as described in Eq. 4. The resulting speedup factor of the FPGA-only part for the complete data set is 19.99 whereby we observed a speedup of 29.02 for chromosome 21.

$$\text{Speedup}_{\text{FPGA}} = \frac{t_{\text{PhEAGLE}} - t_{\text{notacc}}}{t_{\text{PhFPGA}} - t_{\text{notacc}}} \tag{4}$$

Additionally, the introduced time measure of the not accelerated part allowed us to exactly identify the ratio of this part to the complete phasing, and in reverse conclusion, the ratio of the part accelerated by the FPGA. For the complete data set we calculated this ratio to be 48.8%, and thus, according to Amdahl's Law, the theoretical maximum speedup of the phasing part to be $1/(1 - 0.488) = 1.95$, which we have almost reached.

Figure 4 shows the runtimes of the single- and multi-threaded EAGLE2 runs as well as for the single- and multi-threaded FPGA-accelerated runs plotted over the number of phased variants from each input subset. The graphs show a nearly linear behaviour as expected. Tables 1 and 2 show the runtimes and speedups for selected chromosome-wise subsets of our input data including the worst and best runs.

Table 1. Wall-clock runtimes (in s) and speedup of FPGA-accelerated EAGLE2 phasing using a single thread only. t_{notacc} is the time for the phasing part which was not accelerated by the FPGA. Speedup is shown for the total runtime and for the part accelerated by the FPGA according to Eq. 4. Total runtimes include phasing and data IO. The table shows selected chromosome-wise subsets, $\sum$ is the sum of all 22 subsets.

Dataset		EAGLE2		FPGA accel.			% accelerated	Speedup	
chr	# variants	Total	Phasing	Total	Phasing	t_{notacc}	by FPGA	Total	FPGA
chr1	43344	5376.24	5337.55	2893.89	2863.74	2729.07	48.9	1.86	19.37
chr2	42210	5093.09	5052.83	2712.61	2680.65	2551.46	49.5	1.88	19.36
chr6	36892	4237.13	4203.38	2225.90	2200.34	2080.43	50.5	1.90	17.70
chr10	27360	3362.27	3335.68	1844.03	1823.93	1739.58	47.8	1.82	18.92
chr13	17819	2323.75	2303.55	1227.39	1212.45	1159.92	49.6	1.89	21.77
chr17	16612	2119.15	2101.84	1107.58	1094.92	1049.73	50.1	1.91	23.28
chr21	7107	977.28	966.60	506.71	499.26	482.58	50.1	1.93	29.02
chr22	8262	1062.41	1051.59	555.09	546.32	527.11	49.9	1.91	27.30
$\sum$	520789	64366.04	63832.30	34636.55	34231.12	32672.01	48.8	1.86	19.99

Table 2. Wall-clock runtimes (in s) and speedup of FPGA-accelerated EAGLE2 phasing using 32 threads on a system with two Intel Xeon E5-2667v4 and Xilinx Kintex UltraScale KU115 FPGA. Speedup is shown for the total runtime and for the phasing part only. Total runtimes include phasing and data IO. The table shows selected chromosome-wise subsets, $\sum$ is the sum of all 22 subsets.

Dataset		EAGLE2		FPGA accel.		Speedup	
chr	# variants	Total	Phasing	Total	Phasing	Total	Phasing
chr1	43344	338.71	306.30	189.41	158.70	1.79	1.93
chr2	42210	326.69	292.64	183.05	150.70	1.78	1.94
chr6	36892	267.58	239.95	151.93	125.65	1.76	1.91
chr10	27360	213.93	191.44	123.52	102.48	1.73	1.87
chr13	17819	146.11	128.47	85.46	69.65	1.71	1.84
chr17	16612	133.18	118.17	77.27	63.46	1.72	1.86
chr21	7107	64.54	54.71	39.03	30.16	1.65	1.81
chr22	8262	70.05	60.02	41.78	33.07	1.68	1.81
$\sum$	520789	4093.11	3635.05	2360.24	1935.15	1.73	1.88

5 Conclusions and Future Work

With around 50% of the total runtime, data preparation for reference-based phasing with EAGLE2 [9] is the most time consuming part of the software. In this paper, we showed how to accelerate this part by a factor of up to 29 with the help of reconfigurable hardware, leading to a total speedup of almost 2 for the complete process, which is the theoretical maximum according to Amdahl's Law [1]. While preserving the switch error rate of the original software, we reduced genome-wide phasing of 520,000 standard variant markers in 2500 samples from more than 68 min to 39 min on a server-grade computing system.

The single-threaded runtime was reduced in the same manner from more than 18 h to 9.5 h. We used the commonly used and publicly available 1000 Genomes Project [12] phased haplotypes as reference panel.

We are already examining the possibilities of harnessing GPU hardware for supporting the final phasing step by creating the PBWT data structure, which is the second most time consuming step in the phasing process. Furthermore, since phasing is generally only used as preliminary process for genotype imputation, introducing hardware support for corresponding software, such as PBWT [7] or minimac4 [5], as they are used in the Sanger or Michigan Imputation Services [14, 16], is our next goal.

References

1. Amdahl, G.M.: Validity of the single processor approach to achieving large scale computing capabilities. In: Proceedings of the April 18–20, 1967, Spring Joint Computer Conference, AFIPS 1967 (Spring), pp. 483–485. ACM, New York (1967). https://doi.org/10.1145/1465482.1465560
2. Browning, S.R., Browning, B.L.: Haplotype phasing: existing methods and new developments. Nat. Rev. Genet. **12**, 703–714 (2011). https://doi.org/10.1038/nrg3054
3. Chen, G.K., Wang, K., Stram, A.H., Sobel, E.M., Lange, K.: Mendel-GPU: haplotyping and genotype imputation on graphics processing units. Bioinformatics **28**(22), 2979–2980 (2012). https://doi.org/10.1093/bioinformatics/bts536
4. Das, S., Abecasis, G.R., Browning, B.L.: Genotype imputation from large reference panels. Annu. Rev. Genomics Hum. Genet. **19**, 73–96 (2018). https://doi.org/10.1146/annurev-genom-083117-021602
5. Das, S., Forer, L., Schönherr, S., et al.: Next-generation genotype imputation service and methods. Nat. Genet. **48**, 1284–1287 (2016). https://doi.org/10.1038/ng.3656
6. Delaneau, O., Zagury, J.F., Marchini, J.: Improved whole chromosome phasing for disease and population genetic studies. Nat. Methods **10**, 5–6 (2013). https://doi.org/10.1038/nmeth.2307
7. Durbin, R.: Efficient haplotype matching and storage using the positional Burrows-Wheeler transform (PBWT). Bioinformatics **30**(9), 1266–1272 (2014). https://doi.org/10.1093/bioinformatics/btu014
8. Illumina: Infinium Global Screening Array-24 Kit. https://www.illumina.com/products/by-type/microarray-kits/infinium-global-screening.html
9. Loh, P.R., Danecek, P., Palamara, P.F., et al.: Reference-based phasing using the haplotype reference consortium panel. Nat. Genet. **48**, 1443–1448 (2016). https://doi.org/10.1038/ng.3679
10. Loh, P.R., Price, A.L.: EAGLE v2.4.1, 18 November 2018. https://data.broadinstitute.org/alkesgroup/Eagle/downloads/
11. Tewhey, R., Bansal, V., Torkamani, A., Topol, E.J., Schork, N.J.: The importance of phase information for human genomics. Nat. Rev. Genet. **12**, 215–223 (2011). https://doi.org/10.1038/nrg2950
12. The 1000 Genomes Project Consortium: A global reference for human genetic variation. Nature **526**, 68–74 (2015). https://doi.org/10.1038/nature15393

13. The Haplotype Reference Consortium: A reference panel of 64,976 haplotypes for genotype imputation. Nat. Genet. **48**, 1279–1283 (2016). https://doi.org/10.1038/ng.3643
14. US National Institutes of Health: Michigan Imputation Server. https://imputationserver.sph.umich.edu/
15. Wellcome Sanger Institute: SAMtools/BCFtools/HTSlib. https://www.sanger.ac.uk/science/tools/samtools-bcftools-htslib
16. Wellcome Sanger Institute: Sanger Imputation Service. https://imputation.sanger.ac.uk/
17. Williams, A.L., Patterson, N., Glessner, J., Hakonarson, H., Reich, D.: Phasing of many thousands of genotyped samples. Am. J. Hum. Genet. **91**, 238–251 (2012). https://doi.org/10.1016/j.ajhg.2012.06.013

Tree Based Advanced Relative Expression Analysis

Marcin Czajkowski[(⊠)], Krzysztof Jurczuk, and Marek Kretowski

Faculty of Computer Science, Bialystok University of Technology,
Wiejska 45a, 15-351 Bialystok, Poland
{m.czajkowski,k.jurczuk,m.kretowski}@pb.edu.pl

Abstract. This paper presents a new concept for biomarker discovery and gene expression data classification that rises from the Relative Expression Analysis (RXA). The basic idea of RXA is to focus on simple ordering relationships between the expression of small sets of genes rather than their raw values. We propose a paradigm shift as we extend RXA concept to tree-based Advanced Relative Expression Analysis (ARXA). The main contribution is a decision tree with splitting nodes that consider relative fraction comparisons between multiple gene pairs. In addition, to face the enormous computational complexity of RXA, the most time-consuming part which is scoring all possible gene pairs in each splitting node is parallelized using GPU. This way the algorithm allows searching for more tailored interactions between sub-groups of genes in a reasonable time. Experiments carried out on 8 cancer-related datasets show not only significant improvement in accuracy and speed of our approach in comparison to various RXA solutions but also new interesting patterns between subgroups of genes.

Keywords: Relative Expression Analysis · Decision trees · Gene expression data

1 Introduction

High-throughput technologies are generating large volumes of omics data at an unprecedented rate [11]. Traditional machine learning algorithms have been quite successful in automatically identifying complex patterns. Unfortunately, the overwhelming majority of systems focus on complex decision rules that are obstacles to mature applications [2]. Currently developed classification methods to biological data are usually designed for other purposes, such as improving statistical learning or applications to vision and speech, with little emphasis on transparency. The complexity of the decision rules that emerge from standard machine learning impedes biological understanding.

Comprehensive analysis poses new computational challenges and specialized computational approaches are required to effectively and efficiently carry out the predictions using biomedical data. It can be observed that there is a strong

© Springer Nature Switzerland AG 2020
V. V. Krzhizhanovskaya et al. (Eds.): ICCS 2020, LNCS 12139, pp. 496–510, 2020.
https://doi.org/10.1007/978-3-030-50420-5_37

need for such 'white box' models which may actually help in understanding and identifying relationships between specific features and improve biomarker discovery [22]. One of the solutions is Relative Expression Analysis (RXA) which is a powerful collection of easily interpretable computational methods for gene expression data classification. It focuses on finding interactions among a small collection of genes by studying relative ordering of their expressions rather than their raw values.

The most significant novelty in the proposed paper is the new, much more general concept of gene-gene interaction within RXA called Advanced Relative Expression Analysis (ARXA). By introducing relative fraction comparison between multiple gene pairs within a single individual we can detect not only the ordering shifts between the genes but also the percent changes in their relations. In addition, we have applied this strategy to the splitting nodes of the Decision Trees (DTs) in order to detect hierarchical relations as well. The traditional DTs have a long history in predictive modeling [17] but result in insufficient accuracy when applied to gene expression data. By combining and extending these two 'white box' algorithms we managed to significantly improve the classification accuracy on several publicly available gene expression datasets. Finally, to face up the enormous computational complexity which rises from an exhaustive analysis of all possible pairs of genes, we designed and implemented a graphic processing unit (GPU)-based parallelization.

The next section provides our motivations and a brief background on RXA, DTs and GPGPU parallelization. Section 3 describes in detail our concept of tree-based ARXA and its GPU-based implementation. In Sect. 4, experimental validation is performed and in the last section, the paper is concluded and possible future works are outlined.

2   Background

While great progress has been achieved in what entails biodata analysis, most of the research effort tends to focus almost exclusively on the prediction accuracy of core data mining tasks (e.g., classification and regression), and far less effort has gone into the crucial task of knowledge discovery itself. Specifically, the rules generated by nearly all standard, off-the-shelf techniques applied to genomics data [1], such as neural networks, random forests, SVMs, and linear discriminant analysis usually involve nonlinear functions of hundreds or thousands of genes, many parameters, and are therefore too complex to characterize mechanistically. Currently, deep learning approaches have been getting attention [23] as they can better recognize complex features through representation learning with multiple layers, and can facilitate the integrative analysis by effectively addressing the challenges discussed above. However, we know very little about how such results are derived internally.

In contrast to data mining systems, statistical methods for analyzing high-dimensional biomolecular data generated with high-throughput technologies permeate the literature in computational biology. Those analyses have uncovered a great deal of information about biological processes [1], such as important mutations and lists of "marker genes" associated with common diseases and key interactions in transcriptional regulation. However, the analysis is often limited to a relatively small number of features thus a small set of informative variables needs to be identified out of a large number (or dimension) of candidates.

2.1   Relative Expression Analysis

The process of biomarker discovery and characterization provides opportunities for more sophisticated solutions that integrate statistical, data mining and expert knowledge-based approaches. One of the ideas for the gene expression data is the concept of Relative Expression Analysis which focuses on testing relative expression ordering among a small number of transcripts. In the pioneering research from 2004, a Top Scoring Pair (TSP) method is proposed [10] which is a straightforward prediction rule based on the RXA concept that utilizes building blocks of rank-altered gene pairs in case and control comparison. Such pairs of genes can be viewed as "biological switches" which can be directly related to regulatory "motifs" or other properties of transcriptional networks. The discriminating power of each pair of genes i, j was measured by the absolute difference between the probabilities P_{ij} of the event that gene i is expressed more than gene j in the two classes.

Let x_i and x_j be the expression values of two different genes from available set of genes and there are only two classes: *normal* and *disease*. At first, algorithm calculates the probability of the relation $x_i < x_j$ between those two genes in the objects from the same class: $P_{ij}(normal) = Prob(x_i < x_j | Y = normal)$ and $P_{ij}(disease) = Prob(x_i < x_j | Y = disease)$, where Y denotes the class of the objects. Next, the score for this pair of genes (x_i, x_j) is calculated: $\Delta_{ij} = |P_{ij}(normal) - P_{ij}(disease)|$. This procedure is repeated for all distinct pairs of genes and the pair with the highest score becomes titled top scoring pair. In the case of a draw, a secondary ranking that bases on raw genes expression differences in each class is used [24].

The k-TSP algorithm [24] is one of the first extensions of the TSP solution. It focuses on increasing the number of pairs in the prediction model and applies no more than k top scoring disjoint gene pairs with the highest score, where the parameter k is determined by the internal cross-validation. This method was later combined with a top-down induced decision tree in an algorithm called TSPDT [5]. In this hybrid solution, each non-terminal node of the tree divides instances according to a splitting rule that is based on TSP or k-TSP accuracy.

Different approaches for the TSP extension focus on the relationships between more than two genes. Algorithms Top Scoring Triplet (TST) [19] and Top Scoring N (TSN) [21] analyze all possible ordering relationships between the genes, however, the general concept of TSP is retained. One of the first heuristic approaches that applied the RXA concept was the evolutionary algorithm called EvoTSP [6] where the authors proposed an evolutionary search for the TSP-like rules. This approach, later extended with additional features ranking in REHA [7] showed that evolutionary search is a good alternative to the traditional RXA algorithms. Finally, there are many variations of the TSP-family solutions that involve changes in ranking calculations, we can distinguish AUCTSP classifier that uses the ROC curve [15] or VH-k-TSP [12] that focuses on vertical and horizontal genes relations. What's more, the strength and simplicity of RXA has been recognized outside genomics data and is being successfully used in the proteomic and metabolomic analysis.

2.2   Decision Trees

The popularity of Decision trees (DTs) [17] can be explained by its ease of use, speed of classification and effectiveness. In addition, the hierarchical structure of the tree, where appropriate tests are applied successively from one node to the next, closely resembles the human way of making decisions. DT has a knowledge representation structure made up of nodes and branches, where: each internal node is associated with a test on one or more attributes; each branch represents the test result, and each leaf (terminal node) is designed by a class label. Induction of optimal DT is a known NP-complete problem [13]. As a consequence, practical DT learning algorithms must be heuristically enhanced.

DT represents a white-box approach and has considerable potential for bio-data research and scientific modeling of the underlying processes. Unfortunately, there are not so many new solutions in the literature that focus on the classification of genomic data with comprehensive DT models. Existing attempts showed that decision tree algorithms often induce classifiers with the inferior predictive performance [8] and one of the alternatives is combining DTs with evolutionary approaches [18]. However, nowadays, much more interest is given in trees as sub-learners of an ensemble learning approach, such as Random Forests. These solutions alleviate the problem of low accuracy by averaging or adaptive merging of multiple trees. However, when modeling is aimed at understanding basic processes, such methods are not so useful because they generate more complex and less understandable models.

2.3   GPGPU Parallelization

Recent research on the parallelization of various evolutionary computation methods has seemed to focus on GPUs as the implementation platform. The popularity of GPUs results from their general availability, relatively low cost, and high computational power. Parallel evaluation of instances is considered much more scalable with respect to the size of the dataset than a population approach. It focuses on gradually distributing the entire dataset among the local memories of all processors.

In the literature, we can find a few systems where GPU-based parallelization of the induction of DTs was examined. One of the propositions was CUDT [20] that parallelized the top-down induction process. In each internal node, in order to find the best locally optimal splits, the attributes are processed in parallel. With this approach, the authors managed to reduce the induction time of a typical decision tree from 5 to 55 times when compared with the traditional

CPU version. The GPGPU parallelization was also introduced to evolutionary induced DTs [14]. In the case of RXA there exists also research considering GPU parallelization. In [21] authors managed to speed up calculations of basic TSP and TST solutions by two orders of magnitude.

2.4   Motivation and Contribution

Most recently proposed data mining methods for genomic data generate complex rules that constrain the process of uncovering new biological understanding that, after all, is the ultimate goal of data-driven biology. However, it is not enough to simply produce good outcomes but to provide logical reasoning just as clinicians do for medical treatments. In addition, whereas the need for statistical methods in biomedicine continues to grow, the effects on the clinical practice of existing classifiers based on genomic data are widely acknowledged to remain limited. One of the barriers is the study-to-study diversity in reported prediction accuracies, problems with data integration and the unfavorable ratio of the sample size to the number of potential biomarkers. The main TSP advances for gene expression data analysis are:

- the method is non-parametric since the method is constructed based on the relative ranking of gene pairs;
- the method is based on one or a few gene pairs. The biological interpretation of the model and the translational application are more straightforward;
- researchers have repeatedly found that the family of TSP algorithms provides good prediction performance in many transcriptomic data [1].

The main drawback of TSP-family algorithms is that they are focused only on gene expression data and can only be used locally and on a small scale. There are two reasons why: (i) focusing on simple "biological switches" may not work where more advanced relations occur; (ii) exhaustive search performed by TSP-solutions has enormous computational complexity which strongly limits the number of features and inter-relations that can be analyzed [16]. In our previous research, we managed to partially address both issues separately by using decision trees with TSP splits [5] and/or evolutionary algorithms [6,8].

Nonetheless, the true core of the problem (i) still remains as deliberately replacing the raw data values with the ordering relationships between the features obviously causes loss of potentially important information. Let us hypothetically assume that for some tested sample two genes $X1$ and $X2$ can discriminate normal class from cancer one. Figure 1 shows three simple scenarios (a), (b), (c) together with the outcome of RXA. The example (a) shows the ultimate goal of RXA as it illustrates the perfect "top scoring pair". We can observe that the ordering relations between genes X1 and X2 is opposite in different classes among all the instances. Unfortunately, RXA outcome for scenario (a) and scenario (b) would be the same as in both cases "biological switch" occurs, at least in theory. However, when we look at the expression image and the chart axis we see, that in fact, X1 and X2 have low expression values in both classes. Such

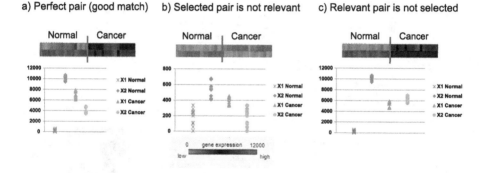

Fig. 1. Comparison of possible relations between two genes X1 and X2 in normal and cancer samples together with RXA outcome

selected pair is not relevant despite the fact that it will be promoted by RXA. An even worse scenario is presented in example (c) of Fig. 1 where undoubtedly relevant pair is not considered by the RXA despite significant variations in the expression values of genes in normal and cancer classes. As the simple ordering relationship between X1 and X2 is not changed between both classes, currently available RXA-family algorithms will never mark these genes as "top scoring pair". It might choose them with together other genes, by making multiple top pairs, but besides potential interpretability problems, lower accuracy issues may also arise. The issue (i) is also aggravated by the second (ii) problem which is computational complexity equals $O(T * k * M * N^Z)$, where T is the size of DT, k is the number of top-scoring groups, M is the number of instances, N is the number of analyzed genes and Z is the size of a group of genes which ordering relationships are searched. Sequential calculation of all possible gene pairs or gene groups strongly limits the number of genes and inter-relations that can be analyzed in a reasonable time and at the same time limits the number of features having similar expression values and being opposite to each other in different classes.

In this paper we propose the comparison of percentage changes of gene expressions in pairs among different classes. Within our approach the algorithm can easily ignore not relevant pairs (scenario (b)), select relevant ones (scenario (a) and (c)) and work even with smaller number of features. It should be noted that our new weight approach is even more computationally demanding than a typical RXA which will be shown in the following section. That is why we designed the GPU parallelization as an alternative to the above-mentioned evolutionary approaches to enable much faster RXA calculations.

3   Tree Based Advanced Relative Expression Analysis

The overall structure of the proposed solution is based on a typical top-down induced [17] binary classification tree. The greedy search starts with the root

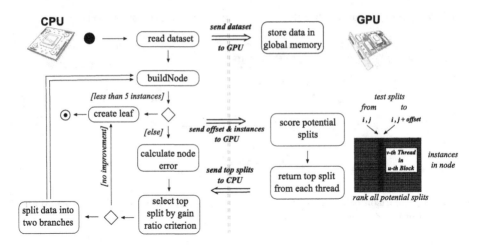

Fig. 2. General flowchart of a GPU-accelerated ARXA

node, where the locally optimal split (test) applies the new rank concept (denoted as ARXA). Then the training instances are redirected to the newly created nodes and this process is repeated for each node until the stop condition is met. Currently, we do not apply any form of post-pruning due to the small sample sizes, however, it should be considered in the future to improve the generalizing power of the predictive model.

The general flowchart of our GPU-accelerated ARXA is illustrated in Fig. 2. Each internal node contains information about a relation of pairs of genes that is later used to constitute the split. The basic idea to analyze relations within a single instance alike in RXA solutions, however, there are fundamental differences in scoring the collections of genes. It can be seen that the DT induction is run in a sequential manner on a CPU, and the most time-consuming operation which is scoring all potential splits is performed in parallel on a GPU. This way, the parallelization does not affect the behavior of the original algorithm.

Let us consider a gene expression microarray dataset consisting of N genes and M samples. Let the data be represented as an $N \times M$ matrix in which an expression value of u-th gene from v-th sample is denoted as x_{uv}. Each row represents observation of a particular gene X_u over M training samples, and each column Y_v represents a sample v described by the N genes. Let's for the simplicity of presentation assume that there are only two classes: C_1 and C_2, and instances with indexes from 1 to M_1 ($M_1 < M$) belong to the first class (C_1) and instances from range $(M_1 + 1, M\rangle$ to the second class (C_2).

At first, the ARXA method focuses on gene pair matching (i, j) ($i, j \in \{1, \ldots, N\}, i \neq j$) for which there is the highest averaged over instances probability p of an event $\frac{x_{im}}{x_{jm}} < \frac{x_{in}}{x_{jn}}$ ($m \in C_1$ and $n \in C_2$). For each pair of genes (i, j) the probability p_{ij} is calculated:

$$p_{ij} = \frac{\sum_{m=1}^{M_1} \sum_{n=M_1+1}^{M} I\left(\frac{x_{im}}{x_{jm}} < \frac{x_{in}}{x_{jn}}\right)}{(|C_1| * |C_2|)}$$

where $|C_1|$ denotes the number of instances from class C_1 and $I(\frac{x_{im}}{x_{jm}} < \frac{x_{in}}{x_{jn}})$ is the indicator function defined as:

$$I(\frac{x_{im}}{x_{jm}} < \frac{x_{in}}{x_{jn}}) = \begin{cases} 1, \text{if } \frac{x_{im}}{x_{jm}} < \frac{x_{in}}{x_{jn}} \\ 0, \text{if } \frac{x_{im}}{x_{jm}} \geq \frac{x_{in}}{x_{jn}} \end{cases}.$$

This computationally expensive calculations performed in each splitting node with complexity equals $O(N^2 * M^2)$ are handled by the GPU. Next, the top ranked pair from each thread is considered in building the splitting node. The threshold are calculated on the CPU and a single test that constitute splitting node has a form e.g. $\frac{x_i}{x_j} < 5$. It denotes that the instances can be divided into two sub-groups (branches or even classes) by simply checking if expression value of gene x_j is greater than 20% of gene x_i. Alike in k-TSP [24] we define maximum number of pairs that can constitute a node (the upper bound denoted as k can be set up before the classification) but instead of minimizing the prediction error we apply the gain ratio criterion. The number of pairs that creates the node may vary due to the internal cross-validation which throws away tests that do not contribute just as it is in k-TSP. The splitting criterion is guided by a majority voting mechanism in which all pair components of the split have the same weight. In the case of a draw, the vote of the primary pair is decisive.

3.1   ARXA Scoring on GPU

We propose two-level scoring due to the performance reasons as the major part of the scoring procedure is performed on the GPU and next the top gathered results are processed by the CPU (see Fig. 2). The RXA methods like TSP and TST exhibit characteristics that make them ideal for a GPU implementation as there is no data dependence between individual scores. As it is illustrated in Fig. 2, the dataset is first copied from the CPU main memory to the GPU device memory so each thread block can access it. It is performed only once before starting the tree induction as later only the indexes of the instances that remain in a calculated node are sent. In each node, possible relations X_i/X_j need to be processed and scored. Each thread on the device is assigned an equal amount of relations (called offset) to compute (see Fig. 2). This way each thread 'knows' which relations of genes it should analyze and where it should store the result.

In addition, number of instances for which the score is calculated varies in each tree node - from the full set of samples in a root to a few instances in the lower parts of the tree. Each thread loops over the instances that reach the node and calculates the scores to the assigned relations. After all thread blocks have finished, the results are copied from the GPU device memory back to the CPU main memory where the top split is established.

3.2   ARXA Scoring on CPU

After letting GPU know which instances residue in a current node and what offset is assigned to each thread, the CPU calculates the gain ratio for the node.

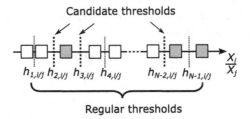

Fig. 3. Candidate thresholds for gene pair x_i/x_j

It is essential to check if potential splits returned from the GPU improves overall gain ratio as otherwise the leaf will be created. ARXA scoring on the CPU starts with sorting the results returned from threads according to their score (calculated on the GPU). Next, the results are filtered, alike in k-TSP solution, to leave only the k (default value: $k = 9$) top-ranked disjoint gene pairs. It should be noted that the GPU returns only the information about the relations and scores which is not enough to constitute a split.

Therefore, in the next step a set of tests is determined for further evaluation. Each test is constituted from a single top pair and has a form: $\frac{x_i}{x_j} < h_{i/j}$, where $h_{i/j}$ is the selected threshold. The search for the threshold only considers the relevant thresholds, called the candidate thresholds, which split instances from different classes as it is illustrated in Fig. 3. This way the algorithm does not consider e.g. $h_{1,i/j}$, $h_{4,i/j}$ and $h_{M-1,i/j}$ as those thresholds are useless for creating new tests because they split two training instances from the same class. The gain ratio criterion is used to determine the best possible threshold $h_{i/j}$, and the midpoint of the interval is applied as the value of this threshold. As an alternative to midpoint, we also performed experiments with smoothed threshold is e.g. an integer value (see enclosed results in Table 3). Finally, the choice of the number of gene pairs (parameter k) that constitute splitting node is determined by internal cross-validation.

4 Experimental Validation

In this section, we present a detailed experimental analysis to evaluate the relative performance of the proposed weight and hierarchical approach in RXA. Using several cancer-related gene expression datasets we have checked ARXA prediction power and confronted its results with popular RXA extensions.

4.1 Algorithms and Datasets

To make a proper comparison with the RXA algorithms, we use the same 8 cancer-related benchmark datasets (see Table 1) that are tested with the EvoTSP solution [6]. Datasets are deposited in NCBI's Gene Expression Omnibus and summarized in Table 1. A typical 10-fold cross-validation is applied and depending on the system, different tools are used:

Table 1. Details of gene expression datasets: abbreviation with name, number of genes and number of instances.

Datasets	Genes	Instances	Datasets	Genes	Instances
(a) GDS2771	22215	192	(e) GSE10072	22284	107
(b) GSE17920	54676	130	(f) GSE19804	54613	120
(c) GSE25837	18631	93	(g) GSE27272	24526	183
(d) GSE3365	22284	127	(h) GSE6613	22284	105

Table 2. Comparison of RXCT with top-scoring algorithms, including accuracy and the size of the classifier's model. The best accuracy for each dataset is bolded.

Dataset	TSP	TST	k-TSP		EvoTSP		TSPDT		ARXA		
	Acc.	Acc.	Acc.	Size	Acc.	Size	Acc.	Node size	Acc.	Node size	Tree size
(a)	57.2	61.9	62.9	10	65.6	4.0	60.1	15.4	**70.9**	5.7	3.6
(b)	88.7	89.4	90.1	6	96.5	2.1	**98.2**	1.0	92.5	1.0	1.0
(c)	64.9	63.7	67.2	10	78.1	2.8	72.3	5.8	**84.7**	7.6	1.4
(d)	93.5	92.8	94.1	10	**96.2**	2.1	88.3	2.0	95.0	3.0	1.0
(e)	56.0	60.5	58.4	14	66.9	3.1	68.1	4.7	**68.3**	6.7	3.4
(f)	47.3	50.1	56.2	18	66.2	2.7	67.2	10.9	**78.5**	8.1	2.2
(g)	81.9	84.2	87.2	14	86.1	4.1	88.6	3.3	**94.4**	6.7	1.0
(h)	49.5	51.7	55.8	10	53.6	6.1	59.6	7.0	**65.6**	5.9	2.4
Average	67.4	69.3	71.5	11.5	76.2	2.7	75.3	6.2	**81.2**	5.6	2.1

- evaluation of TSP, TST, and k-TSP was performed with the AUERA software [9], which is an open-source system for identification of relative expression molecular signatures;
- EvoTSP results were taken from the publication [6];
- original TSPDT and ARXA implementations are used.

Due to the performance reasons concerning other approaches, the Relief-F feature selection was applied and the number of selected genes was arbitrarily limited to the top 1000. In the experiments, we provide results for the proposed ARXA solution as well as its simplified variants which uses e.g. integer percentage split values.

Experiments were performed on a workstation equipped with Intel Core i5-8400 CPU, 32 GB RAM and NVIDIA GeForce GTX 1080 GPU card (8 GB memory, 2 560 CUDA cores). The sequential algorithm was implemented in C++ and the GPU-based parallelization part was implemented in CUDA-C (compiled by nvcc CUDA 10; single-precision arithmetic was applied).

4.2   Accuracy Comparison of ARXA to Popular RXA Counterparts

Table 2 summarizes classification performance for the proposed solution and its competitors. The model size of TSP and TST is not shown as it is fixed and

Table 3. Comparison of ARXA variants results with different model comprehensibility settings. Averaged value through all datasets are shown.

Algorithm	Accuracy	Node size	Tree size
$ARXA_{no\ round}$	**81.2**	5.6	2.1
$ARXA_{round\ 0.5}$	80.7	4.9	3.1
$ARXA_{round\ 1.0}$	79.6	4.6	3.4

equals correspondingly 2 and 3. It is clearly visible that the proposed ARXA solution managed to outperform all popular RXA classifiers in 6 out of 8 datasets. The statistical analysis of the obtained results using the Friedman test and the corresponding Dunn's multiple comparison test (significance level/p-value equals 0.05), as recommended by Demsar [9] showed that the differences in accuracy are significant. We have also performed an additional comparison between the datasets with the corrected paired t-test [24] with the significance level equals 0.05 and 9 degrees of freedom (n-1 degrees of freedom where n = 10 folds). It showed that ARXA significantly outperforms all algorithms on more than half datasets.

However, it should be noticed that improving classification accuracy was not our primary goal. We wanted to make a model in which gene pairs somehow interact with each other more deeply and also to promote finding sub-interactions between co-expressed genes and pairs. Such improvement in terms of classification accuracy was a surprise even for us, however, this may indicate the importance of the founded patterns.

4.3   ARXA Comprehensibility and GPGPU Acceleration

As we mentioned in Sect. 3.2, the fraction value which denotes the relation between two features can be rounded to improve model comprehensibility. Table 3 shows the ARXA average accuracy results from the performed experiments with different roundup of the threshold value. Therefore, in the first row we can see original ARXA version, in second row all the thresholds values in the tests that constituted splits we rounded to 0.5 and in last row the thresholds were rounded to the integer values. From the table we can observe, that, on average, as the thresholds are less precise, the number of tests in internal nodes decreases while the size of the tree increases. This outcome was consistent to all tested datasets.

In Figs. 4 and 5 we show an example decision trees induced by ARXA with and without threshold roundup. In both cases the prediction accuracy is similar but the structure and relations slightly differs. Although, there are a few similarities especially in the top nodes where e.g. two out of three relations in the root node from the DT illustrated in Fig. 5 is the same as in Fig. 4. There are also single genes that appear in both trees but constitute different pairs.

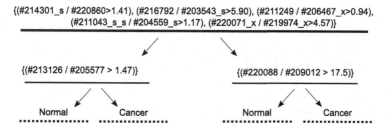

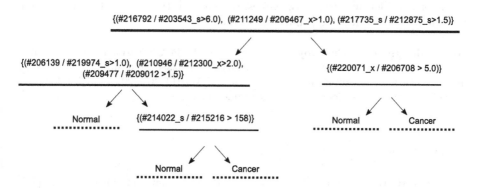

Fig. 4. An example decision tree induced by ARXA with detailed thresholds for GSE6613 Parkinson's disease.

Based on the description of the dataset (GSE6613 series) from GenBank NCBI [3] we performed a brief examination of one of the ARXA output prediction models (see Figs. 4 and 5). To check if founded genes or gene pair have some biological meaning we have decoded gene names from GSE6613 with GPL96 platform provided by NCBI. We found out that most of the founded genes are related with Parkinson's disease, for example #211249 (gene symbol: GPR68) is the top significantly deregulated gene identified through integrated analysis in Parkinson's disease [25] and gene LSM7 (#204559_s) is reported as significant in meta-analysis of genome-wide association studies of Parkinson's disease risk [4]. This is only an example of a fraction of knowledge discovered by ARXA but even the presented model is at some point supported by biological evidence in the literature.

Even with applied feature selection step (to make other algorithms work in a reasonable time), the number of possible relations for which the GPU needs to calculate is very high. For example, for N = 1000 genes and M = 100 instances the number of scores to calculate is over 10^9 in a root node (sub-nodes have fewer instances). However, if we would take the full dataset, this number drastically increases to 10^{13}. With the GPGPU acceleration the score ranking, on average through all datasets, was reduced from 20 s to 0.15 of a second which is over

Fig. 5. An example decision tree induced by ARXA with rounded to 0.5 thresholds for GSE6613 Parkinson's disease.

two magnitude faster for a single run. The time included also the data transfer to and from the GPU. When, the whole data was used, the GPU took up to several minutes where a single sequential run of a single dataset took over a day. Through all the runs number of blocks equals to 1024 and threads equals to 128. In the profiling we noticed that processing too many possible relations by each thread (high load) slows down the parallelization. By decreasing the offset value we managed to improve load balancing and thus the overall ARXA speedup.

5   Conclusions

In this paper, we introduce a hybrid approach to analyze gene expression data which combines the problem-specific methodology with the popular white-box classifier. The Advanced Relative Expression Analysis (ARXA) fundamentally changes the RXA solution in the context of relations and pairs ranking. Our implementation considers involving GPGPU accelerated decision trees in order to open ARXA on finding interesting hierarchical patterns in subgroups of genes in a reasonable time. In addition, experiments show that knowledge discovered by ARXA is accurate, comprehensive and at some point supported by biological evidence in the literature.

We see many promising directions for future research. In particular, we are currently working with biologists and bioinformaticians to better understand the gene relations generated by ARXA. Next, there is still a lot of ways to improve the GPU parallelization of RXCT, e.g. load-balancing of tasks based on the number of instances in each node, simultaneous analysis of two branches, better GPU hierarchical memory exploitation.

It should be noted that most of RXA solutions are not fully detached decision model from the raw values of the dataset. Such an approach may reduce robustness to methodological and technical factors, study-specific biases as well as limits the potential of exploring merged data from different omics, platforms, and experiments. Most of TSP-family solutions use e.g. raw values in their secondary rankings, others mean or variance of a given gene in the data. ARXA does not consider analyzing raw values, therefore in the nearest future we want to validate our approach on multi-omics data.

Acknowledgments. This project was funded by the Polish National Science Center and allocated on the basis of decision 2019/33/B/ST6/02386 (first author). The second and third author were supported by the grant WZ/WI-IIT/3/2020 from BUT founded by Polish Ministry of Science and Higher Education.

References

1. Afsari, B., Braga-Neto, U.M., Geman, D.: Rank discriminants for predicting phenotypes from RNA expression. Ann. Appl. Stat. **8**(3), 1469–1491 (2014)

2. Bacardit, J., et al.: Hard data analytics problems make for better data analysis algorithms: bioinformatics as an example. Big Data **2**(3), 164–176 (2014)
3. Benson, D.A., et al.: GenBank. Nucleic Acids Res. **46**(D1), D41–D47 (2018)
4. Chang, D., Nalls, M.A., et al.: A meta-analysis of genome-wide association studies identifies 17 new Parkinson's disease risk loci. Nat Genet. **49**(10), 1511–1516 (2017)
5. Czajkowski, M., Kretowski, M.: Top scoring pair decision tree for gene expression data analysis. Adv. Exp. Med. Biol. **696**, 27–35 (2011)
6. Czajkowski, M., Kretowski, M.: Evolutionary approach for relative gene expression algorithms. Sci. World J. **2014**, 7 (2014). 593503
7. Czajkowski M., Kretowski M.: Relative evolutionary hierarchical analysis for gene expression data classification. In: GECCO 2019, pp. 1156–1164 (2019)
8. Czajkowski, M., Kretowski, M.: Decision tree underfitting in mining of gene expression data. An evolutionary multi-test tree approach. Expert Syst. Appl. **137**, 392–404 (2019)
9. Earls, J.C., et al.: AUREA: an open-source software system for accurate and user-friendly identification of relative expression molecular signatures. BMC Bioinform. **14**, 78 (2013). (Article 19)
10. Geman, D., et al.: Classifying gene expression profiles from pairwise mRNA comparisons. Stat. Appl. Genet. Mol. Biol. **3**(19) (2004)
11. Huang, S., Chaudhary, K., Garmire, L.X.: More is better: recent progress in multi-omics data integration methods. Front. Genet. **8**(84) (2017)
12. Huang, X., et al.: Analyzing omics data by pair-wise feature evaluation with horizontal and vertical comparisons. J. Pharm. Biomed. Anal. **157**, 20–26 (2018)
13. Hyafil, L., Rivest, R.L.: Constructing optimal binary decision trees is NP complete. Inf. Process. Lett. **5**(1), 15–17 (1976)
14. Jurczuk, K., Czajkowski, M., Kretowski, M.: Evolutionary induction of a decision tree for large scale data. A GPU-based approach. Soft Comput. **21**, 7363–7379 (2017)
15. Kagaris, D., Khamesipour A: AUCTSP: an improved biomarker gene pair class predictor. BMC Bioinform. **19**(244) (2018). (Article 244)
16. Kim, S., Lin, C.W., Tseng, G.C.: MetaKTSP: a meta-analytic top scoring pair method for robust cross-study validation of omics prediction analysis. Bioinformatics **32**(13), 1966–1973 (2016)
17. Kotsiantis, S.B.: Decision trees: a recent overview. Artif. Intell. Rev. **39**(4), 261–283 (2013)
18. Kretowski, M.: Evolutionary Decision Trees in Large-scale Data Mining. Studies in Big Data 59 (2019)
19. Lin, X., et al.: The ordering of expression among a few genes can provide simple cancer biomarkers and signal BRCA1 mutations. BMC Bioinform. **10**(256) (2009)
20. Lo, W.T., et al.: CUDT: a CUDA based decision tree algorithm. Sci. World J. **2014**, 12 (2014). 745640
21. Magis, A.T., Price, N.D.: The top-scoring 'N' algorithm: a generalized relative expression classification method from small numbers of biomolecules. BMC Bioinform. **13**(1), 227 (2012)
22. McDermott, J.E., et al.: Challenges in biomarker discovery: combining expert insights with statistical analysis of complex omics data. Expert Opin. Med. Diagn. **7**(1), 37–51 (2013)
23. Min, S., Lee, B., Yoon, S.: Deep learning in bioinformatics. Brief. Bioinform. **18**(5), 851–869 (2017)

24. Tan, A.C., Naiman, D.Q.: Simple decision rules for classifying human cancers from gene expression profiles. Bioinformatics **21**, 3896–3904 (2005)
25. Wang, J., Liu, Y., Chen, T.: Identification of key genes and pathways in Parkinson's disease through integrated analysis. Mol. Med. Rep. **16**(4), 3769–3776 (2017)

Testing the Significance of Interactions in Genetic Studies Using Interaction Information and Resampling Technique

Paweł Teisseyre[1]([⊠]) [ID], Jan Mielniczuk[1,2] [ID], and Michał J. Dąbrowski[1] [ID]

[1] Institute of Computer Science, Polish Academy of Sciences, Warsaw, Poland
{teisseyrep,miel,m.dabrowski}@ipipan.waw.pl
[2] Faculty of Mathematics and Information Sciences,
Warsaw University of Technology, Warsaw, Poland

Abstract. Interaction information is a model-free, non-parametric measure used for detection of interaction among variables. It frequently finds interactions which remain undetected by standard model-based methods. However in the previous studies application of interaction information was limited by lack of appropriate statistical tests. We study a challenging problem of testing the positiveness of interaction information which allows to confirm the statistical significance of the investigated interactions. It turns out that commonly used chi-squared test detects too many spurious interactions when the dependence between the variables (e.g. between two genetic markers) is strong. To overcome this problem we consider permutation test and also propose a novel HYBRID method that combines permutation and chi-squared tests and takes into account dependence between studied variables. We show in numerical experiments that, in contrast to chi-squared based test, the proposed method controls well the actual significance level and in many situations detects interactions which are undetected by standard methods. Moreover HYBRID method outperforms permutation test with respect to power and computational efficiency. The method is applied to find interactions among Single Nucleotide Polymorphisms as well as among gene expression levels of human immune cells.

Keywords: Interactions · Interaction information · Mutual information · SNP · Gene-gene interaction

1 Introduction

Detection of various types of interactions is one of the most important challenges in genetic studies. This is motivated by the fact that most human diseases are complex which means that they are typically caused by multiple factors, including gene-gene ($G \times G$) interactions and gene-environment ($G \times E$) interactions [1]. The analysis may include binary traits (case-control studies) as well as quantitative traits (e.g. blood pressure or patient survival times). The presence of

© Springer Nature Switzerland AG 2020
V. V. Krzhizhanovskaya et al. (Eds.): ICCS 2020, LNCS 12139, pp. 511–524, 2020.
https://doi.org/10.1007/978-3-030-50420-5_38

gene-gene interactions has been shown in complex diseases such as breast cancer [2] or coronary heart disease [3]. The interactions are closely related to the concept of epistasis [4]. In biology, the epistasis is usually referred to as the modification, or most frequently, blocking of one allelic effect by another allele at a different locus [5]. In this work we focus on interactions of the second order, i.e. interactions between two variables in predicting the third variable, although higher order interactions may also contribute to many complex traits [6]. In our notation, (X_1, X_2) denotes a pair of predictors, whereas Y stands for a response variable. We consider a general situation in which Y can be discrete (e.g. disease status) or quantitative (e.g. blood pressure or survival time), in the latter case Y is discretized.

There are many different concepts of measuring interactions, see e.g. [5]. Informally interaction arises when the simultaneous influence of variables X_1 and X_2 on Y is not additive. The classical approach to analyze interactions is to use ANOVA (in the case of quantitative Y) and logistic regression (in the case of binary Y) [7]. Recently entropy-based methods attracted a significant attention including interaction information (II) [8] which is a very promising measure having many desired properties. It is a non-parametric, model-free measure, which does not impose any particular assumptions on the data, unlike parametric measures of interactions based on e.g. linear or logistic regression. It is based on a very general measure of dependence - mutual information and thus it allows to detect interactions which remain undetected by standard methods based on parametric models, see e.g [9]. Finally, it can be applied to any types of variables, unlike e.g. logistic regression which is restricted to the case of binary response variable. Interaction information has been already used in genetic studies. For example, Moore et al. [10] use II for analysing gene-gene interactions associated with complex diseases. Recently, II was also used to verify existence of interactions between DNA methylation sites/gene expression profiles and gender/age in the context of glioma patients survival prediction [11]. II is applied as a main tool to detect interactions in packages: AMBIENCE [12] and BOOST [13]. Jakulin et al. [14] applied II to detect interactions between variables in classification task and studied how the interactions affect the performance of learning algorithms. Mielniczuk et al. [15] studied properties of II and its modifications in the context of finding interactions among Single Nucleotide Polymorphisms (SNPs). Mielniczuk and Teisseyre [9] have shown that, in context of gene-gene interaction detection, II is on the whole much more discriminative measure than the logistic regression, i.e. it finds certain types of interactions that remain undetected by logistic regression. Here, we provide evidences that II is also more powerful than ANOVA F test, when quantitative trait Y is considered. This is especially pronounced when a posteriori probability of Y given values of predictors is a non-linear function of Y.

Although II has attracted some attention, its application was hindered by lack of appropriate statistical tests. Here we study an important problem of testing positiveness of II. Positive value of II indicates that predictive interaction between X_1 and X_2 exists. The main contribution of the paper is a new test for positiveness of II which takes into account the fact that X_1 and X_2

may be dependent. This occurs frequently, e.g. in Genome Wide Associations Studies when dependence of SNPs in close proximity is due to crossing-over mechanism. The task is challenging as the distribution of II under the null hypothesis that its population value is zero is not known, except the special case when all three variables are mutually independent. In this case it turns out to be chi-squared distribution [16]. We show that indeed dependence matters in this context i.e. when association between X_1 and X_2 is strong the distribution of II can significantly deviate from chi-squared distribution. This means that in such cases a pertaining test based on chi-squared null distribution may not have an assumed level of significance, or equivalently, the calculated p-values may be misleading. In view of this we propose a hybrid method that combines two existing approaches: permutation test and chi-squared test. In brief, we use a chi-squared test when the dependence between the original variables is weak and the permutation test in the opposite case. The experiments show that the combined procedure, in contrast to the chi-squared test, allows to control actual significance level (type I error rate) and has a high power. At the same time it is less computationally expensive than standard permutation test.

2   Interaction Information

Variables denoted X_1, X_2, Y take values K_1, K_2, L respectively, and to simplify definitions are assumed to be discrete. Let $P(x_1, x_2) := P(X_1 = x_1, X_2 = x_2)$, $P(x_1) := P(X_1 = x_1)$ and $P(x_2) := P(X_2 = x_2)$ be joint and marginal probabilities, respectively. The independence between variables X_1 and X_2 will be denoted by $X_1 \perp X_2$. χ_k^2 stands for the chi-squared distribution with k degrees of freedom will be denoted by and $\chi_{k,1-\alpha}^2$ for the corresponding $1 - \alpha$ quantile. Entropy of variable X_1, defined as $H(X_1) := -\sum_{x_1} P(x_1) \log P(x_1)$, is a basic measure of an uncertainty of the variable. Furthermore, conditional entropy, $H(X_1|X_2) := -\sum_{x_1,x_2} P(x_1, x_2) \log P(x_1|x_2)$, quantifies the uncertainty about X_1 when X_2 is given. Mutual information (MI) measures the amount of information obtained about one random variable, through the other random variable. It is defined as $I(X_1, X_2) := H(X_1) - H(X_1|X_2)$. MI is a popular non-negative measure of association and equals 0 only when if X_1 and X_2 are independent. MI can be also interpreted as the amount of uncertainty in one variable which is removed by knowing the other variable. In this context it is often called information gain. In addition define the conditional mutual information as $I(X_1, X_2|Y) := H(X_1|Y) - H(X_1|X_2, Y) = H(X_2|Y) - H(X_2|X_1, Y)$. It is equal zero if and only if X_1 and X_2 are conditionally independent given Y. For more properties of the basic measures above we refer to [17].

In this work the main object of our interest is interaction information (II) [8] that can be defined in two alternative ways. The first definition is

$$II(X_1, X_2, Y) = I((X_1, X_2), Y) - I(X_1, Y) - I(X_2, Y). \tag{1}$$

Observe that $I(X_1, Y)$ and $I(X_2, Y)$ correspond to main effects. In practice we want to distinguish between situations when II is approximately 0 and when

II is large, the latter indicating non-additive influence of both predictors on Y. In view of definition (1), interaction information can be interpreted as a part of mutual information between (X_1, X_2) and Y which is solely due to interaction between X_1 and X_2 in predicting Y i.e. the part of $I((X_1, X_2), Y)$ which remains after subtraction of the main effect terms due to both predictors. Thus the definition of II corresponds to the intuitive meaning of interaction as a situation in which two variables affect a third one in a non-additive manner. Definition (1) also points out to important and challenging fact that existence of interactions is unrelated to existence of the main effects. Thus if SNPs with small main effects are not considered further, this does not necessarily mean that they do not contribute to the trait. The second definition states that

$$II(X_1, X_2, Y) = I(X_1, X_2|Y) - I(X_1, X_2). \tag{2}$$

The equivalence of (1) and (2) follows from basic properties of MI (see e.g. [15]). Definition (2) indicates that II measures the influence of a variable Y on the amount of information shared between X_1 and X_2. In view of (1) and (2) we see that II is a valuable index which can be interpreted as a predictive interaction measure and at the same time as a measure of a deviation of conditional distributions from the unconditional one. This feature corresponds to two main approaches which are used to study interactions. The first one, which quantifies the remaining part of dependence after removing the main effects is exemplified by linear and logistic regression methods and testing significance of an interaction coefficient in such models [13]. The second one is based on measuring the difference of inter-loci associations between cases and controls [18].

Observe that II in contrast to the mutual information can be either positive or negative. In view of (2) positive value of II indicates that variable Y enhances the association between X_1 and X_2. In other words, the conditional dependence is stronger than the unconditional one. The negative value of II indicates that Y weakens or inhibits the dependence between X_1 and X_2. Alternatively, in view of (1), we can assert that positive interaction information means that information about Y contained in (X_1, X_2) is larger than sum of individual informations $I(X_1, Y) + I(X_2, Y)$.

3   Testing the Positiveness of Interaction Information

The main goal of this paper is to propose a novel procedure for testing the positiveness of II. Such a procedure is useful to find pairs of variables (X_1, X_2) that allow to jointly predict Y, even when the main effects are negligible and to confirm the statistical significance of the detected interaction. We state the following proposition which albeit simple, is instrumental for understanding the presented approach.

Proposition 1. *If* $Y \perp (X_1, X_2)$, *then* $II(X_1, X_2, Y) = 0$.

Proof. The independence of (X_1, X_2) and Y implies that $I((X_1, X_2), Y) = 0$ and also $I(X_1, Y) = I(X_2, Y) = 0$. Thus, the assertion follows directly from (1).

Note that although the converse to Proposition 1 is not true i.e. it is possible to have $II(X_1, X_2, Y) = 0$ while $I((X_1, X_2), Y) > 0$ ([19], p. 121) such examples require special constructions and are not typical. Moreover, it follows from (1) that when X_1 and X_2 are individually independent of Y and $II(X_1, X_2, Y) = 0$ then pair (X_1, X_2) is independent of Y. Whence, from the practical point of view hypotheses $II(X_1, X_2, Y) = 0$ and $I((X_1, X_2), Y) = 0$ are approximately equivalent.

Our principal aim is to test the null hypothesis:

$$H_0 : II(X_1, X_2, Y) = 0, \tag{3}$$

against the alternative hypothesis corresponding to the positiveness of $II(X_1, X_2, Y)$:

$$H_1 : II(X_1, X_2, Y) > 0. \tag{4}$$

In view of the above discussion we replace H_0 by:

$$\tilde{H}_0 : Y \perp (X_1, X_2),$$

The main operational reason for replacing H_0 by $\tilde{H}_0$ is that the distribution of a sample version of II under null hypothesis H_0 is unknown and determining it remains an open problem. We note that the sample versions of $I(X_1, X_2)$, $I(X_1, X_2|Y)$ and $II(X_1, X_2, Y)$ are simply obtained by replacing the true probabilities by estimated probabilities (i.e. fractions). They will be denoted by $\widehat{I}(X_1, X_2)$, $\widehat{I}(X_1, X_2|Y)$ and $\widehat{II}(X_1, X_2, Y)$, respectively. In contrast to H_0 scenario, it is possible to determine distribution of $\widehat{II}(X_1, X_2, Y)$ when $\tilde{H}_0$ is true using permutation based approach. We note that the latter allows to calculate the distribution of $\widehat{II}(X_1, X_2, Y)$ with arbitrary accuracy for any sample size n and for fixed sample distribution of Y and (X_1, X_2) while chi-square approximation, even when it is valid, it is accurate only for large sample sizes. In this paper we combine these two approaches: permutation and based on asymptotic distribution. This yields a novel testing method which is computationally feasible (it is not as computationally intensive as permutation based test) and is more powerful than chi-squared test.

3.1   Chi-Squared Test IICHI

The distribution of $\widehat{II}(X_1, X_2, Y)$ under the null hypothesis (3) is not known. However, in a special case of (3) when all three variables are jointly independent and all probabilities $P(X_1 = x_i, X_2 = x_j, Y = y_k)$ are positive, Han [16] has shown that for a large sample size

$$2n\widehat{II}(X_1, X_2, Y) \sim \chi^2_{(K_1-1)(K_2-1)(L-1)}, \tag{5}$$

approximately, where K_1, K_2, L are the number of levels of X_1, X_2 and Y, respectively. Of course, joint independence of (X_1, X_2, Y) is only a special case of $\tilde{H}_0$, which is in turn a special case of (3). Nonetheless the above approximation is informally used to test the positiveness of II under null hypothesis, see e.g. [12]. Thus for this method, we accept the null hypothesis (3) if $2n\widehat{II}(X_1, X_2, Y) <$ $\chi^2_{(K_1-1)(K_2-1)(L-1),1-\alpha}$, where α is a significance level. It turns out that if the dependence between X_1 and X_2 increases the distribution of $2n\widehat{II}(X_1, X_2, Y)$ deviates from χ^2 distribution. Thus the χ^2 test can be used to test the positiveness of $II(X_1, X_2, Y)$ only if there is a weak dependence between X_1 and X_2. It follows from our experiments that if the dependence between X_1 and X_2 is strong, then χ^2 test tends to reject the null hypothesis too often, i.e. its type I error rate may significantly exceed the prescribed level of significance α.

3.2   Permutation Test IIPERM

The distribution of $2n\widehat{II}(X_1, X_2, Y)$ under the null hypothesis $\tilde{H}_0$ can be approximated using a permutation test. Although $\tilde{H}_0$ is a proper subset of (3), the Monte-Carlo approximation is used to test the positiveness of II under hypothesis (3). Observe that permuting the values of variable Y while keeping values of (X_1, X_2) fixed we obtain the sample conforming the null distribution. An important advantage of permutation test is that while permuting the values of Y the dependence between X_1 and X_2 is preserved. We permute the values of variable Y and calculate $2n\widehat{II}(X_1, X_2, Y)$ using the resulting data. This step is repeated B times and allows to approximate the distribution of $2n\widehat{II}(X_1, X_2, Y)$ under the null hypothesis $\tilde{H}_0$.

Figure 1 shows the permutation distribution (for $B = 10000$), χ^2 distribution and a true distribution of $2n\widehat{II}(X_1, X_2, Y)$, under the null hypothesis (3), for artificial data M0 (see Sect. 4.1), generated as follows. The pair (X_1, X_2) is is drawn from distribution described in the Table 1, Y is generated independently from the standard Gaussian distribution and then discretized using the equal frequencies and 5 bins. The true distribution is approximated by calculating $2n\widehat{II}(X_1, X_2, Y)$ for 10000 data generation repetitions (this is possible only for artificially generated data). Since X_1 and X_2 take 3 possible values, we consider χ^2 distribution with $(3 - 1) \times (3 - 1) \times (5 - 1) = 16$ degrees of freedom. In this experiment we control the dependence strength between X_1 and X_2 and analyse three cases: $I(X_1, X_2) = 0$, $I(X_1, X_2) = 0.27$ and $I(X_1, X_2) = 0.71$. Thus in the first case X_1 and X_2 are independent whereas in the last case there is a strong dependence between X_1 and X_2. First observe that the lines corresponding to the permutation distribution and the true distribution are practically indistinguishable, which indicates that the permutation distribution approximates the true distribution very well. Secondly it is clearly seen that the χ^2 distribution deviates from the remaining ones when the dependence between X_1 and X_2 becomes large. Although this nicely illustrates (5) when complete independence occurs, it also underlines that χ^2 distribution is too crude when the dependence between X_1 and X_2 is strong. It is seen that the right tail of χ^2 distribution is thinner than the right tail of the true distribution and thus the uppermost quantiles of the true distribution are underestimated by the corresponding quantiles of χ^2 (Fig. 1). This is the reason why IICHI rejects the null hypothesis too often leading to many false positives. This problem is recognized for other scenarios of interaction detection (cf. [20]). The drawback of the permutation test is its computational cost. This becomes a serious problem when the procedure is applied for thousands of variables, as in the analysis of SNPs.

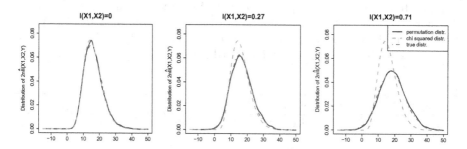

Fig. 1. Probability density functions of chi-squared distribution, permutation distribution and true distribution of $2n\widehat{II}(X_1, X_2, Y)$ under the null hypothesis $(X_1, X_2) \perp Y$. Data is generated from model M0 described by Table 1, for $n = 1000$.

3.3 Hybrid Test

To overcome the drawbacks of a χ^2 test (a significant deviation from the true distribution under the null hypothesis) and a permutation test (high computational cost) we propose a hybrid procedure that combines these two approaches. The procedure exploits the advantages of the both methods. It consists of two steps. We first verify whether the dependence between X_1 and X_2 exists. We use a test for a null hypothesis

$$H_0 : I(X_1, X_2) = 0, \qquad (6)$$

where the alternative hypothesis corresponds to the positiveness of MI:

$$H_1 : I(X_1, X_2) > 0. \qquad (7)$$

It is known (cf e.g. [21]) that under the null hypothesis (6), we approximately have:

$$2n\widehat{I}(X_1, X_2) \sim \chi^2_{(K_1-1)(K_2-1)},$$

for large sample sizes. If the null hypothesis (6) is not rejected, we apply chi-squared test for $II(X_1, X_2, Y)$ described in Sect. 3.1. Otherwise we use a permutation test described in Sect. 3.2. In the case of independence (or weak dependence) of X_1 and X_2 we do not perform, or perform rarely, the permutation test, which reduces the computation effort of the procedure. There are three input parameters. Parameter α is a nominal significance level of the test for interactions. Parameter α_0 is a significance level of the initial test for independence between X_1 and X_2. The larger the value of α_0 is, it is more likely to reject the null hypothesis (6) and thus it is also more likely to use the permutation test. Choosing the small value of α_0 leads to more frequent use of the chi-squared test. This reduces the computational burden associated with the permutation test but can be misleading when chi-squared distribution deviates from the true distribution of $2n\widehat{II}(X_1, X_2, Y)$ under the null hypothesis (3). Parameter B corresponds to the number of loops in a permutation test. The larger the value of B, the more accurate is the approximation of the distribution of $2n\widehat{II}(X_1, X_2, Y)$ under the null hypothesis. On the other hand, choosing large B increases the computational burden. Algorithm for HYBRID method is given below.

Algorithm 1: Hybrid test (HYBRID)

Input   : Sample of size n drawn from a joint distribution of (X_1, X_2, Y)
Parameters: α_0, α, B
calculate $\widehat{I}(X_1, X_2)$ and $\widehat{II}(X_1, X_2, Y)$
if $2n\widehat{I}(X_1, X_2) < \chi^2_{(K_1-1)(K_2-1),1-\alpha_0}$ **then**
$\quad$ # Use the chi-squared test:
$\quad$ **if** $2n\widehat{II}(X_1, X_2, Y) < \chi^2_{(K_1-1)(K_2-1)(L-1),1-\alpha}$ **then**
$\quad\quad$ $\lfloor$ accept the null hypothesis (3)

else
$\quad$ # Use the permutation test:
$\quad$ **for** $b \leftarrow 1$ **to** B **do**
$\quad\quad$ $\lfloor$ Calculate $\widehat{II}^b := \widehat{II}(X_1, X_2, Y^b)$ (Y^b is variable Y with permuted values)
$\quad$ Let $q_{B,1-\alpha}$ be an empirical $1 - \alpha$ quantile based on a sample $\widehat{II}^1, \ldots, \widehat{II}^B$.
$\quad$ **if** $\widehat{II}(X_1, X_2, Y) < q_{B,1-\alpha}$ **then**
$\quad\quad$ $\lfloor$ accept the null hypothesis (3)

---

4 Analysis of the Testing Procedures

In the following we analyse the type I error and the power of the three tests based on II. We present the results of selected experiments, extended results are included in the on-line supplement https://github.com/teisseyrep/Interactions. Although the methods based on II can be applied for any types of variables, in our experiments we focus on the common situation in Genome-Wide Association Studies when X_1 and X_2 are SNPs. For each SNP, there are three genotypes: the homozygous reference genotype (AA or BB), the heterozygous genotype (Aa or Bb respectively), and the homozygous variant genotype (aa or bb). Here A and a correspond to the alleles of the first SNP (X_1), whereas B and b to the alleles of the second SNP (X_2). Moreover we assume that Y is quantitative variable. Experiments for binary trait confirming the advantages of II over e.g. logistic regression are described in [9]. For the comparison we also use standard ANOVA test which is the state-of-the-art method for interaction detection, when

Y is quantitative. In the experiments we compare the following methods: IICHI, IIPERM, HYBRID and as a baseline ANOVA.

4.1 Analysis of Type I Error Rate

In order to analyse the testing procedures, described in the previous sections, we first compare the type I errors rates, i.e. the probabilities of false rejection of the null hypothesis. We consider the following model (called M0) in which Y is independent from (X_1, X_2). The distribution of (X_1, X_2) is given in Table 1. Parameter $p_0 \in [0, 2/9]$ controls the dependence strength between X_1 and X_2. Probabilities of diagonal values $(aa, bb), (Aa, Bb)$ and (AA, BB) are equal $1/9 + p_0$ and increase when p_0 increases. For p_0 in interval $[0, 2/9]$ mutual information $I(X_1, X_2)$ ranges from 0 to 1.1. Value $I(X_1, X_2) = 1.1$ is obtained for $p_0 = 2/9$ and corresponds to the extremal dependence when the probability is concentrated on the diagonal. Variable Y is generated from standard Gaussian distribution independently from (X_1, X_2). To calculate $\widehat{II}(X_1, X_2, Y)$ we discretize Y using the equal frequencies and 5 bins. The type I error rate is approximated by generating data $L = 10^5$ times, for each dataset we perform the tests and then calculate the fraction of simulations for which the null hypothesis is rejected. Number of repetitions in permutation test is $B = 10^4$.

Figure 2 (left panel) shows how the type I error rate for model M0 depends on $I(X_1, X_2)$. For large $I(X_1, X_2)$ the type I error rate of chi-squared interaction test is significantly larger than the nominal level $\alpha = 0.05$. For the other methods, the type I error rate oscillates around α, even for a large $I(X_1, X_2)$. It is also worth noticing that starting from moderate dependence of X_1 and X_2 type I error rates of permutation and hybrid tests are almost undistinguishable. Figure 2 (right panel) shows how the type I error rate for model M0 depends on the sample size n in the case of moderate dependence between X_1 and X_2. For IICHI we observe significantly more false discoveries than for other methods. All methods other than IICHI control type I error rate well for $n \geqslant 300$ when the dependence of predictors is moderate or stronger $(I(X_1, X_2) \geqslant 0.15)$. The above analysis confirms that for a strong dependence between variables, IICHI is not an appropriate test, while the IIPERM and HYBRID work as expected. The supplement contains results for different parameter settings.

4.2 Power Analysis

In the following we analyse the power of the discussed testing procedures. It follows from the previous section that method IICHI based on chi-squared distribution does not control the significance level, especially when the dependence between X_1 and X_2 is strong. Therefore to make a power comparison fair we do not take into account IICHI, which obviously has a largest power as it rejects the null hypothesis too often. For all other methods, IIPERM, HYBRID and ANOVA, which control probability of type I error satisfactorily for considered sample sizes, the distinction between them should be made based on their power and computational efficiency. We use a general framework in which a conditional

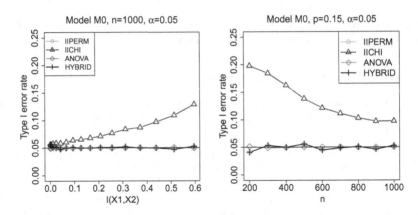

Fig. 2. Type I error rate with respect to the mutual information and n for the simulation model M0, for $\alpha = 0.05$ and $n = 1000$.

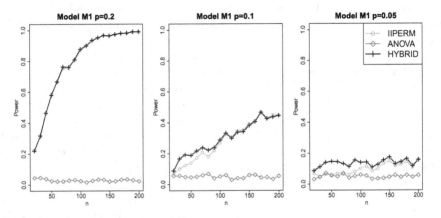

Fig. 3. Power with respect to the sample size n for a simulation model M1. $II(X_1, X_2, Y) = 0.1891, 0.0368, 0.0085$, for $p = 0.2, 0.1, 0.05$, respectively.

distribution of (X_1, X_2) given $Y = y$ is described by Table 1(b). This scenario corresponds to definition (2) of interaction information. Simulation models are designed in such a way to control the interaction strength as well as the dependence between X_1 and X_2 given Y. More precisely, function $s(y)$ controls the value of $I(X_1, X_2|Y)$ and also the value of $II(X_1, X_2, Y)$. In addition we assume that $Y \in \{-1, 0, 1\}$ and $P(Y = -1) = P(Y = 0) = P(Y = 1) = 1/3$. Here we present the result for typical simulation model called M1 in which function $s(y) = p$, when $y = -1$ or $y = 1$ and $s(y) = 0$, when $y = 0$. Other models are described in Supplement. Power is measured as a fraction of simulations (out of 10^4) for which the null hypothesis is rejected.

Observe that the dependence between X_1 and X_2 varies for different values of Y. For example, setting $p \approx 2/9$ we obtain $I(X_1, X_2|Y = 1) = I(X_1, X_2|Y = -1) \approx \log(3)$ and $I(X_1, X_2|Y = 0) \approx 0$. Figure 3 shows how the power depends

Table 1. Distribution of (X_1, X_2) for simulation models M0 (a) and M1 (b).

	$X_2 = bb$	$X_2 = Bb$	$X_2 = BB$
$X_1 = aa$	$\frac{1}{9} + p_0$	$\frac{1}{9} - \frac{p_0}{2}$	$\frac{1}{9} - \frac{p_0}{2}$
$X_1 = Aa$	$\frac{1}{9} - \frac{p_0}{2}$	$\frac{1}{9} + p_0$	$\frac{1}{9} - \frac{p_0}{2}$
$X_1 = AA$	$\frac{1}{9} - \frac{p_0}{2}$	$\frac{1}{9} - \frac{p_0}{2}$	$\frac{1}{9} + p_0$

(a)

	$X_2 = bb$	$X_2 = Bb$	$X_2 = BB$
$X_1 = aa$	$\frac{1}{9} + s(y)$	$\frac{1}{9} - \frac{s(y)}{2}$	$\frac{1}{9} - \frac{s(y)}{2}$
$X_1 = Aa$	$\frac{1}{9} - \frac{s(y)}{2}$	$\frac{1}{9} + s(y)$	$\frac{1}{9} - \frac{s(y)}{2}$
$X_1 = AA$	$\frac{1}{9} - \frac{s(y)}{2}$	$\frac{1}{9} - \frac{s(y)}{2}$	$\frac{1}{9} + s(y)$

(b)

on the sample size n for different values of the parameter p. The larger the value of p, the larger the value of II. Observe that when II is large, it is more likely to reject the null hypothesis. When II is very small, all methods fail to detect interactions for considered sample sizes. The proposed HYBRID procedure is a winner. We should also note that HYBRID is considerably faster than IIPERM. Interestingly, ANOVA does not work for model M1 which is due to the incorrect specification of the linear regression model. For model M1, Y and a pair (X_1, X_2) are non-linearly dependent. The above example shows that there are interesting dependence models in which interactions are detected by II whereas they are undetected by ANOVA. It should be also noted that HYBRID method outperforms IIPERM, which indicates that it is worthwhile to use chi-squared test when the dependence between X_1 and X_2 is weak and the permutation test otherwise. The results for other simulation models are presented in Supplement.

5 Real Data Analysis

5.1 Analysis of Pancreatic Cancer Data

We carried out experiments on publicly available SNP dataset, related to pancreatic cancer [22]. Our aim was to detect interactions associated with the occurrence of cancer. Dataset consists of 230 observations (141 with cancer, 89 controls) and 1189 variables (genetic markers). Each variable takes three possible values (two homozygous variants and one heterozygous). For the considered data we evaluated all pairs of SNPs using the HYBRID procedure described in Sect. 3.3. There were 144158 pairs. We obtained 144158 SNP pairs for which II was computed and out of those 1494 were found significantly positive after Bonferroni correction. In HYBRID method, IICHI option was used for 83% of pairs. In the following we present an analysis of the pair corresponding to the most significant interaction found between SNPs: rs209698, rs2258772. Figure 4 shows the results for this pair. The left hand side plot visualizes the unconditional joint distribution of two SNPs and the remaining two plots correspond to the conditional probabilities. In this case the dependence between SNPs is relatively weak- the mutual information $I(X_1, X_2)$ equals 0.02. On the other hand, the conditional dependence is much stronger which means that the conditional mutual information $I(X_1, X_2|Y)$ being an averaged mutual information of conditional distributions equals 0.14. So interaction information $II(X_1, X_2, Y) = 0.14 - 0.02 = 0.12$ (values rounded to two decimal places). In this example, the main effects are very small. We have that $I(X_1, Y) = 0$, $I(X_2, Y) = 0.02$ and $I((X_1, X_2), Y) = 0.14$, which yields $II = II(X_1, X_2, Y) = 0.14 - 0.02 - 0 = 0.12$. It seems that the information about Y contained in (X_1, X_2) is much larger than information contained in individual variables X_1 and X_2 and joint information about (X_1, X_2) is found to be important in Y prediction. The presented results indicate three important issues: (1) the structure of dependence in subgroups (cases and controls) differs significantly; (2) the conditional dependencies are stronger than the unconditional one; (3) the presence of interaction between variables X_1 and X_2 in predicting Y. Namely, using certain combinations of values of X_1 and X_2 it is possible to predict Y (occurrence of disease) without error. Namely, knowing the values of SNPs in loci X_1 and X_2 it is possible to predict cancer presence (Y) accurately. Interestingly we realize that $X_1 = AA$ and $X_2 = BB$ implies $Y = cancer$. Similarly, $X_1 = aa$ and $X_2 = BB$ implies $Y = no\ cancer$. Such prediction is impossible based on an individual variable X_1 or X_2. Those detected loci could affect transcription factor binding affinity resulting in disregulation of a target gene expression.

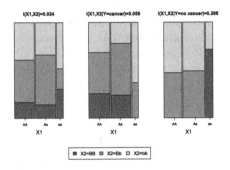

Fig. 4. Distributions of (X_1, X_2). Left figure: joint probabilities for the pair $(X_1, X_2) = (rs209698, rs2258772)$. Middle figure: conditional probabilities given cancer. Right figure: conditional probabilities given no cancer. Mutual information $I(X_1, X_2) = 0.0236$ and conditional mutual information $I(X_1, X_2|Y) = 0.1467$.

5.2   Analysis of Gene Expression Data of CD4+ T Cells

We also applied the proposed method to ImmVar dataset (1250 variables) concerning expression of 236 gene transcripts measured for five stimulation conditions of CD4+ T-cells as well as phenotypic characteristics of 348 doors: Caucasian (183), African-American (91) and Asian (74) ethnicities [23]. We focused on detection of gene-gene interactions that are associated with the specific ethnicity (Y variable). We detected interesting interactions using the HYBRID procedure: (i) Fatty Acid Desaturase 2 (*FADS2*) and Interferon Induced Transmembrane Protein 3 (*IFITM3*); (ii) *IFITM3* and Steroid 5 Alpha-Reductase 3 (*SRD5A3*); (iii) Interferon Induced Transmembrane Protein 1 (*IFITM1*) and *IFITM3*. We have verified that the detected interactions between specific pairs of genes not only make it possible to predict the ethnicity, but also co-participate in biological processes that are known to have various intensity levels in individual ethnicities (see supplement for detailed analysis).

6 Conclusions

In this work we proposed a novel testing procedure which use chi-squared test or permutation test to detect conditional associations, depending on whether the dependence between the variables is weak or not. We showed that the commonly used chi-squared test detects much more false positives than allowed by its nominal significance level. We demonstrated that our method is superior to the standard tests in terms of type I error rate, power and computational complexity. Finally note that standard test IIPERM is computationally expensive and it would be difficult to apply it in the case of really large number of variables. On the other hand, IICHI is fast but it controls type I error rate only for independent X_1 and X_2. In the proposed method HYBRID we use permutation test only when the dependence between independent variables is strong. The experiments on real data indicated that strong dependence occurs relatively rare. Thus the computational complexity of our method is acceptable and unlike IICHI it also controls type I error rate. Future work will include the application of the proposed method on more real datasets related to predicting interesting traits (occurrence of disease, survival times, etc.) using SNPs, gene expression levels and epigenetic regulatory elements. It would be also interesting to compare the proposed method with more model-based approaches. Finally, an interesting challenge is to determine the exact distribution of II under null hypothesis (3) which would allow to avoid using permutation scheme.

References

1. Cordell, H.: Detecting gene-gene interactions that underlie human diseases. Nat. Rev. Genet. **10**(20), 392–404 (2009)
2. Ritchie, M.D., et al.: Multifactor-dimensionality reduction reveals high-order interactions among estrogen-metabolism genes in sporadic breast cancer. Am. J. Hum. Genet. **69**, 138–147 (2001)
3. Nelson, M.R., Kardia, S.L.R., Ferrell, R.R., Sing, C.F.: A combinatorial partitioning method to identify multilocus genotypic partitions that predict quantitative trait variation. Genome Res. **11**, 458–470 (2001)
4. Bateson, W.: Mendel's Principles of Heredity. Cambridge University Press, Cambridge (1909)
5. Moore, J.H., Williams, S. (eds.): Epistasis. Methods and Protocols. Humana Press, New York (2015)
6. Taylor, M.B., Ehrenreich, I.M.: Higher-order genetic interactions and their contribution to complex traits. Trends Genet. **31**(1), 34–40 (2015)
7. Frommlet, F., Bogdan, M., Ramsey, D.: Phenotypes and Genotypes. CB, vol. 18. Springer, London (2016). https://doi.org/10.1007/978-1-4471-5310-8

8. McGill, W.J.: Multivariate information transmission. Psychometrika **19**(2), 97–116 (1954)

9. Mielniczuk, J., Teisseyre, P.: A deeper look at two concepts of measuring gene-gene interactions: logistic regression and interaction information revisited. Genet. Epidemiol. **42**(2), 187–200 (2018)

10. Moore, J.H., et al.: A flexible computational framework for detecting, characterizing, and interpreting statistical patterns of epistasis in genetic studies of human disease susceptibility. J. Theor. Biol. **241**(2), 256–261 (2006)

11. Dabrowski, M.J., et al.: Unveiling new interdependencies between significant DNA methylation sites, gene expression profiles and glioma patients survival. Sci. Rep. **8**(1), 4390 (2018)

12. Chanda, P., et al.: Ambience: a novel approach and efficient algorithm for identifying informative genetic and environmental associations with complex phenotypes. Genetics **180**, 1191–1210 (2008)

13. Wan, X., et al.: A fast approach to detecting gene-gene interactions in genome-wide case-control studies. Am. J. Hum. Genet. **87**(3), 325–340 (2010)

14. Jakulin, A., Bratko, I.: Testing the significance of attribute interactions. In: Proceedings of the Twenty-First International Conference on Machine Learning, ICML 2004, p. 52 (2004)

15. Mielniczuk, J., Rdzanowski, M.: Use of information measures and their approximations to detect predictive gene-gene interaction. Entropy **19**, 1–23 (2017)

16. Han, T.S.: Multiple mutual informations and multiple interactions in frequency data. Inf. Control **46**(1), 26–45 (1980)

17. Cover, T.M., Thomas, J.A.: Elements of Information Theory. Wiley Series in Telecommunications and Signal Processing. Wiley-Interscience, Hoboken (2006)

18. Kang, G., Yue, W., Zhang, J., Cui, Y., Zuo, Y., Zhang, D.: An entropy-based approach for testing genetic epistasis underlying complex diseases. J. Theor. Biol. **250**, 362–374 (2008)

19. Yeung, R.W.: A First Course in Information Theory. Kluwer, New York (2002)

20. Ueki, M., Cordell, H.: Improved statistics for genome-wide interaction analysis studies. PLoS Genet. **8**, e1002625 (2012)

21. Agresti, A.: Categorical Data Analysis. Wiley, Hoboken (2003)

22. Tan, A., et al.: Allele-specific expression in the germline of patients with familial pancreatic cancer: an unbiased approach to cancer gene discovery. Cancer Biol. Theory **7**, 135–144 (2008)

23. Ye, C.J., et al.: Intersection of population variation and autoimmunity genetics in human T cell activation. Science **345**(6202), 1254665 (2014)

Analysis of Ensemble Feature Selection for Correlated High-Dimensional RNA-Seq Cancer Data

Aneta Polewko-Klim[1]([✉]) [ID] and Witold R. Rudnicki[1,2,3] [ID]

[1] Institute of Informatics, University of Białystok, Białystok, Poland
anetapol@uwb.edu.pl
[2] Computational Center, University of Białystok, Białystok, Poland
[3] Interdisciplinary Centre for Mathematical and Computational Modelling,
University of Warsaw, Warsaw, Poland

Abstract. Discovery of diagnostic and prognostic molecular markers is important and actively pursued the research field in cancer research. For complex diseases, this process is often performed using Machine Learning. The current study compares two approaches for the discovery of relevant variables: by application of a single feature selection algorithm, versus by an ensemble of diverse algorithms. These approaches are used to identify variables that are relevant discerning of four cancer types using RNA-seq profiles from the Cancer Genome Atlas. The comparison is carried out in two directions: evaluating the predictive performance of models and monitoring the stability of selected variables. The most informative features are identified using a four feature selection algorithms, namely U-test, ReliefF, and two variants of the MDFS algorithm. Discerning normal and tumor tissues is performed using the Random Forest algorithm. The highest stability of the feature set was obtained when U-test was used. Unfortunately, models built on feature sets obtained from the ensemble of feature selection algorithms were no better than for models developed on feature sets obtained from individual algorithms. On the other hand, the feature selectors leading to the best classification results varied between data sets.

Keywords: Random forest · RNA · Feature selection · Ensemble learning

1 Introduction

The high-throughput DNA sequencing techniques produce data with tens of thousands probes and each of them could be potentially relevant for diagnostics, prognosis and therapeutics.

Feature selection (FS) techniques are indispensable tools for filtering out irrelevant variables and ranking the relevant ones in molecular biological investigations [15,28]. The choice of the FS method is very important for further

© Springer Nature Switzerland AG 2020
V. V. Krzhizhanovskaya et al. (Eds.): ICCS 2020, LNCS 12139, pp. 525–538, 2020.
https://doi.org/10.1007/978-3-030-50420-5_39

investigation because it greatly limits number of features under scrutiny, allowing to concentrate on most relevant ones. On the other hand, FS increases the risk of omitting biological important variables.

FS methods are typically divided into three major groups, namely filters, wrappers, and embedded [1]. The bias in the filtering FS methods does not correlate with the classification algorithms, hence they generalise better than the other methods. Nevertheless, it is well known that individual feature selection algorithms are not robust with respect to fluctuations in the input data [22]. Consequently, application of a single FS algorithm cannot ensure optimal modelling results both in terms of predictive performance and stability. This is particularly evident in the integrative analysis of high-dimensional *omics data [18].

There are numerous FS algorithms that are based on different principles and can generate highly variable results for the same data set. The presence of highly correlated features may result in multiple equally optimal set of features and consequently to the instability of FS method [10]. Such instability reduces the confidence in selected features [22] and their usage as diagnostic or prognostic markers. This variability can be to some extent minimised by application of ensemble methods (EFS) that involve combination of different selectors [1].

1.1 Related Work

The ensemble FS can be broadly assigned to one of two classes: homogeneous (the same base feature selector) and heterogeneous (different feature selectors), [1]. Regardless of the class, the output of ensemble FS is given either in a form of a final feature set or as a ranking of features. Therefore some papers focus on the comparison of different strategies for the ordering of these feature subsets [29]. Other researchers are focused on the evaluation of ensembles. Two quantities of interest are the diversity [25] and stability of the feature selection process [20,22]. And though various methods of feature selection have been developed for high-dimensional data, such as high-throughput genomics data, it is still a big challenge to choose the appropriate method for this type of data [15,30].

The stability of FS algorithms for the classification of this type of data has been investigated for instance by Moulos et al. [20] and Dessi and Pes [6]. It was shown, that stability of ensemble feature selection increase only for these FS methods that are intrinsically weak (in term of stability). Shahrjooihaghighi et al. [26] proposed an ensemble FS based on the fusion of five feature selection methods (rank product, fold change ratio, ABCR, t-test, and PLSDA) for more effective biomarker discovery. The methodology for comparing the outcomes of different FS techniques is presented in [5].

Current study is focused on developing and optimisation of a feature selection protocol aiming at identification of biomarkers important for diagnostic of cancers using the results of high-throughput molecular biology experimental methods. It is based on ensemble of four diverse feature selection methods and application of classification algorithm that is used to evaluate quality of the set of features.

The protocol was applied to analyse four human cancer tumor types from The Cancer Genome Atlas (TCGA, https://www.cancer.gov/tcga).

In particular the following detailed issues were explored:

- whether application of ensemble of FS methods gives more stable results than individual algorithms;
- what is the optimal number of variables for individual algorithms and for ensemble;
- whether models built using features returned by ensemble are better than models built using the same number; of variables returned by individual algorithms;
- which feature selection algorithm returns best sets of variables?

The main contributions of the current study are following:

- we present a novel perspective of optimization and evaluation of the feature selection for correlated high-dimensional RNA-Seq cancer data;
- we compare both the predictive performance of models and the stability of selected feature sets in ensemble feature selection with that of individual FS algorithms;
- we show, that performance of feature selection methods vary between data sets even in for very similar data sets;
- we propose to use an ensemble approach as a reference for selecting the method that works best for a particular data set.

2   Materials and Methods

2.1   Data

Four data sets from The Cancer Genome Atlas database that contain RNA-sequencing data of tumor-adjacent normal tissues for various typed of cancer were used. [3,4,8,12,14,31] These data set all include a large number of highly correlated and potentially informative features [21]. The preprocesing of data involved standard steps for RNA-Seq data. First the log2 transformation was performed. Then features with zero and near zero (1%) variance across patients were removed. After preprocessing the datasets contain:

- the primary BRCA dataset: 1205 samples (112 normal and 1093 tumor), 20223 variables;
- the LUAD dataset: 574 samples (59 normal and 515 tumor), 20172 variables;
- the KIRC dataset: 605 samples (72 normal and 533 tumor), 20222 variables,
- the HNSC dataset: 564 samples (44 normal and 520 tumor), 20235 variables.

All data sets are imbalanced, they contain roughly ten times more cancer than normal samples.

2.2 Methods

Filters Used for Feature Selection. The procedure outlined above was applied to four filter FS methods, namely, Mann-Whitney U-test [17], ReliefF [11,13] and MDFS [19,23] in two variants: one-dimensional (MDFS-1D) and two-dimensional (MDFS-2D). Since only the ranking of variables is used in the procedure outlined above no corrections of p-value due to multiple testing were necessary.

U-test is a robust statistical filter that is routinely used in analysis of *omics data. It assigns probability to the hypothesis that two samples corresponding to two decision classes (normal and tumor tissue) are drawn from populations with the same average value. The U-test use the p-value to select and rank the features.

MDFS is a filter, which is based on information theoretical approach, and which can take synergistic effects between variables into account [19,23]. MDFS also uses p-values of the test to rank features. In the current study, 1D and 2D version of MDFS algorithm were used, referred to as MDFS-1D and MDFS-2D, respectively.

ReliefF is a filter that computes ranking of importance for variables in the information system, based on the distances in the small-dimensional subspaces of the system [13]. Two variants of distance between nearest neighbours, namely, *ReliefFexpRank* and *ReliefFbestK* were tested for the current study. Slightly better results were obtained for the former, hence it was used in all subsequent work. This R implementation of algorithm from *CORElearn* package was used [24].

Filter-Based Feature Selection. The individual prediction models in k-fold cross-validation for each of four filter FS methods and data sets were constructed. The feature selection process and the learning process from RNA-Seq data set were realized by using the Algorithm 1.

This algorithm outlined above was repeated for several values of N and it was repeated multiple times, to minimize the effects of random fluctuations. The stability of feature selection was measured by comparing feature sets obtained in multiple runs of the procedure.

Ensemble Feature Selection. The ensemble set of N-top relevant variables was constructed by a union of top-N variables from each filter FS methods, as it is shown in Algorithm 2. The size of the set may vary between N and $4N$ depending on the similarity of the sets returned by individual FS algorithms. All comparisons between feature sets obtained from ensemble and features sets obtained by individual filters were performed on sets with comparable numbers of total variables. For example if union of four top-5 sets resulted in a set with 20 variables it was subsequently compared with other sets containing 20 variables.

Algorithm 1: FS$(l, f, N, D = \{S_1, \ldots, S_k\})$ the filter feature selection algorithm with Random forest classifier

---

    input : Learning method l
              Feature selection method f
              Number of top features N
              Data set $D = \{(y_i, x_i)\}_{i=1}^{M}$ with $V = \{v_1, \ldots, v_p\}$ features
              and with M instances, randomly partitioned into about
              equally-sized folds S_j

1   **output:** Ranked feature sets F_j, $j = 1, \ldots, k$
              Performance estimation metric E

2 **foreach** S_j **do**

3   Define the training set $D_{\backslash j}(V) \leftarrow D(V) \setminus S_j(V)$

4   Perform feature selection on the training set $R_j \leftarrow f(D_{\backslash j}(V))$

5   Remove highly correlated features with ranked list R_j of features

6   Collect the N highest ranked feature set $F_j = \{v_1, \ldots, v_n\}$ with R_j

7   Build the model on the training set $L_j \leftarrow l(D_{\backslash j}(F_j))$

8   Performance estimation: use the trained model L_j on a test set S_j

9 $E \leftarrow \frac{1}{k}\Sigma E_j$

10 **end**

Algorithm 2: EFS$(l, W = \{F_{11}, \ldots, F_{4k}\}, D = \{S_1, \ldots, S_k\})$ the ensemble feature selection algorithm with Random forest classifier

---

    input : Learning method l
              The $4 \times$ k sets of top-N uncorrelated features with 4 filters W_i
              $F_{i,j}, i = 1, \ldots, 4, j = 1, \ldots, k$ with top-N features
              Data set $D = \{(y_m, x_m)\}_{m=1}^{M}$ described with features $F_{i,j}$
              and with M instances, randomly partitioned into about
              equally-sized folds S_n

1   **output:** Collected feature sets C_p, $p = 1, \ldots, k$
              Performance estimation metric E_p

2 **foreach** S_j **do**

3   Collect the union of feature set $C_p = F_{1j} \cup F_{2j} \cup F_{3j} \cup F_{4j}$

4   Define the training set $D_{\backslash n}(C_p) \leftarrow D(C_p) \setminus S_n(C_p)$

5   Build the model on the training set $L_{n,p} \leftarrow l(D_{\backslash n}(C_p))$

6   Performance estimation: $E_{n,p} \leftarrow L_{n,p}(S_n(C_p))$

7 $E_p \leftarrow \frac{1}{k}\Sigma E_{n,p}$

8 **end**

All applied filters provide their own ranking of the features. The U-test and MDFS algorithms rank features by their statistical significance and, the ReliefF algorithm by their performance in classification. Then the joint set of most important variables is created as a union of top-N sets from individual rankings. The ranking within the combined set was not necessary and it was never performed.

Algorithm 2 was repeated several times for different values of N as in the case Algorithm 1. The stability of feature selection was also estimated.

Classification. The quality of the feature-set was evaluated by building a machine learning model using selected features and measuring its quality. To this end the Random Forest [2] algorithm was used. It has been shown that Random Forest is generally reliable algorithm, that works well out-of the box, rarely fails and in most cases returns results that are very close to best achievable for given problem [7]. The quality of model was evaluated using area under ROC curve (AUC). This measure is independent of the balance of classes in the data and does not need any fitting. The scheme of ensemble feature selection and supervised classification is presented in Fig. 1.

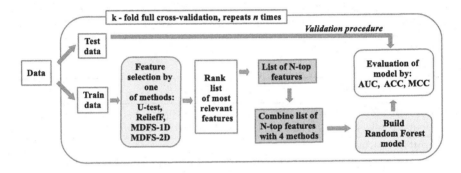

Fig. 1. Pipeline of the ensemble FS method. See notation in text.

Measuring Stability of Feature Selection. The total stability of filter FS method is measured as the average of the pairwise similarity for all pairs of the most informative feature subsets (s_i, s_j) from n runs of a model in full k-fold cross-validation. To this end the Lustgarten's stability measure (ASM) [16], which can be applied to sets of unequal sizes, was used. It is described by the formula:

$$ASM = \frac{2}{c(c-1)} \sum_{i=1}^{c-1} \sum_{j=i+1}^{c} \left(\frac{|s_i \cap s_j| - |s_i| * |s_j| / m}{min(|s_i|, |s_j|) - max(0, |s_i| + |s_j| - m)} \right) \quad (1)$$

where: m is total feature number of dataset and $c = n*k$.

Optimization of Feature Selection. In the first step four threshold levels for defining highly correlated variables were examined to establish threshold leading to best results of classification. Four thresholds levels were tested: $|r| = \{0.7, 0.75, 0.8, 0.9\}$. The subsequent analyses were performed for the optimal threshold level.

The following analyses were performed for each individual FS filter and for ensemble FS filters:

- how many uncorrelated variables should be included in the model to obtain best classification;
- how stable is stability measure for top-N feature subsets;
- whether adding the highly correlated variables to top-N variables influences predictive power.

Entire modelling protocol, including bot feature selection and model building step was performed within $k = 5$ fold cross-validation and was repeated $n = 30$ times, independently for each FS method and data set. Within each cross-validation iteration feature selection algorithm was performed once and then models were trained for all feature set sizes $N = \{5, 10, \ldots, 200\}$.

Analysis was performed using the R (version 3.5) [27] and R/Bioconductor packages [9].

3   Results and Discussion

3.1   Model Accuracy

In the first step, the impact of correlation between informative features on the predictive power of RF model was examined. The results of this analysis are displayed in Fig. 2. It can be seen that squares corresponding to correlation threshold 0.7 in many cases fall bellow other lines on the AUC plots. Therefore threshold for removal of highly correlated variables was set at Spearman's correlation coeffcient r higher than 0.75. This value of coefficient is applied in the subsequent analysis. One may note, that MDFS-1D filter is the most robust with respect to change in the feature level correlation among the applied FS methods.

At the next stage of the analysis, the accuracy of models built using top-N features was examined, see Fig. 3. The number of variables for ensemble model is obtained as the average number of variables in the union of top-N variables from all FS methods averaged over 150 cross-validated sets. Generally, the performance of the models is poor for the smallest sizes of variable sets but increases rapidly with increasing number of variables, reaching plateau after roughly 40 variables are included. However, there are notable exceptions. In particular for BRCA the best model was obtained with 20 variables returned by U-test. Even stronger effect was obtained for KIRC data set. Here the MDFS-2D feature selection leads to clearly best results at 25 variables, whereas the best results for other filters are obtained with 15 variables.

(a) U-test   (b) MDFS-1D   (c) MDFS-2D   (d) ReliefF

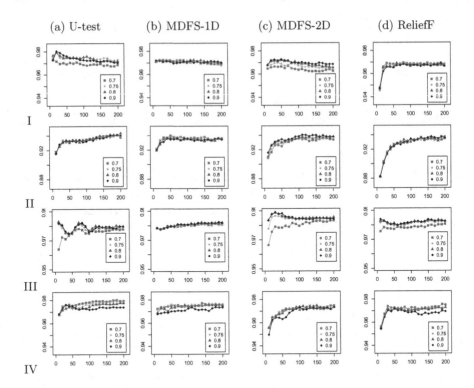

Fig. 2. AUC vs N-top biomarkers for different value of Spearman's rank correlation coefficient. Results for BRCA, HNSC, KIRC and LUAD data sets are displayed in rows I, II, III, and IV, respectively.

Relative performance of models developed using different FS algorithms vary significantly between data sets. For example, the MDFS-2D is clearly the best feature selector for KIRC, and that strongly suggest that non-trivial synergies between variables are present in this data set. MDFS-1D is best feature selector for HNSC data set, U-test is best for BRCA, both algorithms are similarly good for LUAD.

In all cases, the AUC values of models built using variables returned by an ensemble of FS algorithms are comparable to individual models built with a similar number of variables, the AUC curves of the ensemble models (full circle points) are located roughly in the middle of other models as shown in Fig. 3.

Only for BRCA data set for the number of variables larger than 40 the performance of model built on ensemble variables is comparable with the best model for the individual data set, which in this case is a model built using variables from U-test.

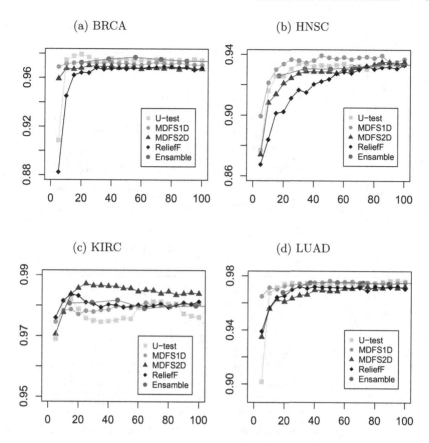

Fig. 3. AUC for models built using top-N variables.

In the next step the effect of adding back redundant variables was examined. To this end the RF models were built for the sets of variables consisting of uncorrelated top-N variables and all informative variables highly correlated with them that were previously removed from feature rank list. The results are displayed in Fig. 4. Clearly adding redundant variables to the main feature set does not improve classification results in most cases. An exception are models built using variables obtained with the MDFS-2D method for BRCA, HNSC and KIRC data. This effect may arise due to inclusion of correlated variables which interact synergistically with other variables in a slightly different way than those previously included.

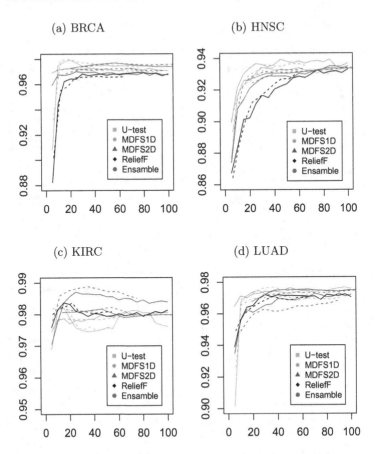

Fig. 4. AUC for models built using top-N variables. Solid lines correspond to models built using top-N uncorrelated variables. Dashed lines correspond to models built using top-N uncorrelated variables and variables correlated with them.

3.2   Stability of Variable Sets

Often the important property of a feature selection method is stability or robustness of the selected features to perturbations in the data. This is particularly important for identification of prognostic or diagnostic markers. Therefore the sensitivity of feature selection algorithms to variations in the training sets that arise in the cross-validation were examined. The similarity between 150 feature subsets obtained in 150 iterations of cross-validation were measured using the Lustgarten's index ASM, see Fig. 5. The highest stability was obtained for variables selected with U-test. For this FS method the value of the ASM index varies between 0.7 and 0.8. The remaining FS methods are much less stable, with least stable MDFS-2D for which the ASM index is generally below 0.2 and most stable ReliefF for which ASM index varies between 0.3 and 0.5. The difference in stability between algorithms is due to the differences in approach used

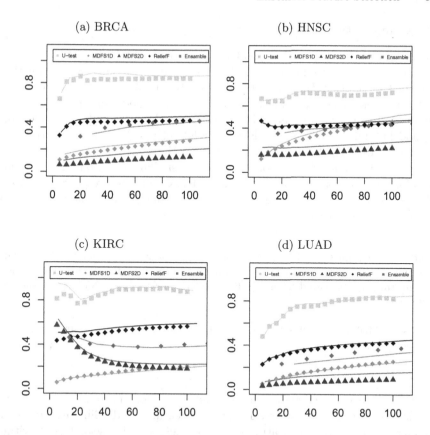

Fig. 5. Clock-wise the average similarity (ASM) between 150 feature subsets for top-N variables. Dotted lines correspond to sets consisting from top-N variables. Solid lines correspond to sets consisting from top-N variables and variables highly correlated with them.

by the algorithm. The U-test is a deterministic algorithm, for which differences arise exclusively due to variation of sample composition. On the other hand all other remaining algorithms rely on randomisation, hence increased variance can be expected. In most cases the stability increases with increasing number of variables. The notable exception is the MDFS-2D algorithm for KIRC data set, where relatively high stability (ASM > 0.5) is achieved for smallest size of feature set, and then it rapidly drops with increasing number of variables. This result is obtained for the same data set, where models built on feature sets returned by MDFS-2D have highest predictive quality. This result strongly suggest existence of a small core of most relevant variables, that must be present in nearly all cases and that strongly contribute to classification. This small core is augmented by a diverse group of loosely correlated relevant but redundant variables. Finally, in most cases adding redundant variables increases the stability of feature subsets, but the difference is small.

3.3   Computational Aspects

The training time does not depend on the type of molecular data, such as microarray gene expression data or DNA copy number data. Execution time of the task depends on the size of the dataset, the number of training iterations, the feature selection algorithm, as well as the CPU model or the GPU model. The most time consuming individual step of the algorithm is feature selection, the model building with Random Forest is relatively quick. However, 150 distinct Random Forest were built using the results of the same feature selection step. Therefore total time of both components was similar. The example execution times for a single iteration of algorithm for KIRC data set are presented in the Table 1. Among feature selection algorithms used in the study, the ReliefF is by far the most time-consuming.

Table 1. Execution times for a single iteration of the algorithm for the KIRC data set. Computations performed on a CPU Intel Xeon Processor E5-2650v2. The MDFS-2D algorithm was executed using a GPU-accelerated version on NVIDIA Tesla K80 co-processor.

U-test	MDFS-1D	MDFS-2D	ReliefF	Ensemble	RF ×1	RF × 100	Total
00 m:37 s	00 m:04 s	00 m:03 s	05 m:41 s	05 m:54 s	00 m:03 s	05 m:19 s	10 m:13 s

The single run of the algorithm involved calling four FS methods, removing correlated features, producing a ranking of the features, and calling RF classification algorithm 150 times (5 feature sets × 20 sizes of feature set). The algorithm was executed 150 times, computations for one data set took about 25 h of CPU time.

4   Conclusions

The current study demonstrates that relying on a single FS algorithm is not optimal. Different FS algorithms are best suited for identification of the most relevant feature sets in various data sets. Combining variables from multiple FS algorithms into a single feature set does not improve the performance in comparison with an equally numerous feature set generated by the individual algorithm that is best suited for the particular data set. On the other hand, application of multiple algorithms increases the chances of identifying the best FS algorithm for the problem under scrutiny. In particular, application of the FS algorithms that can detect synergies in the data can significantly improve the quality of machine learning models.

Interestingly, the stability of a FS algorithm is not required for building a highly predictive machine learning models. This is possible, since biological systems often contain multiple informative variables. Therefore, useful models can be obtained using very diverse combinations of predictive variables.

Notes

Acknowledgements. This work was supported by the National Science Centre, Poland in frame of grant Miniatura 2 No. 2018/02/X/ST6/02571.

References

1. Bolón-Canedo, V., Alonso-Betanzos, A.: Ensembles for feature selection: a review and future trends. Inf. Fus. **52**, 1–12 (2019)
2. Breiman, L.: Random forests. Mach. Learn. **45**(5), 5–32 (2001)
3. Ciriello, G., Gatza, M.L., Beck, A.H., Wilkerson, M.D., et al.: Comprehensive molecular portraits of invasive lobular breast cancer. Cell **163**(2), P506–P519 (2015)
4. Collisson, E., Campbell, J., Brooks, A., et al.: Comprehensive molecular profiling of lung adenocarcinoma. Nature **511**, 543–550 (2014)
5. Dessí, N., Pascariello, E., Pes, B.: A comparative analysis of biomarker selection techniques. BioMed Res. Int. **2013**, 387673 (2013)
6. Dessì, N., Pes, B.: Stability in biomarker discovery: does ensemble feature selection really help? In: Ali, M., Kwon, Y.S., Lee, C.-H., Kim, J., Kim, Y. (eds.) IEA/AIE 2015. LNCS (LNAI), vol. 9101, pp. 191–200. Springer, Cham (2015). https://doi.org/10.1007/978-3-319-19066-2_19
7. Fernández-Delgado, M., Cernadas, E., Barro, S., Amorim, D.: Do we need hundreds of classifiers to solve real world classification problems? J. Mach. Learn. Res. **15**, 3133–3181 (2014)
8. Hammerman, P., Lawrence, M., Voet, D., et al.: Comprehensive genomic characterization of squamous cell lung cancers. Nature **489**, 519–525 (2012)
9. Huber, W., Carey, V., Gentleman, R., Anders, S., et al.: Orchestrating high-throughput genomic analysis with Bioconductor. Nat. Methods **12**(2), 115–121 (2015)
10. Kamkar, I., Gupta, S.K., Phung, D., Venkatesh, S.: Exploiting feature relationships towards stable feature selection. In: 2015 IEEE International Conference on Data Science and Advanced Analytics (DSAA), pp. 1–10 (2015)
11. Kira, K., Rendell, L.: The feature selection problem: Traditional methods and a new algorithm. AAAI **2**, 129–134 (1992)
12. Koboldt, D., Fulton, R., McLellan, M., et al.: Comprehensive molecular portraits of human breast tumours. Nature **490**, 61–70 (2014)
13. Kononenko, I.: Estimating attributes: analysis and extensions of RELIEF. In: Bergadano, F., De Raedt, L. (eds.) ECML 1994. LNCS, vol. 784, pp. 171–182. Springer, Heidelberg (1994). https://doi.org/10.1007/3-540-57868-4_57
14. Lawrence, M., Sougnez, C., Lichtenstein, L., et al.: Comprehensive genomic characterization of head and neck squamous cell carcinomas. Nature **517**, 576–582 (2015)
15. Liang, S., Ma, A., Yang, S., Wang, Y., Ma, Q.: A review of matched-pairs feature selection methods for gene expression data analysis. Comput. Struct. Biotechnol. J. **16**, 88–97 (2018)
16. Lustgarten, J.L., Gopalakrishnan, V., Visweswaran, S.: Measuring stability of feature selection in biomedical datasets. In: AMIA Annual Symposium Proceedings, pp. 406–410. AMIA (2009)
17. Mann, H., Whitney, D.: Controlling the false discovery rate: a practical and powerful approach to multiple testing. Ann. Math. Stat. **18**(1), 50–60 (1947)

18. Meng, C., Zeleznik, O.A., Thallinger, G.G., et al.: Dimension reduction techniques for the integrative analysis of multi-omics data. Briefings Bioinf. **17**(4), 628–641 (2016)
19. Mnich, K., Rudnicki, W.: All-relevant feature selection using multidimensional filters with exhaustive search. Inf. Sci. **524**, 277–297 (2020)
20. Moulos, P., Kanaris, I., Bontempi, G.: Stability of feature selection algorithms for classification in high-throughput genomics datasets. In: 13th IEEE International Conference on BioInformatics and BioEngineering, pp. 1–4 (2013)
21. Peng, L., Bian, X.W., Li, D.K., et al.: Large-scale RNA-Seq transcriptome analysis of 4043 cancers and 548 normal tissue controls across 12 TCGA cancer types. Sci. Rep. **5**(1), 1–18 (2015)
22. Pes, B.: Ensemble feature selection for high-dimensional data: a stability analysis across multiple domains. Neural Comput. Appl. **32**(10), 5951–5973 (2019). https://doi.org/10.1007/s00521-019-04082-3
23. Piliszek, R., Mnich, K., Migacz, S., et al.: MDFS: multidimensional feature selection in R. R J. **11**(1), 197–210 (2019)
24. Robnik-Sikonja, M., Savicky, P.: CORElearn: classification, regression and feature evaluation, R package version 1.54.1 (2018). https://CRAN.R-project.org/package=CORElearn
25. Seijo-Pardo, B., Bolón-Canedo, V., Alonso-Betanzos, A.: Testing different ensemble configurations for feature selection. Neural Process. Lett. **46**(3), 857–880 (2017)
26. Shahrjooihaghighi, A., Frigui, H., Zhang, X., Wei, X., Shi, B., Trabelsi, A.: An ensemble feature selection method for biomarker discovery. In: 2017 IEEE International Symposium on Signal Processing and Information Technology (ISSPIT), pp. 416–421 (2017)
27. Team, R.C.: R: A language and environment for statistical computing. R Foundation for Statistical Computing (2017). https://www.R-project.org/
28. Vanjimalar, S., Ramyachitra, D., Manikandan, P.: A review on feature selection techniques for gene expression data. In: 2018 IEEE International Conference on Computational Intelligence and Computing Research (ICCIC), pp. 1–4. IEEE, ICCIC (2018)
29. Wang, J., Xu, J., Zhao, C., Peng, Y., Wang, H.: An ensemble feature selection method for high-dimensional data based on sort aggregation. Syst. Sci. Control Eng. **7**(2), 32–39 (2019)
30. Wenric, S., Shemirani, R.: Using supervised learning methods for gene selection in RNA-seq case-control studies. Front. Genet. **34**(4), 301–312 (2018)
31. Zhou, Y., Zhou, B., Pache, L., Chang, M., et al.: Metascape provides abiologist-oriented resource for the analysis of systems-level datasets. Nat. Commun. **10**(1), 1–10 (2019)

Biological Network Visualization for Targeted Proteomics Based on Mean First-Passage Time in Semi-Lazy Random Walks

Tomasz Arodz[(✉)]

Department of Computer Science, Virginia Commonwealth University,
Richmond, VA 23284, USA
tarodz@vcu.edu

Abstract. Experimental data from protein microarrays or other targeted assays are often analyzed using network-based visualization and modeling approaches. Reference networks, such as a graph of known protein-protein interactions, can be used to place experimental data in the context of biological pathways, making the results more interpretable. The first step in network-based visualization and modeling involves mapping the measured experimental endpoints to network nodes, but in targeted assays many network nodes have no corresponding measured endpoints. This leads to a novel problem – given full network structure and a subset of vertices that correspond to measured protein endpoints, infer connectivity between those vertices. We solve the problem by defining a semi-lazy random walk in directed graphs, and quantifying the mean first-passage time for graph nodes. Using simulated and real networks and data, we show that the graph connectivity structure inferred by the proposed method has higher agreement with underlying biology than two alternative strategies.

Keywords: Biological networks · Random walks · Node influence

1 Introduction

Profiling experiments involving gene or protein microarrays or assays based on next-generation sequencing have become a standard approach for gaining new knowledge about biological processes and pathologies. Mining the resulting data for patterns of interest, for example differences between phenotypes, can be done with purely data-driven statistical and machine learning methods [11] that perform the discovery *de novo*. But approaches that make use of existing knowledge about biological networks in analyzing profiling data are, in principle, better suited to deal with the complexity of biological systems.

Extensive knowledge has been gathered about physical or functional interactions between biological entities of many types. For example, it may be known

© Springer Nature Switzerland AG 2020
V. V. Krzhizhanovskaya et al. (Eds.): ICCS 2020, LNCS 12139, pp. 539–549, 2020.
https://doi.org/10.1007/978-3-030-50420-5_40

that a certain kinase phosphorylates a specific protein, or that a particular microRNA interacts physically with mRNA transcript of a gene, and in effect the protein encoded by the gene is not expressed. All known interactions of a specific type, taken together, form reference networks, such as a protein-protein interaction network or a gene regulatory network. A reference network describes the interaction potential of a given species. In a specific phenotype, that is, a specific tissue in a specific condition, only some of the interactions actually take place.

Based on the network topology and measurements from profiling experiments, dynamic behavior of the system can be modeled using stochastic Petri nets [12], Boolean networks [14], Bayesian networks [31], or systems of differential equations [5]. Pathways that can discriminating between phenotypes can be discovered by mapping expression data on reference networks and performing a bottom-up [7,13] or top-down [2,6] search. Depicting experimental results visually by mapping the up-regulated or down-regulated genes or proteins can also make the data more interpretable to biologists.

2   Motivation and Problem Statement

Network-based analysis involves mapping the measured experimental endpoints to nodes in the reference network. Often, many nodes will have no corresponding endpoints. This is particularly true in studies that involve targeted assays. For example, only a limited number of antibodies are available and validated for use with the reverse phase protein array (RPPA) immunoassay [22]. Thus, for many nodes in a reference protein-protein interaction network, experimental protein levels will not be measured.

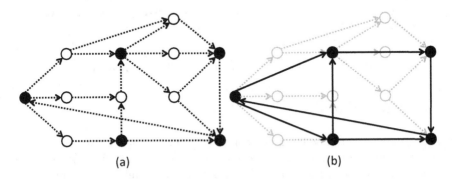

(a)   (b)

Fig. 1. Illustration of the problem. (a) Input: a reference biological network and a subset of nodes (filled black nodes) for which measurements from a targeted assay are available; measurements for the other nodes are not available. Typically the measured nodes are a small fraction of all nodes in the network. (b) Output: an informative and interpretable network (black edges) connecting the measured nodes.

Some network analysis methods can deal with lack of measurements at a subset of nodes. Prize-collecting Steiner tree and related approaches [2] can involve unmeasured nodes in finding discriminative pathways, but these approaches limit the results to a tree or forest structure. They also limit the computational methods used for differential analysis at individual nodes to univariate statistical tests, and are not applicable to directed graphs, such as kinase-substrate network that describes protein signaling. Graphical models such as Bayesian networks [31] can in principle deal with unmeasured nodes, but their applicability is limited by their computational complexity.

Other algorithms for network-oriented analysis typically assume that all nodes in the reference network correspond to measured endpoints or, conversely, that the reference network provides direct edges between the measured nodes and does not leave nodes only connected to unmeasured nodes. Simply eliminating all the unmeasured nodes from consideration is a poor option as it fragments the reference network and leaves many nodes unconnected. The alternative simple approach of adding a direct edge between all pairs of measured nodes that are connected by a path in the reference network will result in a dense graph where connections lose their specificity. Data-driven network inference algorithms commonly used to predict regulatory networks based on gene expression [19] can provide a graph linking the measured nodes, but then the network is grounded in data and not in existing biological knowledge, which prevents its use as an additional source of information to complement the experimental results.

Finding a network of connections between measured nodes based only on a given topology of a larger network consisting of nodes with and without experimental measurements is thus a non-trivial new problem. It can be stated in the following way. Given a reference directed network $G = (V, E)$ and a subset of measured nodes $W \subset V$ that is often much smaller than V, find a new network $G_W = (W, E')$ that captures best the biological processes, such as regulation or signaling, described by G. The new graph G_W should not be based on experimental data, but only on the set of known interactions represented by edges of the original graph G. A graphical illustration of the problem is shown in Fig. 1.

Since the signal in molecular networks can spread through many paths, we want to take them into account in a way that avoids making the graph too dense. In a simple strategy for addressing the problem, we could place a direct edge from measured node $i \in W$ to measured node $j \in W$ in the output network G_W if and only if the underlying network G has a path connecting i to j that does not pass through any other measured node from W. Another strategy could place an edge from i to j in G_W when that unmeasured path is a shortest path. Here, we propose a method that produces networks that are easier to visualize and more interpretable than networks produced by these simple strategies, and at the same time have higher agreement with the underlying biology.

3   Proposed Method

In the proposed method, we treat each measured node in the network as a source and aim to find other measured vertices that would, in the new network being constructed, serve as targets of direct edges from that source measured node. Specifically, for a given source measured node, our goal is to identify a group of measured nodes that are hit first as the signal from the source spreads in the reference network. Those nodes will be connected directly to the source. On the other hand, measured nodes that are reachable from the source but most of the signal passing to them traverses first through other measured nodes will not be connected to the source directly. This intuition leads to a solution that is based on mean first-passage times in a semi-lazy random walk on a directed graph.

3.1 Mean First-Passage Time in Directed Graphs

The mean first-passage time $H(i,j)$, known also as the expected hitting time, from node i to j in a strongly connected, directed graph is defined as the expected number of steps it takes for a random walker starting from node i to reach node j for the first time, where the walk is Markov chain defined by transition probabilities resulting from the graphs connectivity. The average is taken over the number of transitions, that is, lengths L of all paths $s_{(i \to j)}$ from i to j that do not contain a cycle involving j, with respect to probabilities P of the paths:

$$H(i,j) = \sum_{s_{(i \to j)}} P(s_{(i \to j)}) L(s_{(i \to j)}). \tag{1}$$

Compared to the shortest path from i to j, the mean first-passage time includes multiple paths and node degrees into consideration. For example, paths through hub nodes increase H, since the walker has high probability of moving to nodes connected to the hub that are not on the shortest path to the target.

The study of mean first-passage time on various domains has long history in physics [24]. It has also been well-characterized for undirected graphs [10]. Recently, it has been shown that for directed graphs, mean first-passage time $H(i,j)$ can be obtained analytically in close form given the Laplacian matrix and node stationary probabilities in a random walk in the graph [4]. More specifically, let A be the, possibly weighted, adjacency matrix of the input strongly connected, directed graph, D a diagonal matrix of node out-degrees, and I an identity matrix. Then, the expected hitting time can be calculated as [4]:

$$H(i,j) = M(j,j) - M(i,j) + \sum_{k \in V} (M(i,k) - M(j,k))\pi(k). \tag{2}$$

where $\Pi = Diag(\pi)$ is the matrix of node stationary probabilities, $P = D^{-1}A$ captures node transition probabilities, and $M = L^{+}$ is defined as the Moore-Penrose pseudo-inverse of the assymetric graph Laplacian $L = \Pi(I - P)$.

3.2 Semi-Lazy Random Walk and Mean First-Passage Time

Assume we have an unweighted strongly connected directed graph $G = (V, E)$ with two types of nodes, $V = U + W$. Nodes in U are regular nodes, which do not affect the behavior of a random walker in the graph. On the other hand, upon arriving at a node from W, the random walker is trapped. In each subsequent step the walker remains at the node with probability γ. That is, in each time step, the walker has probability $1 - \gamma$ of escaping the trap and continuing with the walk through other nodes[1]. This bears resemblance to a lazy random walk, in which the random walker stays at a node with probability $\frac{1}{2}$ or, more generally, with some fixed probability. Here, we call the walk semi-lazy, since the random walker is lazy only at nodes from W.

In this setting, mean first-passage time no longer depends only on the topology of the graph, but also on whether the paths contain nodes from W or not. We can define the mean first-passage time for a semi-lazy random walk induced by imperfect traps as:

$$
H_{IT}(i,j) = \sum_{s_{(i \to U \to j)}} P(s_{(i \to U \to j)}) L(s_{(i \to U \to j)})
$$
$$
+ \sum_{s_{(i \to M \to j)}} P(s_{(i \to W \to j)})[L(s_{(i \to W \to j)})
$$
$$
+ \Delta(s_{(i \to W \to j)})], \tag{3}
$$

where $s_{(i \to U \to j)}$ is any path from i to j that goes only through regular nodes from U and $s_{(i \to W \to j)}$ is any path that includes at least one trap from set W, and Δ is a stochastic penalty function depending on the number of nodes from the set W on the path. By convention, if $i \in W$ then $H_{IT}(i,j)$ is defined as a walk that starts at the point when random walker escapes the trap i, that is, the first step is always a step to some other node.

We calculate H_{IT} for a directed graph with transition probabilities P separately for each starting node i. We create a new transition probability matrix $P'_i = \gamma I_{W,i} + (I - \gamma I_{W,i})P$, where $I_{W,i}$ is a diagonal matrix that has ones for rows and columns corresponding to $W \setminus \{i\}$ and zeros elsewhere. The Markov chain specified by P'_i is irreducible and aperiodic for a strongly connected graph G. Based on P'_i, we calculate node stationary probabilities and the Moore-Penrose pseudo-inverse of the graph Laplacian, and then use Eq. 2 to obtain $H_{IT}(i,j)$ for each j.

3.3 Connectivity Between Measured Endpoints in Biological Networks

Given a reference biological network and experimental data, we equate the set of traps W with the nodes for which we have experimental measurements and the set U with all other nodes. In this way, if most of the paths from i to j lead through other measured nodes, the mean first-passage time will be much higher than if the paths lead only through non-measured nodes.

First, for every measured node $i \in W$, we calculate $H_{IT}(i,j)$ to all measured nodes $j \in W$. We ignore hitting times from or to non-measured nodes in U.

---

[1] We set the default value of γ to 0.99.

Prior to the calculation of the values $H_{IT}(i, \cdot)$ for starting node i, we eliminate all nodes from U that do not lie on any path from i to any node in W, because either they cannot be reached from i, or they do not have a path to any node in W. Also, if there are dangling nodes, that is, nodes with null out-degree, we add a connection from those nodes to i, allowing the walker to continue the walk.

Once $H_{IT}(i, j)$ is calculated for every $i, j \in W$, we treat H_{IT} as a weighted adjacency matrix and calculate shortest paths $\sigma(i, j)$ for $i, j \in W$. Finally, we create the output graph G_W by keeping edges $i \to j$ for which there is no shorter path in H_{IT} than the direct edge:

$$\forall i, j \in W: \quad G_W(i, j) = 1 \quad \text{iff} \quad H_{IT}(i, j) = \sigma(i, j). \tag{4}$$

In effect, we place a direct edge from i to j if the random walker starting from i tends to avoid other nodes from W on its way towards hitting j. If some other node k is often encountered during the $i \to j$ walk, then $H_{IT}(i, k) + H_{IT}(k, j) < H_{IT}(i, j)$ since the trap at k is not considered when quantifying hitting times $H_{IT}(i, k)$ and $H_{IT}(k, j)$, but it is considered when estimating $H_{IT}(i, j)$. In this way, the new graph G_W will contain only edges between measured nodes, and the edge structure will be based on connectivity in the original reference network in a way that keeps connections through measured nodes explicit and avoids indirect connections.

The computational complexity of the proposed method is $\mathcal{O}(|W||V|^3)$. For each node $i \in W$, calculating $H_{IT}(i, \cdot)$ involves finding the pseudoinverse of the Laplacian and estimating the stationary node probabilities, which both are $\mathcal{O}(|V|^3)$. Calculations for different i can be done independently in parallel, and need to be followed by all-pairs shortest path involving W nodes, which requires $\mathcal{O}(|W|^3)$. In effect, the method can be successfully applied to biological networks, which have on the order of 10^4 nodes or less.

4   Experimental Validation

We evaluated our method by comparing it with two alternative strategies. In the connectivity-based strategy, we place an edge from measured node i to measured node j in the output network if and only if the underlying network has a path connecting i to j that does not pass through any other measured node. In the shortest-paths-based strategy, we place an edge from i to j in the output network when a shortest path from i to j in the underlying network does not pass through any other measured node.

To compare the quality of the network of measured nodes resulting from the proposed method and a network returned by an alternative strategy, we used expression data measured over a set of samples. In both networks each node is associated with vector of expression values of a corresponding gene or protein. For each edge $i \to j$ in both networks, we calculated the p-value of the correlation between expression vector associated with i and the expression vector associated with j. We treat the correlation as an imperfect but easy to obtain surrogate measure of edge quality. We assume that when comparing two graphs inferred

Table 1. Comparison of the proposed approach with two alternative strategies for 5 simulated and 1 real-world dataset. Columns are: $\overline{\mathbf{R}}$: mean -log(p-value) of correlation between endpoints connected by edges from set $\mathbf{R}$, that is, present in results of an alternative strategy and retained by our method; $\overline{\mathbf{F}}$: mean -log(p-value) of correlation between endpoints connected by edges from set $\mathbf{F}$, that is, present in results of an alternative strategy but filtered out by our method; p-value for a test if the means of the negated log-transformed p-values in $\mathbf{R}$ are higher than in $\mathbf{F}$, that is, if the expression profiles for nodes linked by retained edges are more highly correlated than for nodes linked by filtered out edges; #$\mathbf{R}$: number of edges in $\mathbf{R}$; #$\mathbf{F}$: number of edges in $\mathbf{F}$.

Comparison with connectivity-based strategy					
Dataset	$\overline{\mathbf{R}}$	$\overline{\mathbf{F}}$	p-value	#$\mathbf{R}$	#$\mathbf{F}$
DREAM 4 I	2.89	1.28	3.47e−4	1861	165
DREAM 4 II	2.93	0.98	3.78e−6	2071	184
DREAM 4 III	4.52	2.35	5.11e−18	3734	1442
DREAM 4 IV	3.36	1.52	7.10e−39	3602	2345
DREAM 4 V	4.87	2.84	1.69e−8	2482	386
TCGA BRCA	13.52	10.55	0.0110	156	533
Comparison with shortest-path-based strategy					
Dataset	$\overline{\mathbf{R}}$	$\overline{\mathbf{F}}$	p-value	#$\mathbf{R}$	#$\mathbf{F}$
DREAM 4 I	2.89	1.22	4.57e−4	1858	147
DREAM 4 II	2.93	0.98	1.47e−5	2068	161
DREAM 4 III	4.54	2.37	1.44e−14	3706	1058
DREAM 4 IV	3.37	1.46	1.86e−31	3580	1689
DREAM 4 V	4.87	2.88	1.69e−6	2473	283
TCGA BRCA	13.52	10.67	0.0364	156	240

from the same reference network G without looking at expression data, the one that has higher correlation between expression of genes or proteins linked by the graph edges represents the underlying signaling or regulation better.

Since edges detected by our method are a subset of edges detected by the alternative strategies, we partitioned the edge p-values into two groups. In the retained edges group, $\mathbf{R}$, we put the p-values of edges that are found both by the proposed method and by the alternative strategy used for comparison. In the filtered-out edges group, $\mathbf{F}$, we put the p-values of edges detected only by the alternative method but not by the proposed method. Then, we tested if the mean of p-values in the $\mathbf{R}$ group is lower than mean in the $\mathbf{F}$ group, that is, if the proposed method is effective at filtering out low p-value edges.

4.1   Simulated Data

In our evaluation, we used simulated expression data, for which the expression profiles come from known, pre-specified networks connecting genes, and are simulated using a system of differential equations. We used networks and data from GeneNetWeaver [25] available as part of the DREAM 4 In Silico Network Challenge [18]. We used the five multifactorial directed networks from the challenge, each with 100 nodes. Each network is accompanied by simulated expression data for the 100 endpoints in 100 samples. To test our method, we randomly picked 20 nodes, kept their expression data, and ignored the expression data for the remaining 80 nodes. The task for our method is to connect those 20 nodes based on the known network of all 100 nodes, without looking at the data. We repeated the experiment 100 times with different random samples of measured nodes, and grouped the p-values for the discovered edges together.

We carried out the above procedure independently for each of the 5 networks available in DREAM 4. The results are presented in Table 1. In each of the simulated networks, the p-values of the edges retained by our methods are significantly lower on average than those that are filtered out, compared to edges picked by the two alternative strategies.

4.2   Real-World Data

We validated the proposed methods using protein expression data measured using RPPA assays for 410 samples from breast cancer patients gathered from the Cancer Genome Atlas (TCGA) [15]. As the underlying reference network, we used a recently published directed human phosphorylation network that captures protein signaling [20]. The network has 1191 nodes, of which 69 have corresponding protein measured with reverse phase protein array in the TCGA samples. The task for our method is to connect those 69 nodes based on the known network of all 1191 nodes.

We used the same approach as above to compare the connectivity between the 69 nodes resulting from our methods with the connectivity from the alternative strategies. As shown in Table 1, our method performs significantly better. The number of edges returned by the proposed method is only 156, whereas the connectivity-based strategy returns a dense structure of 689 edges for a 69 node graph, and the shortest-path-based strategy returns 396 edges. As seen in Fig. 2, the network returned by the method is much more interpretable than the network resulting from the strategy of placing an edge between all connected nodes. The pairs of proteomic endpoints connected by the edges retained by the proposed method are on average more highly correlated than those connected by the edges from alternative strategies we filtered out.

5   Discussion

We have proposed an approach for visualizing, in a compact way, biological networks in scenarios when only some subset of nodes has measurements available. Our approach is based on theory of random walks [3]. Random walks have been previously used for estimating influence between nodes in biological networks [1,9,16,27–30,32]. The influence has been defined in terms of a diffusion kernel

Fig. 2. Visualization of human kinase phospohrylation network of 1191 proteins for a dataset of 69 proteins measured using reverse phase protein array. The approach of using direct edges between measured proteins results in a network that only presents 46 out of 69 measured proteins and thus leaves 33% of measured proteins out of the picture (left). The alternative approach of placing an edge between all directly or indirectly connected measured proteins as long as there is a path between them that does not pass through any other measured proteins results in an uninterpretable, dense network with 689 edges (center). The proposed algorithm results in a sparse, interpretable network (right) that connects all 69 measured nodes through a set of 156 direct or indirect connections chosen based on mean first-passage time criterion.

[17], diffusion with loss [23] or a heat kernel [8], but these kernels are defined for undirected graphs, which reduces their use for directed networks such as kinase-substrate protein signaling network or gene regulatory networks. These measures of influence also ignore the time progression associated with the spread of signal in the network, since they are based on the stationary state of the random walk.

The progression of the random walk can be quantified using mean first-passage times for individual nodes. In computational biology, it has been used previously for analyzing state transition graphs in probabilistic Boolean networks to identify genes perturbations that lead quickly to a desired state of the system [26]. Here, we proposed to use it to decide if signaling from one measured node to another measured node typically passes through other measured nodes. This task bears similarities to the problem in physical chemistry of finding reaction paths from a reactant to a product. Mean first-passage time has been used as one way of solving that problem for reactions with continuous or discrete reaction coordinates [21], for example to uncover the path of excitation migration after photon absorption in photosynthetic complex. Our approach could be viewed as an exploration of reaction paths on a cellular scale where the reaction coordinates are nodes in a directed graph.

The analogy between the graph problem explored here and the chemical reaction path detection problem indicates that the mean first-passage time could be an effective way of representing paths in the underlying network by direct edges between measured nodes. Experimental validation using simulated and real networks and data show that this is indeed the case. The uncovered connectivity structures better approximate the underlying biology that other strategies we used for comparison. With the proposed approach, for a specific experimental study, one can obtain a dedicated network that includes only nodes for which experimental data is measured in the study, linked by edges representing causal interactions based on known connections from the reference network. The network can then be used as input for network-oriented analyses, or for compact, interpretable visualization of the relationships between measured nodes.

Acknowledgement. TA is supported by NSF grant IIS-1453658.

References

1. Arodz, T., Bonchev, D.: Identifying influential nodes in a wound healing-related network of biological processes using mean first-passage time. New J. Phys. **17**(2), 025002 (2015)
2. Bailly-Bechet, M., et al.: Finding undetected protein associations in cell signaling by belief propagation. Proc. Natl. Acad. Sci. **108**, 882–887 (2011)
3. Berg, H.C.: Random Walks in Biology. Princeton University Press, Princeton (1993)
4. Boley, D., Ranjan, G., Zhang, Z.L.: Commute times for a directed graph using an asymmetric Laplacian. Linear Algebra Appl. **435**, 224–242 (2011)
5. Chen, T., He, H., Church, G.: Modeling gene expression with differential equations. In: Pacific Symposium on Biocomputing, pp. 29–40 (1999)
6. Chowdhury, S.A., Nibbe, R.K., Chance, M.R., Koyutürk, M.: Subnetwork state functions define dysregulated subnetworks in cancer. In: Berger, B. (ed.) RECOMB 2010. LNCS, vol. 6044, pp. 80–95. Springer, Heidelberg (2010). https://doi.org/10.1007/978-3-642-12683-3_6
7. Chuang, H., Lee, E., Liu, Y., Lee, D., Ideker, T.: Network-based classification of breast cancer metastasis. Mol. Syst. Biol. **3**, 140 (2007)
8. Chung, F.: The heat kernel as the pagerank of a graph. Proc. Natl. Acad. Sci. **104**, 19735–19740 (2007)
9. Cowen, L., Ideker, T., Raphael, B.J., Sharan, R.: Network propagation: a universal amplifier of genetic associations. Nat. Rev. Genet. **18**(9), 551 (2017)
10. Göbel, F., Jagers, A.: Random walks on graphs. Stochastic Process. Appl. **2**, 311–336 (1974)
11. Golub, T.R., Slonim, D.K., Tamayo, P., Huard, C., Gaasenbeek, M., et al.: Molecular classification of cancer: class discovery and class prediction by gene expression monitoring. Science **286**, 531–537 (1999)
12. Goss, P., Peccoud, J.: Quantitative modeling of stochastic systems in molecular biology by using stochastic Petri nets. Proc. Natl. Acad. Sci. **95**, 6750–6755 (1998)
13. Ideker, T., Ozier, O., Schwikowski, B., Siegel, A.F.: Discovering regulatory and signalling circuits in molecular interaction networks. Bioinformatics **18**(S1), S233–S240 (2002)
14. Kauffman, S.: Metabolic stability and epigenesis in randomly constructed genetic nets. Journal of Theoretical Biology **22**, 437–467 (1969)
15. Koboldt, D., Fulton, R., McLellan, M., Schmidt, H., Kalicki-Veizer, J., McMichael, J.: Comprehensive molecular portraits of human breast tumours. Nature **490**, 61–70 (2012)
16. Köhler, S., Bauer, S., Horn, D., Robinson, P.N.: Walking the interactome for prioritization of candidate disease genes. Am. J. Hum. Genet. **82**, 949–958 (2008)

17. Kondor, R.I., Lafferty, J.: Diffusion kernels on graphs and other discrete input spaces. In: International Conference on Machine Learning, pp. 315–322 (2002)
18. Marbach, D., Costello, J., Küffner, R., Vega, N., Prill, R., et al.: Wisdom of crowds for robust gene network inference. Nat. Methods **9**, 797–804 (2012)
19. Margolin, A., et al.: ARACNE: an algorithm for the reconstruction of gene regulatory networks in a mammalian cellular context. BMC Bioinf. **7**(S1), S7 (2006)
20. Newman, R.H., Hu, J., Rho, H.S., Xie, Z., Woodard, C., et al.: Construction of human activity-based phosphorylation networks. Mol. Syst. Biol. **9**, 655 (2013)
21. Park, S., Sener, M.K., Lu, D., Schulten, K.: Reaction paths based on mean first-passage times. J. Chem. Phys. **119**, 1313–1319 (2003)
22. Paweletz, C.P., Charboneau, L., Bichsel, V.E., Simone, N.L., Chen, T., et al.: Reverse phase protein microarrays which capture disease progression show activation of pro-survival pathways at the cancer invasion front. Oncogene **20**, 1981–1989 (2001)
23. Qi, Y., Suhail, Y., Lin, Y.Y., Boeke, J.D., Bader, J.S.: Finding friends and enemies in an enemies-only network: a graph diffusion kernel for predicting novel genetic interactions and co-complex membership from yeast genetic interactions. Genome Res. **18**, 1991–2004 (2008)
24. Redner, S.: A Guide to First-Passage Processes. Cambridge University Press, Cambridge (2001)
25. Schaffter, T., Marbach, D., Floreano, D.: GeneNetWeaver: In silico benchmark generation and performance profiling of network inference methods. Bioinformatics **27**, 2263–2270 (2011)
26. Shmulevich, I., Dougherty, E.R., Zhang, W.: Gene perturbation and intervention in probabilistic Boolean networks. Bioinformatics **18**, 1319–1331 (2002)
27. Tsuda, K., Noble, W.S.: Learning kernels from biological networks by maximizing entropy. Bioinformatics **20**(S1), i326–i333 (2004)
28. Valdeolivas, A., et al.: Random walk with restart on multiplex and heterogeneous biological networks. Bioinformatics **35**(3), 497–505 (2018)
29. Vandin, F., Clay, P., Upfal, E., Raphael, B.J.: Discovery of mutated subnetworks associated with clinical data in cancer. In: Pacific Symposium on Biocomputing, pp. 55–66 (2012)
30. Vandin, F., Upfal, E., Raphael, B.J.: Algorithms for detecting significantly mutated pathways in cancer. In: Berger, B. (ed.) RECOMB 2010. LNCS, vol. 6044, pp. 506–521. Springer, Heidelberg (2010). https://doi.org/10.1007/978-3-642-12683-3_33
31. Yu, J., Smith, V.A., Wang, P.P., Hartemink, A.J., Jarvis, E.D.: Advances to Bayesian network inference for generating causal networks from observational biological data. Bioinformatics **20**, 3594–3603 (2004)
32. Zhang, W., Lei, X.: Two-step random walk algorithm to identify cancer genes based on various biological data. In: 2018 IEEE International Conference on Bioinformatics and Biomedicine (BIBM), pp. 1296–1301 (2018)

Bootstrap Bias Corrected Cross Validation Applied to Super Learning

Krzysztof Mnich[1]([✉])[ID], Agnieszka Kitlas Golińska[2][ID], Aneta Polewko-Klim[2][ID], and Witold R. Rudnicki[1,2,3][ID]

[1] Computational Centre, University of Białystok,
ul. Konstantego Ciołkowskiego 1M, 15-245 Białystok, Poland
k.mnich@uwb.edu.pl
[2] Institute of Informatics, University of Białystok,
ul. Konstantego Ciołkowskiego 1M, 15-245 Białystok, Poland
[3] Interdisciplinary Centre for Mathematical and Computational Modelling,
University of Warsaw, ul. Tyniecka 15/17, 02-630 Warsaw, Poland
https://ii.uwb.edu.pl, https://icm.edu.pl

Abstract. Super learner algorithm can be applied to combine results of multiple base learners to improve quality of predictions. The default method for verification of super learner results is by nested cross validation; however, this technique is very expensive computationally.

It has been proposed by Tsamardinos et al., that nested cross validation can be replaced by resampling for tuning hyper-parameters of the learning algorithms. The main contribution of this study is to apply this idea to verification of super learner. We compare the new method with other verification methods, including nested cross validation.

Tests were performed on artificial data sets of diverse size and on seven real, biomedical data sets. The resampling method, called Bootstrap Bias Correction, proved to be a reasonably precise and very cost-efficient alternative for nested cross validation.

Keywords: Super learning · Cross validation · Bootstrap · Resampling

1 Introduction

Numerous machine learning algorithms with roots in various areas of computer science and related fields have been developed for solving different classes of problems. They include linear models, support vector machines, decision trees, ensemble algorithms like boosting or random forests, neural networks etc [1]. There are also many feature selection techniques used to prepare the input data for the predictive algorithm. Different methods can extract and utilise different parts of the information contained in the data set under scrutiny. It has been shown that in many cases an ensemble machine learning model, which combines several different predictions, can outperform the component learners. This can

© Springer Nature Switzerland AG 2020
V. V. Krzhizhanovskaya et al. (Eds.): ICCS 2020, LNCS 12139, pp. 550–563, 2020.
https://doi.org/10.1007/978-3-030-50420-5_41

Algorithm 1: k-fold cross validation $\mathbf{CV}(f, D)$

input : learning method f,
           data set $D = \{y_j, X_j\}$, $j = 1, \ldots, N$

output: N predictions for the response variable ψ_j, $j = 1, \ldots, N$

split randomly the data set into k almost equally-sized folds F_i
foreach F_i **do**
     define a training set as $D_i = D \setminus F_i$
     learn a predictive model using the training set $M_i \leftarrow f(D_i)$
     apply the model to the remaining subset $\Psi_i \leftarrow M_i(F_i)$
     collect the predictions Ψ_i
end

---

be realised in the form of wisdom of crowds [2] or in a more systematic way as a Super Learner [3]. The wisdom of crowds is one of the principles underlying the design of DREAM Challenges, where multiple team contribute diverse algorithms and methodologies for solving complex biomedical problems [4]. Super learning was proposed by van der Laan et al. and implemented as SuperLearner R language package. It utilises cross validation to estimate the performance of the component algorithms and dependencies between them. One may notice that wisdom of crowds can be formally cast as a special example of super learning. The goal of the current study is to examine several methods for estimate the performance of the super learner algorithm. In particular it explores methods for minimising bias with minimal computational effort.

1.1 Super Learning – Basic Notions and Algorithms

The input of any machine learning algorithm is a set of N observations, usually assumed to be independent and identically distributed, of the response variable y_j and a set of p explanatory variables $X_j = \{x_{jm}\}$, $m = 1, \ldots, p$. A predictive algorithm f (that includes also the feature selection algorithm and the set of hyper-parameters) can be trained using a data set $D = \{y_j, X_j\}$ to produce a model M. Then, the model can be applied to another set of variables X' to obtain a vector of predictions $\Psi = M(X')$.

K-fold cross validation is an almost unbiased technique to estimate performance of a machine learning algorithm when applied to unseen data. Algorithm 1 allows to compute a vector of N predictions for all the observations in the data set. The predictions can be compared with the original decision variable Y to estimate the performance of the learning algorithm.

In super learning, we use multiple machine learning algorithms f_l, $l - 1, \ldots, L$. The cross validation procedure leads to L vectors of predictions Ψ_l. The idea is to treat them as new explanatory variables and apply some machine learning algorithm to build a second-order predictive model (see Algorithm 2). The new features are expected to be strongly and linearly connected with the response variable, so linear models with non-negative weights seem to be appropriate super learning algorithms. Indeed they are default methods in SuperLearner R package. Note, however, that any method that selects a weighted subset of the elementary learning algorithms can be formally considered as and example of super learning. This includes also selection of the best-performing algorithm, which is a common application of cross validation, and can be considered as a special case of super learning [3]. Other examples can be unweighted mean of k best performing algorithms, as in the wisdom of crowds, or even a simple mean of all learning algorithms used.

Algorithm 2: Super learning $\mathbf{SL}(f_l, f_c, D, D')$

input : L learning methods f_l,
combining method f_c,
data set $D = \{Y, X\}$,
new data set $D' = \{Y', X'\}$

output: ensemble predictive model M,
predictions for the new response variable Ψ'

compute L vectors of cross validated predictions $\Psi_l \leftarrow \mathbf{CV}(f_l, D)$
build a second-order data set $D_c \leftarrow \{Y, \Psi_l\}$
apply the method f_c to the second-order data set $M_c \leftarrow f_c(D_c)$
build L predictive models using the entire data set $M_l \leftarrow f_l(D)$
apply the models to the new data $\Psi'_l \leftarrow M_l(D')$
compute the ensemble predictions $\Psi' \leftarrow M_c(\Psi'_l)$
return $M = \{M_l, M_c\}$, Ψ'

1.2 Performance Estimation for Super Learning – Nested Cross Validation

Super learning, as every machine learning method, is sensitive to overfitting. Therefore, an unbiased estimate of the performance of ensemble models is necessary. The obvious and most reliable method to obtain it is an external cross validation. The entire procedure is called nested cross validation, as it contains two levels of CV loop (see Algorithm 3).

Nested cross validation is implemented in SuperLearner R package as a default method of performance estimation. The authors of Super learner algorithm recommend to use 10-fold internal cross validation [3]. The external CV should also be at least 10-fold, to avoid a meaningful reduction of the sample size. Even if one restricts himself to a single loop of the external CV, the entire procedure requires all the learners to be run 110 times. Although the routine is easy to parallelise, the computational complexity is very large.

Algorithm 3: Nested cross validation $\mathbf{NCV}(f_l, f_c, D)$

---

input : L learning methods f_l,
combining method f_c,
data set $D = \{Y, X\}$

output: predictions for the response variable Ψ

split randomly the data set into k almost equally-sized folds F_i
foreach F_i **do**
    define a training set as $D_i = D \setminus F_i$
    run a super learning procedure $\{M_i, \Psi_i\} \leftarrow \mathbf{SL}(f_l, f_c, D_i, F_i)$
    collect the predictions Ψ_i
end
compare Ψ with corresponding values of Y

1.3 Bootstrap Bias Correction for Super Learning

An alternative approach to verification of complex machine learning algorithms was proposed by Tsamardinos et al. [5]. It is called the Bootstrap Bias Correction and bases on Efron's bootstrapping technique [6]. The method was originally proposed to estimate the bias caused by a choice of the optimal set of hyper-parameters for a predictive model. However, as it has been mentioned, this task can be considered as a special case of super learning. What is more, the procedure is general and does not involve any actions specific to selection of hyper-parameters. Thus, the Tsamardinos' method can be easily generalised for any kind of super learning.

Main Contribution of the Present Study. The purpose of the current study is to develop and test the algorithm of Bootstrap Bias Correction for an arbitrary super learning method. The idea is shown in Algorithm 4. The cross validated predictions for all the algorithms are computed only once. Then, the combining models are computed many times for samples that are drawn with replacement from the predictions. The predictions of each combination model are then tested on it's out-of-bag observations.

In this method, all base predictions come from the same cross-validation run, hence the entire sample is somewhat co-dependent. Therefore learning on the outcome will be overfitted. On the other hand, the training set contains duplicates what introduces additional noise. The effective data set size is smaller than the sample. This effect should decrease the overfitting and cancel the bias of the performance estimate.

The major advantage of BBC method is its small computational complexity. The elementary predictions are computed only once, multiple runs are required only for the relatively simple combining procedure.

Algorithm 4: Bootstrap bias corrected super learning **BBCSL**(f_l, f_c, D)

---

input : L learning methods f_l,
   combining method f_c,
   data set $D = \{Y, X\}$

output: set of pairs of the response variable and predictions $\{y_j, \Psi_j\}$

foreach f_l **do**
  | compute the cross validated predictions $\Psi_l \leftarrow \mathbf{CV}(f_l, D)$
  | build a second-order data set $D_c \leftarrow \{Y, \Psi_l\}$
end
for $b \leftarrow 1$ **to** B **do**
  | draw a random sample D_b of the prediction set D_c with replacement
  | define the out-of-bag set $D_{\backslash b} \leftarrow D_c \setminus D_b$
  | apply the method f_c to the data set D_b $M_b \leftarrow f_c(D_b)$
  | compute predictions for the out-of-bag set $\Psi_b \leftarrow M_b(D_{\backslash b})$
  | compare the predictions Ψ_b and with the corresponding values of Y
end

In the current study, we apply bootstrap bias corrected super learning algorithm to synthetic and real-world data sets and compare the results with other verification methods, including the nested cross validation.

2   Materials and Methods

The tests were focused on binary classification tasks. The area under ROC curve (AUC) was used as a quality measure of predictions, since it does not depend on the class balance in the data set. The methods, however, can be applied also for multi-class classification or regression tasks, with different quality metrics.

2.1   Data

Artificial Data. The methodology was developed on the synthetic data set. The data set was created as follows:

- First, the two vectors of expected values and two covariance matrices were randomly generated. These parameter sets are common for all the observations.
- The requested number of instances of the binary decision variable was randomly chosen, with the same probability for both classes.
- For each class of the decision variable, the explanatory variables are drawn from a multivariate normal distribution with the corresponding set of parameters.

This procedure allows to create an arbitrary big sample with the same statistical properties and verify directly the predictions on the unseen data. The parameters were tuned to emulate the strength of linear, quadratic and pairwise interactions as well as the dependencies between variables that may appear in the real-world data sets.

The data sets we had generated contained 5000 explanatory variables and 50, 100, 150, 200 observations. Statistic tests indicate from 2 to 20 relevant variables, depending on the sample size.

Biomedical Data Sets. The tests were performed on seven data sets that contain measurements of gene expression and copy number variation for four cancer types. These data sets correspond to biological questions with different levels of difficulty. These are:

– data sets obtained from the CAMDA 2017 Neuroblastoma Data Integration Challenge (http://camda.info):

  • CNV – 39 115 array comparative genomic hybridization (aCGH) copy number variation profiles, data set limited to a cohort of 145 patients,
  • MA – 43 349 GE profiles analysed with Agilent 44 K microarrays, data set limited to a cohort of 145 patients,
  • RNA-seq – 60 778 RNA-seq GE profiles at gene level, data set limited to a cohort of 145 patients.

The data collection procedures and design of experiments were previously described in the original studies [7–11]. Data sets are also accessible in Gene Expression Omnibus [12]. The relevant question for these data sets is predicting the final clinical status of the patient using molecular data. This is difficult problem.

– data sets with The Cancer Genome Atlas database generated by the TCGA Research Network (https://www.cancer.gov/tcga) that contain RNA-sequencing data for various types of cancer [13–18]:

  • BRCA RNA – data set contains 1205 samples and 20223 probes (112 normal and 1093 tumor),
  • BRCA CNV – data set contains 1314 samples and 21106 probes (669 normal and 645 tumor),
  • HNSC – data set contains 564 samples and 20235 probes (44 normal and 520 tumor),
  • LUAD – data set contains 574 samples and 20172 probes (59 normal and 515 tumor).

In this case the relevant question is discerning normal tissue from tumor using data set at hand. It is much easier question since both genetic profile and gene expression patterns are highly modified in cancer cells in comparison with normal tissue.

2.2 Methods

For each data set we applied the full protocol of super learning and nested cross validation. We used the default 10-fold setup for both external and internal cross validation and 100 repeats of resampling procedure. For the artificial data sets, we performed the entire routine for 100 different sets drawn from the same distribution. In the case of real data sets, the protocol was repeated 100 times on the same data for different cross validation splits.

Base Machine Learning Algorithms. Six popular machine learning algorithms were used as base learners:

- Random Forest [19],
- LASSO [20],
- Support Vector Machine (SVM) [21],
- AdaBoost [22],
- Naive Bayesian classifier,
- kNN classifier for $k = 10$.

All the algorithms are implemented from the standard R packages, with the default parameters. The parameters used are obviously not optimal, but the performance optimisation is not the subject of the current study.

For each algorithm, the set of input variables was reduced to the most relevant ones by the feature selection algorithm. To this end, we applied Welch t-test for each explanatory variable and chose 100 variables with the biggest value of the test statistic. This procedure is very sensitive to overfitting, so any bias in the verification methods should be clearly visible.

Methods of Super Learning. Six methods of combining various prediction results via super learner approach were tested:

- two default methods implemented in SuperLearner R package:

  - **NNLS:** non-negative least squares
  - **NNlog:** non-negative logistic regression

- two "toy example" methods, that are, however, commonly used:

  - **Mean:** average of all the base results (the method does not introduce any overfitting)
  - **Best 1:** choice of the best-fitted model (this special case is mentioned in the original van der Laan paper and corresponds directly to the original purpose of the bootstrap method by Tsamardinos)

- **Best** k: average of k best-performing models, where k is optimised on the training set – usually k is set as 3–4
- **RF:** Random Forest algorithm

Methods of Verification. The estimate of quality was performed using the following methods (from the most biased up to unbiased one):

- **Training set:** measure the performance of the combined classifier on the same data that were used to build the combined model (the results are obviously overestimated);
- **Independent CV:** compute the results of base learners only once, then verify the combining algorithm in a second, independent cross validation (due to the common information in the training and validation sets the results are also overestimated);
- **BBC SL:** apply the bootstrap bias correction method (the method proposed in the current study),
- **Nested CV:** apply the nested cross validation (as a gold-standard);
- **New data:** directly measure the performance for new data, drawn from the same distribution (the oracle for artificial data sets).

Table 1. The AUC of base classification algorithms for the artificial data. Comparison between prediction estimate in 10-fold cross-validation (**10CV**), prediction of new data for model trained on entire sample (**100%**) and prediction of new data for model trained on 90% of the sample (**90%**) (the same training set size as in cross-validation). The uncertainty values of AUC were computed as mean square error (MSE) of average of 100 independent measurements.

	10CV	100%	90%	10CV	100%	90%
	50 obs., MSE=0.01			100 obs., MSE=0.01		
RandomForest	0.62	0.64	0.62	0.69	0.71	0.69
LASSO	0.62	0.63	0.62	0.68	0.70	0.69
SVM	0.63	0.64	0.62	0.68	0.71	0.69
AdaBoost	0.38	0.33	0.38	0.68	0.71	0.69
Naive Bayes	0.63	0.65	0.63	0.69	0.72	0.70
kNN	0.56	0.57	0.55	0.61	0.63	0.61
	150 obs., MSE=0.005			200 obs., MSE=0.005		
RandomForest	0.735	0.760	0.745	0.782	0.804	0.788
LASSO	0.744	0.772	0.747	0.817	0.853	0.821
SVM	0.738	0.754	0.744	0.777	0.803	0.787
AdaBoost	0.699	0.731	0.711	0.761	0.789	0.763
Naive Bayes	0.735	0.756	0.744	0.770	0.792	0.779
kNN	0.659	0.678	0.672	0.690	0.711	0.698

3 Results

3.1 Artificial Data Sets

Base Learners – Cross Validation vs. Direct Verification The way of building the artificial data sets allows for comparison between the cross validation results and the actual results obtained for unseen data. This comparison is shown in Table 1.

As it could be expected, the performance of all the machine learning algorithms improves with the sample size. In particular AdaBoost could not cope with 50 observations only.

Interestingly, a considerable negative bias of the cross validated estimations of AUC was observed in comparison with the model trained on the entire sample. To correct this difference, additional models were built using the training sets reduced to 90% of the original sample. Quality of these models, agree much better with the cross-validated estimates. That means that most of the bias of the cross validation procedure comes from the smaller size of training sets. The small remaining bias is most likely due to negative correlations of fluctuations in training and validation sets in cross-validation.

Super Learning. The performance of diverse methods of super learning is shown in Table 2. For this particular data set, the super learning technique needs at least 100 observations to outperform the best single result and at least 200 observations to perform better then a simple average of all the base results. Random forest proves to be a poor super learning method. The non-negative linear models perform the best (surprisingly, the ordinary least squares method was better at classification than the logistic regression, that is specialised for this task). However, the simple best-k method performs almost as well.

One should note the difference between the performance of the "Best 1" method in the Table 2 and the best overall performance from the Table 1. The reason is, that the algorithm, that performed best on a subset of the data will be not necessarily the best one for the entire data set. Everyone, who chooses the most appropriate machine learning algorithm using a single cross validation, should thus expect some decreasing of its performance for new data.

The differences between various performance estimates are small, but some regular patterns are noticeable. As previously, the performance of the models learned on the entire data set and applied to new data is better than measured in the nested cross validation. Moreover, the performance measured for models built using the reduced training set is slightly better as well. The biased performance estimate on the training set is significantly bigger than other estimates. For 100 and more observations, BBC SL method leads to the results very close to the nested cross validation. The results of the independent CV are more unstable, often overestimated.

These results are, however, ambiguous: the bias of cross validation methods due to the smaller size of training sets seems to be stronger than any bias due to overfitting. Nevertheless, the proposed BBC SL algorithm proved better than any other simple verification algorithm and close to the "gold standard" nested cross validation.

3.2   Biomedical Data

The results of super learner procedure obtained for real data are displayed in Tables 3, 4. The direct measurement of performance for unseen data is impossible in this case, hence a nested cross-validation is a reference. The estimated standard deviation of the distribution of results is shown for each data set. The error of the mean value is smaller, but it is hard to estimate, since the measurements are not mutually independent.

As for artificial data, the performance of Random Forest for combining the base results was very poor, and the non-negative logistic regression performed very close to NNLS method. Thus, both were not shown in the tables.

For all the tested data sets ensemble methods outperform the best single classifier. However, in all the cases the best-performing method was a simple average over all the base results. The linear model and k-best proved nearly as good in some cases. This result is obviously not a rule, our artificial data sets give an example for better performance of linear combinations of classifiers.

Table 2. Artificial data – AUC of super learning for 6 diverse methods for three sizes of artificial data sets.

SL method	Training set	Independent CV	BBC SL	Nested CV	New data 100%	New data 90%
50 observations, MSE=0.01						
NNLS	0.66	0.62	0.62	0.59	0.64	0.62
NNlog	0.65	0.62	0.62	0.59	0.59	0.58
RF	0.55	0.57	0.56	0.52	0.56	0.57
Best k	0.67	0.64	0.63	0.61	0.63	0.62
Mean	0.63	0.63	0.63	0.63	0.64	0.62
Best 1	0.67	0.64	0.63	0.61	0.64	0.63
100 observations, MSE=0.01						
NNLS	0.71	0.69	0.69	0.68	0.71	0.70
NNlog	0.71	0.68	0.69	0.68	0.70	0.69
RF	0.65	0.66	0.64	0.63	0.66	0.64
Best k	0.73	0.70	0.69	0.68	0.71	0.70
Mean	0.70	0.70	0.70	0.70	0.72	0.70
Best 1	0.72	0.70	0.68	0.67	0.71	0.69
200 observations, MSE=0.005 for Best 1 and Best k, 0.004 for others						
NNLS	0.829	0.815	0.810	0.814	0.851	0.827
NNlog	0.831	0.816	0.805	0.804	0.854	0.828
RF	0.800	0.802	0.790	0.792	0.832	0.808
Best k	0.830	0.812	0.803	0.799	0.852	0.824
Mean	0.808	0.808	0.807	0.808	0.834	0.816
Best 1	0.822	0.807	0.797	0.794	0.851	0.816

Another interesting point is the stability of the results. The most repeatable values of AUC are produced by mean of all the classifiers, while choice of the best single one and mean of k best ones are the most unstable. The effect is not visible on the training set, but clearly appears when nested CV or BBC SL verification methods are used (see especially Table 4).

The performance measured using the proposed BBC SL method is almost exactly the same as obtained by nested cross validation. Two simpler methods of the performance estimation report inflated results.

Figure 1 shows the performance of three combining methods for the BRCA RNA data set in more detailed way. The "Best single" method is simply a common practice of choosing the best-performing classifier to this particular purpose. In a simple cross validation, the result seems to be better than any ensemble model. However, when the external validation is applied, the performance drops significantly and becomes unstable. It turns out, that as simple operation as choice between six classification algorithms is a considerable source of overfitting. Both nested cross validation and the proposed BBC SL method show this clearly. For this particular data set, the optimal strategy is the average of all the base predictions.

Table 3. The performance of super learning for 4 methods for three Neuroblastoma data sets. Random Forest and NNlog methods were omitted. Note the mean square error for **Best 1** and **Best** k methods, bigger than the typical values.

SL method	Training set	Independent CV	BBC SL	Nested CV
CNV, typical MSE=0.02				
NNLS	0.78	0.77	0.76	0.76
Best k	0.78	0.77	0.75	0.76 ± 0.03
Mean	0.76	0.76	0.76	0.76
Best 1	0.78	0.77 ± 0.03	0.74 ± 0.03	0.75 ± 0.03
MA, typical MSE=0.01				
NNLS	0.89	0.87	0.87	0.88
Best k	0.89	0.88 ± 0.02	0.87 ± 0.02	0.88 ± 0.02
Mean	0.89	0.89	0.89	0.89
Best 1	0.89	0.88 ± 0.02	0.85 ± 0.02	0.87 ± 0.02
RNA, typical MSE=0.01				
NNLS	0.91	0.90	0.89	0.89
Best k	0.91	0.90	0.89 ± 0.02	0.89 ± 0.02
Mean	0.89	0.89	0.89	0.89
Best 1	0.90	0.90 ± 0.02	0.88 ± 0.02	0.88 ± 0.02

Table 4. Real data (2) – the performance of super learning for 4 different methods for TCGA data sets. Note the dependence of the square error on the combining method (the biggest for **Best 1** and **Best** k) and on the verification method (smaller on the training set)

SL method	Training set	Independent CV	BBC SL	Nested CV
BRCA RNA				
NNLS	0.9994 ± 0.0001	0.9992 ± 0.0002	0.9991 ± 0.0002	0.9991 ± 0.0002
Best k	0.9994 ± 0.0001	0.998 ± 0.002	0.998 ± 0.001	0.998 ± 0.003
Mean	0.9993 ± 0.0001	0.9993 ± 0.0001	0.9993 ± 0.0001	0.9993 ± 0.0001
Best 1	0.9994 ± 0.0001	0.998 ± 0.003	0.997 ± 0.002	0.996 ± 0.004
BRCA CNV				
NNLS	0.988 ± 0.001	0.987 ± 0.001	0.987 ± 0.001	0.987 ± 0.001
Best k	0.988 ± 0.001	0.987 ± 0.001	0.987 ± 0.001	0.987 ± 0.001
Mean	0.988 ± 0.001	0.988 ± 0.001	0.988 ± 0.001	0.988 ± 0.001
Best 1	0.983 ± 0.001	0.981 ± 0.003	0.981 ± 0.002	0.981 ± 0.002
HNSC				
NNLS	0.996 ± 0.002	0.992 ± 0.005	0.990 ± 0.002	0.993 ± 0.006
Best k	0.997 ± 0.001	0.988 ± 0.01	0.986 ± 0.005	0.98 ± 0.01
Mean	0.995 ± 0.003	0.995 ± 0.003	0.995 ± 0.002	0.995 ± 0.003
Best 1	0.997 ± 0.001	0.985 ± 0.01	0.983 ± 0.006	0.98 ± 0.01
LUAD				
NNLS	0.9993 ± 0.0002	0.998 ± 0.002	0.9978 ± 0.0007	0.998 ± 0.002
Best k	0.9994 ± 0.0001	0.998 ± 0.003	0.997 ± 0.002	0.997 ± 0.003
Mean	0.9992 ± 0.0001	0.9992 ± 0.0001	0.9991 ± 0.0002	0.9992 ± 0.0001
Best 1	0.9993 ± 0.0001	0.998 ± 0.004	0.997 ± 0.002	0.997 ± 0.003

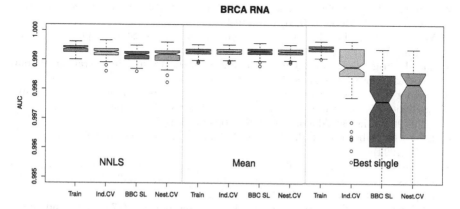

Fig. 1. Box plots of AUC for an example real data set. Note the drop of performance and the instability of results for the best single classifier, when unbiased verification methods are applied.

4   Conclusion

Main Contribution. Super learner as proposed in the [3] is computationally demanding approach that relies on multiple cross-validation and application of multiple learning algorithms. In the original formulation of the algorithm, verification of the quality of the final model involves repeating entire procedure within nested cross-validation, what significantly increases the computational cost of the modelling procedure. The current study shows that the nested cross-validation can be replaced by using resampling protocol, which gives equivalent results.

Additional Remarks. In almost all examined cases, the ensemble of learners gives better results than selection of a single best method for prediction. Interestingly, the simplest method of combining results of all algorithms, namely the the simple average of predictions of all base learners seems to be a very good choice for obtaining a stable non-overfitted estimate. Two simplest methods for assigning weights to base learners, namely a simple linear combination or selection of unweighted average of k-best models have better performance that other methods and should be explored along simple mean of all base methods.

One should note, that Super learner was applied here to merge results of algorithms that are very good predictors themselves. The differences between predictions of these classifiers are concentrated on a handful of difficult cases. The simple average works best, because there are too few independent data point to build reliable model for more advanced methods.

References

1. Fernández-Delgado, M., Cernadas, E., et al.: Do we need hundreds of classifiers to solve real world classification problems? J. Mach. Learn. Res. **15**(1), 3133–3181 (2014)
2. Marbach, D., Costello, J.C., et al.: Wisdom of crowds for robust gene network inference. Nat. Methods **9**(8), 796 (2012)
3. Van der Laan, M.J., Polley, E.C., et al.: Super learner. Stat. Appl. Genet. Mol. Biol. **6**(1) (2007)
4. Saez-Rodriguez, J., et al.: Crowdsourcing biomedical research: leveraging communities as innovation engines. Nat. Rev. Genet. **17**(8), 470 (2016)
5. Tsamardinos, I., Greasidou, E., Borboudakis, G.: Bootstrapping the out-of-sample predictions for efficient and accurate cross-validation. Mach. Learn. **107**(12), 1895–1922 (2018). https://doi.org/10.1007/s10994-018-5714-4
6. Efron, B., Tibshirani, R.J.: An Introduction to the Bootstrap. CRC Press, New York (1994)
7. Zhang, W., Yu, Y., et al.: Comparison of RNA-seq and microarray-based models for clinical endpoint prediction. Genome Biol. **16**, 133 (2015). https://europepmc.org/article/PMC/4506430
8. Theissen, J., Oberthuer, A., et al.: Chromosome 17/17q gain and unaltered profiles in high resolution array-CGH are prognostically informative in neuroblastoma. Genes Chromos. Cancer **53**(8), 639–649 (2014). https://onlinelibrary.wiley.com/doi/abs/10.1002/gcc.22174

9. Stigliani, S., Coco, S., et al.: High genomic instability predicts survival in metastatic high-risk neuroblastoma. Neoplasia **14**(9), 823–910 (2012). http://www.science direct.com/science/article/pii/S1476558612800868

10. Coco, S., Theissen, J., et al.: Age-dependent accumulation of genomic aberrations and deregulation of cell cycle and telomerase genes in metastatic neuroblastoma. Int. J. Cancer **131**(7), 1591–1600 (2012). https://onlinelibrary.wiley.com/doi/abs/10.1002/ijc.27432

11. Kocak, H., Ackermann, S., et al.: Hox-C9 activates the intrinsic pathway of apoptosis and is associated with spontaneous regression in neuroblastoma. Cell Death Dis. **4**, e586 (2013)

12. Edgar, R., Domrachev, M., et al.: Gene expression omnibus: NCBI gene expression and hybridization array data repository. Nucleic Acids Res. **30**(1), 207–210 (2002)

13. Hammerman, P., Lawrence, M., et al.: Comprehensive genomic characterization of squamous cell lung cancers. Nature **489**, 519–525 (2012)

14. Collisson, E., Campbell, J., et al.: Comprehensive molecular profiling of lung adenocarcinoma. Nature **511**, 543–550 (2014)

15. Koboldt, D., Fulton, R., et al.: Comprehensive molecular portraits of human breast tumours. Nature **490**, 61–70 (2014)

16. Ciriello, G., Gatza, M.L., et al.: Comprehensive molecular portraits of invasive lobular breast cancer. Cell **163**(2), P506–519 (2015)

17. Lawrence, M., Sougnez, C., et al.: Comprehensive genomic characterization of head and neck squamous cell carcinomas. Nature **517**, 576–582 (2015)

18. Creighton, C., Morgan, M., et al.: Comprehensive molecular characterization of clear cell renal cell carcinoma. Nature **499**, 43–49 (2013)

19. Breiman, L.: Random forests. Mach. Learn. **45**(1), 5–32 (2001)

20. Friedman, J., Hastie, T., et al.: Regularization paths for generalized linear models via coordinate descent. J. Stat. Softw, **33**(1), 1–22 (2010)

21. Chang, C.C., Lin, C.J.: LIBSVM: a library for support vector machines. ACM Trans. Intell. Syst. Technol. (TIST) **2**(3), 1–27 (2011)

22. Freund, Y., Schapire, R.E.: A desicion-theoretic generalization of on-line learning and an application to boosting. In: Vitányi, P. (ed.) EuroCOLT 1995. LNCS, vol. 904, pp. 23–37. Springer, Heidelberg (1995). https://doi.org/10.1007/3-540-59119-2_166

MMRF-CoMMpass Data Integration and Analysis for Identifying Prognostic Markers

Marzia Settino[1], Mariamena Arbitrio[2], Francesca Scionti[3],
Daniele Caracciolo[3], Maria Teresa Di Martino[3], Pierosandro Tagliaferri[3],
Pierfrancesco Tassone[3], and Mario Cannataro[1(✉)]

[1] Data Analytics Research Center, Department of Medical and Surgical Sciences,
Magna Graecia University, Catanzaro, Italy
marzia.settino@studenti.unicz.it, cannataro@unicz.it

[2] CNR-Institute of Neurological Sciences, UOS of Pharmacology, Catanzaro, Italy
mariamena.arbitrio@cnr.it

[3] Department of Experimental and Clinical Medicine, Medical Oncology Unit,
Mater Domini Hospital, Magna Graecia University, Catanzaro, Italy
{scionti,teresadm,tagliaferri,tassone}@unicz.it,
daniele.caracciolo1@studenti.unicz.it

Abstract. Multiple Myeloma (MM) is the second most frequent haematological malignancy in the world although the related pathogenesis remains unclear. The study of how gene expression profiling (GEP) is correlated with patients' survival could be important for understanding the initiation and progression of MM.

In order to aid researchers in identifying new prognostic RNA biomarkers as targets for functional cell-based studies, the use of appropriate bioinformatic tools for integrative analysis is required.

The main contribution of this paper is the development of a set of functionalities, extending TCGAbiolinks package, for downloading and analysing Multiple Myeloma Research Foundation (MMRF) CoMMpass study data available at the NCI's Genomic Data Commons (GDC) Data Portal. In this context, we present further a workflow based on the use of this new functionalities that allows to i) download data; ii) perform and plot the Array Array Intensity correlation matrix; ii) correlate gene expression and Survival Analysis to obtain a Kaplan–Meier survival plot.

Keywords: TCGABiolinks · MMRF-CoMMpass · Multiple myeloma · TCGA · R · Integrative data analysis

1 Introduction

Multiple myeloma (MM) is a cancer of plasma cell and it is the second most common blood cancer. Myeloma is a heterogeneous disease with great genetic and epigenetic complexity. Therefore, the identification of patient subgroups defined by molecular profiling and clinical features remains a critical need for a

© Springer Nature Switzerland AG 2020
V. V. Krzhizhanovskaya et al. (Eds.): ICCS 2020, LNCS 12139, pp. 564–571, 2020.
https://doi.org/10.1007/978-3-030-50420-5_42

better understanding of disease mechanism, drug response and patient relapse. In this context, the Multiple Myeloma Research Foundation (MMRF-CoMMpass) Study represents the largest genomic data set and the most widely published studies in multiple myeloma.

Transcriptomic studies have largely contributed to reveal multiple myeloma features, distinguishing multiple myeloma subgroups with different clinical and biological patterns. Based on the hypothesis that myeloma invasion would induce changes in gene expression profiles, gene expression profile (GEP) studies constitute a reliable prognostic tool [3,11].

Various studies have identified gene expression signatures capable of predicting event-free survival and overall survival (OS) in multiple myeloma [1,6]. In order to aid researchers in identifying new prognostic RNA biomarkers as well as targets for functional cell-based studies, the use of appropriate bioinformatic tools for integrative analysis can offer new opportunities. Among these tools a promising approach is the use of TCGABiolinks package [2,9,10]. The main contribution of this work is to provide the researchers with a new set of functions extending TCGAbiolinks package that allows to MMRF-CoMMpass database to be investigated. Moreover, a simple workflow for searching, downloading and analyzing RNA-Seq gene level expression dataset from the MMRF-CoMMpass Studies will be described. The same workflow could be in general extended to other MMRF-CoMMpass datasets.

2 Background

Gene expression data from multiple myeloma patients can be retrieved from MMRF-CoMMpass[1] and Gene Expression Omnibus (GEO)[2]. GEO is an international public repository that archives and freely distributes high-throughput gene expression and other functional genomics datasets. The National Cancer Institute (NCI) Genomic Data Commons (GDC) [5] provides the cancer research community with a rich resource for sharing and accessing data across numerous cancer studies and projects for promoting precision medicine in oncology.

The NCI Genomic Data Commons data are made available through the GDC Data Portal[3], a platform for efficiently querying and downloading high quality and complete data. The GDC platform includes data from The Cancer Genome Atlas (TCGA), Therapeutically Applicable Research to Generate Effective Treatments (TARGET) and further studies[4].

Recently, many studies are contributing with additional datasets to GDC platform, including the MMRF CoMMpass Study among others [7]. One of the major goals of the GDC is to provide a centralized repository for accessing data from large-scale NCI programs, however it does not make available a comprehensive toolkit for data analyses and interpretation. To fulfil this need, the

---

[1] https://themmrf.org/we-are-curing-multiple-myeloma/mmrf-commpass-study/.
[2] https://www.ncbi.nlm.nih.gov/geo/.
[3] https://portal.gdc.cancer.gov/.
[4] https://portal.gdc.cancer.gov/projects.

R/Bioconductor package TCGAbiolinks was developed to allow users to query, download and perform integrative analyses of GDC data [2,9,10]. TCGAbiolinks combines methods from computer science and statistics and it includes methods for visualization of results in order to easily perform a complete analysis.

The Cancer Genome Atlas (TCGA): The Cancer Genome Atlas (TCGA) contains data on 33 different cancer types from 11,328 patients and it is the world's largest and richest collection of genomic data. TCGA contains molecular data from multiple types of analysis such as DNA sequencing, RNA sequencing, Copy number, Array-based expression and others. In addition to molecular data, TCGA has well catalogued metadata for each sample such as clinical and sample information.

NCI's Genomic Data Commons (GDC) Data Portal: The National Cancer Institute (NCI) Genomic Data Commons (GDC) is a publicly available database that promotes the sharing of genomic and clinical data among researchers and facilitates precision medicine in oncology. At a high level, data in GDC are organized by project (e.g. TCGA, TARGET, MMRF-CoMMpass). Each of these projects contains a variety of molecular data types, including genomics, epigenomics, proteomics, imaging, clinical and others.

Multiple Myeloma Research Foundation (MMRF) CoMMpass: The MMRF-CoMMpass Study is a collaborative research effort with the goal of mapping the genomic profile of patients with newly diagnosed active multiple myeloma to clinical outcomes to develop a more complete understanding of patient responses to treatments. MMRF-CoMMpass Study identified many genomic alterations that were not previously found in multiple myeloma as well as providing a prognostic stratification of patients leading to advances in cancer care [8]. Recently the MMRF announced new discoveries into defining myeloma subtypes, identifying novel therapeutic targets for drug discovery and more accurately predicting high-risk disease[5]. The NCI announced in 2016 a collaboration with MMRF to incorporate genomic and clinical data about myeloma into the NCI Genomic Data Commons (GDC) platform.

3   Workflow for Downloading and Analysing MMRF-CoMMpass Data

TCGAbiolinks is a R/Bioconductor package that combines methods from computer science and statistics to address challenges with data mining and analysis of cancer genomics data stored at GDC Data Portal. More specifically, a guided workflow [10] allows users to query, download, and perform integrative analyses

---

[5] https://themmrf.org/2018/12/the-mmrf-commpass-study-drives-new-discoveries-in-multiple-myeloma/.

of GDC data. The package provides several methods for analysis (e.g. differential expression analysis, differentially methylated regions, etc.) and methods for visualization (e.g. survival plots, volcano plots and starburst plots, etc.).

TCGAbiolinks was initially conceived to interact with TCGA data through the GDC Data Portal but it can be in principle extended to other GDC datasets if the functions to handle their differences in formats and data availability are properly handled [9]. The GDC API Application Programming Interface (API) provides developers with a programmatic access to GDC functionality.

TCGAbiolinks consists of several functions but in this work we will describe only the main functions used in the workflow described in the Sect. 3. More specifically:

- **GDCquery** uses GDC API for searching GDC data;
- **GDCprepare** allows to read downloaded data and prepare them into an R object;
- **GDCquery_clinic** allows to download all clinical information related to a specified project in GDC;
- **TCGAanalyze_Preprocessing** performs an Array Array Intensity correlation (AAIC). It defines a square symmetric matrix of spearman correlation among samples;
- **TCGAanalyze_SurvivalKM** performs an univariate Kaplan-Meier (KM) survival analysis (SA) using complete follow up taking one gene a time from a gene list.

The *SummarizedExperiment* [4] object is the default data structure used in TCGAbiolinks for combining genomic data and clinical information. A *SummarizedExperiment* object contains sample information, molecular data and genomic ranges (i.e. gene information). MMRF-CoMMpass presents some differences in formats and data respect to TCGA dataset. For example, the sample ID format in MMRF-CoMMpass is "study-patient-visit-source" (e.g. "MMRF-1234-1-BM" means patient 1234, first visit, from bone marrow). Moreover, some fileds in MMRF-CoMMpass *SummarizedExperiment* are lacking or they are named differently respect to TCGA dataset format. In order to fill this gap and to make MMRF-CoMMpass dataset suitable to be handled by previous functions we introduced the following customized functions:

- **MMRF_prepare** adds the sample type information to *SummarizedExperiment* object from *GDCprepare*;
- **MMRF_prepare_clinical** renames the data frame field "submitter_id" of clinical information from *GDCquery_clinic* as the field name found in TCGA dataset (i.e. bcr_patient_barcode);
- **MMRF_prepare_SurvivalKM** makes the MMRF-CoMMpass sample ID format in Gene Expression matrix (dataGE) from *GDCprepare* suitable for using in *TCGAanalyze_SurvivalKM* function.

The following workflow describes the steps for downloading, processing and analyzing MMRF-CoMMpass RNA-Seq gene expression using TCGABiolinks jointly with the new functions before reported.

Search the MMRF-CoMMpass Data: GDCquery uses GDC API to search the data for a given project and data category as well as other filters. A valid data category for MMRF-CoMMpass project can be found using *getProjectSummary* function. The results are shown in Table 1.

Table 1. Results of *getProjectSummary* in the case of the MMRF-CoMMpass project.

Data category	File count	Case count
Simple Nucleotide Variation	10918	959
Sequencing Reads	6577	995
Transcriptome Profiling	4295	787

The following listing illustrates the use of GDCquery for searching gene expression level dataset (HTSeq - FPKM) using the "Trascriptome Profiling" category in the list obtained from *getProjectSummary*.

For simplification purposes just a filtered by barcode subset is downloaded.

```
query.mm.fpkm <- GDCquery(project = "MMRF-COMMPASS",
data.category = "Transcriptome Profiling",
data.type = "Gene Expression Quantification",
workflow.type="HTSeq - FPKM",
barcode = c("MMRF_2473","MMRF_2111","MMRF_2270",
            "MMRF_2238","MMRF_1080","MMRF_2253",
            "MMRF_2119","MMRF_2468", "MMRF_1201",
            "MMRF_2821","MMRF_1957","MMRF_1678"))
```

Listing 1.1. GDCquery function for searching gene expression data in MMRF-CoMMpass. The datset is filtered by barcode.

Download and Prepare the MMRF-COMMPASS Data: The *GDCdownload* function allows to download and save the data in a local folder to be used in *GDCprepare* function that transforms the downloaded data into a *SummarizedExperiment*. The clinical data (e.g. tumor stage, days to last follow up, treatments) can be obtained using the *GDCquery_clinical* function specifying as input project "MMRF-COMMPASS"). At this point, *MMRF_prepare* and *MMRF_prepare_clinical* functions allow to make the output of the previous functions suitable for being handled by TCGABiolinks functions.

Analyse MMRF-COMMPASS Data: Once the data were downloaded and they are prepared, outliers could be discovered through the use of the function *TCGAanalyze_Preprocessing* which performs an Array Array Intensity correlation (AAIC). The plot in Fig. 1 shows an example of heat map of AAIC for MMRF-CoMMpass gene expression data.

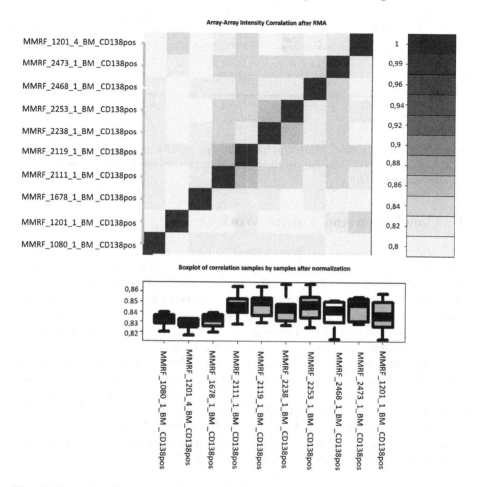

Fig. 1. Example of heat map of the array-array Spearman/Pearson rank correlation coefficients. This plot is useful for detecting outliers.

We used *MMRF_prepare_SurvivalKM* for preparing dataGE from *GDCprepare* to be handled by *TCGAanalyze_SurvivalKM* function.

Finally, we performed a Kaplan-Meier univariate survival analysis (KM-SA) using *TCGAanalyze_SurvivalKM* function. The resulting plot allows to correlate visually gene expression and Survival Analysis. Two thresholds are defined for each gene expression according its level of mean expression in cancer samples. In this example we used the threshold of intensity of gene expression to divide the samples in 2 groups (High, Low). The Fig. 2 shows the correlation between survival and the most high/low expressed gene.

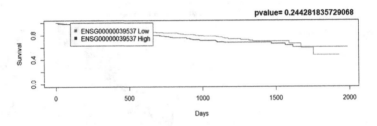

Fig. 2. GDCquery_clinic, MMRF_prepare_clinical, MMRF_prepare_SurvivalKM and TCGAanalyze_SurvivalKM jointly allow to perform an univariate KaplanMeier (KM) survival analysis (SA).

4   Conclusion and Future Work

The MMRF-CoMMpass has proven itself to be a leader in scientific innovation as well as in data sharing when it decided to incorporate their data into the GDC platform. The use of appropriate bioinformatic tools for integrative analysis of MMRF-CoMMpass data can offer great opportunities. In order to take this chance, the TCGAbiolinks package represents a useful tool for data integration and analysis of cancer data. For example, TCGAbiolinks offers the possibility to integrate gene expression data from external sources (e.g GEO) obtaining a merged result that can be used for further analysis such as differential expression analysis. The main contribution of this paper is the extension of TCGABiolinks package with new functions to handle MMRF-CoMMpass data available at the NCI's Genomic Data Commons (GDC) Data Portal. This will allow to MM researchers to better exploit MMRF-CoMMpass data. As future work we plan to make available these new functions as package through a public repository and to extend them to allow further analysis of MMRF-CoMMpass data.

References

1. Chng, W.J., et al.: Gene signature combinations improve prognostic stratification of multiple myeloma patients. Leukemia **30**(5), 1071–1078 (2016)
2. Colaprico, A., et al.: TCGAbiolinks: an R/Bioconductor package for integrative analysis of TCGA data. Nucleic Acids Res. **44**(8) (2016). https://doi.org/10.1093/nar/gkv1507. https://www.ncbi.nlm.nih.gov/pubmed/26704973
3. Gooding, S., et al.: Transcriptomic profiling of the myeloma bone-lining niche reveals BMP signalling inhibition to improve bone disease. Nat. Commun. **10**(1), 4533 (2019)
4. Huber, W., et al.: Orchestrating high-throughput genomic analysis with Bioconductor. Nat. Methods **12**(2), 115–121 (2015)
5. Jensen, M.A., Ferretti, V., Grossman, R.L., Staudt, L.M.: The NCI genomic data commons as an engine for precision medicine. Blood **130**(4), 453–459 (2017). https://doi.org/10.1182/blood-2017-03-735654. https://www.ncbi.nlm.nih.gov/pubmed/28600341
6. Kuiper, R., et al.: A gene expression signature for high-risk multiple myeloma. Leukemia **26**(11), 2406–2413 (2012)

7. Lee, J.S., Kibbe, W.A., Grossman, R.L.: Data harmonization for a molecularly driven health system. Cell **174**(5), 1045–1048 (2018)
8. Liu, Y., et al.: A network analysis of multiple myeloma related gene signatures. Cancers (Basel) **11**(10), 1452 (2019)
9. Mounir, M., et al.: New functionalities in the TCGAbiolinks package for the study and integration of cancer data from GDC and GTEX. PLoS Comput. Biol. **15**(3) (2019). https://doi.org/10.1371/journal.pcbi.1006701. https://www.ncbi.nlm.nih.gov/pmc/articles/PMC6420023/
10. Silva, T.C., et al.: TCGA Workflow: analyze cancer genomics and epigenomics data using bioconductor packages. F1000Res. 5, **1542**, 1–59 (2016)
11. Szalat, R., Avet-Loiseau, H., Munshi, N.C.: Gene expression profiles in myeloma: ready for the real world? Clin. Cancer Res. **22**(22), 5434–5442 (2016)

Using Machine Learning in Accuracy Assessment of Knowledge-Based Energy and Frequency Base Likelihood in Protein Structures

Katerina Serafimova[1], Iliyan Mihaylov[1], Dimitar Vassilev[1(✉)], Irena Avdjieva[1], Piotr Zielenkiewicz[2], and Szymon Kaczanowski[2(✉)]

[1] FMI, Sofia University "St. Kliment Ohridski", 5 James Bourchier Street, 1164 Sofia, Bulgaria
dimitar.vassilev@fmi.uni-sofia.bg
[2] IBB PAN, Warsaw, Poland
szymon@ibb.pan.pl

Abstract. Many aspects of the study of protein folding and dynamics have been affected by the accumulation of data about native protein structures and recent advances in machine learning. Computational methods for predicting protein structures from their sequences are now heavily based on machine learning tools and on approaches that extract knowledge and rules from data using probabilistic models. Many of these methods use scoring functions to determine which structure best fits a native protein sequence. Using computational approaches, we obtained two scoring functions: knowledge-based energy and likelihood of base frequency, and we compared their accuracy in measuring the sequence structure fit. We compared the machine learning models' accuracy of predictions for knowledge-based energy and likelihood values to validate our results, showing that likelihood is a more accurate scoring function than knowledge-based energy.

Keywords: Knowledge-based energy · Statistical potential · Likelihood · Cross-validation · Machine learning · Protein structure prediction

1 Introduction

Proteins are built of one or more linear chains of amino acid residues, which are protein sequences that fold into three-dimensional structures. Correct folding leads to a native structure, and knowledge of the native protein structure is essential for understanding the protein function. A growing amount of structural data in databases such as the Protein Data Bank (PDB) [1] has led to the development of computational approaches for protein structure prediction. However, these approaches are often time-consuming and costly or have low accuracy, so there is a need for effective and accurate computational approaches to protein structure prediction.

Most of the methods for structure prediction use scoring functions to determine which structure best fits a native protein sequence. The native structure generally has a lower free

© Springer Nature Switzerland AG 2020
V. V. Krzhizhanovskaya et al. (Eds.): ICCS 2020, LNCS 12139, pp. 572–584, 2020.
https://doi.org/10.1007/978-3-030-50420-5_43

energy than the other possible structures under the native conditions [2], which means that an accurate free energy function can be applied in the prediction and assessment of protein structures [3], for example, as a scoring function in measuring sequence structure fit. However, calculating the free energy of protein folding (or unfolding) using all-atom coordinates is impractical because it is computationally demanding, so only a small fraction of the available conformational space can be explored in this way. Knowledge-based (KB) approaches that extract knowledge and rules from data are therefore used in the assessment of an ensemble of structural models produced by computational methods to find the correct structure that fits a given sequence.

KB free energy (also known as statistical potential or pseudo-energy potential) is widely used for those purposes. Statistical potential is derived using a mathematical approach, according to which the statistical preferences of interactions between different molecules can be described. However, protein folding is a cooperative process with many driving forces, which means that a residue in a given position has an impact on other residues and, ultimately, on the structure in which the sequence will fold. A way to describe cooperation is by using a likelihood function.

In this study, we used a one-dimensional (1D) representation of the protein structure based on the buried or exposed state of the residues to compare the accuracy of the KB energy (E) and the likelihood of base frequency (L), which are essentially scoring functions and can be used in protein structure prediction.

Machine learning (ML)–based approaches used in the accuracy assessment of sequence-structure fit in proteins provide a large set of models that can contribute to the process by enriching the quantification of different parameters. The ML models can also be efficiently used to validate results related to the accuracy of the functions by assessing the sequence structure fit. The large diversity of models bears some problems related to how and which model is better to choose for a particular case, which can be overcome by different cross-validation approaches.

To validate our results showing that likelihood is a more accurate measure of the sequence-structure fit than KB energy, we used ML models. The application of such ML models to predict likelihood and energy values accordingly provides criteria for assessing the accuracy and predictability of these two approaches.

2   Problem Description

Proteins interact strongly with surrounding solvents, and the exposure of amino acids to solvents is a sensitive parameter that can be used to model energetic features on the protein–solvent boundary [4]. The folding process of soluble proteins also decreases the surface area in contact with the solvent; this is related to the secondary structures of proteins. Accurate knowledge of residue accessibility would thus aid in the prediction of protein structures [5].

The protein residues in a structure are exposed to the solvent to different extents. We applied KB approaches to describe different types of residue preferences for being in a buried or exposed state in the protein structure.

In this work, the model of the protein structure is one-dimensional (1D) and uses only the solvent accessibility of every residue. For simplicity, the solvent accessibility is categorized into two states: buried (0) and exposed (1). There are 20 types of amino acid residues and each of them can be categorized as buried or exposed in a given position in the protein sequence, so the number of residue classes is 40.

The solvent accessible surface area (SASA) of all the proteins in the two sets is calculated, and the threshold of the SASA per residue is selected to classify a residue as buried or exposed. Then, a buried/exposed pattern is assigned to every sequence to construct an **object** that describes the protein using the amino acid (AA) **sequence** and one structural property – a **pattern** of the buried or exposed status of the residues.

The sequence-structure pair objects based on the solvent accessibility were used for optimization of their KB energy or likelihood, accordingly.

The concept of pseudo-energy was introduced to biology by the seminal paper of Tanaka and Scheraga [6]. They assumed that residues behave like molecules interacting in gas, and they used the observed frequencies of the contacts between different types of residues in known X-ray structures. Using these data, they calculated the "free energies" ($\Delta G°$) of the contact between different types of amino acids using a formula exported from statistical chemistry:

$$\Delta G° = -RTX_{ij}X_iX_j \tag{1}$$

where X_{ij} is the frequency of the observed contacts between the residues of type i and the residues of type j, X_iX_j represents a multiplication of these frequencies (statistical expectation of the contact between residue i and residue j), R is the gas constant, and T is temperature.

In this study, using parameters obtained from a set of 200 native protein structures, we calculated the KB energy of proteins, as seen in Eq. 2:

$$kbE(protein) = \sum_1^j E_{i[0\,or\,1]} \tag{2}$$

where i is the type of residue according to the position of the protein sequence, j is the length of the protein, and $E_{i[0\,or\,1]}$ is the KB energy of the i-th residue, which can be buried [0] or exposed [1].

Likelihood is also widely used in biology, for example, in the case of phylogenetics. In this paper, we applied the following formula:

$$L = \log\left(\frac{P_1^{n_{1[0]}} \times P_{2[0]}^{n_{2[0]}} \ldots P_{20[0]}^{n_{20[0]}} \times P_{1[1]}^{n_{1[1]}} \times P_{2[1]}^{n_{2[1]}} \ldots P_{20[1]}^{n_{20[1]}}\left((n_{1[0]} + \ldots n_{20[1]})!\right)}{\left(n_{1[0]}!\right)\left(n_{2[0]}!\right)\left(n_{3[0]}!\right)\ldots\left(n_{20[i]}!\right)}\right) \tag{3}$$

where $P_{i[0\,or\,1]}$ is the observed frequency of the residues in the buried/exposed state in the entire database, $n_{i[0\,or\,1]}$ is the number of residues of a given type in a given protein, and $(n_{1[0]} + \ldots n_{20[1]})$ is equal to the length of the protein.

The object design is highly simplified, and the reduction of 3D structural information to 1D lowers the possible accuracy of the scoring. Therefore, it is possible to optimize a native sequence-structure pair, for example, by changing the pattern to better fit the native sequence according to the selected criteria. The resulting pattern will be different from the native one (because the criteria is imperfect), and the accuracy of the criteria can be assessed by calculating the identity of the resulting pattern to the native pattern.

Using the two criteria – KB energy and likelihood, separately, in two optimization experiments, we were able to conclude that the likelihood is less erroneous than the KB energy as a criterion of the sequence-structure fit.

The concept for this study is to test the accuracy of the two criteria for measuring the sequence-structure fit by using a ML approach.

Machine learning can be applied for multiple purposes in protein folding and structure prediction: measuring the sequence-structure fit, designing energy functions, or analyzing protein simulation data.

In this work, we apply ML models to evaluate the accuracy of the two properties, KB energy and likelihood, that can be used as scoring functions. We have already assessed that likelihood is a more accurate measure of the sequence-structure fit. The goal of this study is to validate that using ML models. We want to check whether the model predictions of the likelihood values will be more accurate than those of KB energy. That will show that the likelihood provides the possibility of better use of structural information in prediction than the KB energy.

A common method to estimate the quality of model predictions is to use cross-validation and calculate the average prediction performance across test samples. Here, we use cross-validation in the context of predictive modeling. This is one of the most widely used data resampling methods to assess the generalizability of a predictive model and to prevent overfitting. To build the final model for the prediction of real future cases, the learning function (or learning algorithm) f is usually applied to the entire learning set. The purpose of cross-validation in the model-building phase is to estimate the performance of the final model on new data.

Cross-validation divides the training data into several disjointed cohorts of approximately equal size. Each cohort is used in turn as testing data, while the remaining cohorts are used as training data. The prediction model built on the training data is then applied to predict the class labels of the testing data. This process is repeated until all cohorts have been used as the testing data once, and then the prediction accuracies of all the blinded tests are combined to produce an overall performance estimate.

3 Related Work

Different bioinformatics and statistical approaches can be used to predict the 3D structure of a protein from its amino acid sequence. Many of these approaches can be viewed as sequence-structure fitness problems. In evaluating a hypothetical structure, such as the fitness of a sequence for a structure, one must be able to distinguish between correct and incorrect structures (to identify the structural states that have a high probability of being observed in given environmental conditions). Success or failure depends crucially on the underlying description of structural states and on the evaluation scheme of sequence-structure fitness [7].

Based on the thermodynamic hypothesis [2], computational studies of proteins, including structure prediction, folding simulation, and protein design, depend on the use of a potential function to calculate the effective energy of the molecule. In protein structure prediction, the potential function is used either to guide the conformational search process or to select a structure from a set of possible sampled candidate structures [8].

Two fundamentally different approaches exist to obtain a potential energy function [9]. The first is an inductive approach [4], a mathematical model that describes the system is assumed without previous knowledge about the physical principles. The resulting potential is directly extrapolated to more complex molecules by assuming that a common behavior will exist in both cases [9]. The second approach is deductive (or KB). In order to obtain an accurate description of the potential energy function, experimental data from large macro-molecular-solvent systems should be used [9]. The parameters of the potential functions are extracted from a database of known protein structures [4]. Because of the deductive nature f this approach, which incorporates many physical interactions (electrostatic, van der Walls, cation interactions) and the extracted potentials do not necessarily reflect true energies, it is often referred to as the "knowledge-based", "empirical", or "statistical" effective potential function or scoring function [8].

Current studies are focused on improving knowledge-based potentials used for: protein structure predictions, [10–12] RNA structure predictions [13, 14], and rational drug design [15].

More complex KB approaches use the advances of ML for protein structure prediction and sequence-structure fit assessment. Theoretically, the implementation of ML can be defined as both supervised learning, where the data includes additional attributes that are expected to be predicted, and unsupervised learning, where the training data consists of a set of input vectors without any corresponding target values. The supervised learning set of models consists of two groups: classification and regression. The large background of standard supervised ML methods provides reasonable results, but the advent of methods based on deep residual networks has shown more promising results in some cases.

Different ML methods have been applied as a tool for protein structure prediction based on KB potentials [16, 17]. It is expected that ML forcefields may soon replace forcefields in protein simulations [18].

Some alternative ML methods for structure prediction, such as probabilistic neural networks and deep learning end-to-end differentiable networks, have shown wider applicability [19].

There have also been attempts to apply likelihood functions as a tool for protein structure predictions using the multiple sequence alignment of related proteins as input data [20, 21]. Multiple sequence alignment shows which residues are evolutionarily related. A likelihood function indicates the probability of contact between different residues.

A significant problem in using ML models for sequence-structure fit is how to validate the results. Very often, this process is based on cross-validation of the outcomes of the applied models. Cross-validation is primarily used in applied ML to assess the potential and the accuracy of certain ML models for certain data. This means that it is possible to use a limited sample to estimate how the model will perform in general when used to make predictions on data not used during training. The cross-validation model can be used to estimate any quantitative measure that is appropriate for the data and the model. The use of cross-validation in sequence to structure fit evaluation models is discussed in [22].

4 Data Description

For the purposes of the study, we used two datasets: 1) a set of 200 protein structures for the calculation of the parameters used in KB energy and likelihood determination, and 2) a set of 45 protein structures for the optimization experiments. The first dataset was extracted from a selection of nonhomologous proteins [23]. The second dataset for testing purposes was obtained from the non-redundant PDB chain set of proteins with a sequence-similarity cut-off BLAST p-value of 10e−7, which is the most non-redundant of the given. The testing set contains 45.*pdb* files that meet the following criteria: having 0% unknown, incomplete, or missing residues or residues with incomplete side-chain; having only one chain (subunit) in the PDB entry; and not containing any heterogens (except for water). The models were determined by X-ray crystallography.

The sequence of every one of the 245 selected proteins is extracted from the .*pdb* file using *Biopython* [24, 25].

The solvent accessible surface area (SASA) of the residues is a geometric measure of exposure to the solvent. SASA is typically calculated by methods involving the in- silico rolling of a spherical probe, which approximates a water molecule, around a full-atom protein model [26]. The SASA of the protein molecule is the surface area traced by the center of the probe. A classical approximation commonly used to calculate SASA is the Lee and Richards (L&R) approximation [27], where the surface is approximated by the outline of a set of slices [28].

In this work, the Python module of FreeSASA, an open source C library [28], is used to calculate the solvent-accessible areas. SASA values for every residue in the protein are obtained by a high precision L&R calculation (probe radius: 1.400; slices: 100) using the default on FreeSASA ProtOr radii [29].

The relative solvent accessibility (RSA) of a residue indicates its degree of burial in a structure. The RSA calculation is important because different amino acids are of different sizes, so they also differ in area. To disregard these differences, the relative exposure (RSA) is calculated by normalizing the surface area of the residue in the structure by the surface area of the same type of residue in some reference state (e.g. the residue X in an extended tripeptide, such as Gly-X-Gly). RSA values are calculated by dividing the absolute SASA by the maximum solvent accessibility (maxSASA). Values for maxSASA based on ProtOr radii were extracted from the default reference values used in the FreeSASA classifier.

The calculated RSA was further divided into two states, using an exposure threshold of 0.1 (10%). Namely, a residue is considered buried (marked as 0) when RSA ≤ 0.1 and exposed (marked as 1) when RSA > 0.1. Each residue in a chain is then assigned to class 0 if it has an RSA lower than or equal to 0.1 and to class 1 if the RSA is higher.

5 Suggested Methodology

For the purposes of this study, we have developed an ML-driven approach for accuracy assessment of knowledge-based energy (E) and frequency base likelihood (L) for protein structure prediction. Both approaches are based on statistics of the buried/exposed properties of residues.

5.1   Data Preparation for ML

The sequence and pattern of every one of the 245 protein objects are transformed into numerical values that can be used as parameters in ML models.

To every type of amino-acid residue a) in the sequence, a corresponding number from b) is assigned:

a) 'A','R','N','D','C','E','Q','G','H','I','L','K','M','F','P','S','T','W','Y','V'
b) 10,11,12,13,14,15,16,17,18,19,20,21,22,23,24,25,26,27,28,29

The sequence of these numbers is specific for every protein and is used as parameter X1 in the ML models. The numbers representing all the residues in one protein sequence are then added to a value that is later used as parameter (p1) in the ML models.

In the patterns, every buried state (before represented as 0) is assigned the coefficient 0.2, and every exposed state (previously represented as 1) is assigned the coefficient 0.5. The sequence of these coefficients is then summed to obtain the second parameter (p2). The purpose of the coefficients is to represent the structural component as a distance.

The values of the KB energy (E) and the likelihood of base frequency (L) are used in the ML and are calculated for every one of the 245 proteins.

Outliers with values greater than five times the mean distance are removed from the study. After this filtering, a dataset generated from 244 native protein structures is used.

The 244 samples of the parameter values X1 and X2 are then individually normalized using the standard normalizer of the *scikit-learn* Python library [30].

To predict the KB energy and likelihood values, we used three supervised regression ML models. The chosen models are from python *scikit-learn* package: 1) Lasso – *linear*_model (alpha = 0.1), which is a regression analysis method that performs both variable selection and regularization to enhance the prediction accuracy and interpretability of the statistical model it produces; 2) Nearest Neighbors Regression (NNR) – *kNeighborsRegressor* (n_neighbors = 5, algorithm ='kd_tree'), which is a non-parametric method used for classification and regression. In both cases, the input consists of the k closest training examples in the feature space. The output depends on whether k-NN is used for classification or regression. We use a regression approach where the output is the property value for the object. This value is the average of the values of k nearest neighbors. 3) Decision tree regression (DTR) – *DecisionTreeRegressor* (max_depth = k). Decision Trees are a non-parametric supervised learning method used for classification and regression. The goal is to create a model that predicts the value of a target variable by learning simple decision rules inferred from the data features.

For every one of the models, a k-fold cross-validation is used to split the set into k smaller sets for better estimation.

As input, we used 244 samples of:

- Two normalized parameters, X1 and X2, that were obtained from data about the protein sequence and the protein structure, respectively.
- The actual values for KB energy or likelihood, obtained from formulas (2) and (3).

For k = 3, 5 and 7:

- The original sample is randomly partitioned into k equal sized subsamples.
- Of the k subsamples, a single subsample is retained as the validation data for testing the model, and the remaining k − 1 subsamples are used as training data.
- The cross-validation process is then repeated k times, with each of the k subsamples used exactly once as the validation data.
- The k train score results are then averaged to produce a single estimation (with a standard deviation).
- The predicted values are plotted against the original data.

The purpose of the suggested methodology is to show the difference in accuracy of prediction performance of the applied ML models based on the values of the two scoring functions: KB energy and likelihood.

6 Results and Discussion

The methodology of this study provides results based on the three above-described ML models and produces scores for comparing the accuracy of these models.

After the data set is normalized, we apply three supervised regression ML models: lasso regression, nearest neighbor regression, and decision tree regression. We test the cross-validation splitting strategy of k = 3, k = 5, and k = 7 folds to compare the models in terms of their accuracy of predicting the scores of KB energy and likelihood.

The graphs in Figs. 1, 2 and 3 show the relatedness of the actual to the predicted values of every particular model used for KB energy and for likelihood respectively with cross-validation (cv) k = 5.

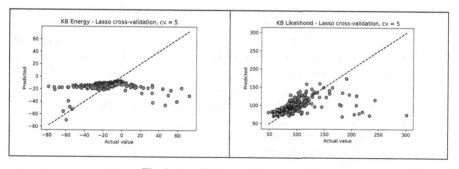

Fig. 1. Lasso cross-validation, k = 5

For the KB energy, lasso has worse predictive results than for likelihood, for which the results are distributed around the fit regression line with very few outliers from the greater actual value of likelihood.

The results of NNR are similar, with the KB energy estimates more dispersed than the values for likelihood.

The DTR produces somewhat similar results for the prediction of KB energy and likelihood values.

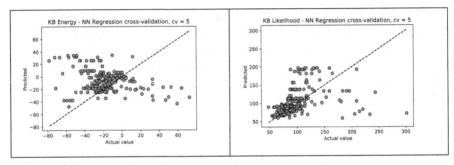

Fig. 2. Nearest neighbor regression cross-validation, k = 5

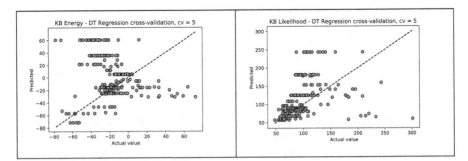

Fig. 3. Decision tree regression cross-validation, k = 5

These results are evidence that the likelihood prediction using is better than the KB energy prediction. These results confirm the analytical computational approach of optimization's finding that likelihood is superior to KB energy as a scoring function.

As a consequence of modelling the relatedness of the predicted to the actual values using the three ML approaches, we can refer to the coefficient of determination resulting from the training scores both for the KB energy and likelihood. The coefficient of determination shows the accuracy of the applied models.

All values given in Figs. 4, 5 and 6 are based on the average values for a particular splitting strategy with different k-fold numbers: 3, 5, and 7.

In Figs. 4, 5 and 6, the greater accuracy of the likelihood prediction ML models over the KB energy prediction ML models is obvious.

In Fig. 4, the NNR (Nearest Neighbor Regression) model with k = 3 produces higher mean values and smaller errors than the other two regression models. Lasso is less accurate model, while DTR (Decision Tree Regression) has an intermediate position.

The accuracy of the applied ML models changed when the splitting training set strategy amounts to five (Fig. 5.) In this case, the NNR and DTR models have very close average mean values and distributions of error values. The lasso regression model is obviously inferior to both NNR and DTR.

The increased accuracy of the DTR model for both KB energy and likelihood predictions is seen in Fig. 6. We can thus infer that, with a higher number of training data

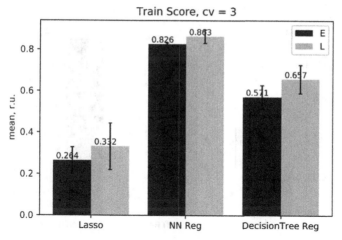

Fig. 4. Mean training scores for the three models with k-fold number cv = 3

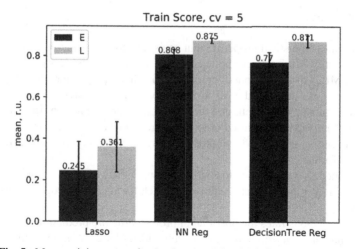

Fig. 5. Mean training scores for the three models with k-fold number cv = 5

sets, we can improve the accuracy of the DTR model. The most important finding is the overall superior accuracy of the likelihood prediction approach.

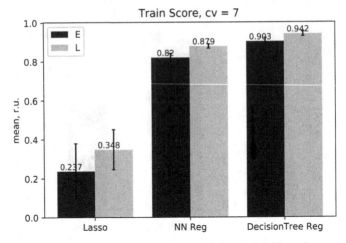

Fig. 6. Mean training scores for the three models with k-fold number cv = 7

7 Conclusions

In bioinformatics, statistical properties can be estimated using likelihood or likelihood function. Recently, ML has been applied as a tool to enhance this classical approach. We showed that ML is more efficient in predicting likelihood parameters than KB energy.

In our study, we developed a ML-driven approach for accuracy assessment of KB energy and frequency base likelihood for protein structure prediction. Both approaches are based on statistics of the buried or exposed properties of residues.

We proposed an approach for model comparison based on cross-validation of the estimated performance.

The ML models were applied to confirm the superiority of the frequency base likelihood approach over the KB based energy approach for assessing sequence-structure fit in proteins.

This study demonstrates the potential of protein structure prediction methods based on ML and indicates that combining ML with frequency base likelihood is more efficient than using KB energy functions.

Acknowledgments. The research presented in this paper was partly supported by the National Scientific Program "Information and Communication Technologies for a Single Digital Market in Science, Education and Security (ICTinSES)", financed by the Bulgarian Ministry of Education and Science. The work planning and conducting the discussed computational experiments was partly supported by the Sofia University SRF within the "A computational approach using knowledge-based energy and entropy for assessment of protein structure prediction" project. The authors are very grateful to the Institute of Biochemistry and Biophysics of the Polish Academy of Sciences for funding the publication of this study.

References

1. Berman, H.M., et al.: The protein data bank. Nucleic Acids Res. **28**(1), 235–242 (2000)

2. Anfinsen, C.B., Haber, E., Sela, M., White, F.H.: The kinetics of formation of native ribonucle-ase during oxidation of the reduced polypeptide chain. Proc. Natl. Acad. Sci. **47**(9), 1309–1314 (1961)

3. Shen, M.-Y., Sali, A.: Statistical potential for assessment and prediction of protein structures. Protein Sci. Publ. Protein Soc. **15**(11), 2507–2524 (2006)

4. Sippl, J.M.: Boltzmann's principle, knowledge-based mean fields and protein folding. An approach to the computational determination of protein structures. J. Comput. Aid. Mol. Des. **7**(4), 473–501 (1993). https://doi.org/10.1007/BF02337562

5. Lins, L., Thomas, A., Brasseur, R.: Analysis of accessible surface of residues in proteins. Protein Sci. **12**, 1406–1417 (2003)

6. Tanaka, S., Scheraga, H.A.: Medium- and long-range interaction parameters between amino acids for predicting three-dimensional structures of proteins. Macromolecules **9**(6), 945–950 (1976)

7. Ouzounis, C., Sander, C., Scharf, M., Schneider, R.: Prediction of protein structure by eval-uation of sequence-structure fitness: aligning sequences to contact profiles derived from three-dimensional structures. J. Mol. Biol. **232**(3), 805–825 (1993)

8. Li, X., Liang, J.: Knowledge-based energy functions for computational studies of proteins. In: Xu, Y., Xu, D., Liang, J. (eds.) Computational Methods for Protein Structure Prediction and Modeling: Volume 1: Basic Characterization, pp. 71–123. Springer, New York (2007). https://doi.org/10.1007/978-0-387-68372-0_3

9. Melo, F., Feytmans, E.: Scoring functions for protein structure prediction. Comput. Struct. Biol. **3**, 61–88 (2008)

10. Ciemny, M.P., Badaczewska-Dawid, A.E., Pikuzinska, M., Kolinski, A., Kmiecik, S.: Mod-eling of disordered protein structures using monte carlo simulations and knowledge-based statistical force fields. Int. J. Mol. Sci. **20**(3), 606 (2019)

11. López-Blanco, J.R., Chacón, P.: KORP: knowledge-based 6D potential for fast protein and loop modeling. Bioinformatics **35**(17), 3013–3019 (2019)

12. Yu, Z., Yao, Y., Deng, H., Yi, M.: ANDIS: an atomic angle- and distance-dependent statistical potential for protein structure quality assessment. BMC Bioinformatics **20**(1), 299 (2019). https://doi.org/10.1186/s12859-019-2898-y

13. Capriotti, E., Norambuena, T., Marti-Renom, M.A., Melo, F.: All-atom knowledge-based potential for RNA structure prediction and assessment. Bioinformatics **27**(8), 1086–1093 (2011)

14. Zhang, T., Hu, G., Yang, Y., Wang, J., Zhou, Y.: All-atom knowledge-based potential for rna structure discrimination based on the distance-scaled finite ideal-gas reference state. J. Comput. Biol. (2019)

15. Chen, P., et al.: DLIGAND2: an improved knowledge-based energy function for protein–ligand interactions using the distance-scaled, finite, ideal-gas reference state. J. Cheminform. **11**(1), 52 (2019). https://doi.org/10.1186/s13321-019-0373-4

16. Pei, J., Zheng, Z., Merz, K.M.: Random forest refinement of the KECSA2 knowledge-based scoring function for protein decoy detection. J. Chem. Inf. Model. **59**(5), 1919–1929 (2019)

17. Xu, J.: Distance-based protein folding powered by deep learning. Proc. Natl. Acad. Sci. **116**(34), 16856–16865 (2019)

18. Noé, F., De Fabritiis, G., Clementi, C.: Machine learning for protein folding and dynamics. Curr. Opin. Struct. Biol. **60**, 77–84 (2020)

19. James, G., Witten, D., Hastie, T., Tibshirani, R.: An Introduction to Statistical Learning with Applications in R. STS, vol. 103. Springer, New York (2013). https://doi.org/10.1007/978-1-4614-7138-7

20. Bywater, R.P.: Prediction of protein structural features from sequence data based on Shannon entropy and Kolmogorov complexity. PLoS ONE **10**(4), e0119306 (2015)

21. Aurell, E.: The maximum entropy fallacy redux? PLoS Comput. Biol. **12**(5), e1004777 (2016)
22. Rashid, S., Saraswathi, S., Kloczkowski, A., Sundaram, S., Kolinski, A.: Protein secondary structure prediction using a small training set (compact model) combined with a complex-valued neural network approach. BMC Bioinform. **17**(1), 1471–2105 (2016). https://doi.org/10.1186/s12859-016-1209-0
23. Zhang, Y., Skolnick, J.: TM-align: a protein structure alignment algorithm based on TM-score. Nucleic Acids Res. **33**(7), 2302–2309 (2005)
24. Hamelryck, T., Manderick, B.: PDB parser and structure class implemented in Python. Bioinformatics **19**, 2308–2310 (2003)
25. Cock, P.J.A., et al.: Biopython: freely available Python tools for computational molecular biology and bioinformatics. Bioinformatics **25**(11), 1422–1423 (2009)
26. Durham, E., Dorr, B., Woetzel, N., Staritzbichler, R., Meiler, J.: Solvent accessible surface area approximations for rapid and accurate protein structure prediction. J. Mol. Model. **15**(9), 1093–1108 (2009). https://doi.org/10.1007/s00894-009-0454-9
27. Lee, B., Richards, F.M.: The interpretation of protein structures: estimation of static accessibility. J. Mol. Biol. **55**, 379–400 (1971)
28. Mitternacht, S.: FreeSASA: An open source C library for solvent accessible surface area calculations. F1000Research (2016)
29. Tsai, J., Taylor, R., Chothia, C., Gerstein, M.: The packing density in proteins: standard radii and volumes. J. Mol. Biol. **290**(1), 253–266 (1999)
30. Pedregosa, F., Varoquaux, G., Gramfort, A., Michel, V.: Scikit-learn: machine learning in python. J. Mach. Learn. Res. **12**, 2825–2830 (2012)

Quantifying Overfitting Potential in Drug Binding Datasets

Brian Davis[1(✉)], Kevin Mcloughlin[2], Jonathan Allen[2], and Sally R. Ellingson[1]

[1] University of Kentucky Markey Cancer Center, Lexington, USA
sel228@uky.edu
[2] Lawrence Livermore National Laboratory, Livermore, USA

Abstract. In this paper, we investigate potential biases in datasets used to make drug binding predictions using machine learning. We investigate a recently published metric called the Asymmetric Validation Embedding (AVE) bias which is used to quantify this bias and detect overfitting. We compare it to a slightly revised version and introduce a new weighted metric. We find that the new metrics allow to quantify overfitting while not overly limiting training data and produce models with greater predictive value.

Keywords: Drug discovery · Machine learning · Data overfitting

1 Introduction

Protein-ligand interactions are important to most processes in the human body, and therefore to regulating disease via drugs. There are an estimated 20,000 different human protein-coding genes[1], and 10^{60} small molecules in the chemical universe [10]. Clearly, exploring all possible protein–drug pairs is not experimentally feasible. Drug discovery programs need accurate computational methods to predict protein–drug binding, and advances in machine learning have improved the accuracy of these predictions used in early stages of drug discovery. For a review of some current trends in making drug binding predictions using machine learning and available datasets see this current review [4].

The primary goal in protein–ligand binding modeling is to produce models that are capable of making accurate predictions on novel protein–drug pairs. Consequently, performance metrics need to reflect expected performance on novel data. This effort is frustrated by benchmark datasets that are not well-sampled from chemical space, so that novel pairs may be relatively far from data available to the modeler. Area under the curve (AUC) scores for various curves are often provided to support suitability of a machine learning model for use in

Supported by Lawrence Livermore National Laboratory and the University of Kentucky Markey Cancer Center.

---

[1] With about one-eighth of the exome containing observable genetic variations [8].

© Springer Nature Switzerland AG 2020
V. V. Krzhizhanovskaya et al. (Eds.): ICCS 2020, LNCS 12139, pp. 585–598, 2020.
https://doi.org/10.1007/978-3-030-50420-5_44

drug binding prediction—in this paper we focus on the Precision–Recall curve and its associated AUC (PR-AUC). Care must be taken in interpreting performance metrics like the PR-AUC, as laboratory experiments reveal that inferring real-world performance from AUC alone is overly-optimistic. This issue of generalizability is common in machine learning applications, but is particularly relevant in settings with insufficient and non-uniformly distributed data, as is the case with drug binding data.

The phenomenon of high performance metrics for low quality models is called overfitting, and is typically combated by using different data for the processes of model training and validation. If the validation data and real-world data for the model application are both distinct from the training data, then we expect the performance metrics on the validation data to be representative of the real-world performance of the model. A common way of splitting the available data into training and validation sets is to select a training ratio and randomly assign that proportion of the data to the training set.

A developing framework to account for overfitting is based on the assumption that the Nearest Neighbor (NN) model has poor generalizability. In the NN model, the test data is classified based on the classification of its nearest neighbor in the training data. Therefore, NN basically memorizes the training data and does not generalize to anything not close to it. Within the context of reporting "fair" performance metrics, this working assumption of poor generalizability of NN models suggests several possibilities for more informative metrics, including:

1. reporting the PR-AUC for a model produced from a training/validation split on which the Nearest Neighbor model has poor performance, and
2. designing a metric which weights each validation molecule according to its relative distance to the binding classes in the training set.

We describe implementations of each of these approaches in this paper. For the first approach, we discuss the efforts presented in the Atomwise paper [14] to produce training/validation splits that are challenging for NN models, hereafter referred to as the Atomwise algorithm. We also describe two variations of the Atomwise algorithm: ukySplit–AVE and ukySplit–VE. As distinct from optimization, we introduce a weighting scheme ω designed to address the second approach, and discuss the consequences of using an ω-weighted PR-AUC versus the traditional PR-AUC with a training/validation split produced by the ukySplit–AVE algorithm.

To summarize the problem at hand, there are too many combinations of potential therapies and drug targets to understand them by experiment alone; simulations still suffer from inaccuracies and high computational costs; with the wealth of biomedical data available, machine learning is an attractive option; current models suffer from data overfitting, where feature sets are too large and patterns can be observed that do not generalize to new data; the current project discusses ways of quantifying and better understanding the potential for overfitting without limiting the predictability of models.

1.1   Current Bias Quantification Methods

For a more articulated description of the original spatial features that inspired this work, including an evaluation of several benchmark datasets with figures to help conceptualize the ideas please see the recent review work [4].

Datasets with a metric feature space can be evaluated using spatial statistics [12] to quantify the dataset topology and better understand potential biases. Of particular interest in the area of drug–binding model generalization are the "nearest neighbor function" G(t) and the "empty space function" F(t). G(t) is the proportion of active compounds for whom the distance to the nearest active neighbor is less than t. F(t) is the proportion of decoy compounds for whom the distance to the nearest active neighbor is less than t. Letting $\sum$G and $\sum$F denote the sum of the values of G and F over all thresholds t, it is reported that large values of $\sum$G indicate a high level of self-similarity and that small values of $\sum$F indicate a high degree of separation. The difference of $\sum$G and $\sum$F gives a quick and interpretable summary of a dataset's spatial distribution, with negative values indicating clumping, near-zero values indicating a random-like distribution, and positive values indicating larger active-to-active distance than decoy-to-active. These spatial statistics were used to develop the Maximum Unbiased Validation (MUV) dataset, with the goal of addressing the reported association of dataset clumping with overly–optimistic virtual screening results [11,12].

Wallach et al. [14] extended the MUV metric, and used it to quantify the spatial distribution of actives and decoys among the training and validation sets. For two subsets V and T of a metric data set with distance function d, define, for each v in V, the function $I_t(v, T)$ to be equal to one if $\min_{w \in T}\{d(v, w)\} < t$ and zero otherwise. For a fixed value of n, define the function $H_{(V,T)}$ by

$$H_{(V,T)} = \frac{1}{n+1} \cdot \frac{1}{|V|} \sum_{v \in V} \left(\sum_{i=0}^{n} I_{i/n}(v, T) \right). \tag{1}$$

Then the Asymmetric Validation Embedding (AVE) bias is defined to be the quantity

$$B(V_A, V_I, T_A, T_I) = H_{(V_A,T_A)} - H_{(V_A,T_I)} + H_{(V_I,T_I)} - H_{(V_I,T_A)}, \tag{2}$$

where the value of n is taken to be 100, and where V_A and V_I are the validation actives and inactives (decoys), respectively, and similarly T_A and T_I are the training actives and inactives. For convenience, we abbreviate $H(V_a, T_a) - H(V_a, T_i)$ and $H(V_i, T_i) - H(V_i, T_a)$ as $(AA - AI)$ and $(II - IA)$, respectively. They are intended to be a quantification of the "clumping" of the active and decoy sets. If the term $(AA - AI)$ is negative, it suggests that, in the aggregate, the validation actives are closer to training decoys than to training actives, with the consequence that the active set is expected to be challenging to classify. If the sum of $(AA - AI)$ and $(II - IA)$ (the AVE bias) is close to zero, it is expected that the data set is "fair", in that it does not allow for easy classification due to clumping. The authors also provide an AVE bias minimization algorithm.

It is a genetic algorithm with breeding operations: merge, add molecule, remove molecule, and swap subset. The algorithm first generates initial subsets through random sampling, measures the bias, and selects subsets with low biases for breeding. The algorithm repeats bias scoring, redundancy removal, and breeding until termination based on minimal bias or maximum iterations.

In their paper, Wallach et al. observe that AUC scores[2] and AVE bias scores are positively correlated for several benchmark data sets, implying that model performance is sensitive to the training/validation split.

In this paper, we present an efficient algorithm for minimizing the AVE bias of training/validation splits. We introduce a variation on the AVE bias, which we call the VE score, and describe its advantages in the context of optimization. We investigate the efficacy of minimizing these metrics for training/validation splits, and conclude by proposing a weighted performance metric as an alternative to the practice of optimizing training/validation splits.

2   Methods

2.1   Dataset

Dekois 2 [2] provides 81 benchmark datasets: 80 with unique proteins, and one with separate datasets for two different known binding pockets in the same protein. The active sets are extracted from BindingDB [6]. Weak binders are excluded, and 40 distinct actives are selected by clustering Morgan fingerprints by Tanimoto similarity. Three datasets are extended by selecting up to 5 actives from each structurally diverse cluster. The decoy set is generated using ZINC [7] and property matched to the actives based on molecular weight, octanol-water partition coefficient (logP), hydrogen bond acceptors and donors, number of rotatable bonds, positive charge, negative charge, and aromatic rings. Possible latent actives in the decoy set are removed using a score based on the Morgan fingerprint and the size of the matching substructures. Any decoy that contained a complete active structure as a substructure is also removed.

2.2   Bias Metrics

Throughout this paper, the term fingerprint refers to the 2048-bit Extended Connectivity Fingerprint (ECFP6) of a molecule as computed by the Python package RDKit [1]. For simplicity, we define

$$d(v, T) := \min_{t \in T}\{d(v, t)\}$$

and

$$\Gamma(v, T) := \frac{\lfloor n \cdot d(v, T) \rfloor}{n + 1},$$

---

[2] They report ROC-AUC scores, as opposed to PR-AUC scores.

where $d(v,t)$ is the Tanimoto distance between the fingerprints of the molecules v and t. We compute the AVE bias via the expression

$$\text{mean}_{v \in V_A}\{\Gamma(v,T_I) - \Gamma(v,T_A)\} + \text{mean}_{v \in V_I}\{\Gamma(v,T_A) - \Gamma(v,T_I)\}, \quad (3)$$

where V_A and V_I are the validation actives and inactives (decoys), respectively, and similarly T_A and T_I are the training actives and inactives. For a derivation of the equivalence of this expression and Expression (2), see the Appendix. Since

$$|d(v,T) - \Gamma(v,T)| < \frac{1}{n+1},$$

for large values of n Expression (3) (and hence the AVE bias) is an approximation of

$$\text{mean}_{v \in V_A}\{d(v,T_I) - d(v,T_A)\} + \text{mean}_{v \in V_I}\{d(v,T_A) - d(v,T_I)\}. \quad (4)$$

We now introduce the VE score, a close relative of the AVE bias:

$$\sqrt{\text{mean}_{v \in V_A}^2\{d(v,T_I) - d(v,T_A)\} + \text{mean}_{v \in V_I}^2\{d(v,T_A) - d(v,T_I)\}}. \quad (5)$$

While the raw ingredients of the AVE bias and the VE score are the same, they are qualitatively different, in particular as the VE score is never negative.

We generate a random training/validation split for each Dekois target and evaluate Expressions (2) through (5) 1,000 times with a single thread on an AMD Ryzen 7 2700x eight-core processor. We compare the mean computation times, as well as the computed values.

2.3 Split Optimization

We implement two custom genetic optimizers, ukySplit-AVE and ukySplit-VE, using the open source DEAP [5] framework. Both ukySplit-AVE and ukySplit-VE optimizers use parameters as described in Table 1. The parameters were chosen after grid-searching for minimum mean-time-to-debias on a sample of the Dekois targets.

Table 1. Evolutionary algorithm parameters

Parameter name	Meaning	Value
POPSIZE	Size of the population	500
NUMGENS	Number of generations in the optimization	2000
TOURNSIZE	Tournament Size	4
CXPB	Probability of mating pairs	0.175
MUTPB	Probability of mutating individuals	0.4
INDPB	Probability of mutating bit of individual	0.005

The optimizer populations consisted of training/validation splits, and the objective functions were given by Expressions (3) and (5), respectively, for *valid* splits, and equal to 2.0 otherwise. We say that a split is valid if

1. the validation set contains at least one active and one decoy molecule,
2. the active/decoy balance in the validation set is within 5% of that in the total dataset,
3. the ratio of training/validation set sizes is $80 \pm 1\%$.

2.4   Modeling

Using scikit-learn [9], we train a random forest classifier ($n\_estimators = 100$) with stratified 5-fold cross-validation and compute the mean PR-AUC for each target of the Dekois data set. We use fingerprints as features, and take the probability of the active class as the output of the model. For each of the folds, we evaluate the Expressions (3) and (5), and report Pearson correlation coefficients with the PR-AUC.

For the training/validation splits produced by an optimizer, we compute PR-AUC of a random forest model and evaluate Expression (3) or (5) as applicable.

2.5   Nearest Neighbor Similarity

An assumption this paper builds upon is that the Nearest Neighbor model does not generalize well since it memorizes the training data. Good metrics come from a Nearest Neighbor model that is only tested using data points very similar to data points in the training data and having the same label. Therefore, we use a similarity measure to the Nearest Neighbor model to show the potential of a model to not generalize.

We gather the binary predictions made by the Nearest Neighbor model, which predicts the class of a validation molecule to be the same as its nearest neighbor (using the metric d) in the training set. Considering the NN predictions as a bit string, we can compare it with the prediction bit string of any other model m using the Tanimoto similarity T:

$$T(NN, m) = \frac{\sum (NN \wedge m)}{\sum (NN \vee m)},$$

with bitwise operations $\wedge$ (and) and $\vee$ (or) and sums over all predictions for the validation set. We take the maximum Tanimoto similarity over all thresholds η for each of the validation folds, and report the mean.

2.6   Weighted PR-AUC

The weighted metric described here gives less of a contribution to the model's performance metric when a tested data point is very similar to a training data point.

For a given model, let TP, TN, FP, and FN be the collections of molecules for which the model predictions are true positive, true negative, false positive, and false negative, respectively. The metrics precision and recall may be easily generalized by assigning a weight $\omega(v)$ to each molecule v, and letting the ω–weighted precision be given by

$$\frac{\sum_{v \in \text{TP}} \omega(v)}{\sum_{v \in \text{TP}} \omega(v) + \sum_{v \in \text{FP}} \omega(v)}$$

and the ω–weighted recall be given by

$$\frac{\sum_{v \in \text{TP}} \omega(v)}{\sum_{v \in \text{TP}} \omega(v) + \sum_{v \in \text{FN}} \omega(v)}.$$

Setting the weight $\omega(v)$ equal to 1 for all molecules v, we recover the standard definitions of precision and recall.

Inspired by Expression (4), we define the ratio $\gamma(v)$ by

$$\gamma(v) = \begin{cases} \frac{d(v,T_A)}{d(v,T_I)} & \text{if } v \text{ is active,} \\[2mm] \frac{d(v,T_I)}{d(v,T_A)} & \text{if } v \text{ is decoy.} \end{cases}$$

When we refer to the weighted PR-AUC in this paper we use the weight ω given by the cumulative distribution function of γ over the validation set for the target protein. Note that the weights ω are between zero and one, and that the weighting de-emphasizes molecules that are closer to training molecules of the same binding class than to training molecules of the opposite class. Thus the larger the contribution of a molecule to the AVE bias, the lower its weight. For further description of the ω–weighted PR-AUC, see the Appendix of [3].

2.7 Generalizability

Inspired by recent work presented on the so-called "far AUC" [13], we attempt to measure the ability of a drug-binding model to generalize. We randomly split the data set for each target 80/20 (preserving the class balance), then remove any molecules in the 80% set that has a distance less than 0.4 from the 20% set. We reserve the 20% set to serve as a proxy for novel data "far" from the training data. We then treat the remainder of the 80% as a data set, running the same analysis as described in the earlier subsections: computing the weighted and unweighted PR-AUC of a random forest trained on random splits, as well as the PR-AUC of random forest models trained on ukySplit-AVE and ukySplit-VE optimized splits.

3 Results

3.1 Computational Efficiency

A naive implementation of Eq. (2) required a mean computation time over all Dekois targets of 7.14 ms, while an implementation of Eq. (3) had a

Table 2. Computational efficiency

Expression	Mean computation time	Relative speedup
(2)	7.14 ms	1
(3)	0.99 ms	7.2
(4)	0.31 ms	23.4
(5)	0.31 ms	23.1

mean computation time of 0.99 ms. The mean computation times for Expressions (4) and (5) were both approximately 0.31 ms.

Evaluations of Expressions (2) through (5) are plotted in Fig. 1. The absolute differences between the computed value of Expression (2) and Expressions (3) and (4) are summarized in Table 3. It is not meaningful to compare the evaluations of Expressions (2) and (5) in this way, as they measure different, though related, quantities.

Table 3. Comparison with Expression (2) over Dekois targets

Expression	Mean abs difference	Max abs difference
(3)	3.1×10^{-4}	4.1×10^{-3}
(4)	9.9×10^{-3}	2.3×10^{-2}

For reference, the AVE paper considered a score of 2×10^{-2} to be "bias free".

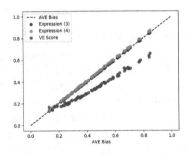

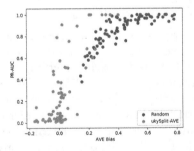

Fig. 1. Comparison of evaluations over Dekois targets

Fig. 2. Mean Split AVE Bias vs. Model PR-AUC.

3.2 Split Bias and Model Performance

In Fig. 2, we plot the mean PR-AUC against the mean AVE bias for 5-fold cross validation on each Dekois target. The Pearson correlation coefficient between mean AVE bias and mean PR-AUC is computed to be 0.80, which is comparable in strength to the correlation reported in the AVE paper for other benchmark datasets. We also plot the AVE bias against the mean PR-AUC for each target after optimization by ukySplit–AVE. Note that, although the optimizer was run with a stopping criterion of 0.02, it is possible for the minimum AVE bias to jump from greater than 0.02 to a negative number (as low as −0.2) in one generation.

We order the target proteins by AVE bias, and plot the two components, AA-AI and II-IA, after optimization by ukySplit–AVE in Fig. 3.

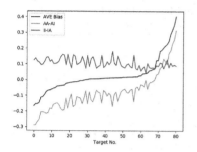

Fig. 3. The two components of the AVE Bias score.

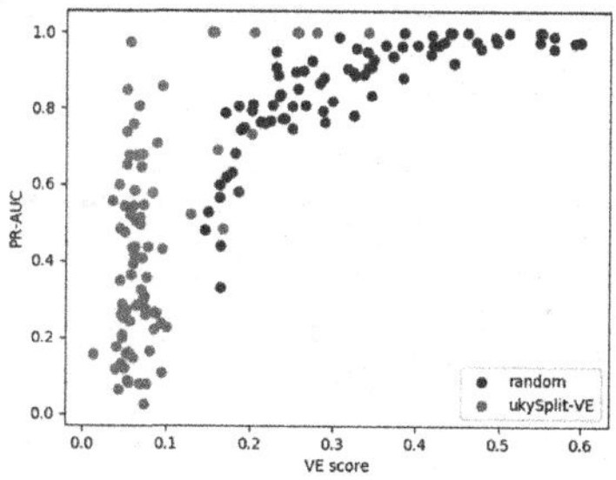

Fig. 4. Mean VE Score vs. Model PR-AUC.

3.3 Optimization by UkySplit–VE

Figure 4 plots the mean VE score against the mean PR-AUC across each fold of the cross-validation split for each target before and after optimization with ukySplit–VE (minimizing VE score as opposed to AVE bias). Figure 5 plots the score components associated to the active and decoy validation molecules after optimizing VE score (for comparison with Fig. 3).

3.4 Weighted Performance

Figure 6 plots the ω–weighted PR-AUC against the mean AVE bias over each fold for each target. Recall that the models and predictions are the same as those represented in Fig. 2 with label "random", but that the contributions of each validation molecule to the precision and recall are now weighted by the weight function ω.

3.5 Nearest Neighbor Similarity

Figure 7 plots the NN- similarity of a random forest model trained on splits produced randomly, by ukySplit-AVE, and by ukySplit-VE. The mean NN-similarities were 0.997, 0.971, and 0.940, respectively.

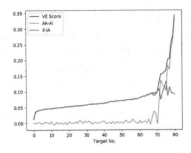

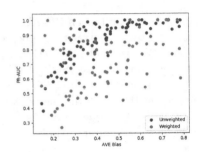

Fig. 5. The two components of the VE Score.

Fig. 6. Mean Split AVE Bias vs. Model Weighted PR-AUC.

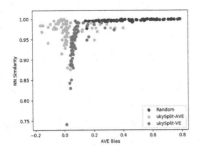

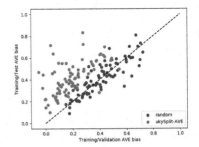

Fig. 7. NN similarity of a random forest model trained on various splits.

Fig. 8. AVE bias of training/validation and training/test splits.

3.6   Model Performance on Distant Data

After reserving 20% of each target's data for the test set, approximately 3% of the remainder was found to be within the 0.4 buffer distance, and was removed before splitting into training and validation sets. Similarly to Sundar and Colwell [13], we find that de-biasing training/validation splits does not lead to increased performance on "distant" test sets: the mean ratio of test PR-AUC before and after split optimization by ukySplit-AVE was 1.010 (1.018 for ukySplit-VE).

Figure 8 plots the AVE bias on the training/test split against the AVE bias on the training/validation split (letting the test set play the role of the validation set in the AVE bias definition). Figure 9 shows the validation and test PR-AUC of a model trained with a training set produced randomly, by ukySplit-AVE, and by ukySplit-VE.

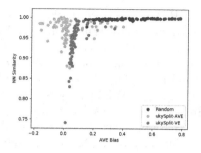

Fig. 9. Model performance on validation vs. test set of a model trained on various splits.

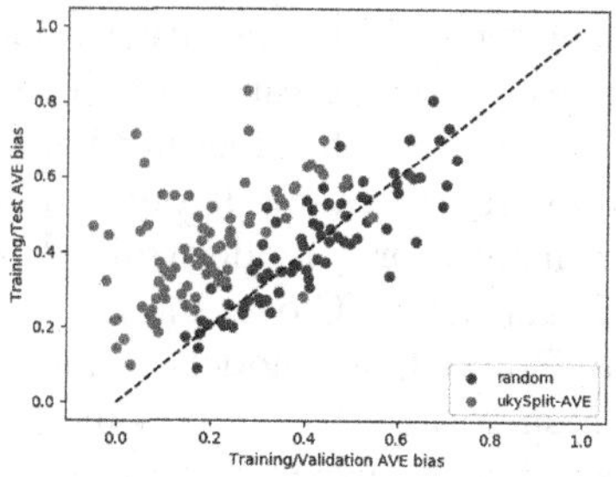

Fig. 10. Comparison of ukySplit-AVE, ukySplit-VE, and weighting methods.

4 Discussion

4.1 Computing Bias

As presented in Table 2, refactoring Expression (2) into Expression (3) yielded speedups of 7x, and the additional use of exact, rather than approximated, values yielded a speedup of roughly 23x for Expression (4). While Expressions (2) and (3) are mathematically equivalent, in practice they yield slightly different results due to machine precision. In the aggregate over the Dekois dataset, the difference is negligible relative to the established definition of "de-biased", as described in Table 3. Expressions (2) and (4) are *not* mathematically equivalent. In light of the equivalence of Expressions (2) and (3), it is clear that AVE bias (Expression (2)) is an *approximation* of Expression (4). Their difference, though slight, is properly interpreted as approximation error in the AVE bias.

4.2 Model Effects of Debiasing

Figures 2 and 4 demonstrate that the process of minimizing bias in the training/validation split risks training a model with little or no predictive value. The expected recall (and precision) for a random guessing model is equal to the balance of active molecules, which for the Dekois dataset is 3%. Of the 81 Dekois targets, 21 (about 26%) had models with below random PR-AUC when trained and validated on a split produced by ukySplit–AVE. This may be understood by considering Fig. 3, which shows that the AVE bias is primarily an indication of the spatial distribution of the (minority) active class in the validation set, with low AVE bias associated with active validation molecules that are closer to training decoys than to training actives. Models trained on such splits are therefore prone to misclassify validation actives, and hence have a low PR-AUC. This phenomenon is less pronounced when splits are optimized for VE score (ukySplit–VE), as it does not allow terms to "cancel out", and so does not incentivize pathological distributions of validation actives. It can be seen when comparing Figs. 3 and 5. When using the AVE bias score (in Fig. 3), AA-AI can get "stuck" at a negative value and then II-IA tends towards the absolute value of AA-AI to result in an overall bias score near zero. When using the VE score (in Fig. 5), this does not happen. Since AA-AI can never be negative, II-IA will not try to cancel it out. Only one Dekois target had worse-than-random performance for a model trained on a split optimized for VE score, and while the mean PR-AUC over models trained with ukySplit–AVE splits was 0.26, the mean PR-AUC for models trained on ukySplit–VE splits was 0.44. Since one of the assumptions built upon in this paper is that the Nearest Neighbor model does not generalize well, we use a measure of similarity to the Nearest Neighbor model as a measure of potential to not generalize well. It can be seen in Fig. 7 that models built on random data can be assumed to not generalize well, that models built on data splits using the AVE bias may sometimes do better but it does not completely alleviate the problem, and that models built on data splits using the VE bias do a better job at diverging from a Nearest Neighbor model.

Therefore, VE may be a better score to debias datasets because datasets debiased with VE are less similar to the NN model and produce a higher PR-AUC.

4.3   Weighted PR-AUC

As described in the Introduction, a weighted metric represents an alternative way of taking into account the assumption of poor generalizability of Nearest Neighbor models. While bias optimization creates training/validation splits that are more challenging for Nearest Neighbor models, they simultaneously result in low quality models, even when using powerful methods like random forests. In Fig. 5, when using an unweighted metric with random data the expected trend of higher bias scores coming from models with better performance metrics can be seen and it can be assumed that this is from data overfitting. However, the trend is not present (or much less pronounced) when using the weighted metric. The assumption here is that the weighted metric gives a better representation of the generalizability of the model without having to limit the training data. The weighted metric ω-PR-AUC discounts the potentially inflated performance of models without degrading the models themselves (see Fig. 10). It is worth noting, as well, that the computational expense of computing the weighting ω is negligible compared with the intensive work performed in optimizing training/validation splits.

The weighted PR-AUC may be used to infer that the standard PR-AUC is inflated due to the spatial distribution of the validation set. In particular, if two trained models have the same performance, the one with the higher weighted performance may be expected to have better generalizability.

4.4   Test Performance

Figure 8 shows that minimizing the AVE bias on the training/validation split does not minimize the AVE bias on the training/test split. Figure 9 demonstrates that even when a split results in a trained model with very low validation PR-AUC, the model still performs fairly well on the test data.

5 Conclusions

In this paper an existing bias metric was evaluated that quantities potential for data overfitting and can be used to optimize splits in order to build models with test data that give a better indication of how generalizable a model may be. An improvement was made to the score by not allowing one of the terms to be negative which was leading to problems during data split optimizations. However, the biggest contribution is the introduction of using a weighted metric instead of optimized data splits. This allows for the training data to not be limited which leads to models with no predictive power while not inflating performance metrics. This will hopefully lead to models that are more predictive on novel real world test data with performance metrics that better represent that potential. Developers of machine learning models for virtual high-throughput screening will have to contend with issues of overfitting as long as drug binding data is scarce. We propose the use of weighted performance metrics as a less computation-intensive alternative to split optimization. If the weighted and un-weighted metrics diverge, it can be concluded that the good performance of a model is concentrated at data points on which a nearest-neighbor model is sufficient.

Future work includes combining a protein distance with the drug distance to represent protein-drug pairs in a dataset. This is to evaluate large datasets to make multi-protein prediction models.

References

1. RDkit, open-source cheminformatics. http://www.rdkit.org
2. Bauer, M.R., Ibrahim, T.M., Vogel, S.M., Boeckler, F.M.: Evaluation and optimization of virtual screening workflows with DEKOIS 2.0-a public library of challenging docking benchmark sets. J. Chem. Inf. Model. **53**(6), 1447–1462 (2013)
3. Davis, B., Mcloughlin, K., Allen, J., Ellingson, S.: Split optimization for protein/ligand binding models. arXiv preprint arXiv:2001.03207 (2020)
4. Ellingson, S.R., Davis, B., Allen, J.: Machine learning and ligand binding predictions: a review of data, methods, and obstacles. Biochimica et Biophysica Acta (BBA)-Gen. Subj. **1846**(6), 129545 (2020)
5. Fortin, F.A., De Rainville, F.M., Gardner, M.A., Parizeau, M., Gagné, C.: DEAP: evolutionary algorithms made easy. J. Mach. Learn. Res. **13**, 2171–2175 (2012)
6. Gilson, M.K., Liu, T., Baitaluk, M., Nicola, G., Hwang, L., Chong, J.: BindingDB in 2015: a public database for medicinal chemistry, computational chemistry and systems pharmacology. Nucleic Acids Res. **44**(D1), D1045–D1053 (2015)
7. Irwin, J.J., Sterling, T., Mysinger, M.M., Bolstad, E.S., Coleman, R.G.: Zinc: a free tool to discover chemistry for biology. J. Chem. Inf. Model. **52**(7), 1757–1768 (2012)
8. Lek, M., et al.: Analysis of protein-coding genetic variation in 60,706 humans. Nature **536**(7616), 285 (2016)

9. Pedregosa, F., et al.: Scikit-learn: machine learning in Python. J. Mach. Learn. Res. **12**, 2825–2830 (2011)

10. Reymond, J.L., Awale, M.: Exploring chemical space for drug discovery using the chemical universe database. ACS Chem. Neurosci. **3**(9), 649–657 (2012)

11. Rohrer, S.G., Baumann, K.: Impact of benchmark data set topology on the validation of virtual screening methods: exploration and quantification by spatial statistics. J. Chem. Inf. Model. **48**(4), 704–718 (2008)

12. Rohrer, S.G., Baumann, K.: Maximum unbiased validation (MUV) data sets for virtual screening based on PubChem bioactivity data. J. Chem. Inf. Model. **49**(2), 169–184 (2009)

13. Sundar, V., Colwell, L.: Debiasing algorithms for protein ligand binding data do not improve generalisation (2019)

14. Wallach, I., Heifets, A.: Most ligand-based classification benchmarks reward memorization rather than generalization. J. Chem. Inf. Model. **58**(5), 916–932 (2018)

Detection of Tumoral Epithelial Lesions Using Hyperspectral Imaging and Deep Learning

Daniel Vitor de Lucena[1,2(✉)], Anderson da Silva Soares[2],
Clarimar José Coelho[3], Isabela Jubé Wastowski[4],
and Arlindo Rodrigues Galvão Filho[3]

[1] Instituto Federal de Educação, Ciências e Tecnologia de Goiás, Luziânia, Brazil
daniel.lucena@ifg.edu.br
[2] Instituto de Informática, Universidade Federal de Goiás, Goiânia, Brazil
anderson@inf.ufg.br
[3] Escola de Informática, Pontifícia Universidade Católica de Goiás, Goiânia, Brazil
clarimarc@gmail.com, argfilho@gmail.com
[4] Pós-Graduação em Ciências Aplicadas a produtos para Saúde,
Universidade Estadual de Goiás, Goiânia, Brazil
wastowski@gmail.com
http://ifg.edu.br, http://inf.ufg.br,
https://www.pucgoias.edu.br, http://www.ueg.br

Abstract. We propose a new method for the analysis and classification of HSI images. The method uses deep learning to interpret the molecular vibrational behaviour of healthy and tumoral human epithelial tissue, based on data gathered via SWIR (short-wave infrared) spectroscopy. We analyzed samples of Melanoma, Dysplastic Nevus and healthy skin. Preliminary results show that human epithelial tissue is sensitive to SWIR to the point of making possible the differentiation between healthy and tumor tissues. We conclude that HSI-SWIR can be used to build new methods for tumor classification.

Keywords: Short-Wave InfraRed · Hyperspectral Imaging · Deep learning · Skin lesions · Dysplastic Nevus · Melanoma

1 Introduction

Skin cancer is the most diagnosed malignant tumor in the whole world [48]. This pathology usually presents in two ways: a) Melanoma, originated from skin cells that produce pigments, called melanocytes and b) the non-melanoma [25].

Although less frequent than other tumors, melanoma is the most aggressive type of skin cancer due to the high possibility of metastasis and high mortality.

Acknowledgments for supporting this research to Instituto Federal de Educação, Ciência e Tecnologia de Goiás for the qualification license and Fundação de Amparo à Pesquisa do Estado de Goiás for the scholarship.

Currently, melanoma accounts of approximately 3% of skin cancer cases and 74% of deaths [5, 11, 20, 25, 39, 49]. In 2015, it was estimated that in the United States alone 73,870 new cases of melanoma would be diagnosed with 9,940 deaths [47]. It was estimated that in 2017 there were 87,110 diagnoses of the disease with about 9,730 deaths [49]. In Brazil it is estimated for each year of 2018–2019 biennium, the occurrence of 6,620 new cases diagnosed [25].

2 Problem

To increase the chances of survival of patients with melanoma, early diagnosis is essential. When detected at early stages, chances of healing are high, but late diagnosis makes treatment ineffective [20]. However, the traditional method of skin cancer detection begins with a visual inspection. If a suspicious stain is identified, the doctor will analyze characteristics such as size, color and texture, besides questions about the stains [48]. Along with the visual inspection, some dermatologists apply a technique called dermoscopy, also known as epiluminescense microscopy (ELM). This technique uses the dermatoscope, a surface microscope with a light source that is held close to the skin for a more detailed view of the lesions [9, 26]. If suspicions remain, then further examinations such as blood tests, biopsies and imaging tests may be performed to confirm or deny the diagnosis [48].

To overcome the difficulties inherent in manual physician visual inspection, the use of an automated method to assist with the task of identifying suspicious stains may increase the effectiveness of the inspection and reduce the subjectivity of the examination.

3 Proposed Solution

As an alternative to the traditional method for investigating skin lesions, in particular, Melanoma, performed by manual dermoscopic inspection, automated inspection by Hyperspectral Imaging (HSI) is proposed. In the context of medical imaging, HSI is an emerging technology that provides, in addition to data such as size and shape, information on the chemical composition of matter analyzed from a set of spatially arranged spectral signals, where each spectral signal corresponds to the electromagnetic interaction of light with the material analyzed in a specific portion of the sample [33]. HSI has been used for the last two decades in medical applications [1, 7, 8, 29, 37, 44, 52] because it offers great potential for the diagnosis of noninvasive diseases, surgical guidance [33] and in particular the diagnosis of tumors [2–4, 10, 16, 18, 19, 22, 28, 32, 34, 35, 38, 41, 42, 45, 46]. Thus, automated inspection employing HSI increases the chances of identifying a suspicious stain of tumor tissue even when the stain is very small and its shape, color and texture are insufficient for accurate dermoscopic identification.

An HSI is composed of n two-dimensional images built from the values measured at a given wavelength [12,40]. Figure 1 exemplifies the organization of the n two-dimensional images on an HSI and the spatial arrangement of the spectrum within the image.

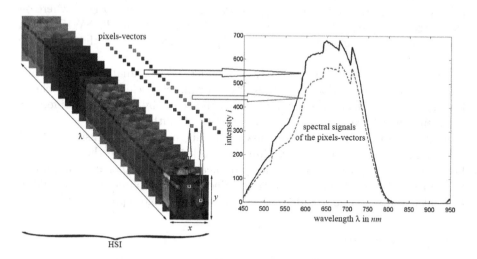

Fig. 1. Representation of the layer structure of an HSI, the spatial arrangement of pixels-vectors and their respective spectral signals. Adapted from Akbari *et al.* [2].

The set of sequentially arranged multicolored frames represent the hypercube. Each frame corresponds to a two-dimensional image of a specific wavelength within the hypercube spectral range. The vertical arrows indicate two distinct sets of pixels selected from the same spatial reference in all images. Each set makes up what is called a pixel-vector. The horizontal arrows indicate the spectral representations of two HSI pixels-vectors in the Cartesian plane, where the λ axis corresponds to the HSI wavelengths and the i axis corresponds to the measured light intensity.

3.1   SWIR Spectroscopy

The construction of an HSI is usually performed employing a particular method of spectroscopy. Each spectroscopic technique acts on a particular region of the light wave spectrum. A region still little explored in medical applications, especially in the diagnosis of tumors is the *Short-Wave Infrared* (SWIR). SWIR comprises the wavelengths between 1000 nm and 3000 nm [51]. Matter when irradiated by electromagnetic waves in the SWIR region, provides a molecular vibrational behavior which intensity is determined by the energy absorption at a given wavelength. [6,23].

Suppose an HSI as illustrated in Fig. 1, obtained by SWIR spectroscopy from an epithelial tissue sample, where the values measured at each wavelength correspond to the energy absorption due to the molecular vibration of this tissue. Each pixel-vector of this HSI will correspond to the molecular vibration of the tissue at that specific location. Thus, the molecular vibration contained in any HSI pixel-vector can be represented as a spectral signal, as shown in Fig. 1, where on the Cartesian plane the λ axis corresponds to wavelengths in the SWIR region and the i axis corresponds to the intensity value measured at the respective wavelength of the λ axis.

Healthy and tumoral epithelial tissues under SWIR radiation may provide different energy absorption intensities at one or more wavelengths due to chemical variations between tissues. Thus, the existence of any measurable difference between the spectral signals of different types of epithelial tissues may determine the feasibility of constructing new tumor diagnosis methods using HSI and SWIR. Therefore, we proposes within this work to investigate the vibrational behavior of melanoma, dysplastic nevi and healthy skin epithelial tissues under SWIR radiation and to employ SWIR-obtained HSI as an alternative method to manual visual inspection by dermoscopy to identify tumor epithelial tissue.

3.2   HSI Acquisition

Hyperspectral images of skin samples were obtained using a high-speed chemical performance analyzer called SisuCHEMA *SWIR*. It uses a Hyperspectral Camera (*HSC*) and combines near-infrared spectroscopy (*NIR*) with high-resolution spectral images with 256 spectral bands. The spectral range comprises wavelengths between 900 and 2500, with a range between 900 and 1700 with a spectral resolution of 10 nm in the *NIR* region and a range between 1000 and 2500 nm with a spectral resolution of 6 nm in *SWIR* region. Image data is automatically calibrated for reflectance, however, the HSC software also provides an estimated absorbance value calculated from the measured reflectance intensity. The calculated absorbance, denoted pseudo-absorbance [50], is the unit registered in the HSI and used by the proposed classifier.

3.3   Epithelial Lesion Classifier

For the task of identifying and classifying the vibrational patterns present in the HSI pixels-vectors as well as the spatial correlation between them, we propose a classifier that uses the concept of deep learning. The neural network used in the experiments was RetinaNet. RetinaNet is a single and unified network composed of a *backbone* and two task-specific subnets. The *backbone* is responsible for computing a convolutional feature map across an entire input image. The first subnet performs the classification of convolutional objects in the *backbone* output, and the second subnet performs convolutional bounding box regression [31]. The rationale for the choice of this approach lies in the heterogeneity of pixels-vectors present in a single HSI. As the primary reference for identifying the sample type in advance is a visual inspection by a microscope, the task of labeling HSI pixels-vectors becomes difficult due to the difference in image precision and scale. Thus, labeling the entire sample and not the pixels-vectors was the way adopted in this paper.

4 Related Works

Many methods have been developed to analyze HSI. However, even with this diversity of methods, exploitation of HSI spatial information for tumor classification is limited. Ding et al. [17] categorized the methods for HSI analysis by the approach used in the classification. These are a) methods based on manual procedures and b) methods based on deep learning (DL).

Recent work has significantly contributed to improving HSI classification by employing deep learning. Hu et al. [24] modeled a CNN architecture with five layers between convolutive layers using basic CNN elements by inserting each pixel-vector with shared weights into the input layer.

Ma, Geng and Wang [36] proposed a CNN architecture, denoted contextual deep learning (CDL) that receives as input each pixel-vector and its neighboring pixels-vectors. This approach allows the extraction of spectral and spatial information providing a fine-tuning in classification. Chen, Zhao and Jia [15] introduced in 2015 a new architecture employing deep belief networks (DBF) and restricted Boltzmann machine (RBM) for the extraction of spectral and PCA characteristics for space extraction. The authors proposed a stacked spectrum-spatial vector as a network input. In 2016, Chen el al. [13] introduced a new network denoted 3-D-CNN that employs multiple convolutive and clustering layers with combined regularization for extraction of HSI spectral and spatial characteristics.

Pan, Shi and Xu [40] implemented a new simplified DL model based on rolling guidance filter (RGF) and vertex component analysis network (R-VCANet), for training a network when there is not an abundance of samples for training. Ding et al. [17] developed an adaptive model employing CNN based on the HSI classification method in which convolutional kernels can be learned automatically from data through clustering, even without knowing the number of clusterings. Similar to Chen et al. [14] proposal, Li, Zhang and Shen [30] proposed a 3D convolutional neural network structure, called 3D-CNN, as a method for analyzing HSI data, but without any preprocessing or postprocessing to extract the combined spectral-spatial resources deeply and effectively.

The most recent work in the context of HSI skin tumor detection employs an approach described as a non-parametric, online and multidimensional probability density estimate [43]. Using the concept of deep learning and HSI, Halicek et al. presented a traditional 6-layer convolutive CNN to classify excised squamous cell carcinoma, thyroid cancer, and normal head and neck tissue samples from 50 (fifty) patients with an accuracy of 80% [22].

This scenario shows that despite advances in HSI and deep learning, this approach is little explored in the context of tumor diagnosis. The contributions of this work are: a) to use SWIR as an acquisition technique of HSI in healthy and tumoral epithelial tissues and to investigate the vibrational behavior of the tissues under this spectroscopy technique, b) to develop a new method for classifying skin lesion samples by object detection through spatial and spectral classification in HSI employing deep learning.

5   Samples

All samples used in the experiments are from human epithelial tissue. They were taken from patients by laboratory procedures performed by physicians and arranged as pathology. Thus, the following sets of samples were defined: C1) Melanoma, containing 12 (twelve) samples divided into 34 (thirty four) parts, C2) Dysplastic Nevi, composed of 18 (eighteen) samples divided into 72 parts and C3) Healthy Skin that has 5 (five) samples divided into 17 (seventeen) parts. These samples were fixed on glass slides without the addition of dyes and without overlapping the sample by coverslipping. Sample thickness is 20 μm.

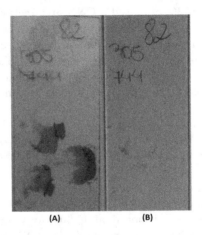

Fig. 2. Skin sample with melanoma fixed to glass slide.

In Fig. 2 are shown two slides referring to the same skin sample with Melanoma. This sample was divided into 3 (three) parts per slide, (A) the slide prepared for microscope viewing and (B) the slide prepared for scanning with SWIR spectroscopy and obtaining the respective HSI.

6   Methodological Procedures

The application of the proposed solution described in Sect. 3 in the analysis of the samples presented in Sect. 5 occurred through the following methodological procedures: A) Sample scanning, B) Annotation, C) Training and D) Detection. Figure 3 illustrates the activities of the Annotation, Training, and Detection procedures represented by a gray-colored bounding box and the flow of activities with their inputs and outputs.

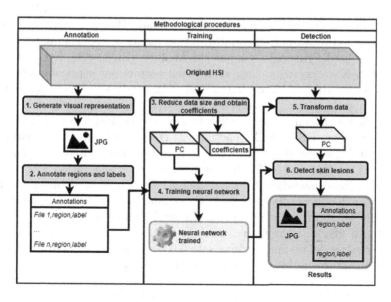

Fig. 3. Methodology of activities with execution flow and inputs and outputs.

6.1   Sample Scanning

The scanning of samples is intended to generate the hyperspectral images by SWIR spectroscopy for each sample. This procedure is fairly simple as it consists of selecting the lens to be used for HSI acquisition, arranging the sample in the reading tray, adjusting the distance from the sensor to the sample, adjusting the lens focus, adjusting the exposure time parameters of the sensor during acquisition and, finally, tray speed during scanning. The images resulting from this procedure are inputs for annotation, training and detection procedures.

6.2   Annotation

Annotation of hyperspectral images is divided into two activities: *1. Generate visual representation*, which consists of transforming one of the layers of the HSI into a visible image to identify the position of the sample within the image. This visual representation is intended to enable the analyst to view the HSI scanning result, to alow the annotation of the images for classifier training and, in the Detection procedure, to view the classification result of the already trained classifier and; *2. Annotate regions and labels*, from the visible image it is possible to delimit the region of the sample within HSI and its respective type.

6.3   Training

The construction of the classifier begins with the reduction of the spectral dimension of each HSI, performed in activity *3. Reduce data size and obtain coefficients*.

This activity aims to minimize information overlays that may exist at different wavelengths and to simplify the mathematical model by reducing data. For this purpose, we used Principal Components Analysis (PCA) [27]. For each training set HSI, $\mathbf{X}' = [\mathbf{x}_1, \mathbf{x}_2, \ldots, \mathbf{x}_p]$ of n pixels-vectors per p possibly correlated variables a new cube of uncorrelated axes with ordered variances is generated $\mathbf{PC}' = [\mathbf{pc}_1, \mathbf{pc}_2, \ldots, \mathbf{pc}_p]$, preserving the spatial arrangement of pixels-vectors. The obtained coefficients of PCA allow to reconstruct the original HSI from $\mathbf{PC}$ or transform a new HSI in $\mathbf{PC}$. In activity *4. Training neural network*, the $\mathbf{PC}$ is used in training a neural network built from a RetinaNet implementation using the TensorFlow and Keras [21] frameworks. A minor adaptation was made to the original implementation to allow training from n layered images and the image resizer has been disabled for not changing the spectral dimension data. The RetinaNet configuration consist in Resnet50 model on backbone, 250 (two hundred and fifty) epochs with 1000 (one thousand) steps each, batch size 4 (four), optimizer Adam and learning rate 0.0001. The code of neural network is avaliable on https://gitlab.com/dvlucena/deep-hsi-swir.

6.4 Detection

Detection is the final procedure of the methodology to evaluate the trained network through object detection and its classification under new HSI. The first activity, *5. Transforming data*, consists in placing each new HSI in the same dimensional space as the samples used in training, using the coefficients obtained in activity 3 of the Training procedure (Sect. 6.3). Finally, each image is subjected to a trained neural network that will detect skin lesions and produce a two-dimensional visible representation of HSI, equivalent to *1. Generate visual representation* activity of the Annotation procedure, however, with the respective demarcation of the region where the lesion is present and its respective label.

7   Results

A HSI for each sample of sets C1, C2, and C3 defined in the Sect. 5 was generated using the *Sample Scanning* procedure described in the Sect. 6.

Shown in Fig. 4 is a two-dimensional visual representation of each HSI of set C1 constructed from conversion of pseudo-absorbance intensity measurements at wavelength λ 1320 nm to whole gray scale values within the range 0–255. Each sample in set C1 is identified by the abbreviations $L1, L2, \ldots, L12$. Results similar to those shown in Fig. 4 were also produced for Dysplastic Nevi (C2) and Healthy Skin (C3) samples. These representations were used in the Annotation and Detection procedures.

For the case study, we separated the sample sets described in the Sect. 5 into training and test samples. The configuration of this separation is presented in Table 1.

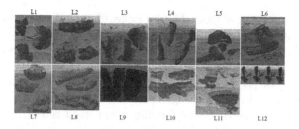

Fig. 4. Two-dimensional representation of HSI from slide-set C1 (Melanoma) sample set.

Table 1. Configuration of training and test sets.

Sample set	Training	Test
C1 - Melanoma	6 samples, 17 parts	6 samples, 17 parts
C2 - Dysplastic Nevus	8 samples, 32 parts	10 samples, 40 parts
C3 - Healthy Skin	3 samples, 9 parts	2 samples, 8 parts

After the classifier training under the training data, we performed the *Detection* procedure using the test samples. The result regarding the classification of Melanoma samples is illustrated in Fig. 5. The orange bounding boxes correspond to the region suggested by the classifier as containing melanoma and the light blue color to the region containing healthy skin.

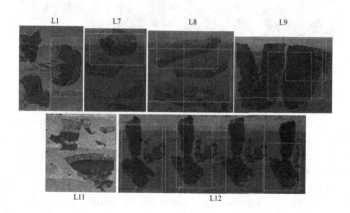

Fig. 5. Melanoma sample classification.

In slide L1 that has three parts of the same sample, only two parts were detected as melanoma. In L7, only one of the three parts was detected. In L8 two regions were suggested, but two of three parts were within one of the suggested regions. In slide L9 despite correctly classifying the sample, there was an overlap of suggested regions, presenting three regions for two parts of samples. In L11 only one region has been suggested and misclassified as Healthy Skin. Finally in L12, three of the four parts were correctly classified and in one part there was a double classification, where overlapping regions received different labels, one correct and one wrong.

To determine the accuracy of the classifier, it was considered correct the correctly suggested and labeled regions on the parts of each sample. Failure to detect or classify with divergent labeling was considered error. Therefore, in numerical terms, the results for the melanoma samples correspond to 11 (eleven) hits, 2 (two) misses and 3 (three) unclassified parts. The accuracy of the classifier for Melanoma samples was 68.8%.

Figure 6 shows ten slides $L13, L14, \ldots, L22$ for the samples of Dysplastic Nevi used in the classifier test. The result of the classification is presented with the demarcation of a dark blue bounding box corresponding to the classifier suggestion for the region with presence of Dysplastic Nevus and, as in Fig. 5, the light blue bounding box corresponds to the region classified as healthy skin.

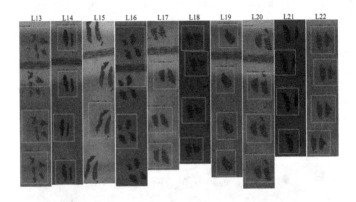

Fig. 6. Classification of Dysplastic Nevi samples.

In all ten slides, at least one part was correctly classified as dysplastic nevus. Of the 40 parts analyzed 29 were classified correctly. In slides L13, L16 and L20 five parts were erroneously classified as Healthy Skin in three suggested regions. Already in slides L15, L16 and L17 five of the twelve parts present were not classified. Therefore, the accuracy of the classifier for Dysplastic Nevi samples was 72.5%. Importantly, for the Dysplastic Nevi samples, the regions suggested by the classifier were well defined, not presenting the problem of overlap or cuts in parts of the samples as occurred with the melanoma samples.

Results for healthy tissue samples were inconclusive because of insufficient samples available in both Training and Detection procedures, so were not presented in this study.

8   Conclusion

We presented a proposal employing HSI obtained by SWIR spectroscopy to identify tumor epithelial tissue using deep learning for classification. The feasibility of using SWIR corroborated with previous studies and confirmed the hypothesis of sensitivity of human epithelial tissue to *SWIR*. This confirmation is most evident when using HSI as a data structure. It has been shown by the construction of the visual representations of each HSI that the morphology of the images coincides with the visible eye shapes of the samples arranged on the slides.

With the implementation of the proposed solution, it was possible to distinguish samples of Melanoma and Dysplastic Nevi by means of the spectral signals and their respective spatial arrangement present in the structure of HSI. Although the pixels-vectors of the epithelial tissues analyzed have a similar spectral profile, there are differences in subtle intensities between the samples that allow them to be distinguished. This result is a strong indication that HSI-SWIR can be used to construct new methods for the classification of epithelial tumors.

While refinements are needed to improve region suggestion for Melanoma samples, labeling suggested regions yielded more assertive than non-assertive results. This shows that deep learning has been able to extract spectral and spatial characteristics from tumor epithelial tissue lesion samples to the extent that they can be distinguished.

It is important to highlight that the samples are not homogeneous, that is, not the entire length of the sample has the pathology. Therefore, we cannot state that in all the length of the Melanoma sample, all pixels-vectors have the pathology. The precise location of tumor cells is most easily determined by using the microscope and preparing the slide. Due to the difference in precision between the microscope and the HSC used in the study, it was not possible to determine in HSI which pixels-vectors correspond to the tumor cells.

We suggest to continue the studies with the following future works: 1) expand the number of samples and perform new experiments to confirm the indicative presented in the results; 2) increase in the *Training* procedure an activity to remove pixels-vectors that do not correspond to the skin sample, performing a semantic segmentation in the sample preceding the neural network training; 3) incorporate semantic segmentation as the final activity of the *Detection* procedure and 4) locate within the sample the pixels-vectors that best match the classified pathology, 5) apply the proposed solution to images acquired from samples *in vivo*. We did not perform acquisition *in vivo* in this study due to the limitations of available HSC.

References

1. Afromowitz, M.A., Callis, J.B., Heimbach, D.M., DeSoto, L.A., Norton, M.K.: Multispectral imaging of burn wounds: a new clinical instrument for evaluating burn depth. IEEE Trans. Biomed. Eng. **35**(10), 842–850 (1988)

2. Akbari, H., et al.: Hyperspectral imaging and quantitative analysis for prostate cancer detection. J. Biomed. Opt. **17**(7), 076005-1–076005-10 (2012). https://doi.org/10.1117/1.JBO.17.7.076005. http://dx.doi.org/10.1117/1.JBO.17.7.076005

3. Akbari, H., Halig, L.V., Zhang, H., Wang, D., Chen, Z.G., Fei, B.: Detection of cancer metastasis using a novel macroscopic hyperspectral method. In: Proceedings of SPIE, vol. 8317, p. 831711. NIH Public Access (2012)

4. Akbari, H., Uto, K., Kosugi, Y., Kojima, K., Tanaka, N.: Cancer detection using infrared hyperspectral imaging. Cancer Sci. **102**(4), 852–857 (2011)

5. Almeida, V.L.D., Leitão, A., Reina, L.D.C.B., Montanari, C.A., Donnici, C.L., Lopes, M.T.P.: Câncer e agentes antineoplásicos ciclo-celular específicose ciclo-celular não específicos que interagem com o dna: uma introdução. Quim. Nova **28**(1), 118–129 (2005)

6. Ball, D.W.: The Basics of Spectroscopy, vol. 49. SPIE Press, Bellingham (2001)

7. Bambery, K.R., Wood, B.R., Quinn, M.A., McNaughton, D.: Fourier transform infrared imaging and unsupervised hierarchical clustering applied to cervical biopsies. Aust. J. Chem. **57**(12), 1139–1143 (2004)

8. Calin, M.A., Parasca, S.V., Savastru, D., Manea, D.: Hyperspectral imaging in the medical field: present and future. Appl. Spectrosc. Rev. **49**(6), 435–447 (2014). https://doi.org/10.1080/05704928.2013.838678. http://dx.doi.org/10.1080/057049 28.2013.838678

9. Carli, P., De Giorgi, V., Soyer, H., Stante, M., Mannone, F., Giannotti, B.: Dermatoscopy in the diagnosis of pigmented skin lesions: a new semiology for the dermatologist. J. Eur. Acad. Dermatol. Venereol. **14**(5), 353–369 (2000)

10. Carrasco, O., Gomez, R.B., Chainani, A., Roper, W.E.: Hyperspectral imaging applied to medical diagnoses and food safety. In: Proceediings of SPIE. vol. 5097, pp. 215–221 (2003)

11. Carvalho, G.C., Alves, F.: Principais marcadores moleculares para os câncceresde pele e mama. NBC-Periódico Científico do Núcleo de Biociências **4**(07), 11–17 (2014)

12. Chang, C.I.: Hyperspectral Imaging: Techniques for Spectral Detection and Classification, vol. 1. Springer, Boston (2003). https://doi.org/10.1007/978-1-4419-9170-6

13. Chen, Y., Jiang, H., Li, C., Jia, X., Ghamisi, P.: Deep feature extraction and classification of hyperspectral images based on convolutional neural networks. IEEE Trans. Geosci. Remote Sens. **54**(10), 6232–6251 (2016)

14. Chen, Y., Lin, Z., Zhao, X., Wang, G., Gu, Y.: Deep learning-based classification of hyperspectral data. IEEE J. Sel. Top. Appl. Earth Obs. Remote Sens. **7**(6), 2094–2107 (2014)

15. Chen, Y., Zhao, X., Jia, X.: Spectral-spatial classification of hyperspectral data based on deep belief network. IEEE J. Sel. Top. Appl. Earth Obs. Remote Sens. **8**(6), 2381–2392 (2015)

16. Dicker, D.T., et al.: Differentiation of normal skin and melanoma using high resolution hyperspectral imaging. Cancer Biol. Ther. **5**(8), 1033–1038 (2006)

17. Ding, C., Li, Y., Xia, Y., Wei, W., Zhang, L., Zhang, Y.: Convolutional neural networks based hyperspectral image classification method with adaptive kernels. Remote Sens. **9**(6), 618 (2017)

18. Fei, B., Akbari, H., Halig, L.V.: Hyperspectral imaging and spectral-spatial classification for cancer detection. In: 2012 5th International Conference on Biomedical Engineering and Informatics (BMEI), pp. 62–64. IEEE (2012)

19. Ferris, D.G., et al.: Multimodal hyperspectral imaging for the noninvasive diagnosis of cervical neoplasia. J. Lower Genital Tract Dis. **5**(2), 65–72 (2001)

20. Figueiredo, L.C., Cordeiro, L.N., Arruda, A.P., Carvalho, M.D.F., Ribeiro, E.M., Coutinho, H.D.M.: Câncer de pele: estudo dos principais marcadores moleculares do melanoma cutâneo. Rev Bras de Cancerologia **49**(3), 179–183 (2003)
21. Fizyr: Keras implementation of retinanet object detection (2019). https://github.com/fizyr/keras-retinanet. Accessed 18 Apr 2019
22. Halicek, M., et al.: Deep convolutional neural networks for classifying head and neck cancer using hyperspectral imaging. J. Biomed. Opt. **22**, 060503 (2017). https://doi.org/10.1117/1.JBO.22.6.060503.   http://dx.doi.org/10.1117/1.JBO.22.6.06 0503
23. Hansen, M.P., Malchow, D.S.: Overview of SWIR detectors, cameras, and applications. In: Proceedings SPIE, vol. 6939, p. 69390I (2008)
24. Hu, W., Huang, Y., Wei, L., Zhang, F., Li, H.: Deep convolutional neural networks for hyperspectral image classification. J. Sens. **2015** (2015)
25. INCA: Estimativa 2018: Incidência de câncer no brasil (2017). https://www.inca.gov.br/sites/ufu.sti.inca.local/files//media/document//estimativa-incidencia-de-cancer-no-brasil-2018.pdf. Accessed 14 Aug 2019
26. Jacques, S.L., Ramella-Roman, J.C., Lee, K.: Imaging skin pathology with polarized light. J. Biomed. Opt. **7**(3), 329–340 (2002)
27. Johnson, R.A., Wichern, D.W.: Applied Multivariate Statistical Analysis, 6th edn. Pearson, London (2014)
28. Kiyotoki, S., et al.: New method for detection of gastric cancer by hyperspectral imaging: a pilot study. J. Biomed. Opt. **18**(2), 026010 (2013)
29. Koh, K.R., Wood, T.C., Goldin, R.D., Yang, G.Z., Elson, D.S.: Visible and near infrared autofluorescence and hyperspectral imaging spectroscopy for the investigation of colorectal lesions and detection of exogenous fluorophores. In: Proceedings of SPIE, vol. 7169, p. 71691E (2009)
30. Li, Y., Zhang, H., Shen, Q.: Spectral-spatial classification of hyperspectral imagery with 3D convolutional neural network. Remote Sens. **9**(1), 67 (2017)
31. Lin, T.Y., Goyal, P., Girshick, R., He, K., Dollár, P.: Focal loss for dense object detection. In: Proceedings of the IEEE International Conference on Computer Vision, pp. 2980–2988 (2017)
32. Lindsley, E.H., Wachman, E.S., Farkas, D.L.: The hyperspectral imaging endoscope: a new tool for in vivo cancer detection. In: Proceedings of SPIE, vol. 5322, pp. 75–82 (2004)
33. Lu, G., Fei, B.: Medical hyperspectral imaging: a review. J. Biomed. Opt. **19**(1), 010901 (2014). https://doi.org/10.1117/1.JBO.19.1.010901. http://dx.doi.org/10.1117/1.JBO.19.1.010901
34. Lu, G., Halig, L., Wang, D., Chen, Z.G., Fei, B.: Spectral-spatial classification using tensor modeling for cancer detection with hyperspectral imaging. In: Proceedings of SPIE, vol. 9034, p. 903413. NIH Public Access (2014)
35. Lu, G., Halig, L., Wang, D., Qin, X., Chen, Z.G., Fei, B.: Spectral-spatial classification for noninvasive cancer detection using hyperspectral imaging. J. Biomed. Opt. **19**(10), 106004 (2014)
36. Ma, X., Geng, J., Wang, H.: Hyperspectral image classification via contextual deep learning. EURASIP J. Image Video Process. **2015**(1), 20 (2015)
37. Malkoff, D.B., Oliver, W.R.: Hyperspectral imaging applied to forensic medicine. In: Proceedings of SPIE, pp. 0277–786X (2000)
38. Martin, M.E., et al.: Development of an advanced hyperspectral imaging (HSI) system with applications for cancer detection. Ann. Biomed. Eng. **34**(6), 1061–1068 (2006)

39. de Moraes Matheus, L.G., Verri, B.H.d.M.A.: Aspectos epidemiológicos do melanoma cutâneo. Revista Ciência e Estudos Acadêmicos de Medicina **1**(03) (2015)
40. Pan, B., Shi, Z., Xu, X.: R-VCANet: a new deep-learning-based hyperspectral image classification method. IEEE J. Sel. Top. Appl. Earth Obs. Remote Sens. **10**(5), 1975–1986 (2017)
41. Panasyuk, S.V., Freeman, J.E., Panasyuk, A.A.: Medical hyperspectral imaging for evaluation of tissue and tumor, US Patent 8,320,996, 27 November 2012
42. Panasyuk, S.V., et al.: Medical hyperspectral imaging to facilitate residual tumor identification during surgery. Cancer Biol. Ther. **6**(3), 439–446 (2007)
43. Pardo, A., Gutiérrez-Gutiérrez, J.A., Lihacova, I., López-Higuera, J.M., Conde, O.M.: On the spectral signature of melanoma: a non-parametric classification framework for cancer detection in hyperspectral imaging of melanocytic lesions. Biomed. Opt. Express **9**(12), 6283–6301 (2018)
44. Schultz, R.A., Nielsen, T., Zavaleta, J.R., Ruch, R., Wyatt, R., Garner, H.R.: Hyperspectral imaging: a novel approach for microscopic analysis. Cytometry Part A **43**(4), 239–247 (2001)
45. Shah, S., Bachrach, N., Spear, S., Letbetter, D., Stone, R., Dhir, R., Prichard, J., Brown, H., LaFramboise, W.: Cutaneous wound analysis using hyperspectral imaging. Biotechniques **34**(2), 408–413 (2003)
46. Siddiqi, A.M., et al.: Use of hyperspectral imaging to distinguish normal, precancerous, and cancerous cells. Cancer Cytopathol. **114**(1), 13–21 (2008)
47. Siegel, R.L., Miller, K.D., Jemal, A.: Cancer statistics. CA Cancer J. Clin. **65**(1), 5–29 (2015)
48. Society, A.C.: Tests for melanoma skin cancer (2016). https://www.cancer.org/cancer/melanoma-skin-cancer/detection-diagnosis-staging/how-diagnosed.html. Accessed 31 July 2017
49. Society, A.C.: Cancer facts and figures 2017 (2017). http://www.cancer.org/acs/groups/content/@editorial/documents/document/acspc-048738.pdf. Accessed 25 July 2017
50. SPECIM: SisuCHEMA - Chemical Imaging Analyzer (2015). http://www.specim.fi/downloads/SisuCHEMA_2_2015.pdf. Accessed 25 July 2017
51. Zevon, M., et al.: CXCR-4 targeted, short wave infrared (SWIR) emitting nanoprobes for enhanced deep tissue imaging and micrometastatic cancer lesion detection. Small **11**(47), 6347–6357 (2015)
52. Zonios, G., et al.: Diffuse reflectance spectroscopy of human adenomatous colon polyps in vivo. Appl. Opt. **38**(31), 6628–6637 (1999)

Statistical Iterative Reconstruction Algorithm Based on a Continuous-to-Continuous Model Formulated for Spiral Cone-Beam CT

Robert Cierniak[(⊠)][iD] and Piotr Pluta

Institute of Computational Intelligence, Czestochowa University of Technology,
Armii Krajowej 36, 42-200 Czestochowa, Poland
robert.cierniak@pcz.pl

Abstract. This paper is closely related to the originally formulated 3D statistical model-based iterative reconstruction algorithm for spiral cone-beam x-ray tomography. The concept proposed here is based on a continuous-to-continuous data model, and the reconstruction problem is formulated as a shift invariant system. This algorithm significantly improves the quality of the subsequently reconstructed images, so allowing a reduction in the x-ray dose absorbed by a patient. This form of reconstruction problem permits a reduction in the computational complexity in comparison with other model-based iterative approaches. Computer simulations have shown that the reconstruction method presented here outperforms standard FDK methods with regard to the image quality obtained and can be competitive in terms of time of calculation.

Keywords: Iterative reconstruction algorithm · Computed tomography · Statistical method

1 Introduction

Recently, the most significant problem in medical CT has been the development of image reconstruction methods which would enable the reduction of the impact of measurement noise on the quality of tomography images and thus decrease the dose of X-ray radiation absorbed by patients during examinations. Some of the most interesting research directions in this area are statistical reconstruction methods, especially those belonging to the MBIR (Model-Based Iterative Reconstruction) approach [1–3], where a probabilistic model of the measurement signals is taken into account. The objective in those solutions was devised

---

Supported by The National Centre for Research and Development in Poland (Research Project POIR.01.01.01-00-0463/17.).

V. V. Krzhizhanovskaya et al. (Eds.): ICCS 2020, LNCS 12139, pp. 613–620, 2020.
https://doi.org/10.1007/978-3-030-50420-5_46

according to a discrete-to-discrete (D-D) data model. Unfortunately, those methods have some very serious drawbacks from the theoretical and practical point of view: for instance, if the image resolution is set to be $I \times I$ pixels, the calculation complexity of the problem is proportional to I^4, the statistical reconstruction procedure based on this methodology necessitates simultaneous calculations for all the voxels in the range of the reconstructed 3D image, the size of the forward model matrix $\mathbf{A}$ is huge, and this makes it often necessary to calculate them in every iteration of the reconstruction algorithm. In this case, the reconstruction problem is extremely ill-conditioned, and it is necessary to introduce an *a priori* term (often referred to in the literature as a regularization term) into the objective, and this leads to the use of the MAP model. The problems connected with the use of a methodology based on the D-D data model can be reduced by using a strategy of reconstructed image processing based on a continuous-to-continuous (C-C) data model. In previous papers we have shown how to formulate reconstruction problems consistent with the ML methodology for parallel scanner geometry [4], and finally for the spiral cone-beam scanner [5,6]. In this paper, we show how to interpret our original statistical reconstruction method as an approach belonging to the C-C mode. We applied very popular and convenient reconstruction strategy, which resemles the FDK-type algorithms, and we present a conception of the direct use of spiral cone-beam projections to a statistical reconstruction algorithm based on the C-C data model.

2   Statistical Reconstruction Algorithm

Our reconstruction method is based on the well-known maximum-likelihood (ML) estimation [8]. In most cases, the objective in those solutions is devised according to a discrete-to-discrete (D-D) data model. We propose here an optimization formula which is consistent with the C-C data model, in the following form:

$$
\mu_{\min} = \arg\min_{\mu} \left(\int_x \int_y \left(\int_{\bar{x}} \int_{\bar{y}} \mu\left(\bar{x}, \bar{y}\right) \cdot h_{\Delta x, \Delta y} d\bar{x} d\bar{y} - \tilde{\mu}\left(x, y\right) \right)^2 dx dy \right), \quad (1)
$$

where $\tilde{\mu}\left(x, y\right)$ is an image obtained by way of a back-projection operation, obtained theoretically in the following way:

$$
\tilde{\mu}\left(x, y\right) \approx \int_0^{2\pi} \int_{-\beta_{max}}^{\beta_{max}} p^h\left(\beta, \alpha^h, z_k\right) \frac{R_{fd}}{\sqrt{R_{fd}^2 + z_k^2}} int_L\left(\Delta\beta\right) d\beta d\alpha, \quad (2)
$$

wherein $p^h\left(\beta, \alpha^h, z_k\right)$ are measurements carried out using a spiral cone-beam scanner, R_{fd} is the SDD (Source-to-Detector Distance), and the coefficients $h_{\Delta i, \Delta j}$ can be precalculated according to the following relation:

$$h_{\Delta x, \Delta y} = \int\limits_{0}^{2\pi} int\left(\Delta x \cos\alpha + \Delta y \sin\alpha\right) d\alpha,\qquad (3)$$

and $int\left(\Delta s\right)$ is a linear interpolation function.

The necessary measurements are performed in a standard helical cone-beam scanner. The mesurement system consists of an x-ray tube and a rigidly coupled screen with a multi-row matrix of detectors. This assembly rotates around the z-axis (the principal axis of the system) and at the same time, the patient table moves into the gantry. Therefore, the moving projection system traces a spiral path around the z-axis. Each ray emitted by the tube at a particular angle of rotation and reaching any of the radiation detectors can be identified by $\left(\beta, \alpha^h, \dot{z}\right)$, as follows: β – the angle between a particular ray in the beam and the axis of symmetry of the moving projection system; α^h – the angle at which the projection is made, i.e. the angle between the axis of symmetry of the rotated projection system and the y-axis; $\dot{z}$ – the z-coordinate relative to the current position of the moving projection system.

In a real spiral cone-beam scanner, the reconstruction algorithm can only make use of projections obtained at certain angles and measured only at particular points on the screen. Let us assume that the beam of x-rays reaches the individual detector rows $k = 1, 2, \ldots, K$, where K is a number of detectors in each row of the array. In every row, selected rays strike the detectors, each of which is described by the index $\eta = -\left(H-1\right)/2, \ldots, 0, \ldots, \left(H-1\right)/2$, where H is a number of detectors in each channel of the array. Detectors are placed on the screen separated by a distance Δ_z in each row, and by an angular distance Δ_η in each channel. Of course, only a limited number of mesurements are performed, each of which is described by the index $\theta = 0, \ldots, \Theta - 1$, where $\Theta - 1$ is the total number of projections made during the examination. Every projection is carried out after rotation by Δ_θ. We can sum up above conditions by saying that the reconstruction algorithm has available to it the projection values $p^h\left(\beta_\eta, \alpha_\theta^h, \dot{z}_k\right)$, in the ranges: $\eta = -\left(H-1\right)/2, \ldots, 0, \ldots, \left(H-1\right)/2$; $\theta = 0, \ldots, \Theta - 1$; $k = 1, 2, \ldots, K$.

According to the originally formulated by us iterative approach to the reconstruction problem, decribed by Eqs. (1)–(3), it is possible to present a practical model-based statistical method of image reconstruction, as follows:

$$\mu_{\min} = \arg\min_{\mu}\left(\sum_{i=1}^{I}\sum_{j=1}^{J}\left(\sum_{\bar{i}}\sum_{\bar{j}}\mu^*\left(x_{\bar{i}}, y_{\bar{j}}\right)\cdot h_{\Delta i, \Delta j} - \tilde{\mu}\left(x_i, y_j\right)\right)^2\right),\qquad (4)$$

and $\tilde{\mu}\left(i, j\right)$ is an image obtained by way of a back-projection operation, in the following way:

$$\tilde{\mu}\left(x_i, y_j\right) = \Delta_{\alpha^h}\sum_{\theta}\dot{p}^h\left(\beta_{ij}, \alpha_\theta^h, \dot{z}_{ij}\right)\qquad (5)$$

It is necessary to use an interpolation to evaluate projections at points β_{ij} based on the measured projections $p^h\left(\beta_\eta, \alpha_\theta^h, \dot{z}_k\right)$. We can obtain an approximations of these projections as follows:

$$\dot{p}^h\left(\beta_{ij}, \alpha_\theta^h, \dot{z}_{ij}\right) =$$

$$\sum_k \sum_\eta p^h\left(\beta_\eta, \alpha_\theta^h, \dot{z}_k\right) \frac{R_{fd}}{\sqrt{R_{fd}^2 + z_k^2}} \cdot int_\beta\left(\beta_{ij} - \eta\Delta_\beta\right) int_z\left(z_{ij} - k\Delta_z\right), \quad (6)$$

where $int_\beta\left(\Delta\beta\right)$ and $int_z\left(\Delta z\right)$ are the interpolation functions, i.e. in the simplest case, linear interpolations:

$$int_\beta\left(\beta\right) = \begin{cases} \frac{1}{\Delta_\beta}\left(1 - \frac{|\beta|}{\Delta_\beta}\right) & \text{for } |\beta| \leq \Delta_\beta \\ 0 & \text{for } |\beta| \geq \Delta_\beta \end{cases}, \quad (7)$$

and

$$int_z\left(z\right) = \begin{cases} \frac{1}{\Delta_z}\left(1 - \frac{|z|}{\Delta_z}\right) & \text{for } |z| \leq \Delta_z \\ 0 & \text{for } |z| \geq \Delta_z \end{cases}. \quad (8)$$

The coefficients $h_{\Delta i, \Delta j}$ are determined according to the following formula:

$$h_{\Delta i, \Delta j} = \frac{1}{\Delta s^2}\Delta_\alpha \sum_{\psi=0}^{\Psi-1} int\left(\Delta i \cos \psi\Delta_\alpha + \Delta j \sin \psi\Delta_\alpha\right), \quad (9)$$

wherein $int\left(\Delta s\right)$ is an interpolation function used in the back-projection operation, and $\Delta_s = R_f * \tan\Delta_\beta$.

The presence of a shift-invariant system in the optimization problem (4) implies that this system is much better conditioned than the least squares problems present in the referential approach [9]. The conception presented above is a full 3D iterative reconstruction algorithm for spiral cone-beam scanner geometry. This algorithm is based on the one of the principal reconstruction methods devised for the cone-beam spiral scanner, i.e. the generalized FDK algorithm. The statistical reconstruction method proposed by us consists of two steps, namely: a back-projection operation described by relations (5)–(8) and an iterative reconstruction procedure according to formula (4). Figure 1 depicts this algorithm after discretization and implementation of FFT which significantly accelerates the calculations (the iterative reconstruction procedure is patented in the United States [10]).

3   Experimental Results

In our experiments, we have used projections obtained from a Somatom Definition AS+ (helical mode) scanner with the following parameters: reference tube potential 120kVp and quality reference effective 200mAs, $R_{fd} = 1085.6\,\text{mm}$,

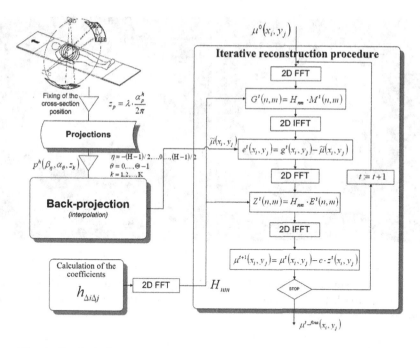

Fig. 1. Statistical reconstruction algorithm for spiral cone-beam scanner.

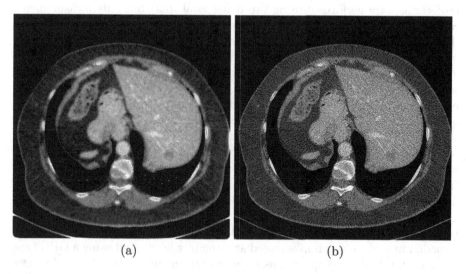

(a)   (b)

Fig. 2. Reconstructed images obtained at 50% x-ray dose reduction, using: (a) the statistical approach presented in this paper obtained after 10000 iterations (b) the standard FDK algorithm.

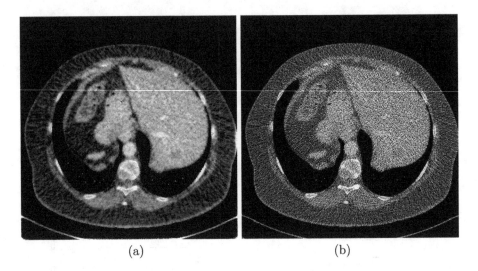

(a)   (b)

Fig. 3. Reconstructed images obtained at 87% x-ray dose reduction, using: (a) the statistical approach presented in this paper obtained after 10000 iterations (b) the standard FDK algorithm.

R_f = 595 mm, number of views per rotation Ψ = 1152, number of pixels in detector panel 736, detector dimensions were 1.09 mm × 1.28 mm. However, these projections were performed using two flying focal spots but only measurements carried out which coincided with the detector's center were used. This means that only every second measurement (theoretically, a 50% reduction of the dose) was useful for the reconstruction algorithm proposed by us. During the experiments, the size of the processed image was fixed at 512 × 512 pixels. A discrete representation of the matrix $h_{\Delta x, \Delta y}$ was established before the reconstruction process was started, and these coefficients were fixed (transformed into the frequency domain) for the whole iterative reconstruction procedure. The image obtained after the back-projection operation was then subjected to a process of reconstruction (optimization) using an iterative procedure. A specially prepared result of an FBP reconstruction algorithm was chosen as the starting point of this procedure. It is worth noting that our reconstruction procedure was performed without any regularization regarding the objective function from (1). The iterative reconstruction procedure was implemented for a computer with 10 cores, i.e. with an Intel i9-7900X BOX/3800MHz processor (the iterative reconstruction procedure was implemented at assembler level), and using a GPU type nVidia Titan V. According to an assessment of the quality of the obtained images by a radiologist, 8000 iterations are enough to provide an acceptable image. The same results were achieved for both hardware implementations after 7.44 s and 7.73 s, for the CPU and GPU implementations, respectively. One can compare the results obtained by assessing the views of the reconstructed images in Figs. 2a and 3a, where the statistical approach presented in this paper was used (image obtained after 10000 iterations), and in Figs. 2b and 3b, where the standard FDK algorithm was applied (with linear interpolation function and Shepp-Logan kernel). Figures 2 and 3 show reconstructed images obtained at 50% x-ray dose reduction, and at 87% x-ray dose reduction, respectively.

4 Conclusion

A statistical iterative reconstruction algorithm which can be used in practice for helical cone-beam scanners has been shown above. We have conducted computer simulations, which proved that our reconstruction method is very fast, above all thanks to the use of FFT algorithms and efficient programming techniques, and it gives satisfactory results regarding the quality of the obtained images at a significantly reduced dose of x-rays absorbed by the patient. If the image resolution is assumed to be I × I pixels, the complexity of the approach implemented here is proportional to $I^2 \log_2 I$, and with the referential approach it is of the level of $I^4 \times number\_of\_cross - sections$. One can note that the iterative reconstruction procedure was performed without introducing any additional regularization term, using only an early stopping regularization strategy. It should be underlined that the price of the hardware used is relatively low (about 5000 USD in both cases) compared with the cost of the equipment necessary in the case of the referential solution.

References

1. Zhou, Y., Thibault, J.-B., Bouman, C.A., Hsieh, J., Sauer, K.D.: Fast model-based x-ray CT reconstruction using spatially non-homogeneous ICD optimization. IEEE Tran. Image Proc. **20**, 161–175 (2011)
2. Ding, Q., Long, Y., Zhang, X., Fessler, J.A.: Modeling mixed Poisson-Gaussian noise in statistical image reconstruction for x-ray CT. In: Proceedings of the 4th International Conference on Image Formation in X-Ray Computed Tomography, Bamberg, Germany, pp. 399–402 (2016)
3. Geyer, L.L., et al.: State of the art: iterative CT reconstruction techniques. Radiology **276**, 339–357 (2017)
4. Cierniak, R.: An analytical iterative statistical algorithm for image reconstruction from projections. Appl. Math. Comput. Sci. **24**, 7–17 (2014)
5. Cierniak, R.: Analytical statistical reconstruction algorithm with the direct use of projections performed in spiral cone-beam scanners. In: Proceedings of the 5th International Meeting on Image Formation in X-Ray Computed Tomography, pp. 293–296, Salt Lake City (2018)
6. Cierniak, R., Pluta, P., Kaźmierczak, A.: A practical statistical approach to the reconstruction problem using a single slice rebinning method. J. Artif. Intell. Soft Comput. Res. **10**, 137–149 (2020)
7. Feldkamp, L.A., Davis, L.C., Kress, J.W.: Practical cone-beam algorithm. J. Opti. Soc. Am. 1(A) **9**, 612–619 (1984)
8. Bouman, C.A., Sauer, K.: A unified approach to statistical tomography using coordinate descent optimization. IEEE Tran. Image Proc. **5**, 480–492 (1996)

9. Cierniak, R., Lorent, A.: Comparison of algebraic and analytical approaches to the formulation of the statistical model-based reconstruction problem for x-ray computed tomography. Comput. Med. Imaging Graph. **52**, 19–27 (2016)

10. Cierniak R.: Fast iterative reconstruction method for 3D computed tomography. Patent US 9.508.164 B2 (2016)

Classification of Lung Diseases Using Deep Learning Models

Matthew Zak[1] and Adam Krzyżak[1,2]([⊠])

[1] Department of Computer Science and Software Engineering, Concordia University, Montreal H3G 1M8, Canada
zak.matthew@yahoo.com, krzyzak@cs.concordia.ca
[2] Department of Electrical Engineering, Westpomeranian University of Technology, 70-313 Szczecin, Poland

Abstract. In this paper we address the problem of medical data scarcity by considering the task of detection of pulmonary diseases from chest X-Ray images using small volume datasets with less than thousand samples. We implemented three deep convolutional neural networks (VGG16, ResNet-50, and InceptionV3) pre-trained on the ImageNet dataset and assessed them in lung disease classification tasks using transfer learning approach. We created a pipeline that segmented chest X-Ray (CXR) images prior to classifying them and we compared the performance of our framework with the existing ones. We demonstrated that pre-trained models and simple classifiers such as shallow neural networks can compete with the complex systems. We also validated our framework on the publicly available Shenzhen and Montgomery lung datasets and compared its performance to the currently available solutions. Our method was able to reach the same level of accuracy as the best performing models trained on the Montgomery dataset however, the advantage of our approach is in smaller number of trainable parameters. Furthermore, our InceptionV3 based model almost tied with the best performing solution on the Shenzhen dataset despite being computationally less expensive.

Keywords: Lung disease classification · Transfer learning · Deep learning

1 Pixel/Voxel-Based Machine Learning

The availability of computationally powerful machines allowed emerging methods like pixel/voxel-based machine learning (PML) breakthroughs in medical image analysis/processing. Instead of calculating features from segmented regions, this technique uses voxel/pixel values in input images directly. Therefore, neither segmentation nor feature extraction is required. The performance

Supported by the Natural Sciences and Engineering Research Council of Canada. Part of this research was carried out by the second author during his visit of the Westpomeranian University of Technology while on sabbatical leave from Concordia University.

© Springer Nature Switzerland AG 2020
V. V. Krzhizhanovskaya et al. (Eds.): ICCS 2020, LNCS 12139, pp. 621–634, 2020.
https://doi.org/10.1007/978-3-030-50420-5_47

of PML's can possibly exceed that of common classifiers [16] as this method is able to avoid errors caused by inaccurate segmentation and feature extraction. The most popular powerful approaches include convolutional neural networks (including shift-invariant neural networks). They resulted in false positive (FP) rates reduction in computer-aided design framework (CAD) for detection of masses and microcalcifications [12] in mammography and in lung nodule detection in chest X-ray CXR images [13], neural filters and massive-training artificial neural networks including massive-training artificial neural networks (MTANNs) including a mixture of expert MTANNs, Laplacian eigenfunction LAP-MTANN and massive-training support vector regression (MTSVR) for classification, object detection and image enhancement in malignant lung modules detection in CT, FP reduction in CAD for polyp detection in CT colonography, bone separation from soft tissue in CXR and enhancement of lung nodules in CT [11].

2   Bone Separation from Soft Tissue in Chest Radiographs Using MTANNs

Chest X-Ray is one of the most frequently used diagnostic modality in detecting different lung diseases such as pneumonia or tuberculosis. Roughly 1 million of adults require hospitalization because of pneumonia, and about 50,000 dies from this disease annually in the US only. Examination of lung nodules in CXR can lead to missing of diseases like lung cancer. However, not all of them are visible in retrospect. Studies show that 82–95% of lung cancer cases were missed due to occlusions (at least partial) by ribs or clavicle. To address this problem researchers examined dual-energy imaging, a technique which can produce images of two tissues, namely "soft-tissue" image and "bone" image. This technique has many drawbacks, but undoubtedly one of the most important ones is the exposure to radiation.

The MTANNs models have been developed to address this problem and serve as a technique for ribs/soft-tissue separation. The idea behind training of those algorithms is to provide them with bone and soft-tissue images obtained from a dual-energy radiography system. The MTANN was trained using CXRs as input and corresponding boneless images. The ribs contrast is visibly suppressed in the resulting image, maintaining the soft tissue areas such as lung vessels.

3   Deep Learning Approaches in Chest X-Ray Analysis

Recent developments in Deep Neural Networks [2] lead to major improvements in medical imaging. The efficiency of dimensionality reduction algorithms like lung segmentation was demonstrated in the chest X-Ray image analysis. Recently researchers aimed at improving tuberculosis detection on relatively small data sets of less than 103 images per class by incorporating deep learning segmentation and classification methods from [4]. We will further explore these techniques in this paper.

3.1   Dataset

In this paper we combine two relatively small datasets containing less than 103 images per class for classification (pneumonia and tuberculosis detection) and segmentation purposes. We selected 306 examples per "disease" class (306 images with tuberculosis and 306 images with pneumonia) and 306 of healthy patients yielding the set of 918 samples from different patients. Sample images from both datasets are shown in Fig. 1.

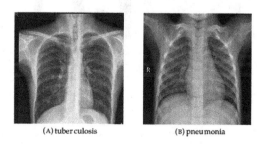

(A) tuberculosis   (B) pneumonia

Fig. 1. Sample Chest X-Ray images containing traces of tuberculosis (A) and pneumonia (B) from Shenzhen dataset [2,6].

The Shenzhen Hospital dataset (SH) [2,6] containing CXR images was created by the People's Hospital in Shenzhen, China. It includes both abnormal (containing traces of tuberculosis) and standard CXR images. Unfortunately, the dataset is not well-balanced in terms of absence or presence of disease, gender, or age. We extracted only 153 samples of healthy patients (153 from both datasets) and 306 of those labeled with traces of tuberculosis. Selecting information about one class from different resources ensures that the model is not contaminated by the features resulting from the method of taking images, e.g., the lens.

Pneumonia is an inflammatory condition of the lung affecting the little air sacs known as alveoli. Standard symptoms comprise of a blend of a dry hacking cough, inconvenience breathing, chest agony, and fever. The Labeled Optical Tomography and Chest X-Ray Images for Classification dataset [9] includes selected images of pneumonia patients from the Medical Center in Guangzhou. It consists of data with two classes - normal and those containing marks of pneumonia. All data come from the patient's routine clinical care. The volume of the complete dataset includes thousands of validated optical coherence tomography (OCT) and X-ray images yet for our analysis we wanted to keep the dataset tiny and evenly distributed thus only 153 images were selected (other 153 images come from the tuberculosis dataset) from the resources labeled as healthy and 306 as pneumonia - both chosen randomly. External segmentation of left and right lung images (exclusion of redundant information: bones, internal organs, etc.) was proven to be effective in boosting prediction accuracy. To extract lungs information and exclude outside regions, we used the manually prepared masks included in the extension of the SH dataset, namely, the segmented SH dataset, see Fig. 2. Due to nonidentical borders and lung shapes, the segmentation data has high variability although its distribution is quite similar to the regular one when compared to image area distribution.

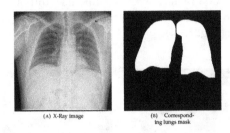

(A) X-Ray image   (B) Corresponding lungs mask

Fig. 2. An X-Ray image and its corresponding lungs mask.

3.2   Image Data Augmentation

Model-based methods greatly improve their predictions when the number of training samples grows. When a limited amount of data is available, some transformations have to be applied to the existing dataset to synthetically augment the training set. Researchers in [10] employed three techniques to augment the training dataset. The first approach was to randomly crop of a 224 × 224 pixel fixed-size window from a 256 × 256 pixel image. The second technique was flipping the image horizontally, which allowed capturing information about reflection invariance. Finally, the third method added randomly generated lighting to capture color and illumination variation.

3.3   Transfer Learning in Lung Diseases Classification

Transfer learning is a very popular approach in computer vision related tasks using deep neural networks when data resources are scarce. Therefore, to launch a new task, we incorporate the pre-trained models skilled in solving similar problems. This method is crucial in medical image processing due to the shortage of real samples. In deep neural networks, feature extraction is carried out but passing raw data through models specialized in other tasks. Here, we can refer to deep learning models such as ResNet, where the last layer information serves as input to a new classifier. Transfer learning in deep learning problems can be performed using a common approach called pre-trained models approach. Reuse Model states that pre-trained model can produce a starting point for another model used in a different task. This involves incorporation of the whole model or its parts. The adopted model may or may not need to be refined on the input-output data for the new task. The third option considers selecting one of the available models. It is very common that research institutions publish their algorithms trained on challenging datasets which may fully or partially cover the problem stated by a new task.

ImageNet [3] is a project that helps computer vision researches in classification and detection tasks by providing them with a large image dataset. This database contains roughly 14 million different images from over 20 thousand classes. ImageNet also provides bounding boxes with annotations for over 1 million images, which are used in object localization problems.

In this work, we will experiment with three deep models (VGG16, ResNet-50, and InceptionV3) pre-trained on the ImageNet dataset.

3.4 Deep Nets in Lung Diseases Classification

The following deep nets have been considered: VGG16, ResNet-50 and InceptionV3. The VGG16 convolutional network is a model with 16 layers trained on fixed size images. The input is processed through a set of convolution layers which use small-size kernels with a receptive field 3×3. This is the smallest size allowing us to capture the notion of up, down, right, left, and center. The architecture also incorporates 1×1 kernels which may be interpreted as linear input transformation (followed by nonlinearity). The stride of convolutions (number of pixels that are shifted in every convolution - step size) is fixed and set to 1 pixel; therefore the spatial resolution remains the same after processing an input through a layer, e.g., the padding is fixed to 1 for 3×3 kernels. Spatial downsizing is performed by five consecutive pooling (max-pooling) layers, which are followed by some convolution layers. However, not all of them are followed by max-pooling. The max-pooling operation is carried over a fixed 2×2 pixel window, with a stride of 2 pixels. This cascade of convolutional layers ends with three fully-connected (FC) layers where the first two consist of 4096 nodes each and the third one of 1000 as it performs the 1000-way classification using softmax. All hidden layers have the same non-linearity ReLU (rectified linear unit) [10].

The ResNet convolutional neural network is a 50-layer deep model trained on more than a million fixed-size images from the ImageNet dataset. The network classifies an input image into one of 1000 object classes like car, airplane, horse or mouse. The network has learned a large amount of features thanks to training images diversity and achieved 6.71% top-5 error rate on the ImageNet dataset. The ResNet-50 convolutional neural network consists of 5 stages, each having convolutions and identity blocks. Every convolution block consists of 3 convolutional layers. ResNet-50 is related to ResNet-34, however, the idea behind its sibling model remains the same. The only difference is in residual blocks; unlike those in ResNet-34 ResNet-50 replaces every two layers in a residual block with a three-layer bottleneck block and 1×1 convolutions, which reduce and eventually restore the channel depth. This allows reducing a computational load when a 3×3 convolution is calculated. The model input is first processed through a layer with 64 filters each 7×7 and stride 2 and downsized by a max-pooling operation, which is carried over a fixed 2×2 pixel window, with a stride of 2 pixels. The second stage consists of three identical blocks, each containing a double convolution with 64 3 × 3 pixels filters and a skip connection block. The third pile of convolutions starts with a dotted line (image not included) as there is a change in the dimensionality of an input. This effect is achieved through the change of stride in the first convolution bloc from 1 to 2 pixels. The fourth and fifth groups of convolutions and skip connections follow the pattern presented in the third stage of input processing, yet they change the number of filters (kernels) to 256 and 512, respectively. This model has over 25 million parameters.

The researchers from Google introduced the first Inception (InceptionV1) neural network in 2014 during the ImageNet competition. The model consisted of blocs called "inception cell" that was able to conduct convolutions using different scale filters and afterward aggregate the results as one. Thanks to 1 × 1 convolution which reduces the input channel depth the model saves computations. Using a set of 1 × 1, 3 × 3, and finally, 5 × 5 size of filters, an inception unit cell learns extracting features of different scale from the input image. Although inception cells use max-pooling operator, the dimension of a processed data is preserved due to "same" padding, and so the output is properly concatenated.

A follow-up paper [17] was released not long after introducing a more efficient InceptionV3 solution to the first version of the inception cell. Large filters sized 5 × 5, and 7 × 7 are useful in extensive spatial features extraction, yet their disadvantage lies in the number of parameters and therefore computational disproportion.

The InceptionV3 model contains over 23 million parameters. The architecture can be divided into 5 modules. The first processing block consists of 3 inception modules. Then, information is passed through the effective grid size reduction and processed through four consecutive inception cells with asymmetric convolutions. Moving forward, information flows to the 17 × 17 pixels convolution layer connected to an auxiliary classifier and another effective grid size-reduction block. Finally, data progresses through a series of two blocs with wider filter banks and consequently gets to a fully-connected layer ended with a Softmax classifier. Visualization of the network architecture can be found in Fig. 3.

4   Experiments

4.1   Image Segmentation Using Deep Neural Networks

Many vision-related tasks, especially those from the field of medical image processing expect to have a class assigned to every pixel, i.e., every pixel is associated with a corresponding class. To conduct this process, we propose so-called U-net neural network architecture described in [18] and in Sect. 4.2. This model works well with very few training image examples yielding precise segmentation. The motivation behind this network is to utilize progressive layers instead of a building system, where upsampling layers are utilized instead of pooling operators, consequently increasing the output resolution. High-resolution features are combined with the upsampled output to perform localization. The deconvolution

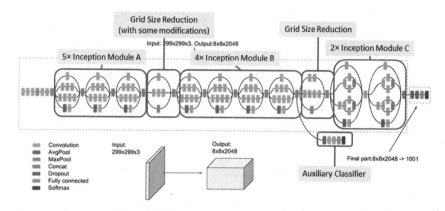

Fig. 3. InceptionV3 architecture. Batch normalization and ReLU units are used after every convolution layer.

layers consist of a high number of kernels, which better propagate information and result in outputs with higher resolution. Owing to the described procedures, the deconvolution path is approximately symmetric to the contracting one and so the architecture resembles the U shape. There are no fully connected layers, therefore, making it possible to conduct the seamless segmentation of relatively large images extrapolating the missing context by mirroring the processed input.

4.2 U-Net Architecture

The network showed in Fig. 4 consists of an expansive path (right) and a contracting one(left). The first part (contracting) resembles a typical convolutional neural network; the repeated 3×3 convolutions followed by a non-linearity (here ReLU), and 2×2 poling with stride 2. Each downsampling operation doubles the number of resulting feature maps. All expansive path operations are made of upsampling of the feature channels followed by a 2×2 deconvolution (or "up-convolution") which reduces the number of feature maps twice. The result is then concatenated with the corresponding feature layer from the contracting path and convolved with 3×3 kernels, and each passed through a ReLU. The final layers apply a 1×1 convolution to map each feature vector to the desired class.

4.3 Lung Segmentation

Following the approaches presented in the literature we wanted to use deep convolutional neural networks to segment lungs [8] before processing it through the classification models mentioned in Sect. 3.4. Researchers in [8] indicate that U-Net architecture and its modifications outperform the majority of CNN-based models and achieve excellent results by easily capturing spacial information about the lungs. Thus, we propose a pipeline that consists of two stages: first segmentation and then classification.

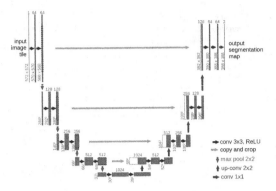

Fig. 4. U-net architecture (example for 32 × 32 pixels in the lowest resolution). Each blue box corresponds to a multichannel feature map. The number of channels is denoted on top of the box. The x-y-size is provided at the lower left edge of the box. White boxes represent copied feature maps. The arrows denote the different operations.

4.4   Dataset

The phase of extracting valuable information (lungs) is conducted with a model presented in Sect. 3.2 Our algorithms trained for 500 epochs on an extension of the SH dataset. The input to our U-shaped deep neural network is a regular chest X-Ray image, whereas the output is a manually prepared binary mask of lung shape, matching the input.

4.5   Software and Hardware

The code for the transfer-learning models is publicly available through a python API, Keras. Our algorithms were trained on servers equipped with GPU provided by Helios Calcul Québec, which consists of fifteen computing nodes each having eight Nvidia K20 GPUs and additionally six computing nodes with eight Nvidia K80 boards each. Every K80 board includes two GPU's and so the total of 216 GPU's in the cluster.

4.6   Training

As mentioned before, our model was trained for 500 epochs using a dataset partitioned into 80%, 10%, and 10% bins, for training, validation and test parts, respectively using the models introduced in Sect. 3.4 using the batch size of 8 samples, augmentation techniques briefed in Sect. 3.2, Adam optimizer and categorical cross-entropy as a loss function for pixel-wise binary classification. The training results are shown in Fig. 5. As we can easily notice, the validation error is slowly falling throughout the whole training, whereas there is no major change after the 100th epoch. The final error on the validation set is right below 0.05 and slightly above 0.06 on the test set.

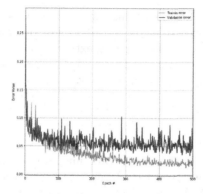

Fig. 5. U-Net training and validation losses change during training.

4.7   Segmentation Results

Our algorithm learns shape-related features typical for lungs and can generalize well further over unseen data. Figure 6 shows the results of our U-Net trained models. It is clear that the network was able to learn chest shape features and exclude regions containing internal organs such as heart. These promising results allowed us to process the whole dataset presented earlier and continue our analysis on the newly processed images.

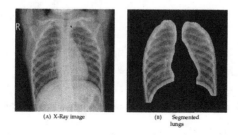

Fig. 6. (A) training X-ray lung image and (B) segmented lung image.

5   Training Deep Learning Models on Segmented Images

We propose a two-stage pipeline for classifying lung pathologies into pneumonia and tuberculosis consisting of two stages: first for chest X-ray image segmentation and second for lung disease classification. The first stage (segmentation) is trained during experiments described in the previous section. The second stage utilizes deep models described in Sect. 3.4, whereas we investigate potential improvements in performance depending on the type of model used. Our classification models were trained using the same setup as described in Sect. 3.4. Here, we conduct our experiments using the data described in Sect. 3.1. The difference is in prior segmentation, which extracts valuable information for the task, namely lungs. Figure 6 shows the training samples; the left and right panels correspond to input and output, respectively.

5.1 Inception Results

We tried all models with three deep net classifiers (VGG16, ResNet-50, InceptionV3) in the task of classification of lung images into two classes: pneumonia and tuberculosis. We observed that InceptionV3 based model performed best and thus due to lack of space we display only its performance results. The confusion matrix in Fig. 8 (A) shows that the new model improved the number of true positives (TP) in all classes in comparison with the VGG16 and ResNet-50 based models. Image Fig. 8 (B) shows that the AUC score for healthy, tuberculosis and pneumonia cases were 90%, 93%, and 99%, respectively.

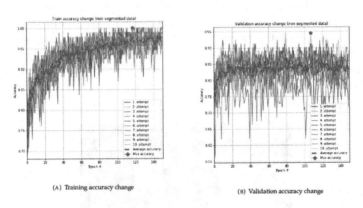

(A) Training accuracy change

(B) Validation accuracy change

Fig. 7. InceptionV3 based model training and validation accuracy versus number of epochs.

5.2 Comparison of Results on Non-segmented and Segmented Images

After comparing the results obtained by models without transfer learning we observe that transfer-learning models perform well in lung diseases classification using segmented images tasks even when the data resources are scarce. In this section, we compare the performance of our models to the results achieved in the literature over different datasets (Fig. 9).

The algorithm that scored the best in the majority results was InceptionV3 trained on the segmented images. What is more, it produced very high scores for the "disease" classes showing that a random instance containing marks of

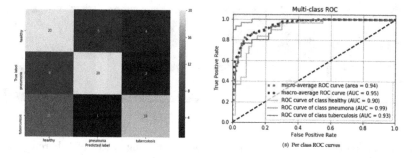

Fig. 8. Classification results obtained by InceptionV3 model on segmented lung images.

	Non segmented CXRs			Segmented CXRs		
	VGG16	ResNet-50	InceptionV3	VGG16	ResNet-50	InceptionV3
Accuracy	0.64	0.72	0.81	0.70	0.75	**0.82**
AUC (healthy)	0.68	0.84	0.89	0.75	0.77	**0.90**
AUC (pneumonia)	0.80	0.84	0.92	0.81	0.91	**0.99**
AUC (tuberculosis)	0.82	0.76	0.87	0.90	0.82	**0.93**
F1 score (healthy)	0.55	0.66	**0.76**	0.55	0.62	**0.76**
F1 score (pneumonia)	0.64	0.82	0.90	0.84	0.89	**0.93**
F1 score (tuberculosis)	0.72	0.66	0.75	**0.79**	0.73	0.78
precision (healthy)	0.49	0.77	**0.80**	0.67	0.66	0.75
precision (pneumonia)	0.87	0.76	0.89	0.92	**0.93**	0.90
precision (tuberculosis)	0.71	0.67	0.74	0.68	0.71	**0.81**
sensitivity (healthy)	0.62	0.60	0.73	0.48	0.62	**0.77**
sensitivity (pneumonia)	0.53	0.88	0.92	0.78	0.85	**0.95**
sensitivity (tuberculosis)	0.77	0.68	0.77	**0.95**	0.78	0.75

Fig. 9. Comparison of all results obtained for segmented and non-segmented data.

tuberculosis or pneumonia has over 90% probability to be classified to the correct class. Although the scores of the healthy class are worse than the diseased ones, its real cost is indeed lower as it is always worse to classify a sick patient as healthy. The InceptionV3 based model scored best, reaching better accuracy than VGG16 algorithms by over 12%. Although the interpretability of our methods is not guaranteed, we can clearly state that using transfer-learning based algorithms on small datasets allows achieving competitive classification scores on the unseen data. Furthermore, we compared the class activation maps shown in Fig. 10 in order to investigate the reasoning behind decision making. The remaining features, here lungs, force the network to explore it and thus make decisions based on observed changes. That behavior was expected and additionally improved the interpretability of our models as the marked regions might bring attention of the doctor in case of sick patients.

5.3 Comparison with Other Works

In this section, we compare performance of our models with the results in the literature using over different datasets. In order to do so we trained our algorithms on the Shenzhen and Montgomery datasets [6] ten times, generated the results for all the models and averaged their scores: accuracy, precision, sensitivity, specificity, F1 score and AUC. Table 1 presents comparison of different deep

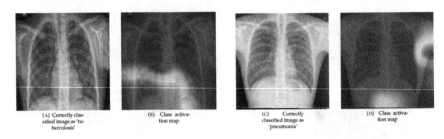

(A) Correctly classified image as 'tuberculosis'   (B) Class activation map   (C) Correctly classified image as 'pneumonia'   (D) Class activation map

Fig. 10. Two pairs of correctly labelled images containing traces of tuberculosis and pneumonia with their class activation maps.

learning models trained on the Shenzhen dataset [6]. Although our approach does not guarantee the best performance, it is always close to the highest even though it is typically less complex. Researchers in [5] used various pre-trained models in the pulmonary disease detection task, and the ensemble of them yields the highest accuracy and sensitivity. To compare, our InceptionV3-based model achieves accuracy smaller by only one percent and has identical AUC, which means that our method gives an equal probability of assigning a positive case of tuberculosis to its corresponding class over a negative sample. Although we could not outperform the best methods, our framework is less complicated. Furthermore, in Table 2 we compared the performance of our framework trained on the Montgomery dataset [6] to the literature. Our InceptionV3-based model tied with [14] in terms of accuracy, yet showed higher value of AUC. ResNet-50 and VGG16 based models performed worse, however not by much as they reached accuracies of 76% and 73% respectively, which is roughly 3 and 6% less than the highest score achieved.

Table 1. Comparison of different deep learning based solutions trained on the Shenzhen dataset [6]. Although our result is not the best, it performs better than any single model (excluding Ensemble). Horizontal line means that corresponding results were not provided in literature.

Model	Accuracy	Precision	Sensitivity	Specificity	F1 score	AUC
[1]	0.82	–	–	–	–	–
[14]	0.84	–	–	–	–	0.90
VGG16 [5]	0.84	–	**0.96**	0.72	–	0.88
ResNet-50 [5]	0.86	–	0.84	0.88	–	0.90
ResNet-152 [5]	0.88	–	0.80	0.92	–	0.91
Ensemble [5]	**0.90**	–	0.88	0.92	–	**0.94**
VGG16 [5]	0.84	0.88	0.80	0.89	0.83	0.86
ResNet-50 [5]	0.85	**0.97**	0.73	**0.98**	0.83	0.92
InceptionV3 [17]	0.89	0.96	0.80	0.97	**0.88**	**0.94**

Table 2. Comparison of different deep learning based solutions trained on the Montgomery dataset [6]. Our average performance is almost identical to [14].

Model	Accuracy	Precision	Sensitivity	Specificity	F1 score	AUC
[14]	0.790	–	–	–	–	0.811
[15]	0.674	–	–	–	–	0.884
[7]	0.783	–	–	–	–	0.869
VGG16 [5]	0.727	0.842	0.581	0.872	0.669	0.931
ResNet-50 [5]	0.764	0.814	0.691	0.836	0.744	0.891
InceptionV3 [17]	0.790	0.822	0.745	0.836	0.779	0.884

6 Conclusions

We created lung diseases classification pipeline based on transfer learning that was applied to small datasets of lung images. We evaluated its performance in classification of non-segmented and segmented chest X-Ray images. In our best performing framework we used U-net segmentation network and InceptionV3 deep model classifier. Our frameworks were compared with the existing models. We demonstrated that models pre-trained by transfer learning approach and simple classifiers such as shallow neural networks can successfully compete with the complex systems.

References

1. Anuj Rohilla, R.H., Mittal, A.: TB detection in chest radiograph using deep learning architecture. Int. J. Adv. Res. Sci. Eng. **6**, 1073–1084 (2017)
2. Candemir, S., et al.: Lung segmentation in chest radiographs using anatomical atlases with nonrigid registration. IEEE Trans. Med. Imaging **33**, 577–590 (2014)
3. Deng, J., Dong, W., Socher, R., Li, L.-J., Li, K., Fei-Fei, L.: ImageNet: a large-scale hierarchical image database. In: IEEE Conference on Computer Vision and Pattern Recognition (CVPR 2009), pp. 248–255 (2009)
4. Gordienko, Y., et al.: Deep learning with lung segmentation and bone shadow exclusion techniques for chest X-ray analysis of lung cancer. Computing Research Repository, vol. abs/1712.07632 (2017)
5. Islam, M.T., Aowal, M.A., Minhaz, A.T., Ashraf, K.: Abnormality detection and localization in chest X-rays using deep convolutional neural networks. ArXiv, vol. abs/1705.09850 (2017)
6. Jaeger, S., Candemir, S., Antani, S., Wang, Y.-X., Lu, P.-X., Thoma, G.: Two public chest X-ray datasets for computer-aided screening of pulmonary diseases. Quant. Imaging Med. Surg. **4**, 475–477 (2014)
7. Jaeger, S., Karargyris, A., Candemir, S., Folio, L., Siegelman, J., Callaghan, F., Xue, Z., Palaniappan, K., Singh, R.K., Antani, S., Thoma, G., Wang, Y., Lu, P., McDonald, C.J.: Automatic tuberculosis screening using chest radiographs. IEEE Trans. Med. Imaging **33**, 233–245 (2014)

8. Kang, Q., Lao, Q., Fevens, T.: Nuclei segmentation in histopathological images using two-stage learning. In: Shen, D., et al. (eds.) MICCAI 2019. LNCS, vol. 11764, pp. 703–711. Springer, Cham (2019). https://doi.org/10.1007/978-3-030-32239-7_78

9. Kermany, K.Z.D., Goldbaum, M.: Large dataset of labeled optical coherence tomography (OCT) and chest X-ray images. Cell **172**, 1122–1131 (2018)

10. Krizhevsky, A., Sutskever, I., Hinton, G.E.: ImageNet classification with deep convolutional neural networks. In: Advances in Neural Information Processing Systems (NIPS 2012), vol. 25, pp. 1097–1105 (2012)

11. Li, F., et al.: Computer-aided detection of peripheral lung cancers missed at CT: ROC analyses without and with localization. Radiology **237**(2), 684–90 (2005)

12. Lo, S.-C.B., Li, H., Wang, Y.J., Kinnard, L., Freedman, M.T.: A multiple circular path convolution neural network system for detection of mammographic masses. IEEE Trans. Med. Imaging **21**, 150–158 (2002)

13. Lo, S.-C.B., Chan, H.-P., Lin, J.-S., Li, H., Freedman, M.T., Mun, S.K.: Artificial convolution neural network for medical image pattern recognition. Neural Netw. **8**, 1201–1214 (1995)

14. Pasa, F., Golkov, V., Pfeiffer, F., Cremers, D., Pfeiffer, D.: Efficient deep network architectures for fast chest X-ray tuberculosis screening and visualization. Sci. Rep. **9**, 1–9 (2019)

15. Hwang, S., Kim, H.-E.: A novel approach for tuberculosis screening based on deep convolutional neural networks. In: Medical Imaging 2016: Computer-Aided Diagnosis, vol. 9785, pp. 1–23 (2016)

16. Suzuki, K.: Pixel-based machine learning in medical imaging. Int. J. Biomed. Imaging **2012**, 1 (2012)

17. Szegedy, C., Vanhoucke, V., Ioffe, S., Shlens, J., Wojna, Z.: Rethinking the inception architecture for computer vision. In: 2016 IEEE Conference on Computer Vision and Pattern Recognition (CVPR 2016), pp. 2818–2826, June 2016

18. Ronneberger, O., Fischer, P., Brox, T.: U-net: convolutional networks for biomedical image segmentation. In: Navab, N., Hornegger, J., Wells, W.M., Frangi, A.F. (eds.) MICCAI 2015. LNCS, vol. 9351, pp. 234–241. Springer, Cham (2015). https://doi.org/10.1007/978-3-319-24574-4_28

An Adaptive Space-Filling Curve Trajectory for Ordering 3D Datasets to 1D: Application to Brain Magnetic Resonance Imaging Data for Classification

Unal Sakoglu[1(✉)], Lohit Bhupati[2], Nazanin Beheshti[3], Nikolaos Tsekos[3], and Lennart Johnsson[3]

[1] Computer Engineering, University of Houston – Clear Lake, Houston, TX 77058, USA
sakoglu@uhcl.edu
[2] Computer Science, University of Houston – Clear Lake, Houston, TX 77058, USA
[3] Computer Science, University of Houston, Houston, TX 77004, USA

Abstract. In this work, we develop an adaptive, near-optimal, 3-Dimensional (3D) to 1D ordering methodology for brain magnetic resonance imaging (MRI) data, using a space-filling curve (SFC) trajectory, which is adaptive to brain's shape as captured by MRI. We present the pseudocode of the heuristics for developing the SFC trajectory. We apply this trajectory to functional MRI brain activation maps from a schizophrenia study, compress the data, obtain features, and perform classification of schizophrenia patients vs. normal controls. We compare the classification results with those of a linear ordering trajectory, which has been the traditional method for ordering 3D MRI data to 1D. We report that the adaptive SFC trajectory-based classification performance is superior than the linear ordering trajectory-based classification.

Keywords: Space-filling curve · Classification · MRI · Neuroimaging · Adaptive compression

1 Introduction

Magnetic resonance imaging (MRI) is a widely used tomography technology which is used to capture the structure of the brain in three dimensions (3D) of space. A specific modality of MRI technology, known as functional MRI (fMRI), can capture hemodynamic response signals from the brain within the order of s, in a repeated manner, albeit in lower spatial resolution. FMRI therefore includes in time-series data for each volume element (voxel) of the brain. Datasets of both structural and functional MRI, which are 3D spatial matrices (e.g. a $64 \times 64 \times 64$ matrix), generally have to be converted to vector arrays of a single dimension (1D), for further analyses (Fig. 1), such as analyses with general linear models, regression, and independent component analysis. Traditionally, a linear ordering/mapping of data from 3D to 1D have been used for this purpose, i.e. 3D data are scanned consecutively along the first, second, third spatial dimensions in

© Springer Nature Switzerland AG 2020
V. V. Krzhizhanovskaya et al. (Eds.): ICCS 2020, LNCS 12139, pp. 635–646, 2020.
https://doi.org/10.1007/978-3-030-50420-5_48

order to obtain a 1D ordering of volumes. Linear ordering introduces large 'jumps' or 'discontinuities' of the recorded signal; for example, if it is applied to structural MRI datasets, it does not preserve the structure of the brain in 1D.

Previously, pre-defined space-filling curves such as Hilbert curve and Z-curve have been suggested for ordering of the 3D MRI datasets [1]. Hilbert curve, specifically, results in better preservation of local features [2–4] when compared to linear ordering; it was shown in [1] that it could result in less discontinuities in brain MRI signals, and it was also applied for classification based on fMRI brain activation maps [5, 6]. In general, space-filling curves are used in a wide range of applications [9–27] in reducing two or higher dimensional spaces to a one-dimensional space. Hilbert [4, 9] and Z [9] orderings are among the most widely used methods focused on preserving spatial locality in the mapping to one dimension. However, as has been noticed by e.g. the image compression community [17–24] context-aware, or adaptive [20–24] orderings taking pixel attributes into account in addition to spatial information leads to better compression.

In [1], using a least-squares signal-difference approach which uses sum of squared signal intensity differences (TSSID) was proposed as a cost function and measure of how adaptive an SFC was to the data/signal being traversed. Since the cost function is the sum of signal 'jumps', the goal is to minimize the cost function to find the most adaptive trajectory. In the context of 3D brain imaging, by traversing 3D volumes using a space-filling curve (SFC) that is adaptive to brain's shape, a 3D MRI image can be ordered into 1D space (e.g. into a $64^3 \times 1$ vector from a $64 \times 64 \times 64$ volume), which can better preserve the brain's structure in 1D. Although this idea was suggested in [1], it was noted that this was inherently a Hamiltonian path problem which also can be formulated as a modified traveling salesman problem (TSP), which is an NP-hard problem to solve. With as many nodes as number of voxels, which is on the order of thousands for fMRI brain activation datasets, it makes the problem computationally intractable [1]. If an approximation of an adaptive SFC could be found and applied for 3D to 1D ordering of the dataset, any dimensionality reduction, smoothing, down-sampling, compression, and

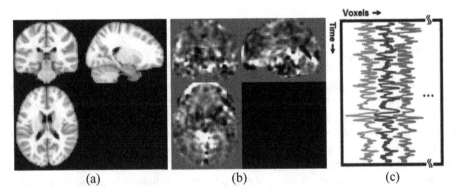

(a)   (b)   (c)

Fig. 1. Three views of a structural T1 MRI dataset (a), and an fMRI brain activation map (b), which is computed from fMRI volumes taken at multiple time-points. Conventionally, fMRI dataset voxels are ordered using linear ordering trajectory into rows of a matrix, as a result, a matrix of *time* × *voxels* is generated for further analyses (c).

feature selection/reduction in 1D could benefit from better preservation of information, when compared with linear ordering, or other predefined orderings such as the Hilbert curve ordering.

In this work, we computed an SFC adaptive to brain's shape (as recorded in a T1 MRI canonical/template image as an example) and developed an algorithm pipeline which uses the computed SFC for ordering of fMRI activation maps from 3D to 1D, obtains features from the 1D orderings, and performs classification of participants; we applied the pipeline to an fMRI study of two groups of participants: schizophrenia patients (SP) and healthy controls (HC), and performed classification of a given brain activation map belonged to SP or HC.

2 Materials and Methods

2.1 Data and Participants

95 schizophrenia patients (SP) and 89 healthy control (HC) participants were scanned using 3T Siemens Trio MRI scanners at four different research sites in the USA. Research protocol was reviewed and approved by the institutional review board of the local institutions where scans were performed, and written consent was obtained from each participant. The parameters for the functional scan were: TR/TE $= 2$ s/30 ms, BW $= \pm 100$ kHz $= 3126$ Hz/pixel, FA $= 90°$, slice thickness 4 mm, slice gap 1 mm, voxel size $= 3.4$ mm $\times$ 3.4 mm $\times$ 4 mm, FOV $= 22$ cm, PACE-enabled, single shot, single-echo EPI pulse sequence, oblique axial slice plane; 64×64 acquisition matrix, 27 slices in ascending sequential acquisition. Participants performed a sensorimotor (SM) task during the scan. After standard pre-processing, MRI volumes were warped to Montreal Neurological Institute (MNI) standard canonical/template T1 MRI volume [7]. The fMRI activation maps computed as standard parametric maps of t-value with the SM task, were resampled to the 3 mm $\times$ 3 mm $\times$ 3 mm standard MNI volume, which resulted in $53 \times 63 \times 46$ data matrix. An adaptive space-filling curve (SFC) using the $53 \times 63 \times 46$ T1 MRI template volume was computed using a graphical processing unit using a greedy search algorithm developed in-house.

2.2 The Adaptive Space-Filling Curve Algorithm for 3D MRI Data

The algorithm proceeds as follows. At any voxel, the signal values of its 26 immediate neighbors in 3D is retrieved and sorted based on their signal intensity difference with the current voxel. The next voxel along the adaptive SFC is selected as the voxel with the minimum absolute signal difference that is not already included in the adaptive SFC. The algorithm traverses back in the list of voxels selected for the adaptive SFC until one voxel to be included in the SFC is found. Using the list for backtracking in the worst case requires a time proportional to the length of the list and hence in the worst case is of $O(n)$ where n is the length of the list. Instead, a hash table is used with key value 1 if a voxel is included in the SFC, otherwise the value is zero. A snapshot of the SFC obtained from the T1 image is presented in Fig. 2 and Table 1.

To make the algorithm faster and reduce memory consumption we remove voxels with zero signal intensity in a preprocessing step before applying the algorithm, since zero signal voxels provide no useful information for feature selection for classification. In a predefined SFC, such as the Hilbert SFC, voxels with zero signal intensity are not consecutive and they are scattered over the SFC, and their indexes have to be traced. In the adaptive SFC, these voxels are consecutive and either at the beginning or at the end of the curve and they can be easily trimmed reducing the number of useless signals in the data. Figure 3 presents this effect. Using a pre-defined space-filling curve such as Hilbert curve results in tracing the full volume (Fig. 3(a)), versus, the adaptive space

Table 1. SFC algorithm execution flow/pseudocode for traversing 3D MRI signal

```
myimage = niftiread('ABC.nii')
SizeVec = size(myimage)
M = sizeVec(0), N = sizeVec(1), T = sizeVec(2);
data = non-zero voxels present in myimage,
 to keep track of non-zero voxels we define NonZeroVoxels_TBL
NonZeroVoxels_TBL[all voxels in data] = 1;
NonZeroVoxels_TBL is used in US_twentysix_neighbors function to find neighboring voxels
whose values are nonzero, helps not to traverse zero voxels around a non-zero voxel.
start = data(1,:);
i = start(1); j = start(2); k = start(3)
hashTbl(i,j,k) = 1
for i_SFC = 0 to size(Data)
    not_found = 0
    traversed_voxels_lst.add([i,j,k])
    neighbors= US_twentysix_neighbors (i,j,k,M,N,T,mat_data)
    signalNeighs = myimage(neighbors)
    [index] = sort(SignalNeighs, 'descending')
    [next_i, next_j, next_k] = neighbors(index(0))
    i = 0;
    while(HashTbl(next_i,next_j,next_k) == 1):
            i = i + 1
            [next_i, next_j, next_k] = neighbors(index(i))
    if (i == len(index))
            not_found = 1
    while(not_found):
            [on_i, on_j, on_k] = traversed_voxels_lst[counter]
            neighbors= US_twentysix_neighbors (on_i, on_j, on_k ,M,N,T,mat_data)
            signalNeighs = myimage(neighbors)
            [index] = sort (SignalNeighs, 'descending')
            [next_i, next_j, next_k] = neighbors (index (0))
            i = 0
            while(hashTbl(next_i,next_j,next_k) == 1)
                    i = i + 1
                    [next_i, next_j, next_k] = neighbors(index(i))
            if (i== len(index))
                    not_found = 1;
    i = next_i, j = next_j, k.= next_k
```

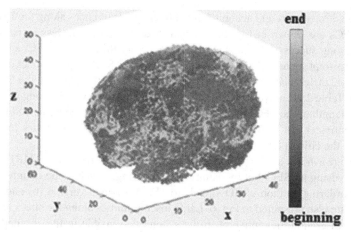

Fig. 2. A color-coded space-filling curve (SFC) trajectory of a template T1 MRI brain in Montreal Neurological Institute (MNI) brain template space. This trajectory was applied to the SFC orderings of the fMRI brain activation maps in this study. Cold (blue): beginning; hot (yellow): end. (Color figure online)

filling curve results in voxels with only non-zero signal values, a 57% reduction in the number of voxels (Fig. 3(b)).

2.3 Using SFC-Ordered FMRI Activation Data as Features for Classification

The 3D fMRI maps from each participant were converted to 1D with linear ordering and also with the computed adaptive SFC. The 1D arrays were down-sampled by a bin size of 100, values were averaged across each bin ("binning"), which constituted raw features. Features were further reduced to 100, and also to 30, by using and support vector machine based sequential forward search feature reduction algorithm [8]. The reduced features were used to train the classification algorithm by using a random 70% of the dataset and were tested on the remaining 30%. The training-testing process was repeated 100 times by using new random selection of training and testing datasets. Average classification accuracies were computed. The results are presented in the next section.

3 Results

To make the algorithm faster and reduce memory consumption we remove voxels with zero signal intensity in a preprocessing step before applying the algorithm. As an example, around 57% of voxels in one of the 3D T1 MRI brain were zero. The impact of removing zero signal voxels is clearly visible in Fig. 3.

Figure 4 presents a sample progression snapshots of the adaptive SFC at the end of the trajectory, where the SFC is traversing mostly along the outer surface of the brain. Part of the SFC with length of 1000, 3000, 5000, and 7000 (in red color) are overlaid on the full-length adaptive SFC (in gray color). Traversal of the MRI signal along the brain's shape, hence adaptive to the brain's shape, is visible.

The signal intensities and absolute signal intensity differences along the Hilbert and adaptive SFCs are shown in Fig. 5. The adaptive SFC results in less signal intensity difference along successive voxels along the trajectory than Hilbert SFC, and hence it results in less total squared signal intensity difference (TSSID). The impact on the total squared signal difference between successive voxels along the SFC and the SFC length is shown in Table 2. Relative total TSSID between successive voxels along the SFC and the relative length of the SFC is computed for the adaptive and the Hilbert SFCs for our data, the canonical 3D T1 MRI image. The adaptive SFC has 50.5 times less relative TSSID than the Hilbert SFC, and it is 34% shorter.

Figure 6 (a) shows a 1D-ordered signal array from a participant's fMRI activation map, not by using an SFC, but by using linear ordering, which is the commonly used, traditional ordering method of 3D signals to 1D. A zoomed version (b) and a binned version (c) are also presented in Fig. 6. Large discontinuities along the slices are visible, and further discontinuities inherent in the scheme are visible in the zoomed version, hence the result of binning includes many signals of zero-values. Figure 7 shows results of the adaptive SFC trajectory-based 1D-ordered signal array. Large clusters can be seen in the adaptive SFC trajectory-based array, whereas linear ordering resulted in a highly disconnected or un-clustered brain activation signal. Large clusters are also visible in the zoomed portion (b), and the binning includes no voxels with zero-values (c); in general, the binned adaptive SFC array includes only few such voxels.

Using the SFC trajectory-based ordering, an SVM classification algorithm resulted in 72.1% (74.6%) average accuracy in classification of SP vs. HC participants, whereas the linear ordering resulted in around 49.9% (50.0%) classification accuracy, using 30 (100) features, employing a sequential forward search algorithm for the reduction of features. SFC-based ordering resulted in significantly higher accuracy, whereas linear ordering resulted in just about chance accuracy of 50%. Performing a $4 \times 4 \times 4$ down-sampling of the brain activation maps directly in 3D, and then ordering the resulting down-sampled brain with linear ordering, using the resulting signal as features, resulted also with chance accuracy of around 50% for participant classification. Overall, SFC-based classification results were significantly higher.

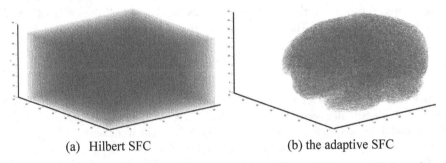

(a) Hilbert SFC   (b) the adaptive SFC

Fig. 3. (a) Using a pre-defined space-filling curve such as Hilbert curve here results in tracing of the full volume (high redundancy), versus, (b) the adaptive space filling curve results in voxels with non-zero signal values, a 57% reduction in the number of voxels (low redundancy).

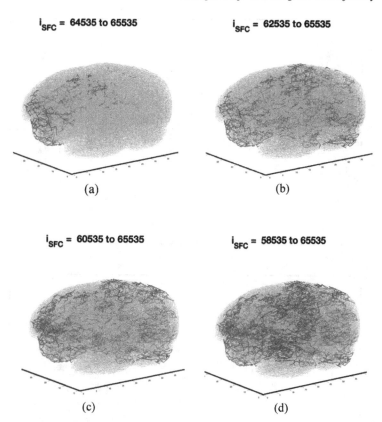

i_{SFC} = 64535 to 65535 i_{SFC} = 62535 to 65535

(a)   (b)

i_{SFC} = 60535 to 65535 i_{SFC} = 58535 to 65535

(c)   (d)

Fig. 4. Representation of fMRI brain activation traversal by the SFC for the last a) 1000, 3000, 5000 and 7000 voxels (red), overlaid on the total SFC trajectory (gray), representing a sample progression of the SFC trajectory during the last 7000-voxel portion of the SFC. (Color figure online)

Table 2. Merit of the adaptive SFC vs. the Hilbert curve for our data set

Relative TSSID between successive voxels along the SFC		Relative length of the SFC	
Adaptive	Hilbert	Adaptive	Hilbert
1	50.5	1	1.53

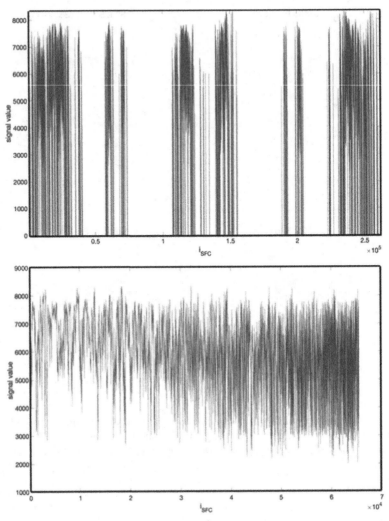

Fig. 5. Signal intensity along the trajectory for successive voxels using Hilbert SFC (top) and the adaptive SFC (bottom).

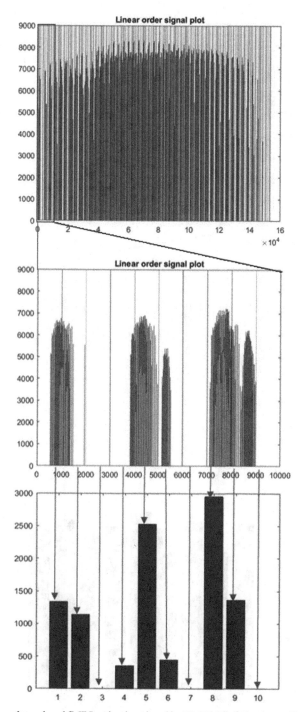

Fig. 6. Top: Linearly-ordered fMRI activation signal in 1D. Middle: a horizontally zoomed portion of the signal. Bottom: bar plot of the zoomed portion after averaging (binning).

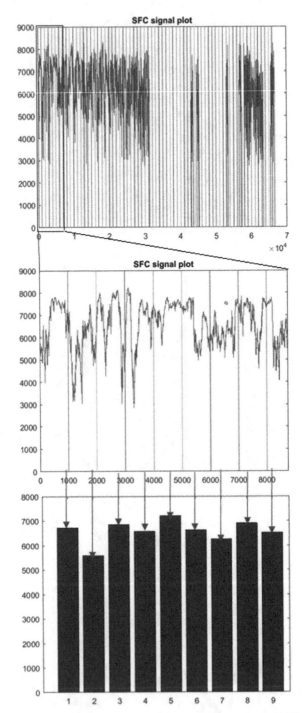

Fig. 7. Top: SFC-ordered fMRI activation signal in 1D. Middle: a horizontally zoomed portion of the signal. Bottom: bar plot of the zoomed portion after averaging (binning).

4 Conclusions and Discussions

An adaptive space-filling curve (SFC) trajectory which is adaptive to brain's shape can be utilized for 1D ordering of 3D MRI data, such as brain activation maps in fMRI, and this ordering proves to be better than using the traditional linear ordering or pre-defined ordering such as Hilbert ordering. We have shown that it reduces the amount of discontinuities and results in shorter signal with less redundancy. In this work, we used the adaptive SFC for feature reduction in classification using fMRI brain activation maps and we showed that SFC ordering resulted in better features from the activation maps, which resulted in higher classification accuracy of two groups of participants, schizophrenia patients and healthy controls.

The results her need to be replicated with larger sample sizes and with other fMRI studies. In addition to using a canonical T1 MRI image, adaptive SFCs can be obtained using T2, DTI, echo-planar imaging (EPI) and many more kinds of different canonical MRI images in the standard imaging space such as MNI space, and they can be applied to traverse participants' brain imaging data in the standard imaging space, and the traversed data can be subsequently compressed, binned, features extracted, and many more applications can be done.

Potential future work also involves using different feature reduction methods and different classification algorithms, and compression of fMRI data based on SFCs.

Overall, data-adaptive SFCs have great potential in adaptive compression of datasets as well, in order to minimize information loss in compression.

Acknowledgements. This research was supported by UHCL College of Science and Engineering, and UHCL Office of Research and Sponsored Programs. Dr. Unal Sakoglu thanks Dr. Vince Calhoun at Tri-institutional Center for Translational Research in Neuroimaging and Data Science (TReNDS), Georgia State University, Atlanta, USA, for providing fMRI activation map data.

References

1. Sakoglu, U., Arslan, A.N., Bohra, K., Flores, H.: In search of optimal space-filling curves for 3-D to 1-D mapping: application to 3-D brain MRI data. In: Proceedings of the 6th International Conference on Bioinformatics and Computational Biology (BICOB), pp. 61–66. International Society for Computers and their Applications (ISCA), Las Vegas (2014)
2. Hilbert, D.: Über die stetige Abbildung einer Linie auf ein Flächenstück (in German). Math. Ann. **38**(3), 459–460 (1891)
3. Griffiths, J.Q.: An algorithm for displaying a class of space-filling curves. Softw. Pract. Exper. **16**, 403–411 (1986)
4. Moon, B., Jagadish, H., Faloutsos, C., Saltz, J.: Analysis of the clustering properties of the Hilbert space-filling curve. IEEE Trans. Knowl. Data Eng. **13**(1), 124–141 (2001)
5. Kontos, D., Megalooikonomou, V., Ghubade, N., Faloutsos, C.: Detecting discriminative functional MRI activation patterns using space filling curves. In: Proceedings of the Engineering in Medicine and Biology Conference, pp. 963–966. IEEE, Hoboken (2003)
6. Wang, Q., Kontos, D., Li, G., Megalooikonomou, V.: Application of time series techniques to data mining and analysis of spatial patterns in 3D images. In: Proceedings of the International Conference on Acoustics, Speech and Signal Processing, pp. 525–528. Institute for EEE, Hoboken (2004)

7. Evans, A.C., Lanke, A.L., Collins, D.L., Baillet, S.: Brain templates and atlases. Neuroimage **62**, 911–922 (2010)
8. Alpaydin, E.: Introduction to Machine Learning, 3rd edn. MIT Press, Cambridge (2014)
9. Sagan, H.: Space-Filling Curves. Springer, Heidelberg (2012)
10. Sasidharan, A., Dennis, J.M., Snir, M.: A general space-filling curve algorithm for partitioning 2D meshes. In: 2015 IEEE 17th International Conference on High Performance Computing and Communications, 2015 IEEE 7th International Symposium on Cyberspace Safety and Security, and 2015 IEEE 12th International Conference on Embedded Software and Systems. IEEE (2015)
11. Sasidharan, A., Snir, M.: Space-filling curves for partitioning adaptively refined meshes. Math. Comput. Sci. (2015)
12. Harlacher, D.F., et al.: Dynamic load balancing for unstructured meshes on space-filling curves. In: 2012 IEEE 26th International Parallel and Distributed Processing Symposium Workshops & PhD Forum. IEEE (2012)
13. Mellor-Crummey, J., Whalley, D., Kennedy, K.: Improving memory hierarchy performance for irregular applications using data and computation reorderings. Int. J. Parallel Prog. **29**(3), 217–247 (2001)
14. Jagadish, H.V.: Linear clustering of objects with multiple attributes. In: Proceedings of the 1990 ACM SIGMOD International Conference on Management of Data (1990)
15. Lawder, J.K.: The application of space-filling curves to the storage and retrieval of multi-dimensional data. Dissertation, University of London, Birkbeck (2000)
16. Asano, T., Ranjan, D., Roos, T., Welzl, E., Widmayer, P.: Space-filling curves and their use in the design of geometric data structures. Theor. Comput. Sci. **181**(1), 3–15 (1997)
17. Moghaddam, B., Hintz, K.J., Stewart, C.V.: Space-filling curves for image compression. In: Automatic Object Recognition. International Society for Optics and Photonics (1991)
18. Liang, J.Y., et al.: Lossless compression of medical images using Hilbert space-filling curves. Comput. Med. Imaging Graph. **32**(3), 174–182 (2008)
19. Lai, Z., et al.: Image reconstruction of compressed sensing MRI using graph-based redundant wavelet transform. Med. Image Anal. **27**, 93–104 (2016)
20. Dafner, R., Cohen-Or, D., Matias, Y.: Context-based space filling curves. In: Computer Graphics Forum. Blackwell Publishers Ltd., Oxford and Boston (2000)
21. Bar-Joseph, Z., Cohen-Or, D.: Hierarchical context-based pixel ordering. In: Computer Graphics Forum. Blackwell Publishing Inc., Oxford (2003)
22. Ouni, T., Abid, M.: Scan methods and their application in image compression. Int. J. Signal Proc. Image Proc. Pattern Recogn. **5**(3), 49–64 (2012)
23. Itani, A., Das, M.: Self-describing context-based pixel ordering. In: Bebis, G., Boyle, R., Koracin, D., Parvin, B. (eds.) ISVC 2005. LNCS, vol. 3804, pp. 413–419. Springer, Heidelberg (2005). https://doi.org/10.1007/11595755_50
24. Ouni, T., Lassoued, A., Abid, M.: Gradient-based space filling curves: application to lossless image compression. In: 2011 IEEE International Conference on Computer Applications and Industrial Electronics (ICCAIE). IEEE (2011)
25. Haverkort, H.: Sixteen space-filling curves and traversals for d-dimensional cubes and simplices. arXiv preprint arXiv:1711.04473 (2017)
26. Lazarus, C., et al.: SPARKLING: variable-density k-space filling curves for accelerated T2*-weighted MRI. Magn. Reson. Med. **81**(6), 3643–3661 (2019)
27. Corcoran, T., et al.: A spatial mapping algorithm with applications in deep learning-based structure classification. arXiv preprint arXiv:1802.02532 (2018)

Correction to: An Agent-Based Simulation of the Spread of Dengue Fever

Imran Mahmood, Mishal Jahan, Derek Groen, Aneela Javed, and Faisal Shafait

Correction to:
Chapter "An Agent-Based Simulation of the Spread of Dengue Fever" in: V. V. Krzhizhanovskaya et al. (Eds.): *Computational Science – ICCS 2020*, **LNCS 12139,** https://doi.org/10.1007/978-3-030-50420-5_8

The original version of this chapter was revised. The missing funding information has been added.

The updated version of this chapter can be found at
https://doi.org/10.1007/978-3-030-50420-5_8

© Springer Nature Switzerland AG 2020
V. V. Krzhizhanovskaya et al. (Eds.): ICCS 2020, LNCS 12139, p. C1, 2020.
https://doi.org/10.1007/978-3-030-50420-5_49

Author Index

Printed in the United States
By Bookmasters